Principles of
Differential Equations

PURE AND APPLIED MATHEMATICS

A Wiley-Interscience Series of Texts, Monographs, and Tracts

A complete list of the titles in this series appears at the end of this volume.

Principles of Differential Equations

Nelson G. Markley
Provost Emeritus
Lehigh University
Bethlehem, PA
and
Professor Emeritus
University of Maryland
College Park, MD

WILEY-INTERSCIENCE
A JOHN WILEY & SONS, INC., PUBLICATION

Published by John Wiley & Sons, Inc., Hoboken, New Jersey.
Published simultaneously in Canada.

Library of Congress Cataloging-in-Publication Data:

Markley, Nelson Groh
Principles of differential equations / Nelson G. Markley.
p. cm — (Pure and applied mathematics: a Wiley-Interscience series of texts, monographs, and tracts)
Includes bibliographical references and index.
ISBN 0-471-64956-2 (cloth)
1. Differential equations. I. Title. II. Pure and applied mathematics (John Wiley & Sons : Unnumbered)

QA371.M27 2004
515'.35—dc22 2004040890

Printed in the United States of America.

10 9 8 7 6 5 4 3 2 1

Contents

Preface

Differential equations is an old but durable subject that remains alive and useful to a wide variety of engineers, scientists, and mathematicians. The purpose of this book is to provide an introductory graduate text for these consumers. It is intended for classroom use or self-study. The goal is to provide an accessible and concrete introduction to the main principles of ordinary differential equations and to present the material in a modern and rigorous way. The intent of this goal is to provide the solid foundation that will enable a reader to learn and understand other parts of the subject easily and to encourage them to learn more about differential equations and dynamical systems.

The study of differential equations began with the birth of calculus, which dates to the 1660s. Part of Newton's motivation in developing calculus was to solve problems that could be attacked with differential equations. For example, an early triumph of differential equations was Newton's demonstration that Kepler's empirical laws of planetary motion could be derived from Newton's laws of motion using differential equations. Now, with over 300 years of history, the subject of differential equations represents a huge body of knowledge including many subfields and a vast array of applications in many disciplines. It is beyond exposition as a whole. Instead, the right question to ask is what are the principles of differential equations that a serious user should know and understand today?

Principles of Differential Equations is my answer to this question. It looks at ordinary differential equations from the viewpoint of important principles. Although the word "principle" is probably overused in the academic world and may be a bit trite, it is used here seriously in the sense of "a basic or essential quality or element determining intrinsic nature or characteristic behavior." Each section presents a coherent picture of a circle of ideas that illustrates a key principle in the study of differential equations. The overarching questions driving the theory are discussed and the value and limitations of results are explained. Throughout, the book a concerted effort is made to tie the pieces together and give the reader a coherent and unified sense of the subject.

Principles of Differential Equations is also largely about the qualitative theory of ordinary differential equations. Qualitative theory refers to the study of the behavior of solutions without determining explicit formulas for the solutions. It originated with Poincaré at the beginning of the twentieth century and, in my judgment, has been the most important theme of ordinary differential equations in that century. Consequently, very little attention is paid to techniques for finding analytic formulas for solutions. The emphasis is on the general properties of the solutions of ordinary differential equations from simple existence of solutions to the remarkable behavior of Hopf bifurcations.

Another important development in the twentieth century was the study of dynamical systems. Since my research has always been in dynamical systems, this book naturally has a dynamical systems perspective. In ordinary differential equations, the dynamical systems approach amounts to a shift in emphasis from finding the solution of a particular problem to studying all the solutions of a differential equation at once, and it is closely linked to the qualitative point of view. Once the existence of a global solution containing all solutions is established at the end of Chapter 1, it plays a central role in the remaining chapters. Furthermore, various branches of modern dynamical systems have roots in ordinary differential equations and are briefly discussed with suggested introductory references at appropriate points in the text.

Since the broad plan of the book is to expose the reader to a range of important ideas and basic results, the focus is on core concepts and theorems that apply to large classes of differential equations and not on being encyclopedic on any topic. This means many things, some more important than others, have been deliberately omitted. There will be, as there should, instructors who strongly disagree with my choices of what to include and what not to include. I would simply invite them to supplement the material in this book with a series of well-prepared lectures on their favorite missing topic and begin expanding their students' horizons.

I have strived to make this volume as complete and self-contained as possible with minimal prerequisites, which are discussed at length in the next three paragraphs. Except for obvious exceptions like the Jordan curve theorem in Chapter 6, stated results are followed by rigorous proofs or left to the reader as straightforward exercises. Theorems and propositions are numbered consecutively in each chapter; lemmas and corollaries are unnumbered. Because the results build on each other, there are many cross-references to help the reader follow the arguments and see how the pieces fit together. There are also approximately 250 exercises that illustrate the material with specific differential equations, fill in gaps, or slightly extend the theory.

To make the book as accessible as possible, the prerequisites have been kept to a minimum. They are primarily undergraduate real analysis of one variable (sometimes called advanced calculus) and introductory linear algebra. For a mathematically capable student one semester of each should suffice, but would require the student to spend more time mastering Chapter 1, which is both challenging and essential. In particular, to gain an understanding of how the fundamentals fit together, some readers may find it beneficial to skip the proofs in a first reading of Chapter 1 and possibly the first two section of Chapter 2 and then go back and study the proofs. A number of advanced topics from both analysis and linear algebra that are less likely to be familiar to a reader are included with proofs when needed in text.

From analysis it is assumed that the reader understands epsilon–delta proofs and knows the standard concepts and results for both the real numbers (density of the rational numbers, Bolzano–Weiestrass theorem, convergence of sequences, Cauchy sequences, etc.) and real-valued function of one variable (limit theorems, continuity, uniform continuity, the intermediate-value theorem, the mean-value theorem, fundamental theorem of calculus, Taylor's formula with remainder, uniform convergence of a sequence of functions, etc.).

It is also assumed that the reader is comfortable working with functions of several variables, their partial derivatives, and integrals. To facilitate the shift from just working with functions of several variables to doing rigorous analysis with them, Section 1.1 and the exercises following it provide a rigorous but brief introduction to the analysis of functions of several variables. The approach to functions of several variables is topological, and depending on a readers background may require more or less time to master.

The prerequisites from linear algebra are a basic knowledge of matrix algebra for real and complex matrices, finite dimensional vector spaces, and linear transformations. From matrix algebra it is assumed that the reader is familiar with matrix calculations including the determinants and inverses of matrices and with systems of linear equations. The vector space prerequisites are subspaces, linear independence, basis, and dimension. Finally, the reader should be familiar with the relationship between matrices and linear transformations and with the nullity, rank, and eigenvalues of a linear transformation, but these concepts are also reviewed when they first occur in the text.

The bibliography consists entirely of books and is longer than would be absolutely necessary. The intent is to provide the reader with a rich list of books that are the next steps toward the frontiers of differential equations and dynamical systems. Many of them have extensive bibliographies of important current and historical research papers in a wide variety of journals. A number of them are excellent introductions to closely related fields. All are appropriately referenced at some point.

This volume grew out of my lecture notes for the introductory graduate course in differential equations at the University of Maryland. The students were typically first or second year graduate students in the mathematics or applied mathematics programs and a few graduate students from the engineering and physics programs. There was a large variation in their backgrounds and the challenge was to engage all of them in the material. This experience more than anything else shaped my thinking about what constitutes a coherent and accessible core to the modern theory of differential equations. My lecture notes over the years contained a variety of different topics that enriched the course but were eventually discarded because they were not really central to the development of the subject. The material was reorganized and the proofs rethought every time I taught the course. Preparing this manuscript was the final distillation.

Over the years I have learned about differential equations from a great variety of books at all levels. I am deeply indebted to the authors of all these books for everything I learned from them. Collectively, they shaped my perspective of the subject and provided a foundation for my lectures, notes, and eventually this book.

I want to thank all my colleagues and friends in the Department of Mathematics at the University of Maryland for affording me the opportunity to regularly teach the graduate course on differential equations in a stimulating mathematical environment during my many years in the department. Numerous discussions about the course content with colleagues in all areas and the semiannual preparation of the qualifying exam in differential equations with the dynamics group were particularly valuable to me as I developed my own approach to the subject. I also want to thank the Department of Mathematics for typing and reproducing an early version of my notes and Jay Alexander for using it as the textbook for the course and for his many insightful comments. Special thanks go to Mary Vanderschoot for reading the final manuscript and doing a wonderful job finding all kinds of little errors that needed to be corrected. Finally, I am particularly grateful to Lehigh University for a very generous sabbatical leave that allowed me to complete this book.

I have tried to write the kind of book I would have enjoyed reading and benefitted from as a graduate student. It is my hope that it will fill that role for others.

Nelson G. Markley

Chapter 1

Fundamental Theorems

The subject of this book is ordinary differential equations of the form $\dot{\mathbf{x}} = \mathbf{f}(t, \mathbf{x})$ where $\mathbf{f}(t, \mathbf{x})$, is a continuous function, $\mathbf{x}$ is a vector, and $\dot{\mathbf{x}}$ denotes the ordinary derivative with respect to the single variable t. The vector $\mathbf{x}$ is often thought of as a space variable and t as time.

Differential equations is an old subject whose long history goes back to Newton and Leibnitz and is tightly interwoven with that of calculus and classical physics. During the nineteenth century, the foundations of differential equations were more rigorously established, and in the twentieth century, it has continued to grow and develop in important new directions. The goal of this book is to provide an accessible concrete introduction to ordinary differential equations that is both modern and rigorous.

The purpose of this first chapter is to prove the basic facts on which the many branches of ordinary differential equations rest. These foundational results—existence, uniqueness, continuation, numerical approximation, and continuity in initial conditions—are akin to the axioms of abstract subjects like group theory. They are always with us and their use in general or specific questions becomes automatic.

To help make the proof of these fundamental results more accessible, the first section of the chapter provides a bridge from the core theoretical ideas of calculus to the analysis of functions of several variables and some specific results needed later in this chapter. Such a bridge cannot meet every reader's needs, but from it most readers should be able to build on their past knowledge to understand better the mathematical framework for studying differential equations.

The most fundamental question is the existence of solutions of an ordinary differential equation, because without solutions there is no subject of differential equations. The question of existence of solutions is addressed by requiring only that $\mathbf{f}(t, \mathbf{x})$ be continuous, although the proofs with more restrictive hypotheses are technically simpler. The advantage of requiring only continuity is that it provides a simple well-understood general context for studying principles of differential equations.

The proof of the main result about the existence of solutions to an ordinary differential equation raises three basic questions. When are solutions to a differential equation uniquely determined? Is there a constructive method for approximating solutions? How far can solutions be extended? These questions will be addressed in subsequent sections on Uniqueness, Numerical Methods, and Continuation. A key hypothesis running through these sections and the final section is that $\mathbf{f}(t, \mathbf{x})$ satisfy a Lipschitz condition. Consequently, understanding this hypothesis and when it holds will be important.

The last section focuses on the collective behavior of all the solutions of a differential equation. The key result, continuity in initial conditions, shows that all the solutions of a differential equation are bound together in one continuous function. In other words, solutions that start sufficiently close stay close over a finite interval of time. This result along with the previous sections provide a good set of fundamental tools for studying the solutions of ordinary differential equations.

1.1 Preliminaries

Before beginning the study of the differential equation $\dot{\mathbf{x}} = \mathbf{f}(t, \mathbf{x})$, it is necessary to set up some notation, review some basic facts, and describe the context for the study of such differential equations. In addition, a key theorem that is needed for existence will be proved at the end of the section.

The real numbers will always be denoted by $\mathbb{R}$, and $\mathbb{R}^m$ will denote m-dimensional Euclidean space, that is, $\mathbb{R}^m$ consists of all m-tuples of real numbers or

$$\mathbb{R}^m = \{(x_1, x_2, \ldots, x_m) : x_i \in \mathbb{R} \text{ for } i = 1, \ldots, m\}.$$

Boldface type will be used consistently to denote elements of $\mathbb{R}^m$, $m > 1$, that is, $\mathbf{x} = (x_1, x_2, \ldots, x_m)$. Elements of $\mathbb{R}^m$ can be thought of as points in Euclidean space or as vectors pointing from the origin to $\mathbf{x}$. Furthermore, boldface type will also be used to denote functions whose values are in $\mathbb{R}^m$ or what are commonly called vector valued functions.

The best approach to the study of $\dot{\mathbf{x}} = \mathbf{f}(t, \mathbf{x})$ is to use the topological ideas of open sets, closed sets, compact sets, and the norms used to define them. Consequently, open, closed, and compact sets of $\mathbb{R}^m$ will be used frequently, and so introducing these topological ideas is a natural starting point.

The *Euclidean norm* on $\mathbb{R}^m$ is defined by

$$\|\mathbf{x}\| = \left(\sum_{i=1}^{m} x_i^2 \right)^{1/2}$$

where $\mathbf{x} = (x_1, x_2, \ldots, x_m)$. The *distance* between $\mathbf{x}$ and $\mathbf{y}$ is defined to be $\|\mathbf{x} - \mathbf{y}\|$. This is the standard *Euclidean distance* between $\mathbf{x}$ and $\mathbf{y}$. In particular, $\|\mathbf{x}\|$ is just the Euclidean distance from $\mathbf{x}$ to $\mathbf{0} = (0, 0, \ldots, 0)$ and satisfies the following conditions for $\mathbf{x}$ and $\mathbf{y}$ in $\mathbb{R}^m$ and α in $\mathbb{R}$:

(a) $\|\mathbf{x}\| = 0$ if and only if $\mathbf{x} = \mathbf{0}$;

(b) $\|\alpha\mathbf{x}\| = |\alpha| \, \|\mathbf{x}\|$; and

(c) $\|\mathbf{x}+\mathbf{y}\| \leq \|\mathbf{x}\| + \|\mathbf{y}\|$.

The last condition is called the *triangle inequality.*

The proof of the triangle inequality and other useful elementary facts that may or may not be familiar to the reader are included in the exercises at the end of this section.

Remark *For* $\mathbf{x}$ *and* $\mathbf{y}$ *in* $\mathbb{R}^m$,

$$\big| \, \|\mathbf{x}\| - \|\mathbf{y}\| \, \big| \leq \|\mathbf{x} - \mathbf{y}\|.$$

Proof. By the triangle inequality $\|\mathbf{x}\| = \|\mathbf{x}-\mathbf{y}+\mathbf{y}\| \leq \|\mathbf{x}-\mathbf{y}\| + \|\mathbf{y}\|$ or $\|\mathbf{x}\| - \|\mathbf{y}\| \leq \|\mathbf{x}-\mathbf{y}\|$. Similarly, $\|\mathbf{y}\| - \|\mathbf{x}\| \leq \|\mathbf{y}-\mathbf{x}\| = \|\mathbf{x}-\mathbf{y}\|$ and the conclusion follows. □ (The symbol □ will be used to indicate the end of a proof.)

A set U in $\mathbb{R}^m$ is an *open set* if for every $\mathbf{x} \in U$ there exists $\varepsilon > 0$ such that

$$\{\mathbf{y} : \|\mathbf{y}-\mathbf{x}\| < \varepsilon\} \subset U.$$

It is easy to show that the *Euclidean ball*

$$\{\mathbf{y} : \|\mathbf{y}-\mathbf{x}\| < r\}$$

of radius r with center at $\mathbf{x}$ is itself an open set. Moreover, the union of open sets is open and the intersection of a finite number of open sets is open.

A subset F of $\mathbb{R}^m$ is a *closed set* if its complement,

$$\mathbb{R}^m \setminus F = \{\mathbf{x} : \mathbf{x} \notin F\}$$

is open. Not surprisingly, the closed Euclidean ball

$$\{\mathbf{y} : \|\mathbf{y}-\mathbf{x}\| \leq r\}$$

is a closed set. It follows from the above remarks about open sets that the intersection of closed sets is closed and the finite union of closed sets is closed. Note that $\mathbb{R}^m$ and the empty set, ϕ, are sets that are both open and closed.

Using the distance function $\|\mathbf{x}-\mathbf{y}\|$ instead of the usual absolute value $|x-y|$, it is easy to define the convergence of sequences in $\mathbb{R}^m$. A sequence of points $\mathbf{x}_k$ in $\mathbb{R}^m$ converges to $\mathbf{y}$ if given $\varepsilon > 0$, there exists $N > 0$ such that $\|\mathbf{x}_k - \mathbf{y}\| < \varepsilon$ when $k \geq N$. Moreover, the standard result from advanced calculus that a sequence of real numbers converges to a real number if and only if it is a Cauchy sequence extends to $\mathbb{R}^m$. (See *Exercise 9.*)

The idea of a compact set is more subtle. A set C is a *compact set* provided that whenever $\{U_\lambda\}$, $\lambda \in \Lambda$, is a family of open sets indexed by Λ such that

$$C \subset \bigcup_{\lambda \in \Lambda} U_\lambda$$

then there exists a finite set of indices $\lambda_1, \ldots, \lambda_k$ such that

$$C \subset \bigcup_{i=1}^{k} U_{\lambda_i}.$$

The essential theorem about compact sets in $\mathbb{R}^m$ is the well-known Heine-Borel theorem.

Theorem 1.1 (Heine-Borel) *A subset C of $\mathbb{R}^m$ is compact if and only if it is both closed and bounded.*

Proof. First assume C is compact. If C is not closed, there exists a sequence $\mathbf{x}_k$ in C converging to $\mathbf{y} \notin C$. (See *Exercise 6.*) For every $r > 0$ set

$$U_r = \mathbb{R}^m \setminus \{\mathbf{x} : \|\mathbf{x} - \mathbf{y}\| \leq r\} = \{\mathbf{x} : \|\mathbf{x} - \mathbf{y}\| > r\}.$$

Then each U_r is open,

$$C \subset \bigcup_{r>0} U_r,$$

but C is not contained in a finite union of the sets U_r, $r > 0$, contradicting the assumed compactness of C. Therefore, C must be closed.

If C is not bounded, there exists a sequence $\mathbf{x}_k \in C$ such that $\|\mathbf{x}_k\|$ goes to infinity. Set $U_r = \{\mathbf{x} : \|\mathbf{x}\| < r\}$ and obtain a contradiction as above. This completes the proof of the first half of the theorem.

Now, assume C is closed and bounded and suppose

$$C \subset \bigcup_{\lambda \in \Lambda} U_\lambda,$$

where each U_λ is open. Every open set is a union of Euclidean balls of the form

$$\{\mathbf{y} : \|\mathbf{y} - \mathbf{x}\| < r\}$$

where r is rational and the center $\mathbf{x}$ has rational coordinates, and there are only countably many such sets. (See *Exercises 11 and 12.*) Consequently, it suffices to consider

$$C \subset \bigcup_{i=1}^{\infty} U_i,$$

where $U_i = \{\mathbf{y} : \|\mathbf{y} - \mathbf{x}_i\| < r_i\}$, r_i is rational, and $\mathbf{x}_i$ has rational coordinates and to show that

$$C \subset \bigcup_{i=1}^{N} U_i,$$

for some N.

Suppose this does not hold. Then for each integer k there exists

$$\mathbf{x}_k \in C \setminus \bigcup_{i=1}^{k} U_i.$$

Consequently, the coordinates of the sequences of $\mathbf{x}_k$, $k = 1, \ldots$, are bounded because C is bounded. A bounded sequence of real numbers has a convergent subsequence by the Bolzano-Weierstrass theorem. (We assume the reader is familiar with the Bolzano-Weierstrass theorem from advanced calculus or real analysis.) In particular, there exists a subsequence with convergent first coordinates. It in turn has a subsequence with both first and second coordinates converging. Do this m-times to get a subsequence $\mathbf{x}_{k_j}$ converging to $\mathbf{y}$. Because C is closed, $\mathbf{y} \in C$. Therefore, $\mathbf{y} \in U_k$ for some k and $\mathbf{x}_{k_j} \in U_k$ for large j. This contradiction completes the proof. □

Consider a function $\mathbf{f} : W \to \mathbb{R}^n$, where W is an open subset of $\mathbb{R}^m$ and m and n are arbitrary positive integers. The function $\mathbf{f}$ is *continuous* at $\mathbf{x} \in W$ if given $\varepsilon > 0$, there exists $\delta > 0$ such that $\|\mathbf{f}(\mathbf{y}) - \mathbf{f}(\mathbf{x})\| < \varepsilon$ whenever $\|\mathbf{y} - \mathbf{x}\| < \delta$. Although continuity is a point property, we will only be using functions that are continuous at every point of their domain. In this context, it is easy to prove the following: $\mathbf{f}$ is continuous at every point of W if and only if $\mathbf{f}^{-1}(U) = \{\mathbf{x} : \mathbf{f}(\mathbf{x}) \in U\}$ is open for every open set U of $\mathbb{R}^n$. The next theorem links compactness and continuity by showing that the continuous image of a compact set is compact.

Proposition 1.2 *Let* $\mathbf{f} : W \to \mathbb{R}^n$ *be a continuous function on the open set* W *of* $\mathbb{R}^m$ *and let* C *be a compact set contained in* W. *Then* $\mathbf{f}(C)$ *is compact.*

Proof. Suppose

$$\mathbf{f}(C) \subset \bigcup_{\lambda \in \Lambda} U_\lambda,$$

where each U_λ is open. It follows that

$$C \subset \bigcup_{\lambda \in \Lambda} \mathbf{f}^{-1}(U_\lambda).$$

Since each $\mathbf{f}^{-1}(U_\lambda)$ is open by the continuity of $\mathbf{f}$, there exist $\lambda_1, \ldots, \lambda_k$ such that

$$C \subset \bigcup_{i=1}^{k} \mathbf{f}^{-1}(U_{\lambda_i})$$

because C is compact. It follows that

$$\mathbf{f}(C) \subset \bigcup_{i=1}^{k} U_{\lambda_i}$$

to complete the proof. □

Knowing that a continuous real-valued function is bounded on a compact set will be a common ingredient in proving theorems about differential equations. In fact, a continuous function assumes its maximum and minimum values on a compact set as the next result establishes.

Proposition 1.3 *Let $f : W \to \mathbb{R}$ be a continuous function on the open set W of $\mathbb{R}^m$ and let C be a compact set contained in W. Then f is bounded on C and there exist $\mathbf{x}_m$ and $\mathbf{x}_M$ in C such that*

$$f(\mathbf{x}_m) \leq f(\mathbf{x}) \leq f(\mathbf{x}_M)$$

for all $\mathbf{x}$ in C.

Proof. Since $f(C)$ is compact by the previous theorem, it is closed and bounded by the Heine-Borel theorem. Hence, $\inf\{f(\mathbf{x}) : \mathbf{x} \in C\}$ and $\sup\{f(\mathbf{x}) : \mathbf{x} \in C\}$ are finite and belong to $f(C)$. So, there exist $\mathbf{x}_m$ and $\mathbf{x}_M$ in C such that

$$f(\mathbf{x}_m) = \inf\{f(\mathbf{x}) : \mathbf{x} \in C\}$$

and

$$f(\mathbf{x}_M) = \sup\{f(\mathbf{x}) : \mathbf{x} \in C\}.$$

Obviously $f(\mathbf{x}_m) \leq f(\mathbf{x}) \leq f(\mathbf{x}_M)$ for all $\mathbf{x}$ in C. □

Let $\mathbf{f} : W \to \mathbb{R}^n$ be continuous function on an open set W of $\mathbb{R}^m$ and let E be a subset of W. The function $\mathbf{f}$ is *uniformly continuous on* E if given $\varepsilon > 0$ there exists $\delta > 0$ such that $\|\mathbf{f}(\mathbf{y}) - \mathbf{f}(\mathbf{x})\| < \varepsilon$ whenever $\|\mathbf{y} - \mathbf{x}\| < \delta$ and both $\mathbf{x}$ and $\mathbf{y}$ are in E. Again there is an important connection with compactness.

Proposition 1.4 *Let $\mathbf{f} : W \to \mathbb{R}^n$ be continuous function on an open set W of $\mathbb{R}^m$. If C is a compact set contained in W, then $\mathbf{f}$ is uniformly continuous on C.*

Proof. Let $\varepsilon > 0$. Because $\mathbf{f}$ is continuous at every point of W, it follows that for each $\mathbf{x} \in C$ there exists $\delta_{\mathbf{x}} > 0$ such that $\|\mathbf{f}(\mathbf{y}) - \mathbf{f}(\mathbf{x})\| < \varepsilon/2$ when $\|\mathbf{y} - \mathbf{x}\| < \delta_{\mathbf{x}}$. Set $U_{\mathbf{x}} = \{\mathbf{y} : \|\mathbf{y} - \mathbf{x}\| < \delta_{\mathbf{x}}/2\}$. Then the sets $U_{\mathbf{x}}$ are a family of open sets such that

$$C \subset \bigcup_{\mathbf{x} \in C} U_{\mathbf{x}}.$$

By the Heine-Borel theorem there exists $\mathbf{x}_1, \ldots, \mathbf{x}_k$ in C such that

$$C \subset \bigcup_{j=1}^{k} U_{\mathbf{x}_j}.$$

Let $\delta = \min\{\delta_{\mathbf{x}_1}/2, \ldots, \delta_{\mathbf{x}_k}/2\}$. Suppose $\|\mathbf{x} - \mathbf{y}\| < \delta$ with $\mathbf{x}$ and $\mathbf{y}$ in C. Then there exists j such that $\|\mathbf{y} - \mathbf{x}_j\| < \delta_{\mathbf{x}_j}/2$. It follows that

$$\|\mathbf{x} - \mathbf{x}_j\| \leq \|\mathbf{x} - \mathbf{y}\| + \|\mathbf{y} - \mathbf{x}_j\| < \delta + \delta_{\mathbf{x}_j}/2 \leq \delta_{\mathbf{x}_j}/2 + \delta_{\mathbf{x}_j}/2 = \delta_{\mathbf{x}_j}.$$

By the choice of $\delta_{\mathbf{x}_j}$, both $\|\mathbf{f}(\mathbf{y}) - \mathbf{f}(\mathbf{x}_j)\| < \varepsilon/2$ and $\|\mathbf{f}(\mathbf{x}) - \mathbf{f}(\mathbf{x}_j)\| < \varepsilon/2$. The triangle inequality implies that $\|\mathbf{f}(\mathbf{x}) - \mathbf{f}(\mathbf{y})\| < \varepsilon$ to complete the proof. □

The entire discussion of open, closed, and compact sets originated from the Euclidean distance between two points. Although we are most familiar with

Euclidean distance, it is not, however, the only distance function on which the discussion could have been based. Many arguments will require estimates of distance or size based on norms, and frequently norms other than the Euclidean norm will be easier to apply. To this end, a discussion of norms in general is worth the time, and it will be helpful to prove that it does not matter which norm is used.

A real-valued function $\|\mathbf{x}\|_a$ on $\mathbb{R}^m$ is called a *norm* if it satisfies the following conditions:

(a) $\|\mathbf{x}\|_a = 0$ if and only if $\mathbf{x} = \mathbf{0}$;

(b) $\|\alpha\mathbf{x}\|_a = |\alpha|\,\|\mathbf{x}\|_a$ for all $\alpha \in \mathbb{R}$ and $\mathbf{x} \in \mathbb{R}^m$

(c) $\|\mathbf{x}+\mathbf{y}\|_a \leq \|\mathbf{x}\|_a + \|\mathbf{y}\|_a$ for all $\mathbf{x}$, $\mathbf{y} \in \mathbb{R}^m$.

As with the Euclidean norm it follows from the triangle inequality that

$$\big|\|\mathbf{x}\|_a - \|\mathbf{y}\|_a\big| \leq \|\mathbf{x}-\mathbf{y}\|_a.$$

Two other simple examples norms on $\mathbb{R}^m$ are

$$|\mathbf{x}| = \sum_{i=1}^{m} |x_i|$$

and

$$\|\mathbf{x}\|_\infty = \max\big\{|x_i| : 1 \leq i \leq m\big\}.$$

The first of these two norms will be particularly useful.

To what extent do the topological ideas of open, closed, and compact sets depend on the norm is now an obvious question. Two norms, $\|\cdot\|_1$ and $\|\cdot\|_2$ are called *equivalent* if there exists positive constants A and B satisfying

$$A\|\mathbf{x}\|_1 \leq \|\mathbf{x}\|_2 \leq B\|\mathbf{x}\|_1$$

for all $\mathbf{x}$. [The dot $\cdot$ in the notation $|\cdot|$ or $\mathbf{g}(\cdot)$ indicates an unnamed variable of a norm or a function.]

If two norms are equivalent, then for a given $\mathbf{x}$

$$\big\{\mathbf{y} : \|\mathbf{y}-\mathbf{x}\|_1 < r/B\big\} \subset \big\{\mathbf{y} : \|\mathbf{y}-\mathbf{x}\|_2 < r\big\} \subset \big\{\mathbf{y} : \|\mathbf{y}-\mathbf{x}\|_1 < r/A\big\},$$

and either norm will define the same family of open sets. The next theorem completely settles the question of which norm to use by establishing that they are all equivalent. This fact will be technically very helpful.

Theorem 1.5 *Any two norms on* $\mathbb{R}^m$ *are equivalent.*

Proof. It suffices to show that the Euclidean norm, $\|\cdot\|$, is equivalent to an arbitrary norm $\|\cdot\|_a$. (See *Exercise 17.*) Let $\mathbf{e}_1, \ldots, \mathbf{e}_m$ be the *standard basis of* $\mathbb{R}^m$, that is, $\mathbf{e}_1 = (1, 0, 0, \ldots, 0)$, $\mathbf{e}_2 = (0, 1, 0. \ldots, 0)$, etc. Set

$$c = \max\big\{\|\mathbf{e}_j\|_a : 1 \leq j \leq m\big\}.$$

Then

$$\mathbf{x} = \sum_{j=1}^{n} x_j \mathbf{e}_j$$

and

$$\|\mathbf{x}\|_a \le \sum_{j=1}^{m} |x_j| \, \|\mathbf{e}_j\|_a \le c \sum_{j=1}^{m} |x_j| \le mc\|\mathbf{x}\|$$

because $|x_j| \le \|\mathbf{x}\|$. This establishes the first required inequalities for the equivalence of norms.

Also note that

$$\big| \|\mathbf{x}\|_a - \|\mathbf{y}\|_a \big| \le \|\mathbf{x} - \mathbf{y}\|_a \le mc\|\mathbf{x} - \mathbf{y}\|$$

implies that $\|\mathbf{x}\|_a$ is a continuous function of $\mathbf{x}$ on $\mathbb{R}^m$. (Given $\varepsilon > 0$, let $\delta = \varepsilon/mc$.)

To establish the second required inequality, it suffices to show that there exists $A > 0$ such that $\|\mathbf{x}\| = 1$ implies $A \le \|\mathbf{x}\|_a$ because then for any $\mathbf{x} \ne \mathbf{0}$,

$$\left\| \tfrac{1}{\|\mathbf{x}\|} \mathbf{x} \right\| = 1$$

and hence

$$A \le \left\| \tfrac{1}{\|\mathbf{x}\|} \mathbf{x} \right\|_a = \frac{1}{\|\mathbf{x}\|} \|\mathbf{x}\|_a$$

or

$$A\|\mathbf{x}\| \le \|\mathbf{x}\|_a.$$

Since $\{\mathbf{x} : \|\mathbf{x}\| = 1\}$ is compact and $\|\mathbf{x}\|_a$ is continuous, by *Proposition 1.3*, there exists $\mathbf{x}_m$ with $\|\mathbf{x}_m\| = 1$ such that

$$\|\mathbf{x}_m\|_a \le \|\mathbf{x}\|_a$$

whenever $\|\mathbf{x}\| = 1$. To complete the proof, set $A = \|\mathbf{x}_m\|_a$, which is positive because $\mathbf{x}_m \ne 0$. □

The above theorem is only true for finite-dimensional vector spaces. In fact, its failure in the infinite-dimensional case is one of the key differences between finite-dimensional and infinite-dimensional normed vector spaces.

Sequences of vector valued functions play a critical role in the study of differential equations, especially uniformly convergent sequences. They are the final preparatory topic in this section. Let $\mathbf{f}_k : W \to \mathbb{R}^n$ be a sequence of continuous functions on an open set $W \subset \mathbb{R}^m$. The sequence of functions $\mathbf{f}_k$ converges to a function $\mathbf{f} : W \to \mathbb{R}^n$ if the sequence $\mathbf{f}_k(\mathbf{x})$ converges to $\mathbf{f}(\mathbf{x})$ for every $\mathbf{x} \in W$. In general, the limit of continuous functions need not be a continuous function unless the convergence is uniform.

The sequence of functions $\mathbf{f}_k$ *converges uniformly on* W to a function $\mathbf{f} : W \to \mathbb{R}^n$ if given $\varepsilon > 0$ there exists $N > 0$ such that $\|\mathbf{f}_k(\mathbf{x}) - \mathbf{f}(\mathbf{x})\| < \varepsilon$ for every $\mathbf{x} \in W$ when $k \ge N$. The proofs of several crucial theorems in this chapter depend on showing that a sequence of functions converges uniformly on an open set and then applying the following result:

Proposition 1.6 *Let* $\mathbf{f}_k : W \to \mathbb{R}^n$ *be a sequence of continuous functions on an open set* $W \subset \mathbb{R}^m$. *If* $\mathbf{f}_k$ *converges uniformly to a function* $\mathbf{f} : W \to \mathbb{R}^n$, *then* $\mathbf{f}$ *is continuous on* W.

Proof. Let $\mathbf{y}$ be a point in W. Given $\varepsilon > 0$, there exists $N > 0$ such that $\|\mathbf{f}_k(\mathbf{x}) - \mathbf{f}(\mathbf{x})\| < \varepsilon/3$ for every $\mathbf{x} \in W$ when $k \geq N$. Choose a $k \geq N$. Since $\mathbf{f}_k$ is continuous at $\mathbf{y}$, there exists $\delta > 0$ such that $\|\mathbf{f}_k(\mathbf{x}) - \mathbf{f}_k(\mathbf{y})\| < \varepsilon/3$ when $\|\mathbf{x} - \mathbf{y}\| < \delta$.

Putting the pieces together

$$\|\mathbf{f}(\mathbf{x})-\mathbf{f}(\mathbf{y})\| \leq \|\mathbf{f}(\mathbf{x})-\mathbf{f}_k(\mathbf{x})\|+\|\mathbf{f}_k(\mathbf{x})-\mathbf{f}_k(\mathbf{y})\|+\|\mathbf{f}_k(\mathbf{y})-\mathbf{f}(\mathbf{y})\| < \frac{\varepsilon}{3}+\frac{\varepsilon}{3}+\frac{\varepsilon}{3} = \varepsilon$$

when $\|\mathbf{x} - \mathbf{y}\| < \delta$, and thus $\mathbf{f}$ is continuous at $\mathbf{x}$. □

The proof of the existence of solutions to $\dot{\mathbf{x}} = \mathbf{f}(t, \mathbf{x})$, the first major result about differential equations, will require a theorem known as Ascoli's Lemma. Since it is not as commonly known as other parts of advanced calculus and real analysis, a complete proof of it is included.

Consider a sequence of functions $\mathbf{f}_m : I \to \mathbb{R}^n$ defined on an interval I. The set of functions $\{\mathbf{f}_m : m \geq 1\}$ is *equicontinuous* if given $\varepsilon > 0$, there exists $\delta > 0$ such that for all $m \geq 1$

$$\|\mathbf{f}(s) - \mathbf{f}(t)\| < \varepsilon,$$

whenever $|s - t| < \delta$.

Theorem 1.7 (Ascoli) *Let* $\mathbf{f}_m : I \to \mathbb{R}^n$ *be a sequence of functions defined on a bounded interval* I. *If the set of functions* $\{\mathbf{f}_m : m \geq 1\}$ *is equicontinuous and for each* $t \in I$, *the sequence* $\mathbf{f}_m(t)$ *is bounded, then there exists a subsequence of* $\mathbf{f}_m$, *which converges uniformly on* I.

Proof. Let $\{r_1, r_2, r_3, \ldots\}$ be an enumeration of the rational numbers in I. (See *Exercise 10.*) Since $\mathbf{f}_m(r_1)$ is a bounded sequence of vectors in $\mathbb{R}^n$, it contains a convergent subsequence by the Bolzano-Weierstrass theorem. Thus there exists a subsequence of $\mathbf{f}_m$ denoted by $\mathbf{f}_{(p,1)}$ such that $\mathbf{f}_{(p,1)}(r_1)$ converges to an element of $\mathbb{R}^n$. Since a subsequence of the sequence $\mathbf{f}_m$ is determined by picking an increasing sequence of integer indices, $(p, 1)$ is just the notation being used for the pth integer in the increasing sequence of integers that determines the first subsequence of $\mathbf{f}_m$.

For the same reason, there exists a subsequence $\mathbf{f}_{(p,2)}$ of $\mathbf{f}_{(p,1)}$ such that $\mathbf{f}_{(p,2)}(r_2)$ converges in $\mathbb{R}^n$. Of course, $\lim_{p\to\infty} \mathbf{f}_{(p,2)}(r_1)$ remains unchanged because a subsequence of a convergent sequence converges to the same limit. Now $(p, 2)$ denotes the pth term of the second subsequence. Because we are selecting a subsequence of a subsequence, the increasing sequence of integers $(p, 2)$ must be selected from the increasing sequence $(p, 1)$. Since this process must be repeated ad infinitum, this notation is not as strange as it might first seem, and $\mathbf{f}_{(p,k)}$ will naturally denote the pth in the kth subsequence of $\mathbf{f}_m$.

Using induction, it follows that for every k there exists a subsequence $\mathbf{f}_{(p,k)}$ of $\mathbf{f}_m$ satisfying:

(a) $\mathbf{f}_{(p,k)}$ is a subsequence of $\mathbf{f}_{(p,j)}$ for $j \leq k$, and

(b) $\mathbf{f}_{(p,k)}(r_j)$ converges for $j \leq k$.

Set $\mathbf{g}_p = \mathbf{f}_{(p,p)}$ and verify that $\mathbf{g}_p$ with $p \geq k$ is a subsequence of $\mathbf{f}_{(p,k)}$. In particular, it follows that $\mathbf{g}_p$ is a subsequence of $\mathbf{f}_m$ and that $\mathbf{g}_p(r_k)$ converges for every rational number r_k in I. The proof will be completed by showing that the sequence $\mathbf{g}_p$ is uniformly convergent on I.

Let $\varepsilon > 0$. By equicontinuity, there exists $\delta > 0$ such that for all p

$$\|\mathbf{g}_p(s) - \mathbf{g}_p(t)\| < \frac{\varepsilon}{3},$$

whenever $|s - t| < \delta$, s, $t \in I$. Because the rational numbers are dense in $\mathbb{R}$ and I is bounded, there exists k such that for every $t \in I$ we have $|t - r_i| < \delta$ for some $i \leq k$. Since the sequences $\mathbf{g}_p(r_i)$, $1 \leq i \leq k$ are all Cauchy sequences of real numbers, there exists N such that for $1 \leq i \leq k$

$$\|\mathbf{g}_p(r_i) - \mathbf{g}_q(r_i)\| < \frac{\varepsilon}{3}$$

whenever p, $q \geq N$. Now, let $t \in I$ and consider p, $q \geq N$. Select an r_i, $1 \leq i \leq k$, such that $|t - r_i| < \delta$, and then by the triangle inequality we have

$$\|\mathbf{g}_p(t) - \mathbf{g}_q(t)\| \leq$$

$$\|\mathbf{g}_p(t) - \mathbf{g}_p(r_i)\| + \|\mathbf{g}_p(r_i) - \mathbf{g}_q(r_i)\| + \|\mathbf{g}_q(r_i) - \mathbf{g}_q(t)\| \leq$$

$$\tfrac{\varepsilon}{3} + \tfrac{\varepsilon}{3} + \tfrac{\varepsilon}{3} = \varepsilon.$$

Therefore, $\mathbf{g}_p(t)$ is a Cauchy sequence, and hence converges to some $\mathbf{g}(t)$ for all t in I. Letting p go to infinity in the above inequality, gives $\|\mathbf{g}(t) - \mathbf{g}_q(t)\| \leq \varepsilon$ for $t \in I$ and $q \geq N$, and proves that $\mathbf{g}_p(t)$ converges uniformly to $\mathbf{g}(t)$ on I. □

EXERCISES

1. Prove the *Cauchy-Schwarz inequality*:

$$|\mathbf{v} \cdot \mathbf{w}| \leq \|\mathbf{v}\| \, \|\mathbf{w}\|$$

Hint: For fixed $\mathbf{v}$ and $\mathbf{w}$ the function $\|t\mathbf{v} - \mathbf{w}\|^2$ is a quadratic in t whose discriminant must be less than or equal to 0 because $0 \leq \|t\mathbf{v} - \mathbf{w}\|^2$. (The *discriminant* of $at^2 + bt + c$ is $b^2 - 4ac$.)

2. Use the Cauchy-Schwarz inequality to show that

$$\|\mathbf{x} + \mathbf{y}\| \leq \|\mathbf{x}\| + \|\mathbf{y}\|.$$

3. Show that for fixed $\mathbf{v}$ and $r > 0$ the set

$$\{\mathbf{x} : \|\mathbf{x} - \mathbf{v}\| < r\}$$

is an open set and the set

$$\{\mathbf{x} : \|\mathbf{x} - \mathbf{v}\| \leq r\}$$

is a closed set.

4. Prove that the union of open sets is open and the intersection of a finite number of open sets is open.

5. Prove that the intersection of closed sets is closed and the finite union of closed sets is closed.

6. Prove that a set C is a closed set if and only if for every convergent sequence $\mathbf{x}_n$ in C, its limit is also in C.

7. Let B be a subset of $\mathbb{R}^m$. Define the *closure* of B by

$$\overline{B} = \{\mathbf{x} : B \cap \{\mathbf{y} : \|\mathbf{y} - \mathbf{x}\| < \varepsilon\} \neq \phi \text{ for all } \varepsilon > 0\}.$$

Prove the following:

(a) The set $\overline{B}$ is a closed set containing B.

(b) If C is a closed set such that $B \subset C$, then $\overline{B} \subset C$.

(c) The set B is closed if and only if $B = \overline{B}$.

(d) $\overline{B} = \cap\{C : B \subset C = \overline{C}\}$.

(e) The closure of an open Euclidean ball is a closed Euclidean ball.

8. Let D be an open subset of $\mathbb{R}^d$ and let C be a compact set contained in D. Set $\rho = \inf\{|\mathbf{x} - \mathbf{y}| : \mathbf{x} \in C \text{ and } \mathbf{y} \notin D\}$.

(a) Show that ρ is positive. Is this true if C is just closed?

(b) Show that $\{\mathbf{y} : |\mathbf{x} - \mathbf{y}| \leq \rho/2$ for some $\mathbf{x} \in C\}$ is a compact set contained in D.

(c) Show that $\{\mathbf{y} : |\mathbf{x} - \mathbf{y}| < \rho/2$ for some $\mathbf{x} \in C\}$ is an open set.

9. A sequence of points $\mathbf{x}_k$ in $\mathbb{R}^m$ is a Cauchy sequence if given $\varepsilon > 0$, there exists $N > 0$ such that $\|\mathbf{x}_j - \mathbf{x}_k\| < \varepsilon$ when $j \geq N$ and $k \geq N$. Using the fact that a sequence of real numbers converges to a real number if and only if it is a Cauchy sequence, show that a sequence $\mathbf{x}_k$ in $\mathbb{R}^m$ converges to a point $\mathbf{y}$ in $\mathbb{R}^m$ if and only if it is Cauchy sequence.

10. Construct a sequence of rational numbers r_n that includes every positive rational number. Modify the construction to include every rational number. Given an interval I, construct a sequence of rational numbers r_n that includes every rational number in the interval I.

11. A set B is *countable* provided there exists a sequence $\mathbf{x}_k$ such that $\mathbf{x}$ is in B if and only if $\mathbf{x} = \mathbf{x}_k$ for some k. The rational numbers are countable by the previous exercise. Prove that the set

$$\{\mathbf{x} = (x_1, x_2, \ldots, x_m) \in \mathbb{R}^m : x_j \text{ is rational for } j = 1, 2, \ldots, m\}$$

is countable.

12. Prove that every open set of $\mathbb{R}^m$ is a union of Euclidean balls of the form $\{\mathbf{y} : \|\mathbf{y} - \mathbf{x}\| < r\}$, where r is rational and the center $\mathbf{x}$ has rational coordinates.

13. Let $\mathbf{f} : \Omega \to \mathbb{R}^n$ be a function, where Ω is an open subset of $\mathbb{R}^m$. Prove that f is continuous at every point of Ω if and only if $f^{-1}(U) = \{\mathbf{x} : f(\mathbf{x}) \in U\}$ is an open set for every open set U of $\mathbb{R}^n$.

14. Let $\mathbf{f} : \mathbb{R}^m \to \mathbb{R}^n$ be a function. Prove that f is continuous at every point of $\mathbb{R}^m$ if and only if $f^{-1}(C) = \{\mathbf{x} : f(\mathbf{x}) \in C\}$ is a closed set for every closed set C of $\mathbb{R}^n$.

15. Let $\mathbf{f} : \Omega \to \mathbb{R}^n$ be a function, where Ω is an open subset of $\mathbb{R}^m$. Prove that f is continuous at $\mathbf{x} \in \Omega$ if and only if for every sequence $\{\mathbf{x}_k\}$ converging to $\mathbf{x}$, the sequence $\{\mathbf{f}(\mathbf{x}_k)\}$ converges to $\mathbf{f}(\mathbf{x})$.

16. Show that

$$|\mathbf{x}| = \sum_{i=1}^{d} |x_i|$$

and

$$\|\mathbf{x}\|_\infty = \max\{|x_i| : 1 \leq i \leq d\}$$

define norms on $\mathbb{R}^d$ and find A and B such that $A|\mathbf{x}| \leq \|\mathbf{x}\|_\infty \leq B|\mathbf{x}|$. For $d = 2$, graph the sets $\{\mathbf{x} : |\mathbf{x}| \leq 1\}$ and $\{\mathbf{x} : \|\mathbf{x}\|_\infty \leq 1\}$.

17. Let $\|\cdot\|_a$, $\|\cdot\|_b$, and $\|\cdot\|_c$ be three norms on a vector space V. Show that if $\|\cdot\|_a$ and $\|\cdot\|_b$ are equivalent norms and if $\|\cdot\|_b$ and $\|\cdot\|_c$ are equivalent norms, then $\|\cdot\|_a$ and $\|\cdot\|_c$ are also equivalent norms.

18. Show by example that Ascoli's Lemma is false when each of the following hypotheses are individually deleted from the statement of the result:

    (a) The interval I is bounded,

    (b) The sequence of functions $\{f_m\}$ is equicontinuous,

    (c) The sequence $\{f_m(t)\}$ is bounded for every $t \in I$.

1.2 Existence

The study of differential equations can now begin in earnest. The class of differential equations to be examined needs to be fully described, the concept of a solution of a differential equation needs to be defined, and the question of whether or not a particular differential equation has solutions needs to be addressed. These things will take place in this section and launch all that follows.

Let D be an open set of $\mathbb{R}^{d+1}$ and let $\mathbf{f} : D \to \mathbb{R}^d$ be a continuous function. As is customary in differential equations, a point of $\mathbb{R}^{d+1}$ will be denoted by $(t, \mathbf{x})$, where $t \in \mathbb{R}$ and $\mathbf{x} \in \mathbb{R}^d$. Moreover, in this notation t will be thought of as time and $\mathbf{x}$ as position in space. In this context, the *differential equation*

$$\dot{\mathbf{x}} = \mathbf{f}(t, \mathbf{x}) \tag{1.1}$$

is the most general differential equation that will be considered, and the notation set out in this paragraph will be used consistently for it.

Actually, equation *(1.1)* is a system of differential equations. Since

$$\mathbf{f}(t, \mathbf{x}) = \big(f_1(t, \mathbf{x}), \dots, f_d(t, \mathbf{x})\big)$$

where the f_i are continuous real-valued functions, $\dot{\mathbf{x}} = \mathbf{f}(t, \mathbf{x})$ can also be written

$$\begin{aligned} \dot{x}_1 &= f_1(t, x_1, x_2, \dots, x_d) \\ \dot{x}_2 &= f_2(t, x_1, x_2, \dots, x_d) \\ \vdots & \quad \vdots \\ \dot{x}_d &= f_d(t, x_1, x_2, \dots, x_d). \end{aligned}$$

Let I be an open interval (possibly infinite). A *curve* is a continuous function $\varphi : I \to \mathbb{R}^d$. The curve $\varphi : I \to \mathbb{R}^d$ is said to be differentiable if $\dot{\varphi} = (\dot{\varphi}_1, \dots, \dot{\varphi}_d)$ exists at every point of I, where as usual $\dot{\varphi}_i$ is the derivative of φ_i, which is a real-valued function of one real variable. A *solution* of 1.1 is simply a differentiable curve $\varphi : I \to \mathbb{R}^d$ such that on I

$$\dot{\varphi}(t) = \mathbf{f}\big(t, \varphi(t)\big).$$

Note it follows that if φ is a solution of *(1.1)*, then $\big(t, \varphi(t)\big)$ is in D for $t \in I$, and the curve $t \to \big(t, \varphi(t)\big)$ lying in D is called a *trajectory* of *(1.1)*. Because $\mathbf{f}$ and φ are continuous, it also follows that $\dot{\varphi}$ is continuous. That is, a solution must be continuously differentiable. Instead of using a different symbol for the curve φ and the point $\mathbf{x}$, we will usually write $\mathbf{x}(t)$ to denote dependency of $\mathbf{x}$ on t in a solution of *(1.1)*.

For differential equations, it is slightly more convenient to use the norm

$$|\mathbf{x}| = \sum_{i=1}^{d} |x_i|$$

and define the distance between $\mathbf{x}$ and $\mathbf{y}$ to be $|\mathbf{x} - \mathbf{y}|$. Using $|\mathbf{x}|$ instead of the more familiar $\|\mathbf{x}\|$ has no effect on the open sets, the continuity of functions, and the convergence of sequences by *Theorem 1.5.* Not only are $\|\cdot\|$ and $|\cdot|$ equivalent, it can easily be verified that

$$\frac{1}{d}|\mathbf{x}| \leq \|\mathbf{x}\| \leq |\mathbf{x}|.$$

Let φ be a curve whose domain includes the closed interval $[a, b]$. Define the integral of φ from a to b by

$$\int_a^b \varphi(t)\, dt = \left(\int_a^b \varphi_1(t)\, dt, \ldots, \int_a^b \varphi_d(t)\, dt \right).$$

When $a < b$, then

$$\begin{aligned} \left| \int_a^b \varphi(t)\, dt \right| &= \sum_{i=1}^{d} \left| \int_a^b \varphi_i(t)\, dt \right| \\ &\leq \sum_{i=1}^{d} \int_a^b \big|\varphi_i(t)\big|\, dt \\ &= \int_a^b \sum_{i=1}^{d} \big|\varphi_i(t)\big|\, dt \\ &= \int_a^b \big|\varphi(t)\big|\, dt. \end{aligned}$$

So in this context, the familiar inequality

$$\left| \int_a^b \varphi(t)\, dt \right| \leq \int_a^b \big|\varphi(t)\big|\, dt$$

when $a < b$ is retained and is one reason for preferring the norm $|\cdot|$ to the Euclidean norm. The next remark shows why the integral of a curve is relevant for differential equations.

Proposition 1.8 *Let* $\mathbf{x}(t)$ *be a continuous function on the open interval* I *such that* $(t, \mathbf{x}(t)) \in D$ *for all* $t \in I$*, and let* $\tau \in I$*. Then* $\mathbf{x}(t)$ *is a solution of* $\dot{\mathbf{x}} = \mathbf{f}(t, \mathbf{x})$ *if and only if*

$$\mathbf{x}(t) = \mathbf{x}(\tau) + \int_{\tau}^{t} \mathbf{f}\big(s, \mathbf{x}(s)\big)\, ds. \tag{1.2}$$

Proof. Just apply the fundamental theorem of calculus to each coordinate. □

The first fundamental question about the differential equation $\dot{\mathbf{x}} = \mathbf{f}(t, \mathbf{x})$ is whether or not any solutions exist, but this question cannot be phrased quite so simply. Since the trajectory of one solution occupies an insignificant portion of D, knowing that *(1.1)* has a solution tells us very little. Furthermore, in many physical situations, the initial data specifies a point in D through which the desired trajectory must pass.

The right existence question to ask is When does there exist a trajectory passing through a specific point of D? The answer to this question is the best possible, and the primary goal of this section is to prove that at least one trajectory passes through each point of D.

Seeking a solution whose trajectory passes through a specified point $(\tau, \boldsymbol{\xi}) \in D$ is called an initial-value problem. Note that $(\tau, \boldsymbol{\xi})$ is on the trajectory of the solution $\mathbf{x}(t)$ if and only if $\big(\tau, \mathbf{x}(\tau)\big) = (\tau, \boldsymbol{\xi})$ or $\mathbf{x}(\tau) = \boldsymbol{\xi}$. Consequently, an *initial-value problem* can always be written in the form

$$\begin{aligned} \dot{\mathbf{x}} &= \mathbf{f}(t, \mathbf{x}) \\ \boldsymbol{\xi} &= \mathbf{x}(\tau). \end{aligned}$$

A solution to the above initial-value problem starts at $\boldsymbol{\xi}$ at time τ and heads in the $\mathbf{f}(\tau, \boldsymbol{\xi})$ direction for just an instant. By actually following the line through $\boldsymbol{\xi}$ in the $\mathbf{f}(\tau, \boldsymbol{\xi})$ direction for a short time we will not stay on the solution, but the error should be small over a short time interval. We can also stop and correct our course by recalculating the direction $\mathbf{f}(t, \mathbf{x})$ at a new time and point, and then follow the line in the corrected direction. Repeating this process for short intervals of time should track an actual solution. The challenge is to show that this intuitive idea really works.

Theorem 1.9 (Peano) *If* $\mathbf{f}(t, \mathbf{x})$ *is continuous on the open set* D*, then for each point* $(\tau, \boldsymbol{\xi}) \in D$*, there exists at least one solution to the initial-value problem*

$$\begin{aligned} \dot{\mathbf{x}} &= \mathbf{f}(t, \mathbf{x}) \\ \boldsymbol{\xi} &= \mathbf{x}(\tau). \end{aligned} \tag{1.3}$$

Before tackling the details of the proof, it is worthwhile outlining the argument to understand how the pieces will fit together. The first step is to define a sequence of approximate solutions

$$\varphi_m : \{t : \tau - \alpha < t < \tau + \alpha\} \to \mathbb{R}^d$$

for a suitable $\alpha > 0$ using line segments. The function $\mathbf{f}(t, \mathbf{x})$ will be used to determine the directions of these line segments. Specifically, starting at the point $\boldsymbol{\xi}$ and at time τ, the curve $\varphi_m(t)$ will move along the line through $\boldsymbol{\xi}$ in the direction $\mathbf{f}(\tau, \boldsymbol{\xi})$ until a time t_1. From t_1 to t_2, it will move along the line through $\varphi_m(t_1)$ in the direction $\mathbf{f}\big(t_1, \varphi_m(t_1)\big)$, and so forth. Of course, as m gets large, the distance between τ and t_1, t_1 and t_2, and so forth, will be sent to zero.

The second step is to show that Ascoli's lemma applies to the sequence φ_m of functions, and thus establishes the existence of a subsequence φ_{m_k}, which converges uniformly on $[\tau - \alpha, \tau + \alpha]$ to a function φ.

The final step is to prove that φ is a solution of $\dot{\mathbf{x}} = \mathbf{f}(t, \mathbf{x})$ satisfying $\varphi(\tau) = \boldsymbol{\xi}$. Carrying out the details of the proof will take several pages and may take more than one reading to understand completely.

Proof. Because D is open, there exists $b > 0$ such that for the given point $(\tau, \boldsymbol{\xi}) \in D$, the set

$$R = \big\{(t, \mathbf{x}) : |t - \tau| \le b \text{ and } |\mathbf{x} - \boldsymbol{\xi}| \le b\big\} \subset D.$$

Since the rectangle R is closed and bounded and hence compact, it follows from *Proposition 1.3* that

$$\sup\big\{\big|\mathbf{f}(t, \mathbf{x})\big| : (t, \mathbf{x}) \in R\big\} = M < \infty.$$

Set $\alpha = \min\{b, b/M\}$ and let I denote the open interval $\{t : \tau - \alpha < t < \tau + \alpha\}$.

For each positive integer m, an approximate solution $\varphi_m : I \to \mathbb{R}^n$ will be defined. Because R is compact, $\mathbf{f}$ is uniformly continuous on R by *Proposition 1.4*. In particular, there exists $\delta_m > 0$ such that

$$\big|\mathbf{f}(t, \mathbf{x}) - \mathbf{f}(s, \mathbf{y})\big| \le \frac{1}{m},$$

whenever $|t - s| < \delta_m$ and $|\mathbf{x} - \mathbf{y}| < \delta_m$. Now, fix m and pick t_i such that

$$\tau - \alpha = t_{-p} < t_{-p+1} < \cdots < t_{-1} < t_0 = \tau < t_1 < \cdots < t_{p-1} < t_p = \tau + \alpha$$

and $|t_i - t_{i-1}| < \min\{\delta_m, \delta_m/M, 1/m\}$. Define $\varphi_m(t)$ on the closed interval $[t_{-1}, t_1]$ by

$$\begin{aligned} \varphi_m(t) &= \boldsymbol{\xi} + (t - t_0) \cdot \mathbf{f}(t_0, \boldsymbol{\xi}) \\ &= \boldsymbol{\xi} + (t - \tau) \cdot \mathbf{f}(\tau, \boldsymbol{\xi}). \end{aligned}$$

Note that this is a line segment with $\varphi_m(t_0) = \varphi_m(\tau) = \boldsymbol{\xi}$. For $t \in [t_{-1}, t_1]$,

$$\big|\varphi_m(t) - \boldsymbol{\xi}\big| = |t - t_0|\,\big|\mathbf{f}(t_0, \boldsymbol{\xi})\big| \le \alpha M \le \frac{b}{M} M = b$$

and $\big(t, \varphi_m(t)\big) \in R$. Extend the definition of φ_m to $[t_1, t_2]$ by

$$\varphi_m(t) = \varphi_m(t_1) + (t - t_1)\,\mathbf{f}\big(t_1, \varphi_m(t_1)\big).$$

For $t \in [t_1, t_2]$,

$$\begin{aligned} \left|\varphi_m(t) - \boldsymbol{\xi}\right| &\le \left|\varphi_m(t) - \varphi_m(t_1)\right| + \left|\varphi_m(t_1) - \boldsymbol{\xi}\right| \\ &\le |t - t_1| \left|\mathbf{f}\big(t_1, \varphi_m(t_1)\big)\right| + |t_1 - t_0| \left|\mathbf{f}(t_0, \boldsymbol{\xi})\right| \\ &\le |t - t_1|M + |t_1 - t_0|M = |t - t_0|M \le b \end{aligned}$$

and $\big(t, \varphi_m(t)\big) \in R$ for $t_0 \le t \le t_2$. It is now clear that this process can be repeated in both directions until it reaches $t_p = \tau + \alpha$ and $t_{-p} = \tau - \alpha$ to define φ_m on I satisfying $\big(t, \varphi_m(t)\big) \in R$ for all $t \in I$. In particular,

$$\varphi_m(t) = \varphi_m(t_k) + (t - t_k)\,\mathbf{f}\big(t_k, \varphi_m(t_k)\big)$$

on $[t_k, t_{k+1}]$ for $k \ge 0$ and on $[t_{k-1}, t_k]$ for $k \le 0$. See *Figure 1.1* for an illustration of a $\varphi_m(t)$ in the plane.

Clearly, φ_m is continuous and

$$\dot{\varphi}(t) = \mathbf{f}\big(t_k, \varphi_m(t_k)\big)$$

when $t_k < t < t_{k+1}$ for $k \ge 0$ and when $t_{k-1} < t < t_k$ for $k \le 0$. This completes the first step in the proof.

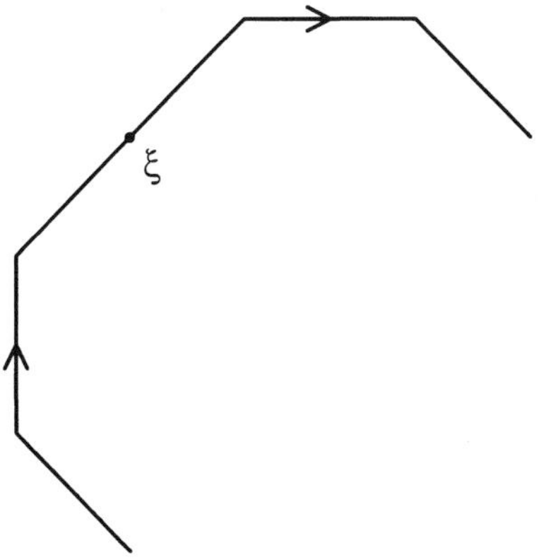

Figure 1.1: Graph of a sample planar $\varphi_m(t)$ with $p = 3$. The arrows indicate the direction of increasing t.

The crucial property of φ_m that must be established is the following: For s and t in I

$$\left|\varphi_m(t) - \varphi_m(s)\right| < |t - s|M.$$

The case when $t_0 \le s < t$ will be established and the other two cases will be left to the reader. First, find j and k, $0 \le j \le k$ such that $t_j \le s < t_{j+1}$ and $t_k \le t < t_{k+1}$. If $j = k$, then

$$\left|\varphi_m(t) - \varphi_m(s)\right| = |t - s|\left|\mathbf{f}\big(t_j, \varphi_m(t_j)\big)\right| \le |t - s|M.$$

If $j < k$, then

$$\begin{aligned}
&\left|\varphi_m(t) - \varphi_m(s)\right| \leq \\
&\left|\varphi_m(t) - \varphi_m(t_k)\right| + \left|\varphi_m(t_k) - \varphi_m(t_{k-1})\right| + \cdots + \left|\varphi_m(t_{j+1}) - \varphi_m(s)\right| \leq \\
&|t - t_k| \left|\mathbf{f}\big(t_k, \varphi_m(t_k)\big)\right| + \cdots + |t_{j+1} - s| \left|\mathbf{f}\big(t_j, \varphi_m(t_j)\big)\right| \leq \\
&|t - t_k|M + |t_k - t_{k-1}|M + \cdots + |t_{j+1} - s|M = \\
&|t - s|M.
\end{aligned}$$

(The last equality holds because $|t - t_k| = t - t_k$, etc.) The proof of the other two cases is similar.

For $t \neq t_i$, $i = -p, \ldots, p$, we have $\dot{\varphi}_m(t) = \mathbf{f}\big(t_k, \varphi_m(t_k)\big)$ for some t_k such that $|t - t_k| < \delta_m$ and $|t - t_k| < \delta_m / M$. Hence $\left|\varphi_m(t) - \varphi_m(t_k)\right| < |t - t_k|M < \delta_m$, and the uniform continuity of $\mathbf{f}$ now implies that

$$\left|\dot{\varphi}_m(t) - \mathbf{f}\big(t, \varphi_m(t)\big)\right| = \left|\mathbf{f}\big(t_k, \varphi_m(t_k)\big) - \mathbf{f}\big(t, \varphi_m(t)\big)\right| < \frac{1}{m},$$

which will be needed for the last step. The above inequality also says that each φ_m is an *approximate solution* of *(1.3)*.

Since $|t - s|M$ is independent of m, it follows from $\left|\varphi_m(t) - \varphi_m(s)\right| \leq |t - s|M$ that $\{\varphi_m : m \geq 1\}$ is an equicontinuous set of functions. Furthermore, φ_m is a uniformly bounded sequence because

$$\begin{aligned}
\left|\varphi_m(t)\right| &\leq \left|\varphi_m(t) - \varphi_m(t_0)\right| + \left|\varphi_m(t_0)\right| \\
&\leq |t - t_0|M + \left|\varphi_m(t_0)\right| \\
&\leq \alpha M + |\boldsymbol{\xi}|.
\end{aligned}$$

Hence, Ascoli's Lemma *Theorem 1.7* applies and there exists a subsequence φ_{m_i}, which is uniformly convergent on I to some function φ. This completes the second step of the proof.

Clearly, $\varphi_m(\tau) = \boldsymbol{\xi}$ for all m implies $\varphi(\tau) = \boldsymbol{\xi}$. To establish that φ is a solution of $\dot{\mathbf{x}} = \mathbf{f}(t, \mathbf{x})$ and complete the proof, it suffices by *Proposition 1.8* to show that *(1.2)* holds or equivalently that

$$\left|\varphi(t) - \varphi(\tau) - \int_\tau^t \mathbf{f}\big(s, \varphi(s)\big)\, ds\right| = 0$$

on I.

By the triangle inequality,

$$\begin{aligned}
&\left|\varphi(t) - \varphi(\tau) - \int_\tau^t \mathbf{f}\big(s, \varphi(s)\big)\, ds\right| \\
&\qquad \leq \left|\varphi(t) - \varphi_{m_i}(t)\right| + \left|\varphi_{m_i}(t) - \varphi(\tau) - \int_\tau^t \mathbf{f}\big(s, \varphi_{m_i}(s)\big)\, ds\right| \\
&\qquad\qquad + \left|\int_\tau^t \mathbf{f}\big(s, \varphi_{m_i}(s)\big) - \mathbf{f}\big(s, \varphi(s)\big)\, ds\right|.
\end{aligned}$$

The first term obviously goes to zero as i goes to infinity. With the observations that

$$\left|\int_\tau^t \mathbf{f}\big(s,\boldsymbol{\varphi}_{m_i}(s)\big) - \mathbf{f}\big(s,\boldsymbol{\varphi}(s)\big)\,ds\right| \le \int_\tau^t \left|\mathbf{f}\big(s,\boldsymbol{\varphi}_{m_i}(s)\big) - \mathbf{f}\big(s,\boldsymbol{\varphi}(s)\big)\right| ds$$

and

$$\left|\mathbf{f}\big(s,\boldsymbol{\varphi}_{m_i}(s)\big) - \mathbf{f}\big(s,\boldsymbol{\varphi}(s)\big)\right| \to 0$$

uniformly as i goes to infinity, it follows that the third term also goes to zero as i goes to infinity.

Because $\dot{\boldsymbol{\varphi}}_m$ is piecewise continuous, it follows that

$$\boldsymbol{\varphi}_m(t) = \boldsymbol{\varphi}_m(\tau) + \int_\tau^t \dot{\boldsymbol{\varphi}}_m(s)\,ds$$

for all m. Now substitute the above expression into the middle term to get

$$\begin{aligned}\left|\int_\tau^t \dot{\boldsymbol{\varphi}}_{m_i}(s) - \mathbf{f}\big(s,\boldsymbol{\varphi}_{m_i}(s)\big)\,ds\right| &\le \int_\tau^t \left|\dot{\boldsymbol{\varphi}}_{m_i}(s) - \mathbf{f}\big(s,\boldsymbol{\varphi}_{m_i}(s)\big)\right| ds \\ &\le \int_\tau^t \frac{1}{m_i}\,ds \le \frac{\alpha}{m_i}\end{aligned}$$

because $\boldsymbol{\varphi}_{m_i}$ is an approximate solution. Thus, the middle term goes to zero as i goes to infinity to complete the proof. □

Corollary *If* $\mathbf{f}$ *is continuous on an open set* D*, then every point in* D *has at least one trajectory of* $\dot{\mathbf{x}} = \mathbf{f}(t,\mathbf{x})$ *passing through it.*

Corollary *If* $\mathbf{f}$ *is continuous on an open set* D *and* C *is a compact subset of* D*, then there exists* $\alpha > 0$ *such that for every* $(t,\boldsymbol{\xi}) \in C$*, the initial-value problem*

$$\begin{aligned}\dot{\mathbf{x}} &= \mathbf{f}(t,\mathbf{x}) \\ \boldsymbol{\xi} &= \mathbf{x}(\tau)\end{aligned}$$

has a solution defined on $\{t : \tau = \alpha < t < \tau + \alpha\}$.

Proof. Exercise.

Peano's theorem supplies a simple general setting in which the existence of solutions can be taken for granted. We will always stay within this context and only consider the differential equations $\dot{\mathbf{x}} = \mathbf{f}(t,\mathbf{x})$, where $\mathbf{f}$ is continuous on an open set.

Peano's theorem, however, has three drawbacks. First, the interval on which the solution is defined in the proof of Peano's theorem may unnecessarily be very short. Second, it leaves open the possibility that an initial-value problem has more than one solution. Third, it is highly non constructive.

EXERCISES

1. Consider the initial-value problem

$$\begin{aligned} \dot{x} &= t^2 + x^2 \\ 0 &= x(0) \end{aligned}$$

on $\mathbb{R}^2$. Determine the longest interval on which the proof of Peano's theorem guarantees a solution. What is the answer to the same question when $t^2 + x^2$ is replaced by $|t|^p + |x|^p$ with $p > 1$?

2. Let $D = \mathbb{R}^2$ and use separation of variables to solve the initial-value problem

$$\begin{aligned} \dot{x} &= 1 + x^2 \\ 0 &= x(0). \end{aligned}$$

(In this problem, $f(t, x) = 1 + x^2$ is independent of t.) Show that the longest interval on which Peano's theorem guarantees a solution is less than one-third of the length of the interval on which there is a known solution.

3. Suppose $\mathbf{f}(t, \mathbf{x})$ is continuous and bounded on $\mathbb{R}^d$. Show that for all $\boldsymbol{\xi}$ the initial-value problem

$$\begin{aligned} \dot{\mathbf{x}} &= \mathbf{f}(t, \mathbf{x}) \\ \boldsymbol{\xi} &= \mathbf{x}(\tau) \end{aligned}$$

has a solution defined on an arbitrarily long interval.

4. Prove the second corollary to Peano's theorem.

5. Let $\varphi_m(t)$ be a sequence of solutions to $\dot{x} = \mathbf{f}(t, \mathbf{x})$ with $\mathbf{f}$ continuous on an open set D. Suppose that

$$R = \{(t, \mathbf{x}) : |t - \tau| \leq c \text{ and } |\mathbf{x} - \boldsymbol{\xi}| \leq c\} \subset D$$

for some $c > 0$. Prove the following: If each $\varphi_m(t)$ is defined on the open interval $I = (\tau - \gamma, \tau + \gamma)$, the sequence $\varphi_m(\tau)$ converges to $\boldsymbol{\xi}$, and $(t, \varphi_m(t)) \in R$ for all $t \in I$, then some subsequence of φ_m converges to a solution of

$$\begin{aligned} \dot{\mathbf{x}} &= \mathbf{f}(t, \mathbf{x}) \\ \boldsymbol{\xi} &= \mathbf{x}(\tau). \end{aligned}$$

6. Suppose $\mathbf{f}(t, \mathbf{x})$ is continuous and satisfies $|f(t, \mathbf{x})| \leq \log(|t| + |\mathbf{x}|)$ on $\mathbb{R}^d$. Show that for all $\boldsymbol{\xi}$ the initial-value problem

$$\begin{aligned} \dot{\mathbf{x}} &= \mathbf{f}(t, \mathbf{x}) \\ \boldsymbol{\xi} &= \mathbf{x}(\tau) \end{aligned}$$

has a solution defined on an arbitrarily long interval.

1.3 Uniqueness

The second fundamental question concerning initial-value problems is to determine when there is exactly one solution. This question is not purely question; it has practical consequences.

For example, suppose the behavior of some physical phenomenon is modelled by the initial-value problem

$$\begin{aligned} \dot{\mathbf{x}} &= \mathbf{f}(t, \mathbf{x}) \\ \boldsymbol{\xi} &= \mathbf{x}(\tau) \end{aligned}$$

Also suppose a specific function $\varphi(t)$ is a solution of this problem. Without knowing that this initial-value problem has a unique solution it cannot be asserted a priori that $\varphi(t)$ describes the behavior of the physical phenomenon being studied. Its behavior may in fact, be governed by a different solution $\psi(t)$. Consequently, simple broadly applicable tests for uniqueness can guarantee that a known solution of an initial-value problem is not extraneous.

Similarly, if a numerical procedure carried out with the aid of a computer is generating a sequence of approximate solutions to an initial-value problem, then without uniqueness the meaning of these computer calculations can be very ambiguous. In fact, uniqueness ensures that solutions can be approximated by numerical procedures. The most elementary of them, Euler's method, will be discussed in the next section.

Continuity of $\mathbf{f}$ is not enough to guarantee that initial-value problems have unique solutions. To construct a counterexample let $D = \mathbb{R}^2$, that is, $d = 1$ and both t, $x \in \mathbb{R}$.

Consider the initial-value problem

$$\begin{aligned} \dot{x} &= \sqrt{|x|} \\ 0 &= x(0). \end{aligned}$$

[Although $f(t, x)$ depends only on $x \in \mathbb{R}$, the t variable is included in the domain so that for now all differential equations are treated consistently in one form.] Clearly, $x(t) \equiv 0$ or $x(t) = 0$ for all $t \in \mathbb{R}$ is a solution. (The symbol $\equiv$ is used to indicate that a function has a particular constant value at each point in its domain.)

The elementary technique of separating variables yields a second solution,

$$x(t) = \begin{cases} 0, & t \leq 0; \\ t^2/4, & t \geq 0, \end{cases}$$

of the same initial-value problem.

The bad behavior of the above example is a result of the cusp of $\sqrt{|x|}$ at 0. However, there are differential equations that exhibit even more pathological behavior. For example, it is possible to construct a continuous function $f(t, x)$ on $\mathbb{R}^2$ such that every initial-value problem has infinitely many solutions. (See [17] page 18.)

The theme of this section is the study of a simple assumption called a Lipschitz condition, which guarantees that initial-value problems have unique solutions. After the Lipschitz condition is defined and its relationship with uniqueness analyzed, it will be shown that continuous first partial derivatives imply the Lipschitz condition. Thus the usually easy to check hypothesis that a function has continuous first partial derivatives will resolve one of the weaknesses of Peano's theorem.

If there exists a constant $L > 0$ such that for every $(t, \mathbf{x})$ and $(t, \mathbf{y})$ in D the following inequality holds:

$$\big|\mathbf{f}(t,\mathbf{x}) - \mathbf{f}(t,\mathbf{y})\big| \leq L|\mathbf{x} - \mathbf{y}|, \tag{1.4}$$

$\mathbf{f}(t, \mathbf{x})$ is said to satisfy a *Lipschitz condition* on D. Technically, this Lipschitz condition is only with respect to the space variable $\mathbf{x}$. Because no other variations of this concept will be used, it will be adequate simply to say that $\mathbf{f}(t, \mathbf{x})$ satisfies a Lipschitz condition and not include the phrase "with respect to $\mathbf{x}$."

Roughly speaking, a Lipschitz condition says that the values of $\mathbf{f}$ cannot separate faster than the distance between $\mathbf{x}$ and $\mathbf{y}$. It can also be thought of as a crude finite derivative condition because

$$\frac{\big|\mathbf{f}(t,\mathbf{x}) - \mathbf{f}(t,\mathbf{y})\big|}{|\mathbf{x} - \mathbf{y}|} \leq L$$

when $\mathbf{x} \neq \mathbf{y}$. Replacing $|\cdot|$ by an equivalent norm will not destroy a Lipschitz condition but will change the constant L.

Theorem 1.10 *Let* $\mathbf{f}$ *satisfy a Lipschitz condition on* D *and let* $\varphi_1(t)$ *and* $\varphi_2(t)$ *be two solutions of the differential equation* $\dot{\mathbf{x}} = \mathbf{f}(t, \mathbf{x})$ *on the domain* D. *Suppose both* $\varphi_1(t)$ *and* $\varphi_2(t)$ *are defined on the open interval* I. *If* $\varphi_1(\tau) = \varphi_2(\tau)$ *for some* $\tau \in I$, *then* $\varphi_1(t) = \varphi_2(t)$ *for every* t *in* I.

Proof. Because $\varphi_1(\tau) = \varphi_2(\tau)$, it follows from

$$\varphi_i(t) = \varphi_i(\tau) + \int_\tau^t \mathbf{f}\big(s, \varphi_i(s)\big)ds,$$

and *(1.4)* that

$$\begin{aligned}
\big|\varphi_1(t) - \varphi_2(t)\big| &= \left|\int_\tau^t \mathbf{f}\big(s,\varphi_1(s)\big) - \mathbf{f}\big(s,\varphi_2(s)\big)\, ds\right| \\
&\leq \int_\tau^t \big|\mathbf{f}\big(s,\varphi_1(s)\big) - \mathbf{f}\big(s,\varphi_2(s)\big)\big|\, ds \\
&\leq \int_\tau^t L\big|\varphi_1(s) - \varphi_2(s)\big|\, ds
\end{aligned}$$

for $t \geq \tau$. Thus

$$\big|\varphi_1(t) - \varphi_2(t)\big| \leq L\int_\tau^t \big|\varphi_1(s) - \varphi_2(s)\big|\, ds$$

for $t \geq \tau$. It remains to show that this inequality forces $\varphi_1(t) = \varphi_2(t)$ for $t \geq \tau$, and then apply the simple technique of reversing time to obtain the result for $t \leq \tau$.

Setting $g(t) = \left|\varphi_1(t) - \varphi_2(t)\right|$ and $G(t) = \int_\tau^t g(s)\,ds$, the previous inequality can be written as

$$g(t) \leq LG(t)$$

or

$$\dot{G}(t) - LG(t) \leq 0.$$

Multiply this inequality by e^{-Lt} and note that the left-hand side is now the derivative of $e^{-Lt}G(t)$. (This is the standard method of integrating factors used to solve a first-order linear differential equations.) Thus

$$\frac{d\left[e^{-Lt}G(t)\right]}{dt} \leq 0.$$

for $t \geq \tau$. Because $e^{-L\tau}G(\tau) = 0$, integrating from τ to t, yields

$$e^{-Lt}G(t) \leq 0$$

or

$$G(t) \leq 0$$

when $t \geq \tau$.

Since $\dot{G}(t) = g(t) \geq 0$ and $G(\tau) = 0$, it also follows by integrating $\dot{G}(t)$ that $G(t) \geq 0$ for $t \geq \tau$. Therefore, $G(t) \equiv 0$ and $\dot{G}(t) = g(t) \equiv 0$ for $t \geq \tau$. It follows that $\varphi_1(t) = \varphi_2(t)$ for $t \geq \tau$.

For $t \leq \tau$ and $i = 1, 2$, let $\boldsymbol{\psi}_i(t) = \boldsymbol{\varphi}_i(-t)$ on $-I = \{-t : t \in I\}$. Then

$$\dot{\boldsymbol{\psi}}_i(t) = -\dot{\boldsymbol{\varphi}}_i(-t) = -\mathbf{f}\big(-t, \boldsymbol{\varphi}_i(-t)\big) = -\mathbf{f}\big(-t, \boldsymbol{\psi}_i(t)\big)$$

and $\boldsymbol{\psi}_i(t)$ is a solution of $\dot{\mathbf{x}} = -\mathbf{f}(-t, \mathbf{x})$ on $D' = \big\{(t, \mathbf{x}) : (-t, \mathbf{x}) \in D\big\}$. Clearly, $\boldsymbol{\psi}_1(-\tau) = \boldsymbol{\psi}_2(-\tau)$ and $\big|-\mathbf{f}(-t, \mathbf{x}) - -\mathbf{f}(-t, \mathbf{y})\big| \leq L|\mathbf{x} - \mathbf{y}|$. The preceding arguments show that $\boldsymbol{\psi}_1(s) = \boldsymbol{\psi}_2(s)$ for $s \geq -\tau$ or $\varphi_1(t) = \varphi_2(t)$ for $t \leq \tau$. □

It is not evident from the previous theorem that uniqueness is really a local issue. By "local" we mean what occurs just near points in D not necessarily throughout D. The next theorem makes this point.

Theorem 1.11 *Let $\varphi_1(t)$ and $\varphi_2(t)$ be two solutions of the differential equation $\dot{\mathbf{x}} = \mathbf{f}(t, \mathbf{x})$ on the domain D and assume they are defined on the same interval I. Suppose that for each point $\mathbf{x}$ of D there exists an open set U containing $\mathbf{x}$ and contained in D such that $\mathbf{f}(t, \mathbf{x})$ satisfies a Lipschitz condition on U. If $\varphi_1(\tau) = \varphi_2(\tau)$ for some $\tau \in I$, then $\varphi_1(t) = \varphi_2(t)$ for all t in I.*

Proof. Assume $\varphi_1(s) \neq \varphi_2(s)$ for some $s > \tau$. (The argument for the case when $s < \tau$ is similar.) Set

$$\beta = \sup\big\{s : \varphi_1(t) = \varphi_2(t) \text{ for } \tau \leq t \leq s\big\}.$$

Clearly, $\beta \in I$ and $\varphi_1(\beta) = \varphi_2(\beta)$ by continuity. Let U be a neighborhood of $(\beta, \varphi_1(\beta))$ in D on which $\mathbf{f}$ satisfies a Lipschitz condition. (A *neighborhood of a point* is an open set containing the specified point.) There exists $\delta > 0$ such that $(t, \varphi_i(t)) \in U$, $i = 1, 2$, when $|t - \beta| < \delta$. Now apply *Theorem 1.10* to $\dot{\mathbf{x}} = \mathbf{f}(t, \mathbf{x})$ restricted to U and φ_i restricted to $\beta - \delta < t < \beta + \delta$ to obtain $\varphi_1(t) = \varphi_2(t)$ when $\tau \leq t < \beta + \delta$, which contradicts the definition of β and completes the proof. □

It is natural to turn the key hypothesis of the previous theorem into a definition. The function $\mathbf{f}(t, \mathbf{x})$ will be called a *locally Lipshitz function* if for each point $\mathbf{x}$ of D there exists a neighborhood U of $\mathbf{x}$ contained in D such that $\mathbf{f}(t, \mathbf{x})$ satisfies a Lipschitz condition on U. Thus the above theorem can be restated as, "If the function $\mathbf{f}(t, \mathbf{x})$ is locally Lipschitz, then the solutions of $\dot{\mathbf{x}} = \mathbf{f}(t, \mathbf{x})$ are unique." Thus for initial-value problems the following corollary holds:

Corollary *Suppose the the function $\mathbf{f}(t, \mathbf{x})$ is locally Lipschitz on the open set D. If $\varphi_1(t)$ and $\varphi_2(t)$ are solutions of the initial-value problem*

$$\begin{aligned} \dot{\mathbf{x}} &= \mathbf{f}(t, \mathbf{x}) \\ \boldsymbol{\xi} &= \mathbf{x}(\tau) \end{aligned}$$

defined on the open intervals I_1 and I_2, then $\varphi_1(t) = \varphi_2(t)$ for all

$$t \in I_1 \cap I_2.$$

Thus far the problem of finding a simple broadly applicable test for uniqueness has only been replaced with the problem of finding a simple broadly applicable test for when a function is locally Lipschitz. The next step is to show that functions with continuous first partial derivatives are locally Lipschitz functions.

As a preliminary step, a vector version of the mean-value theorem is needed. Suppose $g : \Omega \to \mathbb{R}$ has continuous first partial derivatives on some open set Ω of $\mathbb{R}^d$, in other words g is *continuously differentiable* on Ω. Let $\mathbf{x}$ and $\mathbf{y}$ be points in Ω such that the line segment joining them is contained in Ω or

$$\{s\mathbf{x} + (1 - s)\mathbf{y} : 0 \leq s \leq 1\} \subset \Omega.$$

Set $G(s) = g(s\mathbf{x} + (1 - s)\mathbf{y})$ for $0 \leq s \leq 1$. Then $G(0) = g(\mathbf{y})$, $G(1) = g(\mathbf{x})$, $G(\cdot)$ is continuous on $0 \leq s \leq 1$ and differentiable on $0 < s < 1$ with

$$\dot{G}(s) = \nabla g(s\mathbf{x} + (1 - s)\mathbf{y}) \cdot (\mathbf{x} - \mathbf{y}).$$

[As usual

$$\nabla g = \left(\frac{\partial g}{\partial x_1}, \frac{\partial g}{\partial x_2}, \ldots, \frac{\partial g}{\partial x_d}\right)$$

denotes the *gradient* of the function g.] By the mean-value theorem

$$\frac{G(1) - G(0)}{1 - 0} = \dot{G}(\theta)$$

for some θ, $0 < \theta < 1$. This can be written in terms of g to obtain

$$g(\mathbf{x}) - g(\mathbf{y}) = \nabla g\big(\theta\mathbf{x} + (1-\theta)\mathbf{y}\big) \cdot (\mathbf{x} - \mathbf{y}). \tag{1.5}$$

for some θ satisfying $0 < \theta < 1$. This vector version of the mean-value theorem will be used in the next proof of the next theorem along with the usual *Cauchy-Schwarz inequality*

$$|\mathbf{v} \cdot \mathbf{w}| \leq \|\mathbf{v}\| \, \|\mathbf{w}\|$$

for the dot product. (Proving the Cauchy-Schwarz inequality was the first exercise of Section 1 on page 11.)

Theorem 1.12 *If the first partial derivatives of* $\mathbf{f}(t, \mathbf{x})$ *with respect to* $x_1, \ldots,$ x_d *exist and are continuous on* D*, then* $\mathbf{f}(t, \mathbf{x})$ *is locally Lipschitz on* D.

Proof. Consider an arbitrary point $(\tau, \boldsymbol{\xi})$ in D. Because D is open, there exists $r > 0$ such that the compact Euclidean ball

$$B = \big\{(t, \mathbf{x}) : \|(t, \mathbf{x}) - (\tau, \boldsymbol{\xi})\| \leq r\big\} \subset D.$$

Set

$$U = \big\{(t, \mathbf{x}) : \big\|(t, \mathbf{x}) - (\tau, \boldsymbol{\xi})\big\| < r\big\}$$

and note that if $(t, \mathbf{x})$ and $(s, \mathbf{y})$ are in U, then the line segment joining them also lies in U, that is, U is *convex*. [Specifically, a subset C of $\mathbb{R}^m$ is *convex* if the set $\{s\mathbf{x} + (1-s)\mathbf{y} : 0 \leq s \leq 1\}$ is contained in C whenever $\mathbf{x}$ and $\mathbf{y}$ are in C.]

The following notation will be used

$$\nabla_{\mathbf{x}} f_j = \left(\frac{\partial f_j}{\partial x_1}, \ldots, \frac{\partial f_j}{\partial x_d}\right)$$

for the gradient of $f(t, \mathbf{x})$ with respect to the space variable $\mathbf{x}$. Because the function $\nabla_{\mathbf{x}} f_j(t, \mathbf{x})$ is continuous on D and B is compact, $\big\|\nabla_{\mathbf{x}} f_i(t, \mathbf{x})\big\|$ is bounded. Let

$$M_j = \sup\big\{\big\|\nabla_{\mathbf{x}} f_j(t, \mathbf{x})\big\| : (t, \mathbf{x}) \in B\big\}$$

and let

$$M = \sum_{i=1}^{d} M_j.$$

Let $(t, \mathbf{x})$ and $(t, \mathbf{y}) \in U$. For fixed t, the above vector version of the mean-value theorem can be applied to each $f_j(t, \mathbf{x})$, $j = 1, \ldots, d$ to obtain

$$f_j(t, \mathbf{x}) - f_j(t, \mathbf{y}) = \nabla_{\mathbf{x}} f_j\big(t, \theta_j\mathbf{x} + (1-\theta_j)\mathbf{y}\big) \cdot (\mathbf{x} - \mathbf{y})$$

for some θ_j, $0 < \theta_j < 1$.

The following chain of inequalities proves that $\mathbf{f}(t, \mathbf{x})$ satisfies a Lipschitz condition on U:

$$\begin{aligned}
\left|\mathbf{f}(t,\mathbf{x}) - \mathbf{f}(t,\mathbf{y})\right| &= \sum_{j=1}^{d} \left|f_j(t,\mathbf{x}) - f_j(t,\mathbf{y})\right| \\
&= \sum_{j=1}^{d} \left|\nabla_{\mathbf{x}} f_j\big(t, \theta_j \mathbf{x} + (1-\theta_j)\mathbf{y}\big) \cdot (\mathbf{x} - \mathbf{y})\right| \\
&\leq \sum_{j=1}^{d} \left\|\nabla_{\mathbf{x}} f_j\big(t, \theta_j \mathbf{x} + (1-\theta_j)\mathbf{y}\big)\right\| \left\|\mathbf{x} - \mathbf{y}\right\| \\
&\leq \sum_{j=1}^{d} M_j \left\|\mathbf{x} - \mathbf{y}\right\| = M \left\|\mathbf{x} - \mathbf{y}\right\| \leq dM|\mathbf{x} - \mathbf{y}|
\end{aligned}$$

and completes the proof. □

Together *Theorem 1.11* and *Theorem 1.12* imply that when $\mathbf{f}(t, \mathbf{x})$ has continuous first partial derivatives with respect to $x_1, \ldots, x_d$, the solutions to initial-value problems are unique. Thus for a large class of functions $\mathbf{f}(t, \mathbf{x})$ the solutions of $\dot{\mathbf{x}} = \mathbf{f}(t, \mathbf{x})$ are uniquely determined by their value at a single time. Moreover, since it usually easy to determine whether or not $\mathbf{f}$ has continuous first partial derivatives with respect to $\mathbf{x}$, continuous first partial derivatives provides an easily applicable test for uniqueness. Although the proof of this consequence of these two theorems is more like that of a corollary, it is worth stating as a theorem.

Theorem 1.13 *Let $\varphi_1(t)$ and $\varphi_2(t)$ be two solutions of the differential equation $\dot{\mathbf{x}} = \mathbf{f}(t, \mathbf{x})$ on the domain D and assume they are defined on the same interval I. Suppose thats $\varphi_1(\tau) = \varphi_2(\tau)$ for some $\tau \in I$. If the first partial derivatives of $\mathbf{f}(t, \mathbf{x})$ with respect to $x_1, \ldots, x_d$ exist and are continuous on the open set D, then $\varphi_1(t) = \varphi_2(t)$ for all t in I.*

In particular, when $\mathbf{f}$ has continuous first partial derivatives, solutions of initial-value problems are unique in the following sense:

Corollary *Suppose the first partial derivatives of $\mathbf{f}$ with respect to $x_1, \ldots, x_d$ are continuous on D. If $\varphi_1(t)$ and $\varphi_2(t)$ are solutions of the initial-value problem*

$$\begin{aligned}
\dot{\mathbf{x}} &= \mathbf{f}(t, \mathbf{x}) \\
\boldsymbol{\xi} &= \mathbf{x}(\tau)
\end{aligned}$$

defined on the open intervals I_1 and I_2, then $\varphi_1(t) = \varphi_2(t)$ for all

$$t \in I_1 \cap I_2.$$

When $\mathbf{f}$ is locally Lipschitz, the open sets on which the Lipschitz conditions hold can be very small. It will be useful to know that these local Lipschitz conditions at least guarantee that $\mathbf{f}$ satisfies a Lipschitz condition on every compact subset of D even though they do not imply that $\mathbf{f}$ satisfies a Lipschitz condition on D.

Theorem 1.14 *The function $\mathbf{f}(t,\mathbf{x})$ is locally Lipschitz on the open set D if and only if $\mathbf{f}(t,\mathbf{x})$ satisfies a Lipschitz condition on every compact subset of D.*

Proof. First assume that $f(t,\mathbf{x})$ is locally Lipschitz on D and let C be a compact set contained in D. For every point $\mathbf{x}$ in C there exists an open set $U_{\mathbf{x}}$ containing $\mathbf{x}$ on which $\mathbf{f}$ satisfies a Lipschitz condition. Because C is compact, there exist a finite number of these open sets, $U_1, \ldots, U_p$ contained in D such that

$$C \subset \bigcup_{i=1}^{p} U_i$$

and

$$\left|\mathbf{f}(t,\mathbf{x}) - \mathbf{f}(t,\mathbf{y})\right| \le L_i|\mathbf{x}-\mathbf{y}|$$

for $\mathbf{x}$, $\mathbf{y} \in U_i$. Set $M = \sup\left\{\left|\mathbf{f}(t,\mathbf{x})\right| : (t,\mathbf{x}) \in C\right\}$ and $L' = \max_{i \le i \le p} L_i$.

It is easy to prove that there exists $\delta > 0$ such that for $\mathbf{x}$, $\mathbf{y} \in C$, $|\mathbf{x} - \mathbf{y}| < \delta$ implies $\mathbf{x}$, $\mathbf{y} \in U_i$ for some i, $1 \le i \le p$. (See *Exercise 6.*) Set $L = \max\{L', 2M/\delta\}$. For $\mathbf{x}$, $\mathbf{y} \in C$, there are now two cases to check:

(a) if $|\mathbf{x}-\mathbf{y}| < \delta$, then $\left|\mathbf{f}(t,\mathbf{x}) - \mathbf{f}(t,\mathbf{y})\right| \le L_i|\mathbf{x}-\mathbf{y}| \le L|\mathbf{x}-\mathbf{y}|$;

(b) if $|\mathbf{x}-\mathbf{y}| \ge \delta$, then $\left|\mathbf{f}(t,\mathbf{x}) - \mathbf{f}(t,\mathbf{y})\right| \le 2M = \dfrac{2M\delta}{\delta} \le L|\mathbf{x}-\mathbf{y}|$.

Thus $\mathbf{f}$ is Lipschitz on C and locally Lipschitz on D.

For the second half of the proof, assume that $\mathbf{f}$ satisfies a Lipschitz condition on every compact subset of D, and let $\mathbf{x}$ be a point in D. Then there exists $r > 0$ such that

$$B = \left\{(t,\mathbf{x}) : \|(t,\mathbf{x}) - (\tau,\boldsymbol{\xi})\| \le r\right\} \subset D.$$

The closed and bounded Euclidean ball B is compact by *Theorem 1.1*, and hence by hypothesis $\mathbf{f}$ satisfies a Lipschitz condition on B that also holds on the open set

$$U = \left\{(t,\mathbf{x}) : \left\|(t,\mathbf{x}) - (\tau,\boldsymbol{\xi})\right\| < r\right\}.$$

Thus $\mathbf{f}$ is locally Lipschitz on D. □

To summarize, the concept of a Lipschitz condition provides a satisfactory resolution to the question of when the solutions of a differential equation are unique. Consequently, the Lipschitz condition also addresses the nonconstructive nature of the proof of Peano's theorem (*Theorem 1.9*) and turns it into a numerical method, which is is the subject of the next section.

EXERCISES

1. Let $\mathbf{f} : \mathbb{R}^3 \to \mathbb{R}^2$ be given by

$$\mathbf{f}(t, x_1, x_2) = (|x_1 + x_2|, |t|^{1/2}|x_1x_2|)$$

and let $\boldsymbol{\varphi}_1(t)$ and $\boldsymbol{\varphi}_2(t)$ be two solutions of $\dot{x} = \mathbf{f}(t, \mathbf{x})$ defined on an open interval I. Show that if $\boldsymbol{\varphi}_1(\tau) = \boldsymbol{\varphi}_2(\tau)$ for some τ in I, then $\boldsymbol{\varphi}_1 = \boldsymbol{\varphi}_2$.

2. Let $f : \mathbb{R}^2 \to \mathbb{R}$ be continuous and suppose that for every t, the function $f(t, x)$ is nonincreasing in x, that is, $f(t, x) \leq f(t, y)$, when $x > y$.

   (a) Let $\varphi_1(t)$ and $\varphi_2(t)$ be two solutions of $\dot{x} = f(t, x)$ defined on an open interval I. Show that if $\varphi_1(\tau) = \varphi_2(\tau)$ for some τ in I, then $\varphi_1(t) = \varphi_2(t)$ for $t \geq \tau$.

   (b) Show that

$$f(t, x) = \left\{ \begin{array}{lll} |x|^{1/2} & \text{for} & x < 0 \\ 0 & \text{for} & x \geq 0 \end{array} \right\}$$

   satisfies the above condition and the initial-value problem

$$\begin{aligned} \dot{x} &= f(t, x) \\ 0 &= x(0) \end{aligned}$$

   does not have a unique solution.

3. Show that if $\mathbf{f}(\mathbf{x})$ is locally Lipschitz on an open set D, then it is continuous on D.

4. Let Ω be an open subset of $\mathbb{R}^d$ and set

$$V = \{\mathbf{f} : \Omega \to \mathbb{R}^d : \mathbf{f} \text{ is locally Lipschitz } \}$$

Show that V is a vector space of continuous functions and prove the following: If $\mathbf{f}$ is in V and $g : \Omega \to \mathbb{R}$ is locally Lipschitz, then $g\mathbf{f}$ is in V.

5. Let $g : \Omega \to \mathbb{R}$ be a real-valued locally Lipschitz function on an open set $\Omega \subset \mathbb{R}^d$. Show that if $g(\mathbf{x}) > 0$ for every $\mathbf{x} \in \Omega$, then $1/g(\mathbf{x})$ is also locally Lipschitz on Ω.

6. Let C be a compact set in R^d. Prove the following: If $U_1, \ldots, U_p$ are open sets such that

$$C \subset \bigcup_{j=1}^{p} U_j,$$

then there exists $\delta > 0$ such that $|\mathbf{x} - \mathbf{y}| < \delta$ with $\mathbf{x}, \mathbf{y} \in C$ implies that both $\mathbf{x}$ and $\mathbf{y}$ are in U_j for some j.

1.4 Numerical Approximation

Even very innocent looking scalar differential equations need not have solutions in the class of elementary functions, and solutions expressed explicitly in terms of known functions can be unwieldily in calculations. For many applied problems, from the old problem of calculating the trajectory of an artillery shell to the modern one of controlling the orbit of a communications satellite, what is desired is numerical information about the solution of a differential equation. Fortunately, ordinary differential equations lend themselves to numerical solutions. Today's computing technology is a far cry from the mechanical calculators and human calculating teams of the beginning of the twentieth century, but many of the underlying fundamental ideas and problems remain the same.

This book will not pursue in depth the theory of solving differential equations numerically. It is an important and sophisticated subject in its own right. To gain a deeper understanding of the theoretical results presented in subsequent chapters, readers are encouraged to use the best software and computers at their disposal to explore specific examples. Theory and computations do go hand in hand. After reading this section, [13] is a suitable next step for further reading on the theory of numerical approximations to solutions of differential equations, while [23] is a more general and advanced theoretical treatment of numerical analysis.

Euler's method is the original technique for approximating solutions of $\dot{\mathbf{x}} = \mathbf{f}(t, \mathbf{x})$ and yields quite good results, especially with modern equipment, for many simple problems. Many sophisticated methods can trace their roots to Euler's method. In this section, not only will it be shown that Euler's method produces approximate solutions that converge to the actual solution, but their rate of convergence will also be established. To begin the discussion of Euler's method, the ideas in the previous sections on existence and uniqueness are merged.

Consider the usual initial-value problem

$$\begin{aligned} \dot{\mathbf{x}} &= \mathbf{f}(t, \mathbf{x}) \\ \boldsymbol{\xi} &= \mathbf{x}(\tau). \end{aligned}$$

As in the proof of Peano's theorem, choose $b > 0$ so that

$$R = \big\{(t, \mathbf{x}) : |t - \tau| \leq b \text{ and } |\mathbf{x} - \boldsymbol{\xi}| \leq b\big\} \subset D,$$

let $M = \sup\big\{\big|\mathbf{f}(t, \mathbf{x})\big| : (t, \mathbf{x}) \in R\big\}$, let $\alpha = \min\{b, b/M\}$, and let I be the open interval $(\tau - \alpha, \tau + \alpha)$. We will construct a sequence $\boldsymbol{\psi}_k$ of curves defined on I similar to the curves $\boldsymbol{\varphi}_m$ in the proof of *Theorem 1.9*. Specifically, each $\boldsymbol{\psi}_k$ is defined by first picking a finite set of times t_j in I. For convenience, we will assume that $t_0 = \tau$ and an equal number of times, t_i, in I have been selected larger than τ and smaller than τ in I, so that $-p_k \leq j \leq p_k$ and

$$\tau - \alpha = t_{-p_k} < \cdots < t_{-1} < t_0 = \tau < t_1 < \cdots < t_{p_k} = \tau + \alpha.$$

Then, exactly as in the proof of *Theorem 1.9*, define $\boldsymbol{\psi}_k(t)$ so that

$$\boldsymbol{\psi}_k(t) = \boldsymbol{\psi}_k(t_i) + (t - t_i)f\big(t_i, \boldsymbol{\psi}_k(t_i)\big)$$

on $[t_i, t_{i+1}]$ for $i \geq 0$ and on $[t_{i-1}, t_i]$ for $i < 0$ (Note the notation suppresses the fact that the t_i also depend on k.) The next result shows that this is a constructive process for approximating solutions of $\dot{\mathbf{x}} = \mathbf{f}(\mathbf{x})$ when $\mathbf{f}$ is a locally Lipschitz function.

Theorem 1.15 *If* $\mathbf{f}(t, \mathbf{x})$ *is locally Lipschitz on an open set* D *and if*

$$\lim_{k\to\infty} \big[\sup\big\{t_{i+1} - t_i : -p_k \leq i < p_k\big\}\big] = 0,$$

then the sequence of functions $\boldsymbol{\psi}_k(t)$ *defined as above converges uniformly on* I *to the unique solution* $\boldsymbol{\varphi}(t)$ *of the initial-value problem*

$$\begin{aligned} \dot{\mathbf{x}} &= \mathbf{f}(t, \mathbf{x}) \\ \boldsymbol{\xi} &= \mathbf{x}(\tau). \end{aligned}$$

Proof. By *Theorem 1.9* and *Theorem 1.11*, there is a unique solution $\boldsymbol{\varphi}$ to the initial-value problem defined on I. If $\boldsymbol{\psi}_k$ does not converge uniformly to $\boldsymbol{\varphi}$ on I, then there exists $\varepsilon > 0$ such that for any positive integer K, there exists $k > K$ for which

$$\sup\big\{\big|\boldsymbol{\psi}_k(t) - \boldsymbol{\varphi}(t)\big| : t \in I\big\} \geq \varepsilon.$$

The strategy will be to obtain a contradiction by extracting a subsequence of $\boldsymbol{\psi}_k(t)$ to which the proof of Peano's theorem applies and for which no subsequences can converge to a solution.

As in the proof of Peano's theorem, R is compact, $\mathbf{f}$ is uniformly continuous on R, and there exists $\delta_k > 0$ such that

$$\big|\mathbf{f}(t, \mathbf{x}) - \mathbf{f}(s, \mathbf{y})\big| \leq \frac{1}{m}$$

whenever $|t - s| < \delta_m$ and $|\mathbf{x} - \mathbf{y}| < \delta_m$.

Recall that in the proof of Peano's theorem the approximate solutions $\boldsymbol{\varphi}_m$ were chosen so that $t_{i+1} - t_i < \min\{\delta_m, \delta_m/M, 1/m\}$. The hypothesis that $\lim_{k\to\infty}\big[\sup\big\{t_{i+1} - t_i : -p_k \leq i < p_k\big\}\big] = 0$, can be used to ensure that our subsequence meets this required condition. In particular, given m there exists A_m such that $t_{i+1} - t_i < \min\{\delta_m, \delta_m/M, 1/m\}$ for for all $k \geq A_m$.

Now for each integer m choose $k_m \geq A_m$ such that

$$\sup\big\{\big|\boldsymbol{\psi}_k(t) - \boldsymbol{\varphi}(t)\big| : t \in I\big\} \geq \varepsilon$$

and set $\boldsymbol{\varphi}_m(t) = \boldsymbol{\psi}_{k_m}(t)$.

Then $\boldsymbol{\varphi}_m$ is a sequence constructed exactly as in the proof of *Theorem 1.9* (Peano's theorem), and therefore some subsequence converges uniformly on I to a solution of the initial-value problem. But there is only one solution on I, namely, φ, and by the choice of φ_m, no subsequence converges uniformly to φ because

$$\sup\left\{\left|\varphi_m(t)-\varphi(t)\right| : t \in I\right\} \geq \varepsilon,$$

for all m. This contradiction completes the proof. □

The importance of the above theorem is that it provides a simple and surprisingly effective method of numerically approximating solutions. This theorem is the starting point for the development of sophisticated numerical methods for ordinary differential equations.

The remainder of this section will be devoted to exploring two natural questions about this technique, which is known as *Euler's method.* How can it be made to work on a larger interval and how fast will it converge?

For convenience, choose $h > 0$ and set $t_{i+1} - t_i = h$ for all i and only consider $t \geq \tau$. So, $t_i = \tau + ih$, $i = 0, 1, \ldots, p$. If an approximate solution is defined by

$$\boldsymbol{\psi}(t) = \boldsymbol{\psi}(t_i) + (t - t_i)\mathbf{f}\big(t_i, \boldsymbol{\psi}(t_i)\big),$$

then it is only necessary to know the values $\boldsymbol{\xi}_i = \boldsymbol{\psi}(t_i)$ to recapture the entire approximation $\boldsymbol{\psi}$, because it is piecewise linear. Note the points $(t_i, \boldsymbol{\xi}_i)$ of $\mathbb{R}^{n+1}$ can be defined iteratively by

$$\begin{aligned}
(t_0, \boldsymbol{\xi}_0) &= (\tau, \boldsymbol{\xi}) \\
(t_1, \boldsymbol{\xi}_1) &= \big(t_0 + h, \boldsymbol{\xi}_0 + h\mathbf{f}(t_0, \boldsymbol{\xi}_0)\big) \\
&\;\;\vdots \qquad \vdots \\
(t_{i+1}, \boldsymbol{\xi}_{i+1}) &= \big(t_i + h, \boldsymbol{\xi}_i + h\mathbf{f}(t_i, \boldsymbol{\xi}_i)\big).
\end{aligned} \tag{1.6}$$

The only constraint in defining $\boldsymbol{\psi}$ in this way is that $(t_i, \boldsymbol{\xi}_i)$ lies in D.

Letting h go to zero, will produce better and better approximations to the solution of the initial-value problem on $\tau \leq t \leq \alpha$, where $\alpha = \min\{b, b/M\}$ as above. The question is what happens when t goes beyond $\tau + \alpha$. One danger is that the actual solution will go to infinity in finite time and the approximation will not detect it. Consequently, it must be assumed that we have some information about the domain of the solution.

Even when the domain of the solution is known, there is another danger. The accumulation of round off errors might cause the approximation to start tracking a different solution after some interval of time. Thus a single approximation of a solution can be very misleading if the interval of time is too long. By taking h smaller and smaller, however, the approximations are accurate over longer and longer intervals.

The final theorem in this section illustrates how Euler's method can produce a sequence of functions converging to a solution of $\dot{\mathbf{x}} = \mathbf{f}(t, \mathbf{x})$ on a given closed interval. The rate of convergence will be a constant times h. Admittedly, the hypotheses of this theorem were designed to minimize the technical details in the proof, but the theorem still can be applied in many cases.

Theorem 1.16 *Let $\mathbf{f}(t, \mathbf{x})$ have continuous first partial derivatives on D and let D' be an open convex set whose closure is compact and contained in D. Suppose the trajectory of a solution $\boldsymbol{\varphi}(t)$ defined on $\tau \leq t \leq \sigma$ of*

$$\begin{aligned} \dot{\mathbf{x}} &= \mathbf{f}(t, \mathbf{x}) \\ \boldsymbol{\xi} &= \mathbf{x}(\tau). \end{aligned}$$

lies in D'. Let $(t_i, \boldsymbol{\xi}_i)$ be defined by (1.6) for $i = 0, \ldots, p$ so that $t_0 = \tau$ and $t_p = \sigma$. Let $\boldsymbol{\psi}(t) = \boldsymbol{\xi}_i + (t - t_i)\mathbf{f}(t_i, \boldsymbol{\xi}_i)$ when $t_i \leq t < t_{i+1}$. Then there exist constants C and C' depending only on D', $\mathbf{f}$, and $\sigma - \tau$ such that for all i

$$\left|\boldsymbol{\varphi}(t_i) - \boldsymbol{\xi}_i\right| \leq Ch$$

and

$$\left|\boldsymbol{\varphi}(t) - \boldsymbol{\psi}(t)\right| \leq C'h,$$

whenever $(t_i, \boldsymbol{\xi}_i) \in D'$ for $i = 0, \cdots, p$.

Proof. By *Theorem 1.14*, $\mathbf{f}$ satisfies a Lipschitz condition on D'. As usual, let L denote the Lipschitz constant. Since $\dot{\boldsymbol{\varphi}}(t) = \mathbf{f}\big(t, \boldsymbol{\varphi}(t)\big)$ and $\mathbf{f}$ has continuous first partial derivatives, $\boldsymbol{\varphi}(t)$ has a continuous second derivative and the ith coordinate of $\ddot{\boldsymbol{\varphi}}(t)$ is given by

$$\ddot{\varphi}_i(t) = \frac{\partial f_i\big(t, \boldsymbol{\varphi}(t)\big)}{\partial t} + \sum_{j=1}^{n} \frac{\partial f_i\big(t, \boldsymbol{\varphi}(t)\big)}{\partial x_j} f_j\big(t, \boldsymbol{\varphi}(t)\big).$$

Thus, there exists a positive constant B depending only on $\mathbf{f}$ and D' such that for $\tau \leq t \leq \sigma$ and $j = 1, \ldots, n$

$$\left|\ddot{\varphi}_j(t)\right| \leq B.$$

For convenience, let $\beta = \sigma - \tau$, the length of the interval, and

$$\boldsymbol{\varepsilon}_i = \boldsymbol{\varphi}(t_i) - \boldsymbol{\xi}_i,$$

the error at the ith step.

Applying Taylor's formula with the remainder expressed in terms of a derivative at an intermediate value (Some times called the Lagrange form of the remainder.) to each coordinate of $\boldsymbol{\varphi}$, yields

$$\boldsymbol{\varphi}(t_{i+1}) = \boldsymbol{\varphi}(t_i) + h\dot{\boldsymbol{\varphi}}(t_i) + \frac{h^2 \mathbf{r}_i}{2}$$

where the remainder $\mathbf{r}_i \in \mathbb{R}^d$ satisfies $|\mathbf{r}_i| \leq dB$. (An easy to follow proof of Taylor's theorem can be found in [32].) Subtracting

$$\boldsymbol{\xi}_{i+1} = \boldsymbol{\xi}_i + h\mathbf{f}(t_i, \boldsymbol{\xi}_i)$$

from the above, produces

$$\varepsilon_{i+1} = \varepsilon_i + h\big[\mathbf{f}\big(t_i, \boldsymbol{\varphi}(t_i)\big) - \mathbf{f}(t_i, \boldsymbol{\xi}_i)\big] + \frac{h^2 \mathbf{r}_i}{2}.$$

It follows that

$$|\varepsilon_{i+1}| \leq |\varepsilon_i| + hL|\varepsilon_i| + \frac{h^2 dB}{2}.$$

A careful analysis of the right-hand side of the above equation will be used to complete the proof.

Set $a_0 = 0$, and define $a_i \in \mathbb{R}$ inductively by

$$a_{i+1} = (1 + hL)a_i + \frac{h^2 dB}{2}.$$

It follows by induction that

$$|\varepsilon_i| \leq a_i$$

and that

$$a_i = \frac{hdB}{2L}\left[(1+hL)^i - 1\right].$$

Since

$$(1+hL)^i \leq \left(e^{hL}\right)^i = e^{ihL},$$

the following estimates can be made

$$\begin{aligned} |\varepsilon_i| &\leq \frac{hdB}{2L}\left[(1+hL)^i - 1\right] \\ &\leq \frac{hdB}{2L}\left[e^{ihL} - 1\right] \\ &= \frac{hdB}{2L}\left[e^{(t_i-\tau)L} - 1\right] \\ &\leq \frac{hdB}{2L}\left[e^{\beta L} - 1\right]. \end{aligned}$$

Setting

$$C = \frac{dB}{2L}\left[e^{\beta L} - 1\right]$$

completes the proof of the first part.

For the second part, consider t in the closed interval $[t_i, t_{i+1}]$. Then

$$\begin{aligned} \big|\boldsymbol{\varphi}(t) - \boldsymbol{\psi}(t)\big| &\leq \big|\boldsymbol{\varphi}(t) - \boldsymbol{\varphi}(t_i)\big| + \big|\boldsymbol{\varphi}(t_i) - \boldsymbol{\psi}(t_i)\big| + \big|\psi(t_i) - \psi(t)\big| \\ &\leq \int_{t_i}^{t} \big|f\big(s, \boldsymbol{\varphi}(s)\big)\big|\, ds + |\varepsilon_i| + |t - t_i|\,\big|f(t_i, \boldsymbol{\xi}_i)\big| \\ &\leq |t - t_i|M + |\varepsilon_i| + |t - t_i|M \end{aligned}$$

where $\big|\mathbf{f}(t, \mathbf{x})\big| \leq M$ on D'. Therefore,

$$\big|\boldsymbol{\varphi}(t) - \boldsymbol{\psi}(t)\big| \leq 2Mh + Ch,$$

which is the desired result. □

In some cases, the best information available about the solutions of a differential equation comes from numerical studies of them, using more sophisticated numerical methods. For example, numerical studies have shown that the solutions of the relatively simple Lorenz system

$$\begin{aligned}\dot{x} &= \sigma(y-x)\\ \dot{y} &= \rho x - y - xz\\ \dot{z} &= -\beta z + xy\end{aligned}$$

behave in very complicated ways for some values of the parameters that are not yet fully explained by the theory. For an introduction to the use of numerical methods to explore the behavior of solutions see [2] and [28]. The latter includes a software package.

The rate of convergence is a good measure of a numerical algorithm's effectiveness, and one can do much better than the Euler method. For example, the still relatively simple Runge-Kutta method has an error proportional to h^5. The symbolic manipulation programs—Mathematica, Maple, Matlab—all contain sophisticated numerical solvers for differential equations based on Runge-Kutta and other numerical methods. Books like [8], [9], and [10] provide an introduction on using them to explore differential equations.

The exercises for this section are not, however, intended to be done with the full power of Mathematica, Maple, or Matlab. Rather they are designed to illustrate the iterative process of approximating solutions and its pitfalls. These exercises should be done by programming your own calculations and require nothing more than a simple tool like a programable calculator, an Excel spreadsheet, or other appropriate software on a personal computer.

EXERCISES

1. Consider the simple scalar differential equation $\dot{x} = 3x - 2$ with the initial condition $x(0) = 1$.

   (a) Use separation of variables to find the solution of this initial-value problem.

   (b) Use Euler's method to approximate the solution to this initial-value problem for $0 \le t \le 1$ with $h = 0.1,\ 0.05,\ 0.01,$ and 0.005.

   (c) Compare the results of parts (a) and (b) at the points $(t_i, \boldsymbol{\xi}_i)$.

2. Let $x(t)$ be the solution to the initial-value problem $\dot{x} = x^2$ and $x(0.5) = 2$. Use Euler's method with $h = 0.05$ to estimate the value of $x(1.1)$. Then use separation of variables to solve the problem and determine why the estimate for $x(1.1)$ is worthless.

3. Consider the differential equation $\dot{x} = t^2 + x^2$. Use *Theorem 1.15* to approximate the solution $x(t)$ such that $x(0) = 0$ for $-\alpha < t < \alpha$. (See Exercise 1 in Section 2 for the optimal choice of α.) Try pushing the approximations to larger intervals. What tentative conclusion can be made about the behavior of $x(t)$?

4. Consider the differential equation $\dot{x} = x - 3e^{-2t}$ on $D = \mathbb{R}^2$.

   (a) Verify that $x(t) = Ce^t + e^{-2t}$ is a solution for any constant C and determine $\lim_{t\to\infty} x(t)$.

   (b) Use Euler's method to approximate the solution $x(t) = e^{-2t}$ for $0 \le t \le 4$ with $h = 0.1$, 0.05, 0.025, and 0.0125.

   (c) Why are the approximate solutions obtained with Euler's method unreliable toward the end of the range?

5. Solutions to the system of differential equations

$$\begin{aligned} \dot{x} &= -x + y \\ \dot{y} &= -y \end{aligned}$$

   are defined on $\mathbb{R}$. (This will be established in Chapter 3.) Use Euler's method with $h = 0.1$ to approximate and graph the solutions of this system for $0 \le t \le 5$ and the following initial conditions:

   (a) $\big(x(0), y(0)\big) = (1, 0)$

   (b) $\big(x(0), y(0)\big) = (0.5, 0.5)$

   (c) $\big(x(0), y(0)\big) = (0, 1)$

   (d) $\big(x(0), y(0)\big) = (-0.5, 0.5)$.

1.5 Continuation

Although every initial-value problem has a solution, this solution is only known to exist on a limited interval that could be much shorter than necessary. This weakness of Peano's theorem (*Theorem 1.9*) will be addressed in this section. Specifically, the problem of extending solutions to larger intervals will be examined.

Let $\mathbf{x}(t)$ be a solution to $\dot{\mathbf{x}} = \mathbf{f}(t, \mathbf{x})$ defined on the open interval $I = \{t : a < t < b\}$. A *continuation* of $\mathbf{x}(t)$ is a solution $\mathbf{x}_1(t)$ defined on an interval I_1 containing I such that $\mathbf{x}_1(t) = \mathbf{x}(t)$ on I. When $I_1 = \{t : a < t < b_1\}$, we say $\mathbf{x}_1(t)$ is a *continuation to the right*. A *continuation to the left* is defined in the obvious way. The key questions are When does a continuation exist? How far can a given solution be continued? All results will be stated for right continuations. It will be obvious that the corresponding results hold for left continuations.

How far a solution can be continued is governed to some extent by the specified domain for $\mathbf{f}(t,\mathbf{x})$. Given a continuous function $\mathbf{f}(t,\mathbf{x})$ on D, it might be possible to extend $\mathbf{f}(t,\mathbf{x})$ to a continuous function on a domain D' containing D and thereby allow solutions to be defined on larger intervals. To avoid any such ambiguities, it must be understood that all results are in the context of a specified D for a given differential equation. In particular, if $\varphi(t)$ is a solution of the differential equation $\dot{\mathbf{x}} = \mathbf{f}(t,\mathbf{x})$ on the domain D, then the trajectory $\big(t,\varphi(t)\big)$ does not leave the specified D.

Proposition 1.17 *Let $\mathbf{x}(t)$ be a solution of $\dot{\mathbf{x}} = \mathbf{f}(t,\mathbf{x})$ defined on $I = \{t : a < t < b\}$, where $b < \infty$. There exists a continuation of $\mathbf{x}(t)$ to the right if and only if $\lim_{t\to b^-}\mathbf{x}(t) = \boldsymbol{\xi}$ exists and the point $(b,\boldsymbol{\xi})$ lies in D.*

Proof. First, suppose $\lim_{t\to b^-}\mathbf{x}(t) = \boldsymbol{\xi}$ exists and $(b,\boldsymbol{\xi}) \in D$. (The expression $\lim_{t\to b^-}$ denotes a limit from the left and $\lim_{t\to b^+}$ denotes a limit from the right.) By Peano's theorem there exists a solution $\varphi(t)$ defined on $\{t : b-\alpha < t < b+\alpha\}$ such that $\varphi(b) = \boldsymbol{\xi}$. Let

$$\mathbf{x}_1(t) = \begin{cases} \mathbf{x}(t) & \text{when } a < t < b \\ \varphi(t) & \text{when } b \le t < b+\alpha. \end{cases}$$

Clearly, $\mathbf{x}_1(t)$ is continuous on $I_1 = \{t : a < t < b+\alpha\}$. To show that $\mathbf{x}_1(t)$ is a solution, it suffices to show that it satisfies equation *(1.2)*. Select and fix a σ in the interval I. It is immediate that $\mathbf{x}_1(t) = \mathbf{x}_1(\sigma) + \int_\sigma^t f\big(s,\mathbf{x}_1(s)\big)\,ds$ on I. Because both sides are continuous, this also holds for $t = b$. Now, suppose $b < t < b+\alpha$. Then,

$$\begin{aligned}
\mathbf{x}_1(\sigma) + \int_\sigma^t \mathbf{f}\big(s,\mathbf{x}_1(s)\big)\,ds &= \\
\mathbf{x}_1(\sigma) + \int_\sigma^b \mathbf{f}\big(s,\mathbf{x}_1(s)\big)\,ds + \int_b^t \mathbf{f}\big(s,\mathbf{x}_1(s)\big)\,ds &= \\
\mathbf{x}_1(b) + \int_b^t \mathbf{f}\big(s,\mathbf{x}_1(s)\big)\,ds &= \\
\varphi(b) + \int_b^t \mathbf{f}\big(s,\varphi(s)\big)\,ds &= \\
\varphi(t) = \mathbf{x}_1(t).&
\end{aligned}$$

Therefore, $\mathbf{x}_1(t)$ is a solution of $\dot{\mathbf{x}} = \mathbf{f}(t,\mathbf{x})$. The other half is an immediate consequence of the continuity of a trajectory in D. □

The first of several applications of this proposition is that solutions whose trajectories are in a compact subset of D are continuable.

Proposition 1.18 *Let $\mathbf{x}(t)$ be a solution of $\dot{\mathbf{x}} = \mathbf{f}(t,\mathbf{x})$ defined on the interval $I = \{t : a < t < b\}$, where $b < \infty$. If there exists a compact set $C \subset D$ and $\tau \in I$ such that $\big(t,\mathbf{x}(t)\big) \in C$ for all $t \ge \tau$, then there exists a continuation of $\mathbf{x}(t)$ to right.*

Proof. By the previous proposition, it suffices to show that $\lim_{t\to b^-}\mathbf{x}(t)$ exists. To begin, there exists M such that $\left|\mathbf{f}(t,\mathbf{x})\right| \le M$ on C. If $t_1,\ t_2 \ge \tau$, then

$$\left|\mathbf{x}(t_1) - \mathbf{x}(t_2)\right| = \left|\int_{t_1}^{t_2} \mathbf{f}\big(s,\mathbf{x}(s)\big)\,ds\right| \le M\left|t_2 - t_1\right|.$$

By compactness, we can find a sequence s_n, $\tau < s_n < b$, such that

$$\lim_{n\to\infty} s_n = b \text{ and } \lim_{n\to\infty} \mathbf{x}(s_n) = \boldsymbol{\xi}.$$

For $\tau < t < b$, let n go to infinity in the expression

$$\left|\mathbf{x}(t) - \mathbf{x}(s_n)\right| \le M|s_n - t|,$$

to obtain

$$\left|\mathbf{x}(t) - \boldsymbol{\xi}\right| \le M|b - t|.$$

If $|b-t| < \varepsilon/M$ and $t < b$, then $\left|\mathbf{x}(t) - \boldsymbol{\xi}\right| < \varepsilon$ and $\lim_{t\to b^-}\mathbf{x}(t) = \boldsymbol{\xi}$. Clearly, $(b,\boldsymbol{\xi}) \in C \subset D$. □

Theorem 1.19 *Let $\mathbf{x}(t)$ be a solution of $\dot{\mathbf{x}} = \mathbf{f}(t,\mathbf{x})$ defined on the interval $I = \{t : a < t < b\}$, where $b < \infty$. If there exists a compact set $C \subset D$ and $\tau \in I$ such that $\big(t,\mathbf{x}(t)\big) \in C$ for $t \ge \tau$, then $\mathbf{x}(t)$ has a continuation to the right, $\mathbf{x}_1(t)$, such that $\big(t,\mathbf{x}_1(t)\big) \notin C$ for some $t > \tau$.*

Proof. By the second corollary to Peano's theorem on page 19, there exists $\alpha > 0$ such that for any $(\tau,\boldsymbol{\xi}) \in C$ the initial-value problem

$$\begin{aligned} \dot{\mathbf{x}} &= \mathbf{f}(t,\mathbf{x}) \\ \boldsymbol{\xi} &= \mathbf{x}(\tau) \end{aligned}$$

has a solution defined on $\{t : \tau - \alpha < t < \tau + \alpha\}$. By the previous propositions, $\mathbf{x}(t)$ has a continuation $\mathbf{x}_1(t)$ defined on $\{t : a < t < b + \alpha\}$. If $\big(t,\mathbf{x}_1(t)\big) \in C$ for all $t \ge \tau$, there exists a continuation $\mathbf{x}_2(t)$ of $\mathbf{x}_1(t)$, and hence also of $\mathbf{x}(t)$ to $\{t : a < t < b + 2\alpha\}$. This process can be repeated as long as the trajectories do not leave C. Because C is a bounded subset of $\mathbb{R}^{d+1}$, there exists $B > 0$ such that $(t,\mathbf{x}) \in C$ implies $|t| \le B$. Therefore, this process can be repeated at most $(B - b)/\alpha$ times without leaving C and the proof is completed. □

As in the above proof, a continuation to the right of a solution might itself be continuable to the right, and the result is another continuation to the right of the original solution defined on a still larger interval. It is, however, also possible that a solution cannot be continued to the right. The same remarks apply to the left.

A *maximal solution* is one that cannot be continued to either the left or the right. Does an arbitrary solution have a continuation that is a maximal solution? To answer this in the affirmative, it suffices to show that an arbitrary solution has a continuation to the right that cannot be continued any further to the right. Such a solution will be called a *maximal continuation to the right.* Roughly speaking, it will be shown that a solution that cannot be continued to infinity on the right can be continued to the right so that the right end of the trajectory cannot have a limit in D.

Theorem 1.20 *Let* $\mathbf{x}(t)$ *be a solution of* $\dot{\mathbf{x}} = \mathbf{f}(t, \mathbf{x})$ *defined on the open interval* $I = \{t : a < t < b\}$, *where* $b < \infty$. *Then there exists a maximal continuation of* $\mathbf{x}(t)$ *to the right.*

Proof. There exists a sequence of open sets V_m such that $\overline{V}_m$ is compact, $\overline{V}_m \subset V_{m+1}$, and $D = \bigcup_{m=1}^{\infty} V_m$. (The closure of V denoted by $\overline{V}$ was defined in Exercise 7 on page 11.) For example,

$$V_m = \{\mathbf{x} : |\mathbf{x}| < m\} \cap \left\{\mathbf{x} \in D : \inf_{\mathbf{y} \notin D} |\mathbf{x} - \mathbf{y}| > \frac{1}{m}\right\}.$$

Without loss of generality, it can be assumed that $(\tau, \mathbf{x}(\tau)) \in V_1$ for some τ in I.

Now, apply *Theorem 1.19* to $\overline{V}_1$ to obtain a continuation $\mathbf{x}_1(t)$ to $\{t : a < t < b_1\}$ and τ_1, such that $(\tau_1, \mathbf{x}_1(\tau_1)) \notin \overline{V}_1$. [If $(t, x(t)) \notin \overline{V}_1$ for some $t > \tau$, we can use $\mathbf{x}(t)$ for $\mathbf{x}_1(t)$ and b for b_1.] Do the same thing for V_2, V_3, and so forth. This will produce a sequence of continuations $\mathbf{x}_m(t)$ to $\{t : a < t < b_m\}$ such that each $\mathbf{x}_m(t)$ is a continuation of $\mathbf{x}_{m-1}(t)$ and a sequence $\{\tau_m\}$, $\tau < \tau_m < b_m$, such that $(\tau_m, \mathbf{x}_m(\tau_m)) \notin \overline{V}_m$. Set $\boldsymbol{\psi}(t) = \mathbf{x}_m(t)$ if $a < t < b_m$. Clearly, $\boldsymbol{\psi}(t)$ is well defined and a continuation of $\mathbf{x}(t)$ to $\{t : a < t < b'\}$, where $\lim_{m\to\infty} b_m = b'$. If $b' = \infty$, we are done.

Suppose $b' < \infty$. If $\boldsymbol{\psi}(t)$ has a continuation to the right, then it follows that $\lim_{t\to b'} \boldsymbol{\psi}(t) = \boldsymbol{\xi}$ exists and $(b', \boldsymbol{\xi})$ is in D by *Proposition 1.17*. Thus $(b', \boldsymbol{\xi})$ is in V_n for some n. It follows that $\lim_{m\to\infty} \boldsymbol{\psi}(\tau_m) = \boldsymbol{\xi}$ because $b_{m-1} < \tau_m < b_m$ implying that $(\tau_m, \mathbf{x}_m(\tau_m))$ is in V_n for large m, a contradiction. Therefore, $\boldsymbol{\psi}(t)$ is a maximal continuation to the right. □

Corollary *Every solution of* $\dot{\mathbf{x}} = \mathbf{f}(t, \mathbf{x})$ *can be continued to a maximal solution. In particular, every initial-value problem has a maximal solution.*

Maximal solutions that are not defined on all of $\mathbb{R}$ are easy to exhibit. For example, consider $\dot{x} = x^2$ on $D = \mathbb{R}^2$. By separation of variables, $x(t) = -1/(t - c)$, where c is a constant of integration. Because solutions are required to be continuous on an open interval, this must be viewed as two solutions, namely, $\varphi_1(t) = -1/(t-c)$ on $-\infty < t < c$ and $\varphi_2(t) = -1/(t-c)$ on $c < t < \infty$. Clearly, $\lim_{t\to c^-} \varphi_1(t) = \infty$, $\lim_{t\to c^+} \varphi_2(t) = -\infty$, and both $\varphi_1(t)$ and $\varphi_2(t)$ are maximal solutions.

The *boundary of a set* E is defined by

$$\partial E = \overline{E} \cap \overline{(\mathbb{R}^d \setminus E)}.$$

If E is open, then $\partial E = \overline{E} \cap (\mathbb{R}^d \setminus E)$ because $\mathbb{R}^d \setminus E$ is closed. The previous example shows that a maximal solution need not be defined on $\mathbb{R}$ even when the boundary of D is empty. When the boundary of D is not empty a maximal solution can approach the boundary of D in a complicated way as the next example shows. It is easy to check that for $t > 0$

$$\varphi(t) = \big(\cos(1/t), \sin(1/t)\big)$$

defines a maximal solution of the system

$$\begin{aligned} \dot{x} &= y/t^2 \\ \dot{y} &= -x/t^2 \end{aligned}$$

on $D = \big\{(t, x, y) : t > 0\big\} \subset \mathbb{R}^3$. Clearly, $\varphi(t)$ spirals around the origin infinitely often as t decreases to 0 and $\lim_{t\to 0} \varphi(t)$ does not exist. [In $\mathbb{R}^2$, the notation is usually more convenient to write a system of scalar differentiable equations using the variables x and y than to use the vector differential equation $\dot{\mathbf{x}} = \mathbf{f}(t, \mathbf{x})$ and $\mathbf{x} = (x_1, x_2)$. We shall usually do so.]

There is, however, a common behavior that these maximal solutions exhibit near the ends of their domains. Let $\mathbf{x}(t)$ be a solution of $\dot{\mathbf{x}} = \mathbf{f}(t, \mathbf{x})$ defined on $\{t : a < t < b\}$. The trajectory $\big(t, \mathbf{x}(t)\big)$ *goes to the boundary* of D as t increases to b, if given any compact subset K of D there exists $\tau \in \{t : a < t < b\}$ such that $t > \tau$ implies $\big(t, \mathbf{x}(t)\big) \notin K$. In other words, the trajectory escapes forever from each compact set contained in D.

Theorem 1.21 *If $\mathbf{x}(t)$ is a maximal solution of $\dot{\mathbf{x}} = \mathbf{f}(t, \mathbf{x})$ defined on $I = \{t : a < t < b\}$, then the trajectory $\big(t, \mathbf{x}(t)\big)$ goes to the boundary of D as t increases to b.*

Proof. If $b = \infty$, the conclusion is clear. Assume $b < \infty$. Let K be a compact subset of D. Suppose there does not exist $\tau \in I$ such that $t > \tau$ implies $\big(t, \mathbf{x}(t)\big) \notin K$.

First, note that there exists N such that

$$B = \left\{(t, \mathbf{x}) : \big\|(t, \mathbf{x}) - (s, \mathbf{y})\big\| \le \frac{1}{N} \text{ for some } (s, \mathbf{y}) \in K\right\} \subset D$$

Clearly, B is compact. If $\big(t, \mathbf{x}(t)\big) \in K$, then by maximality and *Theorem 1.19*, there exists $s > t$ such that $\big(s, \mathbf{x}(s)\big) \notin B$. Since we are assuming the conclusion of the theorem fails for K, there must exist $t' > s$ such that $\big(t', \mathbf{x}(t')\big) \in K$. Repeating this argument produces sequences $\{t_m\}$ and $\{s'_m\}$ such that $t_m < s'_m < t_{m+1} < b$, $\big(t_m, \mathbf{x}(t_m)\big) \in K$, $\big(s'_m, \mathbf{x}(s'_m)\big) \notin B$. Let

$$s_m = \sup\big\{s \ge t_m : \big(t, \mathbf{x}(t)\big) \in B \text{ for all } t \in [t_m, s]\big\}.$$

It is easy to see that $\big(t, \mathbf{x}(t)\big) \in B$ for all $t \in [t_m, s_m]$ and

$$\big\|\big(s_m, \mathbf{x}(s_m)\big) - \big(t_m, \mathbf{x}(t_m)\big)\big\| \ge \frac{1}{N}.$$

Let

$$M = \sup\Big\{\big\|\mathbf{f}(t,x)\big\| : (t,\mathbf{x}) \in B\Big\}.$$

Clearly, the arc length of the trajectory $\big(t, \mathbf{x}(t)\big)$ between t_m and s_m is at least as big as the Euclidean distance between $\big(s_m, \mathbf{x}(s_m)\big)$ and $\big(t_m, \mathbf{x}(t_m)\big)$. Therefore, using the formula for arc length, it follows that

$$\begin{aligned}
\frac{1}{N} &\le \Big\|\big(s_m, \mathbf{x}(s_m)\big) - \big(t_m, \mathbf{x}(t_m)\big)\Big\| \\
&\le \int_{t_m}^{s_m} \sqrt{1 + \|\dot{\mathbf{x}}\|^2}\, dt \\
&= \int_{t_m}^{s_m} \sqrt{1 + \big\|\mathbf{f}\big(t, \mathbf{x}(t)\big)\big\|^2}\, dt \\
&\le |s_m - t_m|\sqrt{1 + M^2}
\end{aligned}$$

which goes to zero because the sequences $\{t_m\}$ and $\{s_m\}$ both converge to b, which is finite. This contradiction completes the proof. □

An initial-value problem can, of course, have more than one maximal solution. The final example shows that an initial-value problem can also have maximal solutions defined on very different intervals.

Let $D = \big\{(t,x) : -1 < x < 1\big\} \subset \mathbb{R}^2$ and consider the initial-value problem

$$\begin{aligned}
\dot{x} &= \frac{\sqrt{|x|}}{1 - \sqrt{|x|}} \\
0 &= x(0).
\end{aligned}$$

Clearly, $x(t) \equiv 0$ is one maximal solution defined on $\mathbb{R}$. Separating variables produces

$$\frac{1 - \sqrt{|x|}}{\sqrt{|x|}}\, dx = dt,$$

which for $x > 0$ is

$$\big(x^{-1/2} - 1\big)\, dx = dt.$$

Thus

$$2x^{1/2} - x = t + c.$$

To solve this, let $y = x^{1/2}$ so that

$$y^2 - 2y + t + c = 0$$

and

$$y = \frac{2 \pm \sqrt{4 - 4(t+c)}}{2} = 1 \pm \sqrt{1 - t - c}.$$

Since $0 \le x < 1$ and $0 \le y < 1$, the minus sign is the correct choice and $0 \le t + c < 1$. It follows that

$$x(t) = \big(1 - \sqrt{1 - t - c}\big)^2, \qquad 0 \le t + c < 1$$

is a solution. Similarly for $x \leq 0$

$$x(t) = -\left(1 - \sqrt{1+t+c}\right)^2, \qquad -1 < t+c \leq 0$$

is a solution. Thus with $c = 0$

$$x(t) = \begin{cases} \left(1 - \sqrt{1-t}\right)^2 & \text{for } 0 \leq t < 1 \\ -\left(1 - \sqrt{1+t}\right)^2 & \text{for } -1 < t \leq 0 \end{cases}$$

is another maximal solution, but only defined for $-1 < t < 1$. Notice that as t increases to 1, the trajectory $\big(t, \mathbf{x}(t)\big)$ approaches the point (1,1), which is on the boundary of D.

The designation of t as the independent variable in $\dot{\mathbf{x}} = \mathbf{f}(t, \mathbf{x})$ is also an important aspect of this differential equation. The previous example can be considered with x or a new parameter s as the independent variable. In each of these cases, solutions lie on the curves of the form

$$2x^{1/2} - x - t = C,$$

but the parameterizations of the solutions are quite different, as will be evident in the final paragraphs of this section

If we let x be the independent variable instead of t, then we would be considering

$$\frac{dt}{dx} = \frac{1 - \sqrt{|x|}}{\sqrt{|x|}}$$

on $D = \big\{(t, x) : x \neq 0\big\}$ and $t = 2x^{1/2} - x + c$ would be a solution defined on $\{x : x > 0\}$.

We could also let x and t depend on a new independent variable s, and considering the system

$$\begin{aligned} \frac{dx}{ds} &= \sqrt{|x|} \\ \frac{dt}{ds} &= 1 - \sqrt{|x|} \end{aligned}$$

on $\mathbb{R}^2$. Now

$$x(s) = \begin{cases} s^2/4 & \text{for } s \geq 0 \\ -s^2/4 & \text{for } s < 0 \end{cases}$$

$$t(s) = \begin{cases} s - s^2/4 & \text{for } s \geq 0 \\ s + s^2/4 & \text{for } s < 0 \end{cases}$$

is a solution defined on $\mathbb{R}$. Solving for s in terms of t or x would recover solutions to the two previous versions. Even though the theory applies equally well in all these cases and the tangent lines remain the same, the solutions differ because the context changes.

EXERCISES

1. Find all maximal solutions of

$$\dot{x} = \frac{\sqrt{|x|}}{1 - \sqrt{|x|}}$$
$$0 = x(0)$$

2. Suppose $f(t, x)$ is locally Lipschitz with respect to x on $\mathbb{R}^2$ and there exist a and b with $a < b$ such that $f(t, a) = f(t, b) = 0$ for all t. Show that if $a < \xi < b$, then the solution of

$$\dot{x} = f(t, x)$$
$$x(\tau) = \xi$$

is defined on $\mathbb{R}$.

3. Consider $\dot{\mathbf{x}} = \mathbf{f}(t, \mathbf{x})$ on $D = \mathbb{R}^{d+1}$ and let $x(t)$ be a maximal solution defined on the interval $a < t < b$. Show that if $b < \infty$, then

$$\lim_{t \to b^-} |\mathbf{x}(t)| = \infty.$$

4. Suppose $|\mathbf{f}(t, x)| \leq M$ on $D \neq \mathbb{R}^{d+1}$ and let $\mathbf{x}(t)$ be a maximal solution defined on $a < t < b < \infty$. Show that

$$\lim_{t \to b^-} (t, \mathbf{x}(t))$$

exists and lies in the boundary of D.

5. Consider $\dot{\mathbf{x}} = \mathbf{f}(t, \mathbf{x})$ on $\mathbb{R}^{d+1}$ satisfying the following condition: For every compact interval I of the real line there exists a positive real number M_I such that

$$\mathbf{f}(t, \mathbf{x}) \cdot \mathbf{x} < 0$$

when $\|\mathbf{x}\| \geq M_I$ and $t \in I$. Prove that every maximal solution is defined on an interval of the form $a < t < \infty$.

1.6 Continuity in Initial Conditions

Given a continuous function $\mathbf{f}(t, \mathbf{x})$ defined on $D \subset \mathbb{R}^{d+1}$, for every $(\tau, \boldsymbol{\xi}) \in D$ there is an initial-value problem, namely,

$$\begin{aligned} \dot{\mathbf{x}} &= \mathbf{f}(t, \mathbf{x}) \\ \boldsymbol{\xi} &= \mathbf{x}(\tau). \end{aligned}$$

It is natural to ask how the solution of this problem changes as the initial time τ and initial position $\boldsymbol{\xi}$ are varied. The first specific behavior to look for is some sort of continuous dependency on the initial conditions. After all, in many physical problems, it is apparent that a very small change in starting time or starting position will cause only a correspondingly small change in the outcome after a finite amount of time. Of course without uniqueness of solutions there is no hope of proving a general result of this type.

Throughout this section it will be assumed that $\mathbf{f}$ is continuous and locally Lipschitz on D. Consequently, by Peano's theorem, every initial-value problem has at least one solution; by the corollary to *Theorem 1.11*, there is at most one solution to an initial-value problem on a given interval; and by the corollary to *Theorem 1.20*, for each initial-value problem there is a largest interval to which the solution can be continued. Therefore, we can let $\mathbf{x}(t, \tau, \boldsymbol{\xi})$ denote the maximal solution of

$$\begin{aligned} \dot{\mathbf{x}} &= \mathbf{f}(t, \mathbf{x}) \\ \boldsymbol{\xi} &= \mathbf{x}(\tau), \end{aligned}$$

and know that $\mathbf{x}(t, \tau, \boldsymbol{\xi})$ is a well-defined function on some subset $D_{\mathbf{f}}$ of $\mathbb{R}^{d+2}$ with values in $\mathbb{R}^d$. In other words, $\mathbf{x}(t, \tau, \boldsymbol{\xi})$ is the solution at time t, which at time τ passes through the point $\boldsymbol{\xi}$, and as a function of three variables t, τ, and $\boldsymbol{\xi}$, the function $\mathbf{x}(t, \tau, \boldsymbol{\xi})$ can be thought of as the *global solution* or *general solution* of $\dot{\mathbf{x}} = \mathbf{f}(t, \mathbf{x})$ because every solution can be retrieved from it by choosing a suitable τ and $\boldsymbol{\xi}$.

The goal is to prove that $\mathbf{x}(t, \tau, \boldsymbol{\xi})$ is a continuous function on its domain in $\mathbb{R}^{d+2}$. The main step is to see what happens near a specific solution on a finite time interval.

Theorem 1.22 *Fix $(\sigma, \boldsymbol{\zeta}) \in D$. If the domain of the solution $\mathbf{x}(t, \sigma, \boldsymbol{\zeta})$ contains the closed interval $\{t : a \le t \le b\}$, then there exists $\delta > 0$ such that for any point $(\tau, \boldsymbol{\xi})$ in the set*

$$U = \big\{(\tau, \boldsymbol{\xi}) : a < \tau < b \text{ and } \big|\boldsymbol{\xi} - \mathbf{x}(\tau, \sigma, \boldsymbol{\zeta})\big| < \delta\big\}$$

the domain of the solution $\mathbf{x}(t, \tau, \boldsymbol{\xi})$ contains the open interval $I = \{t : a < t < b\}$, and the function $\mathbf{x}(t, \tau, \boldsymbol{\xi})$ is continuous on the set

$$W = \big\{(t, \tau, \boldsymbol{\xi}) : a < t < b \text{ and } (\tau, \boldsymbol{\xi}) \in U\big\}.$$

The idea of the proof is to define an iterative procedure to produce approximations of $\mathbf{x}(t, \tau, \boldsymbol{\xi})$, which converge uniformly to $\mathbf{x}(t, \tau, \boldsymbol{\xi})$ on W. As a by-product of this approach we will obtain another method beside Euler's method for approximating the solution of an initial-value problem.

Proof. For ease of notation, let $\boldsymbol{\psi}(t)$ denote $\mathbf{x}(t, \sigma, \boldsymbol{\zeta})$ on I. Choose $\delta_1 > 0$ such that

$$C = \big\{(t, \mathbf{x}) : a \le t \le b \text{ and } \big|\mathbf{x} - \boldsymbol{\psi}(t)\big| \le \delta_1\big\} \subset D.$$

Clearly, C is closed and bounded and hence compact. Because we are assuming that $\mathbf{f}$ is locally Lipschitz, *Theorem 1.14* applies and $\mathbf{f}$ satisfies a Lipschitz condition on C for some $L > 0$. Next, choose δ so that $\delta < e^{-L(b-a)}\delta_1 < \delta_1$.

For $(t, \tau, \boldsymbol{\xi}) \in W$, set

$$\boldsymbol{\varphi}_0(t, \tau, \boldsymbol{\xi}) = \boldsymbol{\psi}(t) + \boldsymbol{\xi} - \boldsymbol{\psi}(\tau).$$

Note that

$$\left|\boldsymbol{\varphi}_0(t, \tau, \boldsymbol{\xi}) - \boldsymbol{\psi}(t)\right| = \left|\boldsymbol{\xi} - \mathbf{x}(\tau, \sigma, \boldsymbol{\zeta})\right| < \delta < \delta_1$$

and $\big(t, \boldsymbol{\varphi}_0(t, \tau, \boldsymbol{\xi})\big) \in C$ for all $(t, \tau, \boldsymbol{\xi}) \in W$. Set

$$\boldsymbol{\varphi}_1(t, \tau, \boldsymbol{\xi}) = \boldsymbol{\xi} + \int_\tau^t \mathbf{f}\big(s, \boldsymbol{\varphi}_0(s, \tau, \boldsymbol{\xi})\big)\, ds.$$

Because $\boldsymbol{\psi}(t) - \boldsymbol{\psi}(\tau) = \int_\tau^t \mathbf{f}\big(s, \boldsymbol{\psi}(s)\big)\, ds$, it follows that

$$\begin{aligned} \left|\boldsymbol{\varphi}_1(t, \tau, \boldsymbol{\xi}) - \boldsymbol{\varphi}_0(t, \tau, \boldsymbol{\xi})\right| &= \left|\int_\tau^t \mathbf{f}\big(s, \boldsymbol{\varphi}_0(s, \tau,)\big) - \mathbf{f}\big(s, \boldsymbol{\psi}(s)\big)\, ds\right| \\ &\leq L\delta|t - \tau| \end{aligned}$$

and by the triangle inequality

$$\left|\boldsymbol{\varphi}_1(t, \tau, \boldsymbol{\xi}) - \boldsymbol{\psi}(t)\right| < \delta\big(1 + L(b-a)\big) < \delta e^{L(b-a)} < \delta_1$$

because $L(b-a) > 0$ and

$$e^w = \sum_{m=0}^{\infty} w^m/(m!).$$

Thus $\big(t, \boldsymbol{\varphi}_1(t, \tau, \boldsymbol{\xi})\big) \in C$ for all $(t, \tau, \boldsymbol{\xi}) \in W$. Arguing inductively, suppose

$$\boldsymbol{\varphi}_j(t, \tau, \boldsymbol{\xi}) = \boldsymbol{\xi} + \int_\tau^t \mathbf{f}\big(s, \boldsymbol{\varphi}_{j-1}(s, \tau, \boldsymbol{\xi})\big)\, ds$$

is defined on W for $1 \leq j \leq n$ so that $\big(t, \boldsymbol{\varphi}_j(t, \tau, \boldsymbol{\xi})\big) \in C$ for all $(t, \tau, \boldsymbol{\xi}) \in W$ and

$$\left|\boldsymbol{\varphi}_j(t, \tau, \boldsymbol{\xi}) - \boldsymbol{\varphi}_{j-1}(t, \tau, \boldsymbol{\xi})\right| \leq \frac{L^j|t - \tau|^j\delta}{j!}$$

for $(t, \tau, \boldsymbol{\xi}) \in W$ and $1 \leq j \leq n$. Set

$$\boldsymbol{\varphi}_{n+1}(t, \tau, \boldsymbol{\xi}) = \boldsymbol{\xi} + \int_\tau^t \mathbf{f}\big(s, \boldsymbol{\varphi}_n(s, \tau, \boldsymbol{\xi})\big)\, ds.$$

It follows that

$$\left|\boldsymbol{\varphi}_{n+1}(t, \tau, \boldsymbol{\xi}) - \boldsymbol{\varphi}_n(t, \tau, \boldsymbol{\xi})\right| \quad \leq$$

$$\begin{aligned}\int_\tau^t \left|\mathbf{f}\big(s,\varphi_n(s,\tau,\boldsymbol{\xi})\big)-\mathbf{f}\big(s,\varphi_{n-1}(s,\tau,\boldsymbol{\xi})\big)\right| ds &\le \\ \int_\tau^t L\delta\,\frac{L^n|s-\tau|^n}{n!}\,ds &= \\ \frac{\delta L^{n+1}}{n!}\cdot\frac{|t-\tau|^{n+1}}{n+1} &= \\ \frac{\delta L^{n+1}|t-\tau|^{n+1}}{(n+1)!}&.\end{aligned}$$

(The integration is obvious for $t > \tau$, and a matter of carefully keeping track of minus signs when $t < \tau$.) Thus

$$\left|\varphi_{n+1}(t,\tau,\boldsymbol{\xi})-\boldsymbol{\psi}(t)\right| \le \delta\sum_{j=0}^{n+1}\frac{L^j|t-\tau|^j}{j!} \le \delta e^{L(b-a)} < \delta_1,$$

which implies $\big(t,\varphi_{n+1}(t,\tau,\boldsymbol{\xi})\big) \in C$ for all $(t,\tau,\boldsymbol{\xi}) \in W$. Clearly each φ_n is continuous on W and for $(\tau,\boldsymbol{\xi}) \in U$, $\varphi_n(t,\tau,\boldsymbol{\xi})$ is defined for all $t \in I$. It now suffices to show that φ_n converges uniformly to $\mathbf{x}(t,\tau,\boldsymbol{\xi})$ on W.

Since for $n > m$,

$$\left|\varphi_n(t,\tau,\boldsymbol{\xi})-\varphi_m(t,\tau,\boldsymbol{\xi})\right| \le \delta\sum_{j=m+1}^{n}\frac{L^j(b-a)^j}{j!},$$

$\varphi_n(t,\tau,\boldsymbol{\xi})$ is a Cauchy sequence at every point in W and therefore converges to a function $\varphi(t,\tau,\boldsymbol{\xi})$ on W. Furthermore, letting n go to infinity in the above inequality, shows that

$$\left|\varphi(t,\tau,\boldsymbol{\xi})-\varphi_m(t,\tau,\boldsymbol{\xi})\right| \le \delta\sum_{j=m+1}^{\infty}\frac{L^j(b-a)^j}{j!}.$$

Hence, the convergence is uniform and φ is continuous on W. To show that $\varphi(t,\tau,\boldsymbol{\xi}) = \mathbf{x}(t,\tau,\boldsymbol{\xi})$ for $a < t < b$ and complete the proof, it must be shown that

$$\varphi(t,\tau,\boldsymbol{\xi}) = \boldsymbol{\xi} + \int_\tau^t \mathbf{f}\big(s,\varphi(s,\tau,\boldsymbol{\xi})\big)\,ds$$

Let $\varepsilon > 0$. By adding $\boldsymbol{\xi}+\int_\tau^t \mathbf{f}\big(s,\varphi_{n-1}(s,\tau,\boldsymbol{\xi})\big)\,ds$ and subtracting $\varphi_n(t,\tau,\boldsymbol{\xi})$ which are different ways of writing the same thing, we get

$$\begin{aligned}\left|\varphi(t,\tau,\boldsymbol{\xi})-\boldsymbol{\xi}-\int_\tau^t \mathbf{f}\big(s,\varphi(s,\tau,\boldsymbol{\xi})\big)\,ds\right| &\le \\ \left|\varphi(t,\tau,\boldsymbol{\xi})-\varphi_n(t,\tau,\boldsymbol{\xi})\right| &+ \\ \left|\int_\tau^t \mathbf{f}\big(s,\varphi_{n-1}(s,\tau,\boldsymbol{\xi})\big)-\mathbf{f}\big(s,\varphi(s,\tau,\boldsymbol{\xi})\big)\,ds\right| &< \\ \varepsilon+\int_\tau^t L\left|\varphi_{n-1}(s,\tau,\boldsymbol{\xi})-\varphi(s,\tau,\boldsymbol{\xi})\right| ds &< \\ \varepsilon+L\varepsilon(b-a)&\end{aligned}$$

for n large. Because ε was arbitrary, it follows that

$$\left|\varphi(t,\tau,\boldsymbol{\xi}) - \boldsymbol{\xi} - \int_\tau^t \mathbf{f}\big(s,\varphi(s,\tau,\boldsymbol{\xi})\big)\,ds\right| = 0$$

or

$$\varphi(t,\tau,\boldsymbol{\xi}) = \boldsymbol{\xi} + \int_\tau^t \mathbf{f}\big(s,\varphi(s,\tau,\boldsymbol{\xi})\big)\,ds.$$

Finally,

$$\varphi(\tau,\tau,\boldsymbol{\xi}) = \lim_{n\to\infty} \varphi_n(\tau,\tau,\boldsymbol{\xi}) = \lim_{n\to\infty} \boldsymbol{\xi} = \boldsymbol{\xi}$$

and it follows that $\varphi(t,\tau,\boldsymbol{\xi}) = \mathbf{x}(t,\tau,\boldsymbol{\xi})$ for all $(t,\tau,\boldsymbol{\xi}) \in W$ by uniqueness. This completes the proof. □

The previous theorem established the continuity of $\mathbf{x}(t,\tau,\boldsymbol{\xi})$ on an open set containing an arbitrarily long compact piece of the trajectory of $\mathbf{x}(t,\sigma,\boldsymbol{\zeta})$ for each $(\sigma,\boldsymbol{\zeta}) \in D$ and makes it easy to prove the general continuity in initial conditions.

Theorem 1.23 (Continuity in Initial Conditions) *The set $D_{\mathbf{f}}$ is open in $\mathbb{R}^{d+2}$ and $\mathbf{x}(t,\tau,\boldsymbol{\xi})$ is a continuous function on $D_{\mathbf{f}}$ with values in $\mathbb{R}^d$.*

Proof. Let $(s,\sigma,\boldsymbol{\zeta})$ be an arbitrary point in $D_{\mathbf{f}}$. Then s is in the domain of $\mathbf{x}(t,\sigma,\boldsymbol{\zeta})$, and there exist $a < s < b$ such that the closed interval $[a,b]$ is contained in the domain of $\mathbf{x}(t,\sigma,\boldsymbol{\zeta})$. By the *Theorem 1.22*, the function $\mathbf{x}(t,\tau,\boldsymbol{\xi})$ is continuous on an open set W containing $(t,\sigma,\boldsymbol{\zeta})$ for $a < t < b$. In particular, $(s,\sigma,\boldsymbol{\zeta}) \in W \subset D_{\mathbf{f}}$. Therefore, $D_{\mathbf{f}}$ is the union of open sets on which $\mathbf{x}(t,\tau,\boldsymbol{\xi})$ is continuous. In particular, $D_{\mathbf{f}}$ is open and $\mathbf{x}(t,\tau,\boldsymbol{\xi})$ is continuous on $D_{\mathbf{f}}$. □

An alternative description of $D_{\mathbf{f}}$ is the following: $(t,\tau,\boldsymbol{\xi}) \in D_{\mathbf{f}}$ if and only if

(a) $(\tau,\boldsymbol{\xi}) \in D$;

(b) t is in the domain of the maximal solution $\mathbf{x}(t,\tau,\boldsymbol{\xi})$.

The technique used to prove *Theorem 1.22* can also be used to approximate the solution of an initial-value problem as *Theorem 1.24* shows.

Theorem 1.24 (Picard-Lindelöf) *Let $\mathbf{f}$ be locally Lipschitz on D and consider the initial-value problem*

$$\begin{aligned} \dot{\mathbf{x}} &= \mathbf{f}(t,\mathbf{x}) \\ \boldsymbol{\xi} &= \mathbf{x}(\tau). \end{aligned}$$

There exists $\alpha > 0$ such that every function in the sequence

$$\begin{aligned} \varphi_0(t) &= \boldsymbol{\xi} \\ \varphi_m(t) &= \boldsymbol{\xi} + \int_\tau^t \mathbf{f}\big(s,\varphi_{m-1}(s)\big)\,ds \end{aligned}$$

is defined on the interval $I = \{\tau - \alpha < t < \tau + \alpha\}$ and the sequence $\{\varphi_m\}$ converges uniformly on I to the solution of this initial-value problem.

Proof. Exercise.

Suppose in addition to varying the initial conditions the function $\mathbf{f}(t, x)$ is allowed to vary. This corresponds to varying the force in Newtonian mechanics. Again, for physical reasons, we would expect to find continuous behavior of the solutions. The simplest way to vary the force is with an external parameter, say $\boldsymbol{\mu}$. Consider a continuous function $\mathbf{f}(t, \mathbf{x}, \boldsymbol{\mu})$ on D, an open subset of $\mathbb{R}^{1+d+k}$. As before $t \in \mathbb{R}$, $\mathbf{x} \in \mathbb{R}^d$ and now $\boldsymbol{\mu} \in \mathbb{R}^k$. For each fixed $\boldsymbol{\mu}$, the set $E = \{(t, \mathbf{x}) : (t, \mathbf{x}, \boldsymbol{\mu}) \in D\}$ is open in $\mathbb{R}^{d+1}$ and consider $\dot{\mathbf{x}} = \mathbf{f}(t, \mathbf{x}, \boldsymbol{\mu})$ on E. When $\mathbf{f}(t, \mathbf{x}, \boldsymbol{\mu})$ is assumed to be locally Lipschitz, there is a function $\mathbf{x}(t, \tau, \boldsymbol{\xi}, \boldsymbol{\mu})$ defined by $\mathbf{x}(t, \tau, \boldsymbol{\xi}, \boldsymbol{\mu})$, which is the maximal solution of $\dot{\mathbf{x}} = \mathbf{f}(t, \mathbf{x}, \boldsymbol{\mu})$ such that $\mathbf{x}(\tau, \tau, \boldsymbol{\xi}, \boldsymbol{\mu}) = \boldsymbol{\xi}$. As before, let $D_{\mathbf{f}}$ denote the domain of $\mathbf{x}(t, \tau, \boldsymbol{\xi}, \boldsymbol{\mu})$. The set $D_{\mathbf{f}}$ consists of the points $(t, \tau, \boldsymbol{\xi}, \boldsymbol{\mu}) \in \mathbb{R}^{1+d+k}$ satisfying the following conditions:

(a) $(\tau, \boldsymbol{\xi}, \boldsymbol{\mu}) \in D$;

(b) t is in the domain of the maximal solution $\mathbf{x}(t, \tau, \boldsymbol{\xi}, \boldsymbol{\mu})$.

Theorem 1.25 *If $\mathbf{f}(t, \mathbf{x}, \boldsymbol{\mu})$ is locally Lipschitz with respect to the variable $\mathbf{y} = (\mathbf{x}, \boldsymbol{\mu})$ on D, then $\mathbf{x}(t, \tau, \boldsymbol{\xi}, \boldsymbol{\mu})$ is a continuous function on the open set $D_{\mathbf{f}}$ of $\mathbb{R}^{d+k+2}$ with values in $\mathbb{R}^d$.*

Proof. Let $\mathbf{f}(t, \mathbf{x}, \boldsymbol{\mu}) = \big(f_1(t, \mathbf{x}, \boldsymbol{\mu}), \ldots, f_n(t, \mathbf{x}, \boldsymbol{\mu})\big)$. Define $\mathbf{F} : D \to \mathbb{R}^{d+k}$ by $F_i(t, \mathbf{y}) = f_i(t, \mathbf{x}, \boldsymbol{\mu})$ for $1 \le i \le d$ and $F_i(t, y) = 0$ for $d + 1 \le i \le d + k$. Consider $\dot{\mathbf{y}} = \mathbf{F}(t, \mathbf{y})$ on D and note that $\mathbf{F}(t, \mathbf{y})$ is locally Lipschitz. Let $\boldsymbol{\theta} = (\boldsymbol{\xi}, \boldsymbol{\mu})$. Clearly, $\mathbf{y}(t, \tau, \boldsymbol{\theta}) = \big(\mathbf{x}(t, \tau, \boldsymbol{\xi}, \boldsymbol{\mu}), \boldsymbol{\mu}\big)$. By *Theorem 1.23*, $\mathbf{y}(t, \tau, \boldsymbol{\theta})$ is continuous on an open set $D_{\mathbf{F}}$. Clearly, $D_{\mathbf{F}} = D_{\mathbf{f}}$ and hence $\mathbf{x}(t, \tau, \boldsymbol{\xi}, \boldsymbol{\mu})$ is also continuous on an open set. □

The global solution $\mathbf{x}(t, \tau, \boldsymbol{\xi}, \mu)$ is not as mysterious as the proof might suggest. In many simple cases, this function is familiar. For example, if $\dot{x} = \mu x$, x, $\mu \in \mathbb{R}$, then $x(t, \tau, \xi, \mu) = \xi e^{\mu(t-\tau)}$ on $\mathbb{R}^4$. The fact that in this case $x(t, \tau, \xi, \mu)$ also has continuous partial derivatives of all orders in all variables is no accident, but rather a consequence of $f(t, x, \mu) = \mu x$ having the same property. However, the proof of this general result is a level deeper than the continuity of $\mathbf{x}(t, \tau, \xi, \mu)$ and will not be considered until Chapter 7.

The function $\mathbf{x}(t, \tau, \boldsymbol{\xi})$ will play a major role in every chapter. It provides a tool for simultaneously investigating the behavior of all the solutions and their interrelations instead of studying a particular solution to a specific initial-value problem. Using the function $\mathbf{x}(t, \tau, \boldsymbol{\xi})$ to understand the behavior of the solutions of a differential equation is the essence of dynamical systems.

EXERCISES

1. For the initial-value problem

$$\dot{x} = x$$

$$x(0) = 0$$

with x real construct the sequence of functions $\{\varphi_m(t)\}$ as defined in the statement of *Theorem 1.24*.

2. Using α as defined in the proof of Peano's theorem on page 15 prove *Theorem 1.24*.

3. Determine the function $x(t, \tau, \xi)$ and its domain for each of the following separable equations:

   (a) $\dot{x} = t^2/x$ on $D = \{(t, x) : x > 0\}$;

   (b) $\dot{x} = x/t^2$ on $D = \{(t, x) : t > 0\}$;

   (c) $\dot{x} = t^2/x^2(1 - t^3)$ on $D = \{(t, x) : t < 1 \text{ and } x > 0\}$.

4. Find the function $x(t, \tau, \xi, \mu)$ for $\dot{x} = x\cos(\mu t)$.

5. Find the function $x(t, \tau, \xi, \mu, \nu)$ for

$$\dot{x} = e^{\mu t - \nu x}$$

   on $D = \{(t, x, \mu, \nu) : \mu > 0 \text{ and } \nu > 0\}$.

6. Show that $D_{\mathbf{f}} = \mathbb{R}^{d+2}$ for $\dot{\mathbf{x}} = \mathbf{f}(t, \mathbf{x})$, if $\mathbf{f}(t, \mathbf{x})$ is locally Lipschitz and bounded on $\mathbb{R}^{d+1}$.

Chapter 2

Classical Themes

The focus of this chapter is three themes that have their origins at the beginning of the twentieth century or earlier and in that sense can be called classical. But they are also part of the modern study of differential equations and provide a basis for broadly understanding a major paradigm shift from finding formulas for solutions to describing the long-term behavior of solutions. This paradigm shift is arguably the most important distinction between differential equations in the nineteenth and twentieth centuries.

Before Poincaré initiated the qualitative paradigm, the common approach to investigating and solving differential equations was to find integrals, that is, functions that are constant along solutions. The method of separation of variables fits this mold along with exact equations and the conservation of quantities like energy and angular momentum. These topics are the subject of the first section of this chapter.

The first section begins with a careful analysis of the separation of variables method of solving simple differential equations that leads naturally to the concept of an integral. It is also helpful to understand how elementary methods like separation of variables fit into the general theory. In subsequent sections and chapters, a variety of other elementary techniques will be revisited from a rigorous perspective and incorporated into the larger theoretical picture of the subject.

The solutions of most ordinary differential equations, however, cannot be understood using integrals, let alone solved analytically. The qualitative viewpoint created by Poincaré seeks instead to describe the behavior of all the solutions of a differential equation without solving the equation in the traditional sense. It has been a major force in the subject during the twentieth century and is the subject of the second section.

In the spirit qualitative approach, there are general estimates on how fast solutions can grow. These estimates lead to the idea of a differential inequality. The last section covers differential inequalities and lays the foundation for the study of linear differential equations in Chapter 3.

2.1 Integrals

The most common elementary technique for solving an ordinary differential equation is separation of variables. Consider the scalar differential equation

$$\dot{x} = f(t, x)$$

on D, a subset of $\mathbb{R}^2$, and suppose that after some algebraic manipulation, $f(t, x)$ can be put in the form $h(t)/g(x)$ where $g(x)$ and $h(t)$ are continuous and $g(x) \neq 0$. Without loss of generality, it can be assumed that

$$D = \{(t, x) : \alpha < t < \beta \text{ and } \gamma < x < \delta\}.$$

Such a differential equation is called *separable.* To solve

$$\frac{dx}{dt} = \frac{h(t)}{g(x)}$$

by the separation of variables procedure, one writes it as

$$g(x)\, dx = h(t)\, dt,$$

integrates both sides, and solves for x in terms of t and a constant of integration. The constant of integration is then determined by specifying initial conditions. In other words, one finds indefinite integrals $G(x)$ and $H(t)$ for $g(x)$ and $h(t)$, respectively, that is,

$$\dot{G}(x) = g(x) \text{ and } \dot{H}(t) = h(t),$$

and then solves

$$G(x) = H(t) + C$$

for x. The first and only theorem in this section puts this elementary technique in the form of a rigorous result there-by explaining why it works and launching the discussion of integrals.

Theorem 2.1 *Consider a separable differential equation*

$$\dot{x} = \frac{h(t)}{g(x)}$$

on $D = \{(t, x) : \alpha < t < \beta \text{ and } \gamma < x < \delta\}$ such that $h(t)$ is continuous on $\alpha < t < \beta$ and $g(x)$ is nonzero and continuous on $\gamma < x < \delta$. If $G(x)$ and $H(t)$ satisfy

$$\frac{dG}{dx} = g \text{ and } \frac{dH}{dt} = h,$$

then the following hold:

(a) *If $G(\xi) = H(\tau) + C$ for some (τ, ξ) in D and real number C, then the equation $G(x) = H(t) + C$ can be solved for x as a function of t on some interval and the result is a solution of the separable differential equation.*

(b) *Every solution of the separable differential equation arises in this way.*

(c) *If φ_1 and φ_2 are two solutions of the separable differential equation defined on an open interval I such that $\varphi_1(\tau) = \varphi_2(\tau)$ for some τ in I, then $\varphi_1 = \varphi_2$.*

Proof. Since it is assumed that $g(x) \neq 0$ on $\gamma < x < \delta$, the function $g(x)$ can not change sign when $\gamma < x < \delta$. Hence, $G(x)$ is either increasing or decreasing with range $\{y : c < y < d\}$. It follows that G and has a differentiable inverse $G^{-1} : \{y : c < y < d\} \to \{x : \gamma < x < \delta\}$ and

$$\frac{dG^{-1}}{dy} = \frac{1}{g(G^{-1}(y))}.$$

Because $c < G(\xi) = H(\tau) + C < d$, by continuity there exists an interval $I = \{t : a < t < b\}$ containing τ such that

$$c < H(t) + C < d,$$

when t is in I. For t in I we can use G^{-1} to solve

$$G(x) = H(t) + C$$

for x and express x as the function t given by

$$x(t) = G^{-1}\big(H(t) + C\big).$$

It must now be shown that $x(t)$ is a solution of the separable equation.

Using the chain rule to calculate the derivative of $x(t)$ yields

$$\dot{x}(t) = \frac{1}{g\big(G^{-1}\big(H(t) + C\big)\big)} h(t) = \frac{h(t)}{g\big(x(t)\big)}$$

and proves the first statement.

Suppose $\varphi(t)$ is a solution of $\dot{x} = h(t)/g(x)$ defined on the open interval I. Set $E(t, x) = G(x) - H(t)$ and calculate

$$\begin{aligned} \frac{dE}{dt} &= g\big(\varphi(t)\big)\frac{h(t)}{g\big(\varphi(t)\big)} - h(t) \\ &= 0. \end{aligned}$$

Hence, $E\big(t, \varphi(t)\big) = C$ for some real C or

$$G\big(\varphi(t)\big) = H(t) + C$$

for all t in I. In particular, $H(t) + C$ is in the range of G and

$$\varphi(t) = G^{-1}\big(H(t) + C\big)$$

which proves part (b).

Finally, if $\varphi_1(\tau) = \varphi_2(\tau)$, then $E(\tau, \varphi_1(\tau)) = E(\tau, \varphi_2(\tau))$. From part (b), $E(t, \varphi_1(t))$ and $E(t, \varphi_2(t))$ are constant for all t in I. Since they are equal at τ, it follows that $E(t, \varphi_1(t)) = E(t, \varphi_2(t)) = C$ for all $t \in I$ and $\varphi_1(t) = G^{-1}(H(t) + C) = \varphi_2(t)$ on I. $\square$

The point of these arguments is that separation of variables has a simple rigorous foundation. Furthermore, understanding it can help locate the places where more careful analysis is needed. For example,

$$\dot{x} = \sqrt{|x|} = \frac{1}{1/\sqrt{|x|}}$$

is separable on $\{(t, x) : x \neq 0\}$ but not on $\mathbb{R}^2$. Recall that the initial-value problem

$$\begin{aligned} \dot{x} &= \sqrt{|x|} \\ \xi &= x(\tau). \end{aligned}$$

fails to have a unique solution precisely when $\xi = 0$ and the equation is not separable.

Frequently, the value of $\mathbf{f}(t, \mathbf{x})$ in the differential equation $\dot{\mathbf{x}} = \mathbf{f}(t, \mathbf{x})$ is independent of the variable t. This is evident in some of the examples in Chapter 1. Thinking of t as time, $\mathbf{x}$ as position, $\mathbf{f}(t, \mathbf{x})$ independent of t means the tangential direction of a solution through $\mathbf{x}$ is independent of the time the solution passes through $\mathbf{x}$. Such differential equations are said to be *autonomous* and possess some elementary but basic properties that distinguish them from the general *nonautonomous* variety.

Let Ω be an open subset of $\mathbb{R}^d$ and let $\mathbf{f} : \Omega \to \mathbb{R}^d$ be a continuous function. Set $D = \mathbb{R} \times \Omega$ and for $(t, \mathbf{x})$ in D set $\mathbf{f}(t, \mathbf{x}) = \mathbf{f}(\mathbf{x})$. In this way, the autonomous differential equation

$$\dot{\mathbf{x}} = \mathbf{f}(\mathbf{x})$$

fits the model of the previous sections. In particular, given $\boldsymbol{\xi} \in \Omega$, consider the initial-value problem

$$\begin{aligned} \dot{\mathbf{x}} &= \mathbf{f}(\mathbf{x}) \\ \boldsymbol{\xi} &= \mathbf{x}(\tau) \end{aligned}$$

for any real τ. By Peano's theorem, it has at least one solution defined on $\tau - \alpha < t < \tau + \alpha$ for some $\alpha > 0$.

Conversely, any differential equation $\dot{\mathbf{x}} = \mathbf{f}(t, \mathbf{x})$ can always be thought of as an autonomous equation as follows: let $\mathbf{y} = (t, \mathbf{x})$, $\Omega = D$, and define $\mathbf{g} : \Omega \to \mathbb{R}^{d+1}$ by

$$g_i(\mathbf{y}) = \begin{cases} 1 & \text{for} \quad i = 1 \\ f_{i-1}(t, \mathbf{x}) & \text{for} \quad 2 \leq i \leq d + 1. \end{cases}$$

Then a curve $\mathbf{y}(t) = (t, \varphi(t))$ is a solution of $\dot{\mathbf{y}} = \mathbf{g}(\mathbf{y})$ if and only if it is a trajectory of $\dot{\mathbf{x}} = \mathbf{f}(t, \mathbf{x})$. This is a unifying idea but one that disguises the special role that t often plays.

Consider the general autonomous differential equation $\dot{\mathbf{x}} = \mathbf{f}(\mathbf{x})$ where, as usual, $\mathbf{f} : \Omega \to \mathbb{R}^d$ is continuous. A continuous function $E : \Omega \to \mathbb{R}$ is called an *integral* of $\dot{\mathbf{x}} = \mathbf{f}(\mathbf{x})$ if $E\big(\mathbf{x}(t)\big)$ is a constant function of t for every solution $\mathbf{x}(t)$ of $\dot{\mathbf{x}} = \mathbf{f}(\mathbf{x})$. The term *first integral* is also commonly used.

Remark *Suppose E has continuous first partial derivatives on Ω. Then E is an integral for $\dot{\mathbf{x}} = \mathbf{f}(\mathbf{x})$ if and only if*

$$\nabla E \cdot \mathbf{f} \equiv 0. \tag{2.1}$$

Proof. Use the chain rule to differentiate $E(\mathbf{x}(t))$ with respect to t for an arbitrary solution $\mathbf{x}(t)$. Because $E(\mathbf{x}(t))$ is a differentiable real-valued function, it is constant if and only if its derivative is identically zero. □

The function $E(t, x)$ in the proof of the theorem on separation of variables (*Theorem 2.1*) is an integral of the separable equation thought of as an autonomous differential equation. Since there were only two variables t and x, solving for one in terms of the other was manageable. Solving $E(\mathbf{x}) = C$ as a function of just one of the coordinates is not possible when there are three or more coordinates. The best that can be done is to use the implicit function theorem to express one coordinate in terms of all the others. In other words, an integral reduces the number of space variables by 1, at least locally.

The strategy of solving an autonomous differential equation by integrals is to find sufficiently many integrals and completely determine a solution by the values of its integrals. To put it another way, $d - 1$ independent integrals for a d-dimensional problem would reduce the problem to an autonomous scalar differential equation and then ordinary integration in the remaining variable would at least determine the solution locally. Such systems are often called *completely integrable* .

To more easily picture how several integrals can determine a solution, consider an autonomous differential equation with $d = 3$ and suppose $E_1(x, y, z)$ is an integral. Then the set of points satisfying $E_1(x, y, z) = C_1$ is a surface provided the gradient of $E_1(x, y, z)$ is nonzero. Thus each solution lies on a known surface and that is valuable descriptive information about the solutions. Now suppose $E_2(x, y, z)$ is a second integral with a nonzero gradient. If the gradients of $E_1(x, y, z)$ and $E_2(x, y, z)$ are not parallel, then the intersection of the surfaces $E_1(x, y, z) = C_1$ and $E_2(x, y, z) = C_2$, is a curve. More importantly, a solution starting at a point on that curve would remain on it because $E_1(x, y, z)$ and $E_2(x, y, z)$ are constants on solutions. Thus the two constants C_1 and C_2 would determine the image of the solution in $\mathbb{R}^3$ but not its parameterization by t.

Many equations do not even have one nonconstant integral, but integrals do occur regularly as conservation laws in problems coming from mechanics. The typical differential equation coming from a problem in mechanics starts from Newton's law, $\mathbf{F} = m\mathbf{a}$ or $\mathbf{F}(\mathbf{x}) = m\ddot{\mathbf{x}}$, because acceleration is the second derivative of position. Since $\mathbf{F}(\mathbf{x}) = m\ddot{\mathbf{x}}$ is a second-order differential equation, it is not automatically part of the theory developed in Chapter 1 for first-order differential equations. The generic first-order differential equation $\dot{\mathbf{x}} = \mathbf{f}(t, \mathbf{x})$ is, however, general enough to include higher order equations by rewriting them as systems of equations.

To understand how to rewrite a higher order differential equation as a first-order equation, start with the general *second-order* scalar differential equation $\ddot{y} = g(t, y, \dot{y})$. By using the variable $\mathbf{x} \in \mathbb{R}^2$ instead of $y \in \mathbb{R}$, it can be written in the form

$$\begin{aligned} \dot{x}_1 &= x_2 \\ \dot{x}_2 &= g(t, x_1, x_2) \end{aligned}$$

or

$$\dot{\mathbf{x}} = \mathbf{f}(t, x_1, x_2) = \big(x_2, g(t, x_1, x_2)\big).$$

The corresponding initial-value problem is

$$\begin{aligned} \dot{x}_1 &= x_2 \\ \dot{x}_2 &= g(t, x_1, x_2) \\ x_1(\tau) &= \xi_1 \\ x_2(\tau) &= \xi_2 \end{aligned}$$

or more compactly

$$\begin{aligned} \dot{\mathbf{x}} &= \mathbf{f}(t, x_1, x_2) \\ \boldsymbol{\xi} &= \mathbf{x}(\tau), \end{aligned}$$

where $\boldsymbol{\xi} = (\xi_1, \xi_2)$. In other words, both an initial position and initial derivative must be specified because $\xi_2 = x_2(\tau) = \dot{x}_1(\tau)$. Furthermore, a solution is now of the form $\mathbf{x}(t) = \big(x_1(t), x_2(t)\big) = \big(y(t), \dot{y}(t)\big)$, not just $y(t)$.

Similarly an *nth-order* scalar differential equation

$$y^{(n)} = g\big(t, y, \ldots, y^{(n-1)}\big), \tag{2.2}$$

where $y^{(i)}$ equals the ith derivative can be written

$$\begin{aligned} \dot{x}_1 &= x_2 \\ \dot{x}_2 &= x_3 \\ &\vdots \\ \dot{x}_n &= g(t, x_1, x_2, \ldots, x_n) \end{aligned}$$

with initial conditions of the form

$$\begin{aligned}
x_1(\tau) &= \xi_1 = y(\tau) \\
x_2(\tau) &= \xi_2 = y^{(1)}(\tau) \\
&\vdots \\
x_n(\tau) &= \xi_n = y^{(n-1)}(\tau)
\end{aligned}$$

and with solutions

$$\mathbf{x}(t) = \big(y(t), y^{(1)}(t), \ldots, y^{(n-1)}(t)\big).$$

Of course, using the same device, an nth order system of differential equations can also be expressed in the form $\dot{\mathbf{x}} = \mathbf{f}(t, \mathbf{x})$.

As an example of a second-order differential equation occurring naturally in Newtonian mechanics, consider a particle of mass m moving in $\mathbb{R}^3$ under the action of a force $\mathbf{F}$, which is continuous and depends only on its position. Then the differential equation governing its motion is Newton's second law, $\mathbf{F} = m\ddot{\mathbf{x}}$. Letting

$$\mathbf{F}(x, y, z) = \big(F_1(x, y, z), F_2(x, y, z), F_3(x, y, z)\big),$$

$\mathbf{F} = m\ddot{\mathbf{x}}$ becomes

$$m(\ddot{x}, \ddot{y}, \ddot{z}) = \big(F_1(x, y, z), F_2(x, y, z), F_3(x, y, z)\big),$$

and applying the previous device with $x = x_1$, $y = x_2$, $z = x_3$, $\dot{x} = x_4$, and so forth, it can be written as $\dot{\mathbf{x}} = \mathbf{f}(\mathbf{x})$ or as the system of differential equations:

$$\begin{aligned}
\dot{x}_1 &= x_4 \\
\dot{x}_2 &= x_5 \\
\dot{x}_3 &= x_6 \\
\dot{x}_4 &= F_1(x_1, x_2, x_3)/m \\
\dot{x}_5 &= F_2(x_1, x_2, x_3)/m \\
\dot{x}_6 &= F_3(x_1, x_2, x_3)/m.
\end{aligned}$$

Notice that an initial-value problem for $\mathbf{F} = m\ddot{\mathbf{x}}$ must include an initial velocity as well as an initial position and solutions should be thought of as including both position and velocity as functions of time. The general theory developed in the Chapter 1 can now be applied. For example, if $\mathbf{F}$ has continuous first partial derivatives, then given an initial position and velocity the associated initial-value problem has a unique maximal solution and continuity in initial conditions holds.

Conservation of energy is just an integral of a differential equation of the form $\mathbf{F} = m\ddot{\mathbf{x}}$. A force $\mathbf{F}(x, y, z)$ depending as above on position only is said to be *conservative* if $\mathbf{F} = -\nabla G$ for some $G : \mathbb{R}^3 \to \mathbb{R}$. The *total energy* is defined by

$$E = \frac{1}{2}m\|\mathbf{v}\|^2 + G \tag{2.3}$$

where $\mathbf{v} = (\dot{x}, \dot{y}, \dot{z})$ is the velocity of the particle of mass m moving under the action of $\mathbf{F}$. In terms of the variables, $x_1, \ldots, x_6$ used above,

$$E(x_1, \ldots, x_6) = \frac{1}{2} m \left(x_4^2 + x_5^2 + x_6^2\right) + G(x_1, x_2, x_3).$$

The following easy calculation shows that E is an integral of $\mathbf{F} = m\ddot{\mathbf{x}}$ when $\mathbf{F}$ is conservative:

$$\begin{aligned}
\nabla E \cdot \mathbf{f} &= \left(\frac{\partial G}{\partial x_1}, \frac{\partial G}{\partial x_2}, \frac{\partial G}{\partial x_3}, m x_4, m x_5, m x_6\right) \cdot \mathbf{f} \\
&= \left(-F_1, -F_2, -F_3, m x_4, m x_5, m x_6\right) \cdot \mathbf{f}(x_1, \ldots, x_6) \\
&= -F_1 x_4 - F_2 x_5 - F_3 x_6 + m x_4 F_1/m + m x_5 F_2/m + m x_6 F_3/m \\
&\equiv 0.
\end{aligned}$$

The condition that $\mathbf{F} = -\nabla G$ is very restrictive for $d > 1$, but for $d = 1$, it is not a restriction by the fundamental theorem of calculus. This observation can be exploited to obtain information about the solutions of $\ddot{y} = g(y)$, a problem that will be discussed in the next section.

We will close this section with a brief discussion of perhaps the most famous classical problem in ordinary differential equations, namely, the *n-body problem.* Even when $n = 3$, it is largely unsolved.

The problem is to describe the behavior of n bodies in three-dimensional Euclidean space where the only forces are the gravitational forces they exert on each other. Let $m_1, \ldots, m_n$ denote their masses, let $\mathbf{x}_1, \ldots, \mathbf{x}_n$ denote their positions, and let $\mathbf{v}_1, \ldots, \mathbf{v}_n$ denote their velocities. Since each $\mathbf{x}_i$ is an arbitrary point of $\mathbb{R}^3$, the natural space in which to set up the second-order equation is $\left(\mathbb{R}^3\right)^n = \mathbb{R}^{3n}$. Then $(\mathbf{x}_1, \ldots, \mathbf{x}_n) = \mathbf{x}$ gives the position of all n bodies simultaneously. Note the ith and jth bodies cannot occupy the same position unless there is a collision that would alter the system depending on the elasticity of the bodies in a collision. For this reason, points at which collisions would occur are removed from the domain. For $i \neq j$, set

$$\Delta(i, j) = \{\mathbf{x} \in \mathbb{R}^{3n} : \mathbf{x}_i = \mathbf{x}_j\}$$

and

$$U = \{\mathbf{x} \in \mathbb{R}^{3n} : \mathbf{x} \notin \Delta(i, j) \text{ for } 1 \leq i < j \leq n\}.$$

Clearly, U is open. Let $\mathbf{F}_i(x)$ be the sum of the forces the masses m_j at $\mathbf{x}_j$, $j \neq i$, exert on m_i at $\mathbf{x}_i$. It will be shown that $\mathbf{F}(\mathbf{x}) = (\mathbf{F}_1(\mathbf{x}), \ldots, \mathbf{F}_n(\mathbf{x}))$ is conservative and hence the system of differential equations for the n-body problem

$$\begin{aligned}
\dot{\mathbf{x}}_i &= \mathbf{v}_i \\
\dot{\mathbf{v}}_i &= \mathbf{F}_i(\mathbf{x})/m_i
\end{aligned}$$

defined on $D = \left\{(\mathbf{x}, \mathbf{v}) : \mathbf{x} \in U \text{ and } \mathbf{v} \in \mathbb{R}^{3n}\right\}$ has an integral, namely, the total energy.

Define $G : U \to \mathbb{R}$ by

$$-G(\mathbf{x}) = \gamma \sum_{i<j} \frac{m_i m_j}{||\mathbf{x}_i - \mathbf{x}_j||},$$

where γ is a universal coefficient of gravity. Let $\nabla_k G(\mathbf{x}) \in \mathbb{R}^3$ denote the gradient of G with respect to the three variables forming $\mathbf{x}_k$. Observe that

$$\begin{aligned} -\nabla_k G(\mathbf{x}) &= \gamma \nabla_k \left(\sum_{i<j} \frac{m_i m_k}{||\mathbf{x}_i - \mathbf{x}_j||} \right) \\ &= \gamma \sum_{i \neq k} m_i m_k \nabla_k \left(\frac{1}{||\mathbf{x}_i - \mathbf{x}_k||} \right). \end{aligned}$$

An easy calculation shows that

$$\begin{aligned} \nabla_k \left(\frac{1}{||\mathbf{x}_i - \mathbf{x}_k||} \right) &= \frac{1}{||\mathbf{x}_i - \mathbf{x}_k||^3} (\mathbf{x}_i - \mathbf{x}_k) \\ &= \frac{1}{||\mathbf{x}_i - \mathbf{x}_k||^2} \left(\frac{\mathbf{x}_i - \mathbf{x}_k}{||\mathbf{x}_i - \mathbf{x}_k||} \right). \end{aligned}$$

Since

$$\frac{\mathbf{x}_i - \mathbf{x}_k}{||\mathbf{x}_i - \mathbf{x}_k||}$$

is a unit vector pointing from $\mathbf{x}_k$ toward $\mathbf{x}_i$,

$$\gamma\, m_i m_k \nabla_k \left(\frac{1}{||\mathbf{x}_i - \mathbf{x}_k||} \right) = \frac{\gamma\, m_i m_k}{||\mathbf{x}_i - \mathbf{x}_k||^2} \frac{\mathbf{x}_i - \mathbf{x}_k}{||\mathbf{x}_i - \mathbf{x}_k||}$$

is a force toward the ith body that is proportional to the product of the masses and inversely proportional to the square of the distance between them. Thus by Newton's law of gravitation, it is precisely the force the gravity exerted on the kth body by the ith body. It follows that

$$-\nabla_k G(\mathbf{x}) = \gamma \sum_{i \neq k} \frac{m_i m_k}{||\mathbf{x}_i - \mathbf{x}_k||^2} \left(\frac{\mathbf{x}_i - \mathbf{x}_k}{||\mathbf{x}_i - \mathbf{x}_k||^2} \right)$$

is the total forces on the mass m_k and $-\nabla G = \mathbf{F}$ as promised and that energy is an integral for the n-body problem. The n-body problem has other important integrals that will be described in the context of Hamiltonian systems in Chapter 7.

The primary motivation for studying the n-body is, of course, celestial mechanics. For an expanded but elementary discussion of the n-body in the context of celestial mechanics see [30]. There are also modern mathematical treatises devoted to mechanics in general. One such book is [1].

EXERCISES

1. Let $f(x)$ be a continuous real-valued function on the interval $c - 2\alpha < x < c + 2\alpha$ with $f(x) = 0$ precisely, when $x = c$. Show that the scalar initial-value problem

$$\begin{aligned} \dot{x} &= f(x) \\ c &= \mathbf{x}(0) \end{aligned}$$

has a unique solution if and only if the improper integrals

$$\int_{c-\alpha}^{c} \frac{dx}{f(x)} \quad \text{and} \quad \int_{c}^{c+\alpha} \frac{dx}{f(x)}$$

are both infinite.

2. Consider the scalar autonomous differential equation $\dot{x} = f(x)$, where $f(x)$ has a continuous first derivative on $\mathbb{R}$. Suppose $f(c) = 0$ and $f(x) > 0$, when $x > c$. For $\xi > c$, show that $x(t, 0, \xi)$ is defined on the interval

$$-\infty < t < \int_{\xi}^{\infty} \frac{dw}{f(w)}.$$

3. The scalar differential equation $\dot{x} = \mathbf{f}(t, x)$ on a domain $D \subset \mathbb{R}^2$ is said to be *exact* if there exists a function $E : D \to \mathbb{R}$ such that

$$f(t, x) = -\frac{\frac{\partial E}{\partial t}}{\frac{\partial E}{\partial x}}.$$

   (a) Show that separable equations are exact.

   (b) Let $\varphi(t)$ be a solution of an exact differential equation and show that $E(t, \varphi(t))$ is a constant.

   (c) Find the solutions of

$$\dot{x} = -\frac{bx + ct}{ax + bt}$$

   on $D = \{(t, x) : ax + bt \neq 0\}$.

4. Let $E(\mathbf{x})$ be an integral for $\dot{\mathbf{x}} = \mathbf{f}(\mathbf{x})$ on an open set Ω and assume that $\mathbf{f}$ is locally Lipschitz on Ω. Show that if the set $\{\mathbf{x} : E(\mathbf{x}) = C\}$ is compact and $E(\boldsymbol{\xi}) = C$, then the solution $\mathbf{x}(t, 0, \boldsymbol{\xi})$ is defined on $\mathbb{R}$.

5. Let $\dot{\mathbf{x}} = \mathbf{f}(\mathbf{x})$ be a locally Lipschitz autonomous differential equation on $\Omega \subset \mathbb{R}^3$ and let $\boldsymbol{\xi}$ be a element of Ω. Suppose $E : \Omega \to \mathbb{R}$ is an integral for $\dot{\mathbf{x}} = \mathbf{f}(\mathbf{x})$ with continuous first partial derivatives and with

$$\frac{\partial E}{\partial x_3}(\boldsymbol{\xi}) \neq 0.$$

So by the implicit function theorem there exits $g(x_1, x_2)$ with continuous first partial derivatives on a neighborhood of (ξ_1, ξ_2) such that $g(\xi_1, \xi_2) = \xi_3$ and $E(x_1, x_2, g(x_1, x_2)) = E(\boldsymbol{\xi})$. Show that if $(\varphi_1(t), \varphi_2(t))$ is a solution of

$$\begin{aligned} \dot{x}_1 &= f_1(x_1, x_2, g(x_1, x_2)) \\ \dot{x}_2 &= f_2(x_1, x_2, g(x_1, x_2)), \end{aligned}$$

then $(\varphi_1(t), \varphi_2(t), g(\varphi_1(t), \varphi_2(t)))$ is a solution of $\dot{\mathbf{x}} = \mathbf{f}(\mathbf{x})$.

2.2 The Qualitative Point of View

Given an autonomous differential equation $\dot{\mathbf{x}} = \mathbf{f}(x)$, it is usually not possible to express the solutions in terms of well-known functions. Nevertheless, these solutions exist and specify curves that can be thought of as the paths along which the solutions move with time. The goal of the qualitative theory of ordinary differential equations is to identify, describe, and when possible, classify the kinds of behaviors these solution curves can exhibit. In particular, it aims to understand the behavior of the solutions of $\dot{\mathbf{x}} = \mathbf{f}(\mathbf{x})$ without explicitly determining them.

The analogy with motion is particularly apt here. Think of a solution $\mathbf{x}(t)$ as describing the motion of a point through space. The motion over a short interval of time is not particularly interesting; the point will move a commensurate distance and by continuity in initial conditions, nearby points will remain near it. The real questions arise analyzing the limiting behavior of a solution. What eventually happens to a point whose motions is governed by $\dot{\mathbf{x}} = \mathbf{f}(\mathbf{x})$? In other words, what are the asymptotic properties of a solution as t goes to infinity.

The study of differential equations from the view point of describing the main qualitative features of solutions originated with Poincaré around the turn of the century. Today, it lies at the heart of what is generally referred to as the study of dynamical systems. In this section, the elementary definitions and principles are introduced, and some examples illustrating the spirit of the qualitative point of view are presented.

In the study of autonomous differential equations like $\dot{\mathbf{x}} = \mathbf{f}(\mathbf{x})$, the function $\mathbf{f} : \Omega \to \mathbb{R}^d$ can be thought of as assigning a vector $\mathbf{f}(\mathbf{x})$ in $\mathbb{R}^d$ to each point $\mathbf{x}$ in $\Omega \subset \mathbb{R}^d$. See *Figure 2.1.* For this reason such a function is often called a *vector field.* Finding a curve whose tangent vector is the value of the vector field at each point is exactly the same as finding a solution of $\dot{\mathbf{x}} = \mathbf{f}(\mathbf{x})$.

The vector field point of view emphasizes the function $\mathbf{f}$ and is easy to represent pictorially as arrows emanating from points in Ω. For example, an autonomous differential equation with an integral can be pictured as a vector field that is tangent to the level surfaces of a real-valued function. The vector field point of view also can be carried over to manifolds, spaces that at least

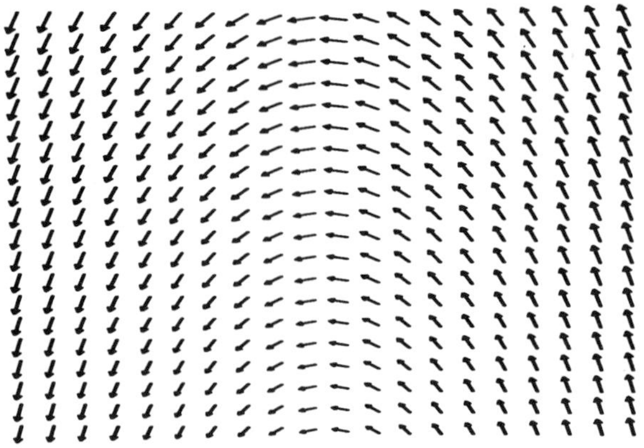

Figure 2.1: A vector field turning a corner.

locally appear to be Euclidean. Investigating vector fields on manifolds uses the principles of differential equations in $\mathbb{R}^d$ as a foundation.

Throughout this section, it will be understood that $\mathbf{f}(\mathbf{x})$ is a continuous function on Ω, an open subset of $\mathbb{R}^d$, with values in $\mathbb{R}^d$. Most of the time, it will also be assumed that $\mathbf{f}(\mathbf{x})$ is locally Lipschitz on Ω. The next two remarks are easy to prove and contain properties of solutions of autonomous differential equations that will be used again and again.

Remark *If $\varphi(t)$ is a solution of $\dot{\mathbf{x}} = \mathbf{f}(\mathbf{x})$ defined on the open interval I, then $\psi(t) = \varphi(t+\tau)$ is a solution defined on $I - \tau$.*

Proof. $\dot{\psi}(t) = \dot{\varphi}(t+\tau) = \mathbf{f}\big(\varphi(t+\tau)\big) = f\big(\psi(t)\big)$. □

Remark *If $\mathbf{f}(\mathbf{x})$ is locally Lipschitz on Ω, then the solutions of $\dot{\mathbf{x}} = \mathbf{f}(\mathbf{x})$ satisfy:*

(a) $\mathbf{x}(t, \tau, \boldsymbol{\xi}) = \mathbf{x}(t - \tau, 0, \boldsymbol{\xi})$,

(b) $\mathbf{x}(t, 0, \mathbf{x}(s, 0, \boldsymbol{\xi})) = \mathbf{x}(t + s, 0, \boldsymbol{\xi})$.

Proof. The two solutions in part (a) agree when t equals τ, and the two solutions in part (b) agree when t equals 0. Hence, both formulas hold by the uniqueness of solutions. □

One implication of part (a) is that the initial time variable is superfluous in $\mathbf{x}(t, \tau, \boldsymbol{\xi})$ when $\mathbf{f}(\mathbf{x})$ is locally Lipschitz. Consequently, in some contexts it is convenient to simply write $\mathbf{x}(t, \boldsymbol{\xi})$ for $\mathbf{x}(t, 0, \boldsymbol{\xi})$ and recapture $\mathbf{x}(t, \tau, \boldsymbol{\xi})$ when needed with $\mathbf{x}(t - \tau, \boldsymbol{\xi})$.

The next special property of locally Lipschitz autonomous differential equations is that the maximal solutions decompose Ω into distinct pieces called orbits.

Proposition 2.2 *If* $\mathbf{f}(\mathbf{x})$ *is locally Lipschitz on* Ω *and if* $\mathbf{x}_1(t)$ *and* $\mathbf{x}_2(t)$ *are maximal solutions of* $\dot{\mathbf{x}} = \mathbf{f}(\mathbf{x})$ *defined on open intervals* I_1 *and* I_2*, respectively, then either*

(a) $\{\mathbf{x}_1(t) : t \in I_1\} \cap \{\mathbf{x}_2(t) : t \in I_2\} = \phi$; *or*

(b) *there exists* $\tau \in \mathbb{R}$ *such that* $I_1 + \tau = I_2$ *and* $\mathbf{x}_1(t) = \mathbf{x}_2(t+\tau)$ *for all* $t \in I_1$*. It follows in this case that*

$$\{\mathbf{x}_1(t) : t \in I_1\} = \{\mathbf{x}_2(t) : t \in I_2\}.$$

Proof. If (a) does not hold, then there exists $t_i \in I_i$ such that $\mathbf{x}_1(t_1) = \mathbf{x}_2(t_2)$. Set $\tau = t_2 - t_1$ and $\varphi(t) = \mathbf{x}_2(t+\tau)$ on $I_2 - \tau$. Now $\varphi(t_1) = \mathbf{x}_2(t_1+\tau) = \mathbf{x}_2(t_2) = \mathbf{x}_1(t_1)$, and by uniqueness, $\varphi(t) = \mathbf{x}_1(t)$ on $I_1 \cap (I_2 - \tau)$. Since $\mathbf{x}_1(t)$ is maximal, $I_2 - \tau \subset I_1$ or $I_2 \subset I_1 + \tau$. Repeating the same argument with $\tau' = t_1 - t_2$ and $\psi(t) = \mathbf{x}_1(t+\tau')$, shows that $I_1 + \tau = I_1 - \tau' \subset I_2$. □

Furthermore, there are three distinct types of solutions to a locally Lipschitz autonomous differential equation.

Proposition 2.3 *If* $\mathbf{f}(\mathbf{x})$ *is locally Lipschitz on* Ω *and* $\mathbf{x}(t)$ *is a maximal solution of* $\dot{\mathbf{x}} = \mathbf{f}(\mathbf{x})$ *defined on the open interval* I*, then exactly one of the following hold:*

(a) $\mathbf{x}(t)$ *is one-to-one on* I;

(b) $I = \mathbb{R}$ *and there exists a smallest positive* ω *such that* $\mathbf{x}(t+\omega) = \mathbf{x}(t)$ *for all* $t \in \mathbb{R}$; *or*

(c) $I = \mathbb{R}$ *and* $\mathbf{x}(t) = \mathbf{x}(0)$ *for all* $t \in \mathbb{R}$.

Proof. Suppose $\mathbf{x}(t)$ is not one-to-one. So there exists t_1 and t_2 in I such that $\mathbf{x}(t_1) = \mathbf{x}(t_2)$. Assume $t_1 < t_2$ and set $\omega = t_2 - t_1$. Then $\varphi(t) = \mathbf{x}(t+\omega)$ is also a maximal solution defined on $I - \omega$ and $\varphi(t_1) = \mathbf{x}(t_1+\omega) = \mathbf{x}(t_2) = \mathbf{x}(t_1)$. By the previous proposition $I - \omega = I$ and $\varphi(t) = \mathbf{x}(t)$ for all t in I. Since $\omega \neq 0$, clearly I must equal $\mathbb{R}$.

Either there exists a smallest $\omega > 0$ such that $\mathbf{x}(t+\omega) = \mathbf{x}(t)$ for all t and (b) holds or there exists a sequence ω_m of positive real numbers such that $\lim_{m\to\infty} \omega_m = 0$ and $\mathbf{x}(t+\omega_m) = \mathbf{x}(t)$ for all t. In the latter case, it follows that $\mathbf{x}(k\omega_m) = \mathbf{x}(0)$ for all $k \in \mathbb{Z}$ and $m \in \mathbb{Z}^+$. (As usual $\mathbb{Z}$ denotes the integers and $\mathbb{Z}^+$ the positive integers.) Since $\{k\omega_n : k \in \mathbb{Z} \text{ and } n \in \mathbb{Z}^+\}$ is dense in $\mathbb{R}$, it follows by continuity that $\mathbf{x}(t) = \mathbf{x}(0)$ for all $t \in \mathbb{R}$ to complete the proof. □

Note that if $\mathbf{x}(t) = \mathbf{x}(0)$ for every t in $\mathbb{R}$, then $\mathbf{0} = \dot{\mathbf{x}}(t) = \mathbf{f}(\mathbf{x}(0))$. Conversely, if $\mathbf{f}(\boldsymbol{\xi}) = \mathbf{0}$, then $\mathbf{x}(t) \equiv \boldsymbol{\xi}$ is a solution. This proves the following:

Remark *Let* $\boldsymbol{\xi} \in \Omega$*. Then* $\mathbf{x}(t) \equiv \boldsymbol{\xi}$ *on* $\mathbb{R}$ *is a solution of* $\dot{\mathbf{x}} = \mathbf{f}(\mathbf{x})$ *if and only if* $\mathbf{f}(\boldsymbol{\xi}) = \mathbf{0}$.

The set $\mathcal{F} = \{\mathbf{x} : \mathbf{f}(\mathbf{x}) = \mathbf{0}\}$ is called the set of *fixed points*, *equilibrium points*, or *critical points* of the equation $\dot{\mathbf{x}} = f(\mathbf{x})$. Our clear preference will be to use the fixed-point terminology because of its dynamical connotation of no motion. The more interesting fixed points are the isolated fixed points. A fixed point is an *isolated fixed point*, if it has a neighborhood that contains no other fixed points.

For autonomous differential equations, we think in terms of orbits instead of trajectories. Specifically, if $\mathbf{x}(t)$ is a maximal solution of $\dot{\mathbf{x}} = \mathbf{f}(\mathbf{x})$ defined on I, then the subset $\{\mathbf{x}(t) : t \in I\}$ of Ω is called an *orbit*. When $\mathbf{f}(\mathbf{x})$ is locally Lipschitz, it follows from *Proposition 2.2* that every point of Ω lies in exactly one orbit. In this case, for example, the orbit of a fixed point is just the fixed point itself.

Naturally, a solution $\mathbf{x}(t)$ is said to be a *periodic solution* if there exists a smallest positive ω, called the *period*, such that $\mathbf{x}(t + \omega) = \mathbf{x}(t)$ for all real t. If $\mathbf{x}(t)$ is a periodic solution of period ω, then as t ranges from τ to $\tau + \omega$, the solution $\mathbf{x}(t)$ traverses its entire orbit exactly once and the orbit of a periodic point is a simple closed curve in Ω .

The space Ω is often called the *phase space*, and a *phase portrait* of $\dot{\mathbf{x}} = \mathbf{f}(\mathbf{x})$ is a sketch of representative orbits showing all possible types of behavior of the solutions for a specific autonomous differential equation. Since most differential equations cannot be solved explicitly, a phase portrait is a very useful device containing the essential descriptive information. In fact, the whole idea of the qualitative theory of ordinary differential equations is to describe the behavior of solutions without actually determining them, and a phase portrait is an effective way of presenting a qualitative solution of $\dot{\mathbf{x}} = \mathbf{f}(\mathbf{x})$.

These ideas can be illustrated by examining a few simple planar examples with $\Omega = \mathbb{R}^2$. Consider the following system of differential equations:

$$\begin{aligned} \dot{x} &= -y \\ \dot{y} &= x. \end{aligned}$$

Clearly, $\mathcal{F} = \{(0, 0)\}$ and $\varphi(t) = (r\cos t, r\sin t)$ is a solution for $r > 0$. Thus, except for one fixed point, every orbit is periodic. Its phase portrait is shown in *Figure 2.2* and is an example of a *center*.

In phase portraits, dots indicate fixed points and arrows show the direction in which the orbit is traced out as t increases. Typically, just one or two orbits of each type are shown and the region in which each type of orbit occurs is usually self-evident. In *Figure 2.2*, three periodic orbits are shown to emphasize the fact that every orbit except the fixed point at the origin is periodic. We omit the axes in most phase portraits because the dynamics being portrayed is more important than the coordinate system. *Figure 2.2* shows a fixed point surrounded by periodic orbits extending to infinity, and whether the fixed point is at the origin or $(-37, 5.36)$ is less important than the dynamical behavior it illustrates.

Notice that a solution that starts close to $(0, 0)$ stays close to $(0, 0)$. In other words, a slight change in initial position near the origin has only a slight

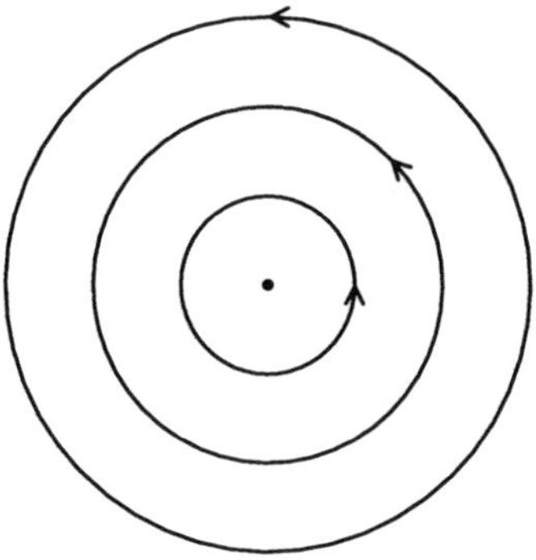

Figure 2.2: Phase portrait for $\dot{x} = -y$ and $\dot{y} = x$.

long term effect on the behavior of the solution. The next example is strikingly different in its behavior near the origin.

Deleting the minus sign from the above system of differential equations, produces the system

$$\begin{aligned} \dot{x} &= y \\ \dot{y} &= x \end{aligned}$$

with a very different phase portrait. Most solutions passing near the origin do not stay near the origin, and the behavior near $(0,0)$ is rather unstable. It is still true that $\mathcal{F} = \{(0,0)\}$, but for $r > 0$, the following are now solutions:

$$\begin{aligned} \varphi_1(t) &= (r\cosh t, r\sinh t), \\ \varphi_2(t) &= (-r\cosh t, -r\sinh t), \\ \varphi_3(t) &= (r\sinh t, r\cosh t), \\ \varphi_4(t) &= (-r\sinh t, -r\cosh t). \end{aligned}$$

In addition to these hyperbolae, the functions

$$\begin{aligned} \varphi_5(t) &= (e^t, e^t), \\ \varphi_6(t) &= (-e^t, -e^t), \\ \varphi_7(t) &= (e^{-t}, -e^{-t}), \\ \varphi_8(t) &= (-e^{-t}, e^{-t}), \end{aligned}$$

are also solutions.

Figure 2.3 shows the phase portrait of this system, which is called a *saddle.* The solutions φ_5 through φ_8 divide the plane into four quadrants. (Although each of these solutions approaches the origin either as t goes to plus or minus infinity, in the phase portrait a little space is left between the solution and the fixed point to indicate that they are distinct orbits.) Each of these quadrants

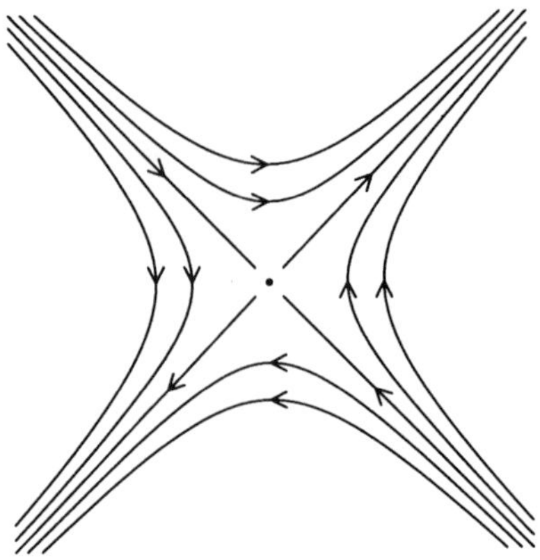

Figure 2.3: Phase portrait for $\dot{x} = y$ and $\dot{y} = x$.

is filled with hyperbolic orbits given by one of the solution types $\varphi_j(t)$ with $j = 1, 2, 3, 4$ and $r > 0$. Notice how different the qualitative behavior of the solutions is near the origin in *Figure 2.3* compared with those in *Figure 2.2*.

When $d = 1$ and $f(x)$ is locally Lipschitz, it is easy to describe all possible phase portraits. For simplicity, assume $\Omega = \mathbb{R}$. In this case, $f(x)$ is real-valued and $\mathcal{F}$ is a closed subset of $\mathbb{R}$. The complement of $\mathcal{F}$ is a finite or a countably infinite union of disjoint open intervals. On each one, f is either positive or negative and $\dot{x} = f(x)$ is separable. It is not hard to show that each of these complementary open intervals constitutes an entire orbit and the sign of f determines the direction of the motion on it.

For $\mathcal{F}$ finite, a sample phase portrait is shown *Figure 2.4*. Notice that the behavior of the solutions near each of the three fixed points is different. In fact, they exhibit the three possible behaviors on the real line—both arrows point toward the fixed point, both point away, and one of each.

Figure 2.4: Sample phase portrait on $\mathbb{R}$.

The basic properties of the solutions to an autonomous locally Lipschitz differential equation are the motivation for the definition of a flow. This abstraction of autonomous ordinary differential equations occurred in the 1940s in both Russia ([27]) and the United States ([14]), and the study of flows has been one of the mainstays of dynamical systems. Part Two of [27] is the classical introduction to flows on metric spaces while [14] is a very abstract treatment of the subject. Differences in terminology between the Russian and American schools unfortunately persist to this day. For a more current introduction to dynamical systems see [6] or [18].

Because many of the systems appearing in subsequent chapters define flows on their phase spaces, it is worth both understanding the connection between differential equations and flows and using the terminology of flows when appropriate.

A *flow* is given by a function $\Phi : \mathbb{R} \times \Omega \to \Omega$ that satisfies the following conditions:

(a) Φ is continuous on $\mathbb{R} \times \Omega$,

(b) $\Phi(0, \boldsymbol{\xi}) = \boldsymbol{\xi}$ for every $\boldsymbol{\xi} \in \Omega$,

(c) $\Phi(t+s, \boldsymbol{\xi}) = \Phi(t, \Phi(s, \boldsymbol{\xi}))$ for every $\boldsymbol{\xi} \in \Omega$ and $s, t \in \mathbb{R}$.

The proof of the following remark is now transparent:

Remark *If* $\mathbf{f}(\mathbf{x})$ *is locally Lipschitz on* Ω *and every solutions to* $\dot{\mathbf{x}} = \mathbf{f}(\mathbf{x})$ *is defined on* $\mathbb{R}$*, then* $\Phi(t, \boldsymbol{\xi}) = \mathbf{x}(t, 0, \boldsymbol{\xi}) = \mathbf{x}(t, \boldsymbol{\xi})$ *defines a flow on* Ω.

Invariant sets play a major role in the study of flows. A subset Y of Ω is an *invariant set* of the flow Φ on Ω if $\Phi(t, \mathbf{x})$ is in Y for all $t \in \mathbb{R}$ whenever $\mathbf{x}$ is in Y. Saying that Y is a *positively invariant set* simply means that $\Phi(t, \mathbf{x}) \in Y$ for all $t > 0$ when $\mathbf{x}$ is in Y. Obviously, the union of invariant sets is an invariant set and the intersection of invariant sets is an invariant set.

If $\mathbf{f}(\mathbf{x})$ is locally Lipschitz on Ω and $\Phi(t, \boldsymbol{\xi}) = \mathbf{x}(t, 0, \boldsymbol{\xi}) = \mathbf{x}(t, \boldsymbol{\xi})$ defines a flow on Ω, then $\mathcal{F}$ is always an invariant set. If $E : \Omega \to \mathbb{R}$ is an integral of $\dot{\mathbf{x}} = \mathbf{f}(\mathbf{x})$, then each set of the form $\{\mathbf{x} \in \Omega : E(\mathbf{x}) = C\}$ is an invariant set. Invariant sets can be useful when describing the phase portrait of a complicated autonomous differential equation by providing macro pieces of the picture.

The requirement in the previous remark that the solutions be defined on $\mathbb{R}$ is not as critical as it might appear because $\dot{\mathbf{x}} = \mathbf{f}(\mathbf{x})$ can usually be modified without changing the orbits so that the solutions are defined on $\mathbb{R}$. This is accomplished by changing the speed of solutions without changing their direction and is discussed in the next several paragraphs.

Let $\boldsymbol{\varphi} : I \to \mathbb{R}^d$ be a continuously differentiable curve. The *velocity* of $\boldsymbol{\varphi}(t)$ is its derivative, the vector $\dot{\boldsymbol{\varphi}}(t)$, and the *speed* of $\boldsymbol{\varphi}(t)$ is the Euclidean length of its velocity, $\|\dot{\boldsymbol{\varphi}}(t)\|$. Qualitative questions about solutions are usually insensitive to the solution's speeds. Intuitively, it is clear that traversing the orbits at different positive speeds should not change the phase portrait. The next few propositions make precise the idea that only the direction of $\mathbf{f}(\mathbf{x})$ is important in determining the phase portrait of $\dot{\mathbf{x}} = \mathbf{f}(\mathbf{x})$.

Proposition 2.4 *If* $\boldsymbol{\varphi} : I \to \Omega$ *is continuously differentiable on the open interval* I*, the velocity* $\dot{\boldsymbol{\varphi}}(t) \neq \mathbf{0}$ *for all* t *in* I*, and* $\dot{\boldsymbol{\varphi}}(t) = \gamma(t) f\big(\boldsymbol{\varphi}(t)\big)$ *for some* $\gamma : I \to \mathbb{R}$*, then there exists a continuously differentiable function* λ *mapping an open interval* I' *onto* I *such that* $\boldsymbol{\varphi}\big(\lambda(t)\big)$ *is a solution of* $\dot{\mathbf{x}} = \mathbf{f}(\mathbf{x})$ *(not necessarily a maximal solution).*

Proof. Clearly, $\gamma(t)$ is never zero and must be continuous. Let $\psi(t) = \int_\tau^t \gamma(s)\, ds$, where $\tau \in I$. Then $\psi(t)$ is either increasing or decreasing and has a continuous derivative, namely, $\gamma(t)$. Let I' be the range of ψ and set $\lambda = \psi^{-1}$. Letting $\mathbf{x}(t) = \varphi\big(\lambda(t)\big)$,

$$\begin{aligned}
\dot{\mathbf{x}}(t) &= \dot{\varphi}\big(\lambda(t)\big)\dot{\lambda}(t) \\
&= \gamma\big(\lambda(t)\big)\mathbf{f}\Big(\varphi\big(\lambda(t)\big)\Big)\dot{\lambda}(t) \\
&= \gamma\big(\lambda(t)\big)\mathbf{f}\big(\mathbf{x}(t)\big)\frac{1}{\dot{\psi}\big(\lambda(t)\big)} \\
&= \gamma\big(\lambda(t)\big)\mathbf{f}\big(\mathbf{x}(t)\big)\frac{1}{\gamma\big(\lambda(t)\big)} \\
&= \mathbf{f}\big(\mathbf{x}(t)\big)
\end{aligned}$$

as required. □

The point of the previous proposition is that a curve whose tangent vectors are in the right direction, but not necessarily of the right magnitude to solve $\dot{\mathbf{x}} = \mathbf{f}(\mathbf{x})$, at least determines part of an orbit, and hence part of the phase portrait. The next proposition shows that continuously changing the magnitude of $\mathbf{f}(\mathbf{x})$ without adding any critical points does not change the phase portrait.

Proposition 2.5 *Let $\mathbf{f}(\mathbf{x})$ be locally Lipschitz on Ω. If $h : \Omega \to \mathbb{R}^+$ is continuous and locally Lipschitz, then $\dot{\mathbf{x}} = \mathbf{f}(\mathbf{x})$ and $\dot{\mathbf{x}} = h(\mathbf{x})\mathbf{f}(\mathbf{x})$ have precisely the same orbits.*

Proof. First note that $h(\mathbf{x})\mathbf{f}(\mathbf{x})$ is continuous and locally Lipschitz because $h(\mathbf{x})$ and $\mathbf{f}(\mathbf{x})$ are continuous and locally Lipschitz. (The assumption that h is locally Lipschitz makes the proof easier but is not essential.) If $\varphi(t)$ is a solution of $\dot{\mathbf{x}} = h(\mathbf{x})\mathbf{f}(\mathbf{x})$, then set

$$\gamma(t) = h(\varphi(t))$$

and apply *Proposition 2.4*. Therefore, every orbit of $\dot{\mathbf{x}} = h(\mathbf{x})f(\mathbf{x})$ is contained in an orbit of $\dot{\mathbf{x}} = f(\mathbf{x})$. To reverse the inclusion, let $\psi(t)$ be a solution of $\dot{\mathbf{x}} = \mathbf{f}(\mathbf{x})$ and $\gamma(t) = 1/h(\psi(t))$. □

Even when $h(\mathbf{x}) = 0$ has solutions, the phase portrait of $\dot{\mathbf{x}} = h(\mathbf{x})\mathbf{f}(\mathbf{x})$ is readily determined by that of $\dot{\mathbf{x}} = \mathbf{f}(\mathbf{x})$. One simply expands the fixed-point set to include $\{\mathbf{x} : h(\mathbf{x}) = 0\}$, reverses the arrows where h is negative, and leaves the remainder of the phase portrait unchanged. This will, of course, cut some of the old orbits apart. For example, consider the system

$$\begin{aligned}
\dot{x} &= -y\left((x-1)^2 + y^2\right), \\
\dot{y} &= x\left((x-1)^2 + y^2\right),
\end{aligned}$$

which comes from multiplying

$$\begin{aligned}
\dot{x} &= -y, \\
\dot{y} &= x
\end{aligned}$$

by $(x-1)^2+y^2$. Now, $\mathcal{F}=\big\{(0,0),(1,0)\big\}$ and all the orbits are unchanged except the orbit consisting of the unit circle described by the solution $\boldsymbol{\varphi}(t)=(\cos t,\sin t)$.

Proposition 2.4 applies to $\boldsymbol{\varphi}(t)=(\cos t,\sin t)$ for $0<t<2\pi$. The unit circle splits into two orbits, the point $(1,0)$ and the rest of the circle. Consequently, the phase portrait is unchanged except for a dot at $(1,0)$ representing the additional fixed point. It follows, however, from continuity in initial conditions that the period of $\boldsymbol{\xi}$ goes to infinity as $\|\boldsymbol{\xi}\|$ approaches 1. Thus although the orbits inside the unit circle have not changed, their periods have been drastically altered.

In the plane, the level curves of an integral for $\dot{\mathbf{x}}=\mathbf{f}(\mathbf{x})$ usually completely determine the phase portrait without calculating the actual solutions. The next result, which will be useful analyzing some standard examples, is in this spirit.

Proposition 2.6 *Let $E:\Omega\to\mathbb{R}$ be an integral with continuous first partial derivatives for $\dot{\mathbf{x}}=\mathbf{f}(\mathbf{x})$ on $\Omega\subset\mathbb{R}^2$, and let $\mathbf{x}(t)$ be a maximal solution with $E\big(\mathbf{x}(t)\big)=C$. Suppose $\mathbf{f}$ is locally Lipschitz on Ω and E has a continuous nonvanishing gradient on Ω.*

(a) If $\mathbf{x}(t_m)$ converges to $\boldsymbol{\xi}\in\Omega$, then either $\boldsymbol{\xi}\in\mathcal{F}$ or $\boldsymbol{\xi}=\mathbf{x}(\tau)$ for some τ.

(b) If $\big\{\mathbf{x}:E(\mathbf{x})=C\big\}$ is a compact subset of Ω and does not intersect $\mathcal{F}$, then $\mathbf{x}(t)$ is periodic.

Proof. The proof of part (a) proceeds by showing that if $\boldsymbol{\xi}\notin\mathcal{F}$, then $\boldsymbol{\xi}=\mathbf{x}(\tau)$ for some τ. Clearly, $E(\boldsymbol{\xi})=C$. Since $\nabla E(\boldsymbol{\xi})\neq 0$, the implicit function theorem can be applied. Without loss of generality there exists $g:\big\{s:|s-\xi_1|<\delta\big\}\to\mathbb{R}$ such that $E\big(s,g(s)\big)=C$ and $g(\xi_1)=\xi_2$. Furthermore, there exists $\varepsilon>0$ such that $E(\mathbf{x})=C$ and $\|\mathbf{x}-\boldsymbol{\xi}\|<\varepsilon$ implies $\mathbf{x}=\big(s,g(s)\big)$ for some s. Because $\big\|\mathbf{x}(t_m)-\boldsymbol{\xi}\big\|<\varepsilon$, for large m, the solution $\mathbf{x}(t_m)=\big(s_m,g(s_m)\big)$ for some s_m. By *Proposition 2.2*, it now suffices to show that $\big\{\big(s,g(s)\big):|\xi_1-s|<\delta\big\}$ is contained in the orbit determined by $\boldsymbol{\xi}$.

Let $\boldsymbol{\varphi}(s)=\big(s,g(s)\big)$ and note that $\nabla E\big(\boldsymbol{\varphi}(s)\big)\cdot\dot{\boldsymbol{\varphi}}(s)\equiv 0$ because $E\big(\boldsymbol{\varphi}(s)\big)\equiv C$. Since $E(\mathbf{x})$ is an integral, $\nabla E(\mathbf{x})\cdot\mathbf{f}(\mathbf{x})=0$. Because the dimension is two, it follows that $\dot{\boldsymbol{\varphi}}(s)=\gamma(s)f\big(\boldsymbol{\varphi}(s)\big)$ as long as $\mathbf{f}\big(\boldsymbol{\varphi}(s)\big)\neq 0$. Using the assumption that $\boldsymbol{\xi}\notin\mathcal{F}$, apply *Proposition 2.4* to conclude that there exists λ mapping $\big\{t:|t|<\varepsilon\big\}$ onto $\big\{s:|\xi_1-s|<\delta\big\}$ with $\lambda(0)=\xi_1$, such that $\boldsymbol{\varphi}\big(\lambda(t)\big)$ is a solution of $\dot{\mathbf{x}}=\mathbf{f}(\mathbf{x})$ and $\boldsymbol{\varphi}\big(\lambda(0)\big)=\boldsymbol{\xi}$. In particular, for large m, we have $\mathbf{x}(t_m)=\boldsymbol{\varphi}\big(\lambda(\tau_m)\big)$. Therefore, the orbits of $\boldsymbol{\xi}$ and $\mathbf{x}(t)$ intersect and are equal by *Proposition 2.2*, which proves the first part.

Now assume that $\big\{\mathbf{x}:E(\mathbf{x})=C\big\}$ is a compact subset of Ω and

$$\big\{\mathbf{x}:E(\mathbf{x})=C\big\}\cap\mathcal{F}=\phi.$$

Then $\mathbf{x}(t)$ cannot approach the boundary of Ω and must be defined on all of $\mathbb{R}$ by *Theorem 1.21*. By compactness, there exists a sequence $\mathbf{x}(t_m)$ with t_m going to infinity such that $\mathbf{x}(t_m)$ converges to a point $\boldsymbol{\xi}$ in Ω. From the above argument, it follows that for m large there exists a sequence of bounded real numbers τ_m

such that $\mathbf{x}(\tau_m, \boldsymbol{\xi}) = \mathbf{x}(t_m)$ or $\mathbf{x}(t_m - \tau_m) = \boldsymbol{\xi}$. Clearly, $t_m - \tau_m$ goes to infinity and hence $\boldsymbol{\xi} = \mathbf{x}(t)$ has infinitely many solutions. Hence, $\mathbf{x}(t_1) = \mathbf{x}(t_2)$ for some $t_1 \neq t_2$. Since $\mathbf{x} \notin \mathcal{F}$, it must be periodic by *Proposition 2.3*. $\square$

The classical example of an application of ordinary differential equations to the life sciences is Voltera's predator-prey model. It models the populations of a simple predator and its only prey. As with most mathematical models simplifying assumptions are needed to build Voltera's predator-prey model, but the results are realistic in special cases. The previous proposition can be used to construct its phase portrait without obtaining formulas for the solutions as functions of t.

Let x denote the population of the prey and y that of the predator. Assuming the prey has adequate food, its population should grow exponentially and satisfy $\dot{x} = ax$ for some $a > 0$. When xy is large, we would expect the predator and prey to have frequent contact and a proportional decrease in the population of the prey to occur. This suggests

$$\dot{x} = ax - cxy,$$

and the reverse analysis for the predator leads to

$$\dot{y} = -by + dxy,$$

where a, b, c, and d are all positive.

The next few paragraphs are devoted to a qualitative analysis of the solutions of the system in the first quadrant:

$$\begin{aligned} \dot{x} &= ax - cxy, \\ \dot{y} &= -by + dxy. \end{aligned}$$

An easy calculation shows that $\mathcal{F} = \big\{(0,0), (b/d, a/c)\big\}$ and the positive x and y axis are the orbits of the solutions $(e^{at}, 0)$ and $(0, e^{-bt})$. It is natural to restrict the problem to the first quadrant because populations are not negative. When $x = b/d$ and $y = a/c$ the populations are in equilibrium. What happens if this equilibrium is disturbed?

Because the problem is set in the plane, it might be possible to solve for y in terms of x along a piece of an orbit. Arguing heuristically,

$$\frac{dy}{dx} = \frac{y(-b + dx)}{x(a - cy)},$$

which is separable and leads to

$$a \log y - cy = -b \log x + dx + C.$$

Set

$$E(x, y) = a \log y + b \log x - cy - dx.$$

Now that this function has been discovered, it is easy to verify that it is an integral for

$$\begin{aligned} \dot{x} &= ax - cxy, \\ \dot{y} &= -by + dxy \end{aligned}$$

by showing that $\nabla E \cdot \mathbf{f} \equiv 0$.

Next fix α, a positive real number, and consider the function $E(\alpha, y)$. A little calculus shows that $E(\alpha, y)$ goes to $-\infty$ as y goes to both 0 and ∞ and that the maximal value of $E(\alpha, y)$ is

$$a \log(c/a) - a + b \log \alpha - c\alpha,$$

which also goes to $-\infty$ as α goes to both 0 and ∞. A similar analysis applies to $E(x, \beta)$. Therefore, the level curves of $E(x, y)$ are closed and bounded subsets in the first quadrant. *Proposition 2.6* can now be applied to conclude that every solution in the first quadrant except the fixed point $(b/d, a/c)$ is periodic and the fixed point $(b/d, a/c)$ is a center. In particular, if the populations are not in equilibrium, they would oscillate periodically forever.

The phase portrait for the Voltera predator-prey model with $a = b = 2$ and $c = d = 1$ is shown in *Figure 2.5.* The positive axes are the orbits of the solutions $\varphi_1(t) = (e^{2t}, 0)$ and $\varphi_2(t) = (0, e^{-2t})$. They are included to provide perspective.

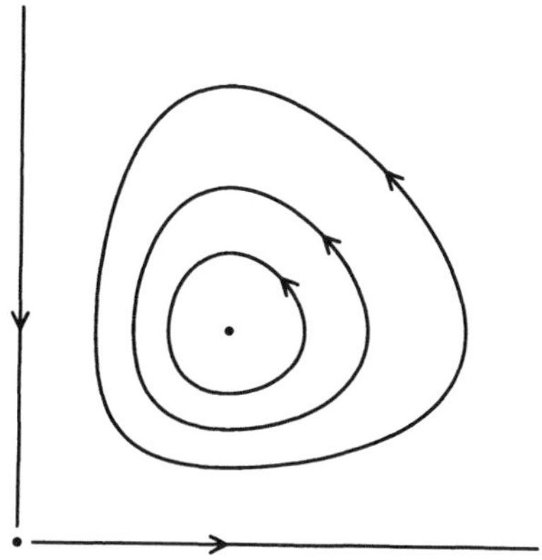

Figure 2.5: Phase portrait of the Voltera predator-prey model.

Thus without explicitly solving the Voltera equation it was possible to completely describe the dynamical behavior of these populations. Euler's method, *Theorem 1.16*, is now applicable to produce good approximations to the actual solutions and to determine their shape and periods to any desired accuracy. From the qualitative point of view, this is a very satisfactory state of affairs.

Another class of problems that fit the mold of *Proposition 2.6* arises when an object of mass m is moving on the x-axis under the action of a force $F : \mathbb{R} \longrightarrow \mathbb{R}$

that is locally Lipschitz on $\mathbb{R}$. Thus its motion is governed by the second-order scalar differential equation $m\ddot{x} = F(x)$, which can be written as the first-order system,

$$\begin{aligned} \dot{x} &= y \\ \dot{y} &= F(x)/m. \end{aligned}$$

Note that $\mathbf{f}(x,y) = \big(y, F(x)/m\big)$ is locally Lipschitz. Set $V(x) = -\int_0^x F(w)\,dw$. Then $F(x)$ is conservative and the total energy

$$E(x,y) = \frac{1}{2}my^2 + V(x)$$

is an integral. *Proposition 2.6* can be used to determine the phase portrait for any problem of this type. In fact, all that is needed to determine the phase portrait is the graph of $V(x)$.

Changing variables is a common technique in differential equations. Although the underlying idea is simple, discovering a change of variables to make an equation more tractable often appears to be magical. A given change of variables can always be implemented ad hoc, but the next proposition provides a unified way of understanding the process.

Proposition 2.7 *Let $\dot{\mathbf{x}} = \mathbf{f}(\mathbf{x})$ be a given autonomous differential equation on $\Omega \subset \mathbb{R}^d$ where $\mathbf{f}$ is continuous on Ω. Suppose $\mathbf{G} : \Omega' \to \Omega$ has continuous first partial derivatives on $\Omega' \subset \mathbb{R}^d$ and the $d \times d$ matrix of partial derivatives*

$$\left[\frac{\partial G_i}{\partial y_j}\right]$$

is nonsingular at every $\mathbf{y} \in \Omega'$. Let $\varphi : I \to \Omega'$ be a continuously differentiable curve. Then $\mathbf{G}\big(\varphi(t)\big)$ is a solution of $\dot{\mathbf{x}} = \mathbf{f}(\mathbf{x})$ if and only if $\varphi(t)$ is a solution of

$$\dot{\mathbf{y}} = \left[\frac{\partial G_i}{\partial y_j}(\mathbf{y})\right]^{-1} \mathbf{f}\big(\mathbf{G}(\mathbf{y})\big).$$

Proof. By the chain rule, the derivative of $\mathbf{G}\big(\varphi(t)\big)$ is given by the matrix product

$$\left[\frac{\partial G_i}{\partial \mathbf{y}_j}\big(\varphi(t)\big)\right] \dot{\varphi}(t)$$

where $\dot{\varphi}(t)$ is thought of as a $d \times 1$ matrix. (The chain rule will be proved in Section 1 of Chapter 7 as part of a larger discussion of differentiable functions of several variables.) Hence $\mathbf{G}\big(\varphi(t)\big)$ is a solution of $\dot{\mathbf{x}} = \mathbf{f}(\mathbf{x})$ if and only if

$$\dot{\varphi}(t) = \left[\frac{\partial G_i}{\partial \mathbf{y}_j}\big(\varphi(t)\big)\right]^{-1} \mathbf{f}\Big(\mathbf{G}\big(\varphi(t)\big)\Big). \ \square$$

The final example in this section has a more interesting phase portrait that in the spirit of the qualitative approach will be obtained without determining explicit solutions or an integral. In fact, there are no integrals for this system except the constant functions, and they are useless as integrals.

Let $\Omega = \mathbb{R}^2$ and consider

$$\begin{aligned} \dot{x} &= -y + \mu x(1 - r^2) \\ \dot{y} &= x + \mu y(1 - r^2), \end{aligned}$$

where $r^2 = x^2 + y^2$ and μ is a real parameter. This example also provides a good illustration of how the right change of variables—polar coordinates—can make the construction of a phase portrait much easier.

Letting $x = r\cos\theta$ and $y = r\sin\theta$, and putting things in the context of the previous proposition,

$$\begin{aligned} \Omega' &= \{(r, \theta) : r > 0\} \\ G(r, \theta) &= (r\cos\theta, r\sin\theta), \end{aligned}$$

and the matrix of first partial derivatives is

$$\begin{bmatrix} \cos\theta & -r\sin\theta \\ \sin\theta & r\cos\theta \end{bmatrix}.$$

The inverse of this matrix is

$$\begin{bmatrix} \cos\theta & \sin\theta \\ -r^{-1}\sin\theta & r^{-1}\cos\theta \end{bmatrix}$$

and it times $\mathbf{f}(\mathbf{G}(r, \theta))$ written as a column vector is

$$\begin{bmatrix} \cos\theta & \sin\theta \\ -r^{-1}\sin\theta & r^{-1}\cos\theta \end{bmatrix} \begin{pmatrix} -r\sin\theta + \mu r(1 - r^2)\cos\theta \\ r\cos\theta + \mu r(1 - r^2)\sin\theta \end{pmatrix} =$$

$$\begin{pmatrix} -r\cos\theta\sin\theta + \mu r(1 - r^2)\cos^2\theta + r\cos\theta\sin\theta + \mu r(1 - r^2)\sin^2\theta \\ \sin^2\theta + \mu(1 - r^2)\cos\theta\sin\theta + \cos^2\theta - \mu(1 - r^2)\cos\theta\sin\theta \end{pmatrix} =$$

$$\begin{pmatrix} \mu r(1 - r^2) \\ 1 \end{pmatrix}.$$

So the new system of differential equations after the change of variables is

$$\begin{aligned} \dot{r} &= \mu r(1 - r^2) \\ \dot{\theta} &= 1, \end{aligned}$$

and $\big(r(t), \theta(t)\big)$ is solution of it if and only if $\big(r(t)\cos\theta(t), r(t)\sin\theta(t)\big)$ is a solution of the original differential equation. (Usually the above system is derived by differentiating $x = r\cos\theta$ and $y = r\sin\theta$ with respect to t, solving for $\dot{r}$ and $\dot{\theta}$, and substituting. Solving for $\dot{r}$ and $\dot{\theta}$ is equivalent to finding the inverse of the matrix of partial derivatives, and substituting for $\dot{x}$ and $\dot{y}$ is the same as computing the above matrix product.)

Hence, $\theta = t + C$, and solutions move around the origin counterclockwise at a uniform rate. It is also easy to determine the phase portrait of $\dot{r} = \mu r(1 - r^2)$ because it is an autonomous locally Lipschitz scalar differential equation. The only critical points are -1, 0, and 1. For convenience, assume $\mu > 0$. Then $\mu r(1 - r^2)$ is positive between $-\infty$ and -1 and between 0 and 1. It is negative between -1 and 0 and between 1 and ∞. In particular, if $r(t)$ is a maximal solution with $0 < r(\tau) < 1$, then by uniqueness of solutions $0 < r(t) < 1$ for all t. Clearly, $r(t)$ is increasing and defined on $\mathbb{R}$.

Let $b = \lim_{t \to \infty} r(t)$ and let $g(r) = \mu r(1 - r^2)$. By the mean-value theorem, there exists β_n, $n < \beta_n < n + 1$, such that $\dot{r}(\beta_n) = r(n + 1) - r(n)$. Because $\lim_{n \to \infty} r(n) = b$ and $\lim_{n \to \infty} r(\beta_n) = b$,

$$\begin{aligned} g(b) &= \lim_{n \to \infty} g\big(r(\beta_n)\big) \\ &= \lim_{n \to \infty} \dot{r}(\beta_n) \\ &= \lim_{n \to \infty} r(n + 1) - r(n) \\ &= 0 \end{aligned}$$

and it follows that $b = 1$. Similarly, $\lim_{t \to -\infty} r(t) = 0$ and, for a solution with $r(\tau) > 1$, we have $\lim_{t \to \infty} r(t) = 1$. The phase portrait of $\dot{x} = \mu r(1 - r^2)$, $\mu > 0$, consist of the seven orbits pictured in *Figure 2.6*. The dot in the center is, of course, at 0.

Figure 2.6: Phase portrait of $\dot{r} = \mu r(1 - r^2)$.
with $\mu > 0$.

For the original equation, it is now clear that orbits inside the unit circle spiral toward the center as t goes to $-\infty$ and toward the unit circle as t goes to ∞. Note $x(t) = (\cos t, \sin t)$ is a solution, so the unit circle is a periodic orbit. Orbits outside the unit circle spiral toward it as t goes to ∞ and toward infinity as t goes to negative infinity as shown in *Figure 2.7*. The slightly darker orbit with two arrows on it is the periodic orbit. The figure was constructed using $\mu = 1/8$.

EXERCISES

1. Let $\mathbf{f}$ be a continuous vector field on Ω an open subset of $\mathbb{R}^d$ and let $\mathbf{x}_m$ be a sequence of fixed points. Show that if $\mathbf{x}_m$ converges to $\mathbf{x}$ in Ω, then $\mathbf{x}$ is also a fixed point of $\mathbf{f}$.

2. Let $\varphi(t)$ be a solution of the autonomous differential equation $\dot{\mathbf{x}} = \mathbf{f}(\mathbf{x})$ and assume $f(\mathbf{x})$ is locally Lipschitz on an open subset Ω of $\mathbb{R}^d$. Show that $\mathbf{f}(\mathbf{q}) = 0$, if $\varphi(t)$ is defined on $a < t < \infty$ and

$$\lim_{t \to \infty} \varphi(t) = \mathbf{q} \in \Omega.$$

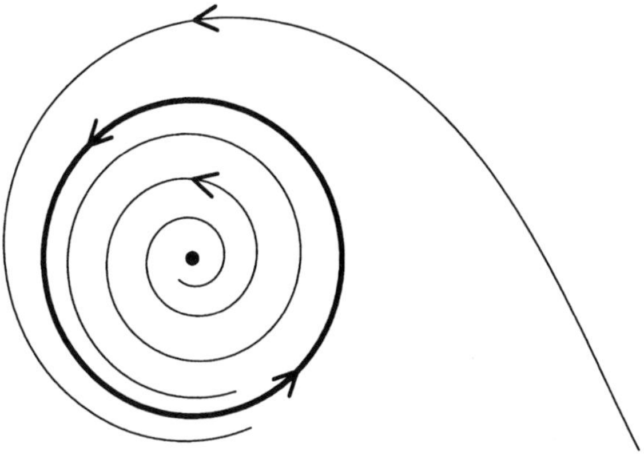

Figure 2.7: Phase portrait showing orbits spiraling towards a periodic orbit.

3. Let $\mathbf{f}(\mathbf{x})$ be locally Lipschitz on the open set $\Omega \subset \mathbb{R}^d$, and suppose that $\mathbf{p}_m$ is a sequence of periodic points for $\dot{\mathbf{x}} = \mathbf{f}(\mathbf{x})$ such that

$$\lim_{m \to \infty} \mathbf{p}_m = \mathbf{q} \in \Omega.$$

Let ω_m be the period of $\mathbf{p}_m$. Prove the following:

   (a) If $\omega_m \to 0$ as $m \to \infty$, then $\mathbf{f}(\mathbf{q}) = 0$.

   (b) If there exist positive constants α and β such that $\alpha < \omega_m < \beta$ for every m, then $\mathbf{q}$ is a periodic point or a fixed point.

   (c) Show by example that $\mathbf{q}$ can be a fixed point even when $\omega_m \to T \neq 0$ as $m \to \infty$.

4. Draw a phase portrait for each of the following autonomous systems of differential equations:

   (a)

$$\begin{aligned} \dot{x} &= xy^2 \\ \dot{y} &= x^2 y \end{aligned}$$

   (b)

$$\begin{aligned} \dot{x} &= -y(x^2 + y^2 - 1) \\ \dot{y} &= x(x^2 + y^2 - 1) \end{aligned}$$

   (c)

$$\begin{aligned} \dot{x} &= -y\big((x-1)^2 + y^2\big) \\ \dot{y} &= x\big((x-1)^2 + y^2\big). \end{aligned}$$

5. Find an integral for the system

$$\begin{aligned}\dot{x} &= y\\ \dot{y} &= x + x^2\end{aligned}$$

and sketch its phase portrait.

6. Consider the scalar second-order differential equation $\ddot{x} = F(x)$, where $F : \mathbb{R} \to \mathbb{R}$ is locally Lipschitz. Suppose that $F(x)$ is a restoring force, $xF(x) < 0$ for $x \neq 0$ and $F(0) = 0$. Show that $V(x) = -\int_0^x F(w)dw$ is increasing for $x > 0$ and decreasing for $x < 0$. Let $\alpha = \lim_{x\to-\infty} V(x)$ and $\beta = \lim_{x\to\infty} V(x)$. Write $\ddot{x} = F(x)$ as a system of differential equations on $\mathbb{R}^2$ and sketch the phase portrait in each of the following situations:

   (a) $\alpha = \beta = \infty$,

   (b) $\alpha < \beta = \infty$,

   (c) $\alpha < \beta < \infty$,

   (d) $\alpha = \beta < \infty$.

7. Each of the scalar autonomous differential equations shown below contains a real parameter μ. Sketch all possible phase portraits for them as μ varies through $\mathbb{R}$.

   (a) $\dot{x} = x^2 + 2\mu x + 1$,

   (b) $\dot{x} = x(x^2 - \mu)$,

   (c) $\dot{x} = (x^2 - \mu)(x^2 - \mu^3)$.

8. Suppose $\Phi : \mathbb{R} \times \Omega \to \Omega$ defines a flow on $\Omega \subset \mathbb{R}^d$ and let $\mathbf{x}$ be a point in Ω. Show that the orbit closure of $\mathbf{x}$ defined by

$$\overline{\mathcal{O}}(\mathbf{x}) = \overline{\{\Phi(t, \mathbf{x}) : t \in \mathbb{R}\}}$$

is an invariant set.

9. Suppose $\Phi : \mathbb{R} \times \Omega \to \Omega$ defines a flow on $\Omega \subset \mathbb{R}^d$ and let Y be an invariant subset of the flow. Show that $\Omega - Y$ and $\overline{Y} \cap \Omega$ are also invariant sets.

10. Prove the following: If $\Phi : \mathbb{R} \times \Omega \to \Omega$ defines a flow on $\Omega \subset \mathbb{R}^d$, then for each $\mathbf{x} \in \Omega$ exactly one of the following hold:

   (a) $t \to \Phi(t, \mathbf{x})$ is one-to-one on $\mathbb{R}$;

   (b) There exists a smallest positive ω such that $\Phi(t + \omega, \mathbf{x}) = \Phi(t, \mathbf{x})$ for all $t \in \mathbb{R}$; or

   (c) $\Phi(t, \mathbf{x}) = \mathbf{x}$ for all $t \in \mathbb{R}$.

11. Prove that if $\mathbf{f}(\mathbf{x})$ is locally Lipschitz on $\mathbb{R}^d$, then there exists a realvalued locally Lipschitz function $h : \mathbb{R}^d \to \mathbb{R}$ such that $\dot{\mathbf{x}} = h(\mathbf{x})\mathbf{f}(\mathbf{x})$ defines a flow on $\mathbb{R}^d$ with the same orbits as $\dot{\mathbf{x}} = \mathbf{f}(\mathbf{x})$.

12. Recall that the arc length of a continuously differentiable curve $\varphi(t)$ on the interval $a \le t \le b$ is given by

$$\int_a^b \|\dot{\varphi}(t)\| dt.$$

Let $\mathbf{f}(\mathbf{x})$ be a vector field with continuous first partial derivatives on an open set $\Omega \subset \mathbb{R}^d$, and let $\mathbf{p}_m$ be a sequence of periodic points for $\dot{\mathbf{x}} = \mathbf{f}(\mathbf{x})$ with periods T_m. Suppose the sequence $\mathbf{p}_m$ converges to $\mathbf{q}$ a fixed point of $\dot{\mathbf{x}} = \mathbf{f}(\mathbf{x})$. Prove the following: If the ω_m is bounded sequence, then the arc lengths of the periodic points $\mathbf{p}_m$ converges to 0.

13. Consider the scalar second-order differential equation $\ddot{x} + 4(x - x^3) = 0$. First, rewrite it as a first-order system and find an integral for the system. Then draw a phase portrait for the system. For what values of ξ, if any, is the solution of $\ddot{x} + 4(x - x^3) = 0$ with initial velocity 0 periodic?

14. Sketch the phase portrait for the undamped pendulum $\ddot{x} + k^2 \sin x = 0$.

15. Show that if $E(x, y)$ is an integral of the system

$$\begin{aligned} \dot{x} &= -y + \mu x(1 - r^2) \\ \dot{y} &= x + \mu y(1 - r^2) \end{aligned}$$

for some $\mu \neq 0$, then $E(x, y)$ is a constant function.

16. Sketch the phase portrait for the system:

$$\begin{aligned} \dot{x} &= -y + x(1 - x^2 - y^2 - z^2) \\ \dot{y} &= x + y(1 - x^2 - y^2 - z^2) \\ \dot{z} &= 0. \end{aligned}$$

17. Sketch the phase portrait for the Voltera predator-prey model in all four quadrants with $a = b = 2$ and $c = d = 1$.

2.3 Differential Inequalities

The results in this section surrounding differential inequalities are also classical, although less familiar than the material in the preceding two sections. The key ideas were known to Peano in the late 1800s. These results can be used to determine the interval on which solutions exist, to prove more general uniqueness theorems, to prove differentiability in initial conditions, and to study the solutions of a particular differential equation. The presentations of the material in this section was influenced by and parallels the discussion of these ideas in Hale's book on differential equations [15].

Solving the differential equation $\dot{\mathbf{x}} = \mathbf{f}(t, \mathbf{x})$ can be thought of as integration, because we are given $\dot{\mathbf{x}}(t)$ and want to determine $\mathbf{x}(t)$. For real-valued functions the integral from a to t of a continuous real-valued function is greater than or equal to the integral from a to t of any function it dominates. More specifically, if $g(t)$ and $h(t)$ have continuous derivatives satisfying $g'(t) \leq h'(t)$ on $a \leq t \leq b$ and $g(a) \leq h(a)$, then

$$g(t) = g(a) + \int_a^t g'(s)\, ds \leq h(a) + \int_a^t h'(s)\, ds = h(t). \tag{2.4}$$

A simple example shows how the inequality *(2.4)* can be used to control solutions and obtain information about them. Consider the scalar equation $\dot{x} = g(t) - x^3$ where $g(t)$ is continuous and positive in $\mathbb{R}$. Clearly, $g(t) - x^3$ is locally Lipschitz. The goal is to show that $x(t) = x(t, \tau, \xi)$ is defined for all $t \geq \tau$ when ξ is positive.

Suppose $x(s) < 0$ for some $s > \tau$. Then there exists σ, $\tau < \sigma < s$ such that $x(\sigma) = 0$ and $x(t) < 0$ for $\sigma < t < s$. By the mean-value theorem

$$\frac{x(s) - x(\sigma)}{s - \sigma} = \dot{x}(\theta) = g(\theta) - x(\theta)^3$$

for some θ between σ and s. Note that the left side of this equation is negative and the right side is positive because $x(\theta) < 0$. Therefore, $x(s) < 0$ is impossible for $s \geq \tau$.

Since $g(t) - x^3 \leq g(t)$ when $x \geq 0$, it follows from *(2.4)* that

$$x(t) = \xi + \int_\tau^t g(s) - x(s)^3\, ds \leq \xi + \int_\tau^t g(s)\, ds$$

when $t \geq \tau$ because $x(t) \geq 0$ for $t \geq \tau$. Thus, if $x(t)$ is defined on $\tau \leq t < b < \infty$, its trajectory is trapped in the compact set

$$\left\{(t, x) : \tau \leq t \leq b \text{ and } 0 \leq x \leq \xi + \int_\tau^t g(x)\, ds\right\}$$

and by *Theorem 1.19*, $x(t)$ can be continued to values of t greater than b. Therefore, $x(t)$ must be defined for $\tau \leq t < \infty$. This example captures much of the flavor of this entire section.

To begin the more general discussion of differential inequalities, let $g(t)$ be a real-valued function defined on an interval I. When

$$\lim_{h \to 0+} \frac{g(\tau + h) - g(\tau)}{h}$$

exists (and is not infinite), the limit is called the *right-hand derivative of* $g(t)$ *at* τ and is denoted by $\mathrm{D_r g}(\tau)$. Note that τ could be the left-end point of I, but not the right-end point.

Let D be an open subset of $\mathbb{R}^2$ and let $g(t,y)$ be a continuous real-valued function defined in D. The differential equation $\dot{y} = g(t,y)$ will be used to study the *differential inequality*

$$\mathrm{D_r w} \leq \mathrm{g(t, w)}. \tag{2.5}$$

A *solution of the differential inequality (2.5)* is a continuous function $w(t)$ defined on an open interval I such that $\mathrm{D_r w(t)}$ exists and satisfies

$$\mathrm{D_r w(t)} \leq \mathrm{g\big(t, w(t)\big)},$$

for all t in I.

The first result is an analogue of *(2.4)* for scalar differential inequalities. The proof of this theorem makes effective use of continuity in initial conditions and leads to theorems that set the stage for Chapter 3.

Theorem 2.8 *Let $g(t,y)$ be continuous and locally Lipschitz on D, an open subset of $\mathbb{R}^2$. Let $w(t)$ be a solution of the differential inequality (2.5) defined on $I = \{t : a \leq t < b\}$, and let $y(t)$ be a solution of $\dot{y} = g(t,y)$ also defined on the open interval I. If $w(a) \leq y(a)$, then $w(t) \leq y(t)$ for all $t \in I$.*

Proof. Consider the parameterized system of differential equations

$$\dot{y} = g(t,y) + \mu$$

on $D \times \mathbb{R}$, and let $\xi = y(a)$. Clearly, continuity in initial conditions (*Theorem 1.23*) applies here. Thus, given $\varepsilon > 0$, the solution $y(t,a,\xi,\mu)$ is defined on $a \leq t \leq b - \varepsilon$ for small μ, and

$$\lim_{\mu \to 0} y(t,a,\xi,\mu) = y(t).$$

Set

$$y_m(t) = y(t,a,\xi,1/m).$$

Since ε can be arbitrarily small, it suffices to show that $w(t) \leq y_m(t)$ when $a \leq t \leq b - \varepsilon$ and m is large. Suppose $w(t) \leq y_m(t)$ does not hold for some m such that $y_m(t)$ is defined for $a \leq t < b - \varepsilon$. Then there exist τ and τ', $a \leq \tau < \tau' \leq b - \varepsilon$, such that $w(\tau) = y_m(\tau)$ and $w(t) > y_m(t)$ for $\tau < t < \tau'$. Hence,

$$w(\tau + h) - w(\tau) > y_m(\tau + h) - y_m(\tau)$$

for small positive h, and it follows that

$$\begin{aligned} \mathrm{D_r w}(\tau) &\geq \dot{y}_m(\tau) = g\big(\tau, y_m(\tau)\big) + 1/m \\ &> g\big(\tau, y_m(\tau)\big) = g\big(\tau, w(\tau)\big), \end{aligned}$$

contradicting $\mathrm{D_r w(t)} \leq \mathrm{g}\big(\tau, \mathrm{w}(\tau)\big)$. □

Corollary *A continuous real-valued function $h(t)$ satisfying* $\mathrm{D_r h(t)} \leq 0$ *on an open interval I is nonincreasing on I.*

The stage is now set for proving the key result in this section.

Theorem 2.9 (Gronwall's Inequality) *Let $\varphi(t)$, $\alpha(t)$, and $\beta(t)$ be continuous real-valued functions on the interval $\{t : a \le t < b\}$ with $\beta(t) \ge 0$. If*

$$\varphi(t) \le \alpha(t) + \int_a^t \beta(s)\varphi(s)\,ds$$

for $a \le t < b$, then

$$\varphi(t) \le \alpha(t) + \int_a^t \alpha(s)\beta(s)\exp\left(\int_s^t \beta(w)\,dw\right)\,ds$$

for $a \le t < b$.

Proof. First rewrite the inequality

$$\varphi(t) \le \alpha(t) + \int_a^t \beta(s)\varphi(s)\,ds$$

as

$$\varphi(t) - \alpha(t) \le \int_a^t \beta(s)\alpha(s)\,ds + \int_a^t \beta(s)\big(\varphi(s) - \alpha(s)\big)\,ds.$$

Letting $\psi(t) = \varphi(t) - \alpha(t)$, this can be written as

$$\psi(t) \le \int_a^t \beta(s)\alpha(s)\,ds + \int_a^t \beta(s)\psi(s)\,ds. = w(t)$$

Note that the right-hand side, which is being denoted by $w(t)$, has a continuous derivative and

$$\begin{aligned}\dot{w}(t) &= \alpha(t)\beta(t) + \beta(t)\psi(t)\\ &\le \alpha(t)\beta(t) + \beta(t)w(t).\end{aligned}$$

Thus $w(t)$ is a solution to a differential inequality and the previous theorem can be applied by first finding a solution to the differential equation $\dot{y} = \beta(t)y + \alpha(t)\beta(t)$.

For convenience, set $\alpha(t) \equiv \alpha(a)$ and $\beta(t) \equiv \beta(a)$ for $t < a$, so that $\alpha(t)$ and $\beta(t)$ are continuous for all $t < b$. It is easy to check that

$$\begin{aligned}y(t) &= \int_a^t \alpha(s)\beta(s)\exp\left(\int_s^t \beta(u)\,du\right)\,ds\\ &= \exp\left(\int_a^t \beta(u)\,du\right)\int_a^t \alpha(s)\beta(s)\exp\left(\int_a^s -\beta(u)\,du\right)\,ds\end{aligned}$$

is a solution of $\dot{y} = g(t,y) = \beta(t)y + \alpha(t)\beta(t)$ on $D = \{(t,y) : t < b\}$. (This type of differential equation will be fully discussed in the beginning of Chapter 3.)

Since $w(a) = 0 = y(a)$ the *Theorem 2.8* applies. Therefore,

$$\psi(t) \le w(t) \le y(t)$$

for $a < t < b$ or

$$\varphi(t) - \alpha(t) \le \int_a^t \alpha(s)\beta(s) \exp\left(\int_s^t \beta(u)\, du\right) ds,$$

which is the desired inequality. □

Corollary *Let $\varphi(t)$ and $\beta(t)$ be continuous real-valued functions on the interval $\{t : a \le t < b\}$ with $\beta(t) \ge 0$ and let C be a constant. If*

$$\varphi(t) \le C + \int_a^t \beta(s)\varphi(s)\, ds$$

for $a \le t < b$, then

$$\varphi(t) \le C \exp\left(\int_a^t \beta(s)\, ds\right).$$

Proof. Applying Gronwall's inequality with $\alpha(t) = C$ yields

$$\begin{aligned}
\varphi(t) &\le C + \int_a^t C\beta(s) \exp\left(\int_s^t \beta(w)\, dw\right) ds \\
&= C - C\left(\int_a^t -\beta(s) \exp\left(\int_t^s -\beta(w)\, dw\right) ds\right) \\
&= C - C\left(1 - \exp\left(\int_a^t \beta(s)\, ds\right)\right) \\
&= C \exp\left(\int_a^t \beta(s)\, ds\right). \quad \square
\end{aligned}$$

Gronwall's inequality has many applications. One of them that will be essential in Chapter 3 is that solutions can be continued as far as possible when $\mathbf{f}(t, \mathbf{x})$ has sublinear growth in $\mathbf{x}$.

Theorem 2.10 *If $\mathbf{f}$ is continuous on $\{(t, \mathbf{x}) : c < t < d$ and $\mathbf{x} \in \mathbb{R}^n\}$, and if there exist continuous nonnegative real-valued functions $\beta(t)$ and $\gamma(t)$ on $\{t : c < t < d\}$ such that*

$$\left|\mathbf{f}(t, \mathbf{x})\right| \le \gamma(t) + \beta(t)|\mathbf{x}|$$

for $(t, \mathbf{x}) \in D$, then every maximal solution of the differential equation $\dot{\mathbf{x}} = \mathbf{f}(t, \mathbf{x})$ is defined on $\{t : c < t < d\}$.

Proof. Let $\varphi(t)$ be a maximal solution defined on the interval $I = \{t : c' < t < d'\} \subset \{t : c < t < d\}$ and let $a \in I$. Note that

$$\left|\varphi(t)\right| = \left|\varphi(a) + \int_a^t \mathbf{f}\left(s, \varphi(s)\right) ds\right|$$

$$\begin{aligned}
&\le \ |\varphi(a)| + \int_a^t \left|\mathbf{f}\big(s, \varphi(s)\big)\right| ds \\
&\le \ |\varphi(a)| + \int_a^t \gamma(s) + \beta(s)|\varphi(s)|\, ds \\
&\le \ |\varphi(a)| + \int_a^t \gamma(s)\, ds + \int_a^t \beta(s)|\varphi(s)|\, ds.
\end{aligned}$$

If $d' < d$, then there exists C and M such that for $a \le t \le d'$, $|\varphi(a)| + \int_a^t \gamma(s)\, ds \le C$ and $\beta(t) \le M$. Thus

$$|\varphi(t)| \le C + \int_a^t \beta(s)|\varphi(s)|\, ds.$$

So the corollary to Gronwall's inequality applies, and it follows that

$$|\varphi(t)| \le C \exp\left(\int_a^t \beta(s)\, ds\right) \le Ce^{M(d'-a)}.$$

Hence, for $a \le t < d'$, the trajectory $\big(t, \varphi(t)\big)$ is contained in the following compact subset of D:

$$\left\{(t, \mathbf{x}) : a \le t \le d' \text{ and } |\mathbf{x}| \le Ce^{M(d'-a)}\right\},$$

which contradicts the maximality of $\varphi(t)$. Therefore, $d' = d$. To prove that $c' = c$, reverse the time and apply the above. □

Theorem 2.8 is limited to scalar differential equations and to use it to study more general ones it must be applied to the norm of a solution. It was for this purpose that right-hand derivatives were introduced. First, it will be shown that any norm has right-hand directional derivatives in all directions, and then this fact will be used to prove that $\mathrm{D_r}\|\mathbf{x}(t)\|_a$ exists when $\mathbf{x}(t)$ is a solution of $\dot{\mathbf{x}} = \mathbf{f}(t, \mathbf{x})$. The general norm inequality

$$-\|\mathbf{u} - \mathbf{v}\|_a \le \|\mathbf{u}\|_a - \|\mathbf{v}\|_a \le \|\mathbf{u} - \mathbf{v}\|_a$$

will be needed several times. It is a consequence of the triangle inequality. (See the Remark on page 3.)

Proposition 2.11 *Let $\|\cdot\|_a$ be any norm on a vector space V and let $\mathbf{x}$, $\mathbf{y} \in V$. Then,*

$$\lim_{h\to 0^+} \frac{\|\mathbf{x} + h\mathbf{y}\|_a - \|\mathbf{x}\|_a}{h}$$

exists and has absolute value at most $\|\mathbf{y}\|_a$.

Proof. It suffices to show that

$$\frac{\|\mathbf{x} + h\mathbf{y}\|_a - \|\mathbf{x}\|_a}{h}$$

is a nondecreasing function of h for $h > 0$, because

$$-\|\mathbf{y}\|_a \le \frac{\|\mathbf{x} + h\mathbf{y}\|_a - \|\mathbf{x}\|_a}{h} \le \|\mathbf{y}\|_a.$$

Given $0 < h' < h$, we can write $h' = \theta h$ for some θ such that $0 < \theta < 1$ and work with $h > 0$ and $0 < \theta < 1$ instead of h and h'. Starting with

$$\|\mathbf{x} + \theta h\mathbf{y}\|_a - \|\theta\mathbf{x} + \theta h\mathbf{y}\|_a \le \|\mathbf{x} - \theta\mathbf{x}\|_a = (1-\theta)\|\mathbf{x}\|_a$$

and rearranging the terms, yields

$$\|\mathbf{x} + \theta h\mathbf{y}\|_a - \|\mathbf{x}\|_a \le \theta\|\mathbf{x} + h\mathbf{y}\|_a - \theta\|\mathbf{x}\|_a.$$

Then dividing by θh to get

$$\frac{\|\mathbf{x} + \theta h\mathbf{y}\|_a - \|\mathbf{x}\|_a}{\theta h} \le \frac{\|\mathbf{x} + h\mathbf{y}\|_a - \|\mathbf{x}\|_a}{h},$$

completes the argument. □

Theorem 2.12 *Let $\|\cdot\|_a$ be any norm on $\mathbb{R}^d$. If $\varphi : I \to \mathbb{R}^d$ is a continuously differentiable curve defined on an open interval I, then $\mathrm{D_r}\|\varphi(\mathrm{t})\|_\mathrm{a}$ exists and*

$$\left|\mathrm{D_r}\left\|\varphi(\mathrm{t})\right\|_\mathrm{a}\right| \le \left\|\dot\varphi(\mathrm{t})\right\|_\mathrm{a}.$$

Proof. From the previous lemma,

$$\lim_{h\to 0^+} \frac{\left\|\varphi(t) + h\dot\varphi(t)\right\|_a - \left\|\varphi(t)\right\|_a}{h}$$

exists and has absolute value at most $\|\dot\varphi(t)\|_a$. The squeeze principle can be applied to show that the above limit also equals $\mathrm{D_r}\|\varphi(\mathrm{t})\|_\mathrm{a}$.

For small positive h,

$$\begin{aligned}
0 &\le \left|\frac{\left\|\varphi(t+h)\right\|_a - \left\|\varphi(t)\right\|_a}{h} - \frac{\left\|\varphi(t) + h\dot\varphi(t)\right\|_a - \left\|\varphi(t)\right\|_a}{h}\right| \\
&= \left|\frac{\left\|\varphi(t+h)\right\|_a - \left\|\varphi(t) + h\dot\varphi(t)\right\|_a}{h}\right| \\
&\le \left\|\frac{\varphi(t+h) - \varphi(t)}{h} - \dot\varphi(t)\right\|_a .
\end{aligned}$$

Since the last term goes to 0 as h goes to 0, it follows that

$$\begin{aligned}
\mathrm{D_r}\left\|\varphi(\mathrm{t})\right\|_\mathrm{a} &= \lim_{h\to 0^+} \frac{\left\|\varphi(t+h)\right\|_a - \left\|\varphi(t)\right\|_a}{h} \\
&= \lim_{h\to 0^+} \frac{\left\|\varphi(t) + h\dot\varphi(t)\right\|_a - \left\|\varphi(t)\right\|_a}{h}
\end{aligned}$$

and the proof is finished. □

The next theorem applies *Theorems 2.8 and 2.12* with the usual norm $|\cdot|$ to prove that a Lipschitz condition guarantees that even approximate solutions can at worst move apart at an exponential rate.

Theorem 2.13 *Let $\mathbf{f}(t,\mathbf{x})$ be continuous and Lipschitz on $D \subset \mathbb{R}^{d+1}$ with Lipschitz constant L. Suppose φ_1 and φ_2 are two continuously differentiable curves defined on an open interval I satisfying $\left|\dot{\varphi}_i(t) - \mathbf{f}\big(t, \varphi_i(t)\big)\right| \le \varepsilon_i$ for positive constants ε_1 and ε_2. If $\left|\varphi_1(\tau) - \varphi_2(\tau)\right| \le \delta$ for some τ in I, then for all t in I*

$$\left|\varphi_1(t) - \varphi_2(t)\right| \le \delta e^{L|t-\tau|} + \frac{\varepsilon_1 + \varepsilon_2}{L}\left(e^{L|t-\tau|} - 1\right).$$

Proof. Set $\varphi(t) = \varphi_1(t) - \varphi_2(t)$ and $\varepsilon = \varepsilon_1 + \varepsilon_2$. Then, by *Theorem 2.12*

$$\begin{aligned} \mathrm{D_r}\left|\varphi(t)\right| &\le \left|\dot{\varphi}(t)\right| \\ &= \left|\dot{\varphi}_1(t) - \dot{\varphi}_2(t)\right| \\ &\le \left|\mathbf{f}\big(t, \varphi_1(t)\big) - \mathbf{f}\big(t, \varphi_2(t)\big)\right| + \varepsilon_1 + \varepsilon_2 \\ &\le L\left|\varphi_1(t) - \varphi_2(t)\right| + \varepsilon \\ &= L\left|\varphi(t)\right| + \varepsilon \end{aligned}$$

and $\left|\varphi(t)\right|$ is a solution of the differential inequality $\dot{w} \le Lw + \varepsilon$. Set $g(t,y) = Ly + \varepsilon$ and check that

$$y(t) = \delta e^{L(t-\tau)} + \frac{\varepsilon}{L}\left(e^{L(t-\tau)} - 1\right)$$

is a solution of $\dot{y} = Ly + \varepsilon$. For $t \ge \tau$, the conclusion follows from *Theorem 2.8* and to complete the proof for $t \le \tau$, apply the above argument with time reversed. □

Corollary *Let $\mathbf{f}(t,\mathbf{x})$ be continuous and locally Lipschitz on $D \subset \mathbb{R}^{n+1}$, let φ_1 and φ_2 be solutions of $\dot{\mathbf{x}} = \mathbf{f}(t,\mathbf{x})$ defined on I. Given a, b, and τ in I with $a \le \tau \le b$ and $\left|\varphi_1(\tau) - \varphi_2(\tau)\right| \le \delta$, there exists $L > 0$ such that for $a \le t \le b$*

$$\left|\varphi_1(t) - \varphi_2(t)\right| \le \delta e^{L|t-\tau|}.$$

Proof. The set

$$K = \left\{\varphi_i(t) : a \le t \le b \text{ and } i = 1,2\right\}$$

is compact. Hence, there exists a Lipschitz constant L for $\mathbf{f}(t,\mathbf{x})$ on K, and the above argument can be repeated. □

EXERCISES

1. Suppose $|\mathbf{f}(t, \mathbf{x})| \leq g(t, |x|)$, where $\mathbf{f}(t, \mathbf{x})$ is locally Lipschitz on D in $\mathbb{R}^d$ and $g(t, y)$ is locally Lipschitz on $\mathbb{R}^2$. Let $y(t)$ be a positive solution of $\dot{y} = g(t, y)$ defined on an open interval $a < t < b$ containing τ. Show that if $|\boldsymbol{\xi}| \leq y(\tau)$, then $\mathbf{x}(t, \tau, \boldsymbol{\xi})$ is defined on $\tau < t < b$.

2. Consider the initial-value problem

$$\begin{aligned} \dot{\mathbf{x}} &= \mathbf{f}(t, \mathbf{x}) \\ \mathbf{0} &= \mathbf{x}(0) \end{aligned}$$

where $\mathbf{f}(t, \mathbf{x})$ continuous on $\mathbb{R}^{d+1}$ and satisfies the following condition:

$$\sqrt{|t|}\,\big|\,\mathbf{f}(t, \mathbf{x}) - \mathbf{f}(t, \mathbf{y})\big| \leq |\mathbf{x} - \mathbf{y}|.$$

Use the Corollary to Gronwall's inequality to show that this initial-value problem has a unique solution for $t \geq 0$.

3. Let $\varphi(t)$ defined on the interval $a < t < b$ be the maximal solution of the initial- value problem $\dot{x} = t^2 + x^2$ and $x(0) = 1$. Use *Theorem 2.8* to show that $b < 1$.

4. Show that if $\mathbf{f}(\mathbf{x})$ is locally Lipschitz on $\mathbb{R}^d$ and satisfies $\|\mathbf{f}(\mathbf{x})\| \leq \|\mathbf{x}\|^2$, then the domain of the general solution $\mathbf{x}(t, \boldsymbol{\xi})$ for $\dot{\mathbf{x}} = \mathbf{f}(\mathbf{x})$ includes the interval $0 \leq t < \|\boldsymbol{\xi}\|^{-1}$.

5. Let $\mathbf{f}(\mathbf{x})$ be a locally Lipschitz function on $\mathbb{R}^d$ satisfying

$$\|\mathbf{f}(\mathbf{x})\| \leq e^{-\|\mathbf{x}\|}.$$

Show that the solution $\mathbf{x}(t, \boldsymbol{\xi})$ of $\dot{\mathbf{x}} = \mathbf{f}(\mathbf{x})$ is defined for all t and satisfies

$$\|\mathbf{x}(t, \boldsymbol{\xi})\| \leq \log\left(|t| + e^{\|\boldsymbol{\xi}\|}\right).$$

6. Let $\mathbf{f} : \mathbb{R}^{d+1} \to \mathbb{R}^d$ be a continuous vector-valued function, and let $\beta : \mathbb{R} \to \mathbb{R}$ be a continuous non-negative function such that

$$\int_0^\infty \beta(s)ds < \infty.$$

Prove that if

$$\|\mathbf{f}(t, \mathbf{x})\| \leq \beta(t)\|\mathbf{x}\|,$$

then the solutions of $\dot{\mathbf{x}} = \mathbf{f}(t, \mathbf{x})$ have the following properties:

(a) The solution $\mathbf{x}(t, \tau, \boldsymbol{\xi})$ is defined for all $t \geq \tau$.

(b) Given τ and $\boldsymbol{\xi}$, there exists a constant C such that $\|\mathbf{x}(t, \tau, \boldsymbol{\xi})\| \leq C$.

(c) The limit as t goes to infinity of $\mathbf{x}(t, \tau, \boldsymbol{\xi})$ exists.

Chapter 3

Linear Differential Equations

As a rule in mathematics, problems that have a linear structure are easier to analyze than those that are nonlinear. Differential equations are no exception to this rule. Linearity has significant consequences when it occurs in the space variable. The key fact in this context is that there is a vector space of solutions, and this vector space structure can be exploited.

Although the linearity provides a good means for understanding the space of solutions of a linear differential equation, determining an explicit solution is often out-of-reach when the equation is nonautonomous. Euler's method and more refined numerical methods, however, work very well on linear differential equations because the theory will show that the solutions are defined on the longest possible interval.

By proving that this vector space of solutions is finite dimensional, the problem of determining all solutions is reduced to finding a finite basis of solutions. This is precisely what is accomplished when one calculates the general solution of the familiar second-order linear differential equation with constant coefficients,

$$a\ddot{x} + b\dot{x} + cx = 0.$$

The elementary techniques used to solve these equations will be revisited at the end of Chapter 4 , which is devoted to linear differential equations with constant coefficients.

After the basic structure of the vector space of all solutions is established in Sections 3.1 and 3.2, it will be applied to higher order linear differential equations in Section 3.3 and then to complex linear differential equations in the final section. The concept of a matrix differential equation will play a key role throughout the study of the vector space of solutions of a linear differential equation both in Chapter 3 and Chapter 4.

3.1 Elementary Properties

To begin the study of linear differential equations, consider the familiar *linear scalar differential equation*

$$\dot{x} = a(t)x + h(t). \tag{3.1}$$

In this equation, x is a real number and both $a(t)$ and $h(t)$ are continuous real-valued functions on some open interval I. In other words, $d = 1$, $D = I \times \mathbb{R}$, and $f(t, x) = a(t)x + h(t)$. This equation was already encountered in the proof of Gronwall's inequality but was not fully analyzed there. It is worth some additional attention because the general goal is to extend its properties to higher dimensional linear differential equations.

Clearly, $a(t)x + h(t)$ is continuous and locally Lipschitz on D, so solutions exist and are unique. Furthermore, the global solution $x(t, \tau, \xi)$ can be explicitly determined for this equation.

Let $g(t)$ be any function satisfying $\dot{g}(t) = a(t)$ on I. For example, $g(t) = \int_\tau^t a(s)\,ds$ is such a function when τ is fixed in I. Set $\mu(t) = e^{-g(t)}$. Since $\mu(t) > 0$ on I, the function $x(t)$ is a solution of *(3.1)* if and only if it is a solution of

$$\mu(t)\dot{x} - \mu(t)a(t)x = \mu(t)h(t).$$

Let $x(t)$ be a maximal solution of *(3.1)*. Then the left-hand side of the above equation is the derivative of $\mu(t)x(t)$. Hence, given a τ at which $x(t)$ is defined, there exists a constant C such that

$$\mu(t)x(t) = C + \int_\tau^t \mu(s)h(s)\,ds$$

and

$$x(t) = \frac{1}{\mu(t)}\left(C + \int_\tau^t \mu(s)h(s)\,ds\right)$$

in the interval on which $x(t)$ is defined. Note, however, the right-hand side of the above equation is defined and differentiable on all of I. Moreover, one can easily check that it is always a solution of *(3.1)*. This proves that every maximal solution of *(3.1)* is defined on all of I and is given by the previous equation. Furthermore, the initial condition $x(\tau) = \xi$ uniquely determines C, namely, $C = \xi\mu(\tau)$. The right-hand side of the above formula for $x(t)$ can now easily be rewritten in the following form:

$$x(t, \tau, \xi) = \exp\left(\int_\tau^t a(s)\,ds\right)\xi + \int_\tau^t \exp\left(\int_s^t a(w)\,dw\right)h(s)\,ds. \tag{3.2}$$

In this section, the generalization of *(3.1)* to higher dimensions will be studied. The property of $f(t, x) = a(t)x + h(t)$ that is used to generalize *(3.1)* to $\mathbb{R}^d$ is the linearity in x of $f(t, x) - f(t, 0) = a(t)x$ for fixed t. In this more general setting, an analogue of *(3.2)* will be established in the next section.

Recall that a function $T : \mathbb{R}^n \to \mathbb{R}^m$ is called a *linear transformation* or *linear mapping* if for all $\mathbf{x}$, $\mathbf{y} \in \mathbb{R}^n$ and α, $\beta \in \mathbb{R}$

$$T(\alpha\mathbf{x} + \beta\mathbf{y}) = \alpha T(\mathbf{x}) + \beta T(\mathbf{y}).$$

For example, if A is an $m \times n$ matrix, then

$$T(\mathbf{x}) = A\mathbf{x}$$

defines a linear transformation from $\mathbb{R}^n$ to $\mathbb{R}^m$, where $A\mathbf{x}$ denotes the usual product of an $m \times n$ matrix times an $n \times 1$ matrix (column vector). There is a slight abuse of notation here because the vector $\mathbf{x}$ has been written horizontally in expressions like $T(\mathbf{x})$ and it now must be written vertically in the expression $A\mathbf{x}$. It would, however, be a nuisance to think of a vector as a $1 \times n$ matrix and always take the transpose of it when the $n \times 1$ form of it was used by writing $\mathbf{x}^t$. Instead, we will let this abuse persist, since it will usually be clear from the context whether a vector is being viewed as a horizontal or vertical vector.

Conversely, if $T : \mathbb{R}^n \to \mathbb{R}^m$ is a linear transformation, then $T(\mathbf{x})$ can be expressed as

$$T(\mathbf{x}) = A\mathbf{x}$$

by letting A be the $m \times n$ matrix whose columns are the coordinates of the images in the standard basis of $\mathbb{R}^n$. In other words, let

$$A = [T(\mathbf{e}_1), \ldots, T(\mathbf{e}_d)] = \begin{bmatrix} a_{11} & a_{12} & \cdots & a_{1n} \\ a_{21} & a_{22} & \cdots & a_{2n} \\ & & \vdots & \\ \cdots & \cdots & & \cdots \\ a_{m1} & a_{m2} & \cdots & a_{mn} \end{bmatrix},$$

where $T(\mathbf{e}_j)$ is the image of the standard basis vector $\mathbf{e}_j$ written as a column vector. It follows that

$$T(\mathbf{x}) = \sum_{j=1}^{n} x_j T(\mathbf{e}_j) = \sum_{j=1}^{n} x_j \begin{pmatrix} a_{1j} \\ a_{2j} \\ \cdots \\ a_{mj} \end{pmatrix}$$

and the range of T is the span of the columns of A (the column space of A). The ith coordinate of $T(\mathbf{x})$ is then given by

$$\left(T(\mathbf{x})\right)_i = \sum_{j=1}^{n} a_{ij} x_j = (A\mathbf{x})_i$$

and $T(\mathbf{x}) = A\mathbf{x}$ as claimed. In this paragraph, the entries of column vectors had parentheses around them and the entries of the matrix A were between brackets. To help distinguish between a matrix and a column vector, this notational device will be used consistently henceforth.

The set of all $m \times n$ matrices is nothing more than $\mathbb{R}^{mn}$, where the coordinates have been arranged in a rectangle instead of a row or column. Consequently, there are open, closed, and compact sets of matrices; functions of matrices; distance between matrices; and norms of matrices. The first ingredient needed to apply these topological ideas to matrices is a convenient norm for matrices.

Proposition 3.1 *If $\|\cdot\|_a$ and $\|\cdot\|_b$ are norms on $\mathbb{R}^m$ and $\mathbb{R}^n$, respectively, then*

$$\|A\| = \sup\left\{\|Ax\|_a : \|\mathbf{x}\|_b = 1\right\}$$

defines a norm on the space of $m \times n$ matrices and satisfies.

$$\|A\mathbf{x}\|_a \leq \|A\|\,\|\mathbf{x}\|_b.$$

Proof. First, note that $A = 0$ if and only if $A\mathbf{x} = 0$ for all $\mathbf{x}$ such that $\|\mathbf{x}\|_b = 1$. It follows that $A = 0$ if and only if $\|A\| = 0$.

For $\alpha \in \mathbb{R}$,

$$\begin{aligned} \|\alpha A\| &= \sup\left\{\|\alpha A\mathbf{x}\|_a : \|\mathbf{x}\|_b = 1\right\} \\ &= \sup\left\{|\alpha|\,\|A\mathbf{x}\|_a : \|\mathbf{x}\|_b = 1\right\} \\ &= |\alpha| \sup\left\{\|A\mathbf{x}\|_a : \|\mathbf{x}\|_b = 1\right\} \\ &= |\alpha|\,\|A\|. \end{aligned}$$

Let B be another $m \times n$ matrix. Then,

$$\begin{aligned} \|A+B\| &= \sup\left\{\left\|(A+B)\mathbf{x}\right\|_a : \|\mathbf{x}\|_b = 1\right\} \\ &\leq \sup\left\{\|A\mathbf{x}\|_a + \|B\mathbf{x}\|_a : \|\mathbf{x}\|_b = 1\right\} \\ &\leq \sup\left\{\|A\mathbf{x}\|_a : \|\mathbf{x}\|_b = 1\right\} + \sup\left\{\|B\mathbf{x}\|_a : \|\mathbf{x}\|_b = 1\right\} \\ &= \|A\| + \|B\|. \end{aligned}$$

Hence, $\|A\| = \sup\left\{\|A\mathbf{x}\|_a : \|\mathbf{x}\|_b = 1\right\}$ defines a norm.

Finally, when $\mathbf{x} \neq 0$,

$$\begin{aligned} \|A\mathbf{x}\|_a &= \|\mathbf{x}\|_b \left\|A\left(\frac{\mathbf{x}}{\|\mathbf{x}\|_b}\right)\right\|_a \\ &\leq \|\mathbf{x}\|_b\,\|A\| \\ &= \|A\|\,\|\mathbf{x}\|_b \end{aligned}$$

because

$$\left\|\frac{\mathbf{x}}{\|\mathbf{x}\|_b}\right\|_b = 1,$$

and the proof is complete. □

When the usual norm $|\mathbf{x}| = \sum_i |x_i|$ is being used on both $\mathbb{R}^m$ and $\mathbb{R}^n$, the resulting matrix norm will be denoted by $|A|$. The following fact about $|A|$ will be occasionally useful.

Proposition 3.2 *Let A be a real $m \times n$ matrix and let $\mathbf{e}_1, \ldots, \mathbf{e}_n$ be the standard basis for $\mathbb{R}^n$. Then,*

$$|A| = \max\left\{|A\mathbf{e}_j| : 1 \le j \le n\right\}.$$

Proof. Let $M = \max\left\{|A\mathbf{e}_j| : 1 \le j \le n\right\}$ and note that $M \le |A|$ because $|\mathbf{e}_j| = 1$. If $|\mathbf{x}| = 1$, then

$$\begin{aligned} |A\mathbf{x}| &= \left|\sum_{j=1}^{n} x_j A\mathbf{e}_j\right| \\ &\le \sum_{j=1}^{n} |x_j|\,|A\mathbf{e}_j| \\ &\le M\sum_{j=1}^{n} |x_j| = M, \end{aligned}$$

proving that $|A| \le M$. □

It is easy to see that the set of linear maps from $\mathbb{R}^n$ to $\mathbb{R}^m$ also form a vector space by setting $(S+T)(x) = S(x) + T(x)$ and $(\alpha T)(x) = \alpha T(x)$. The above process for associating a matrix with a linear map T is a natural one-to-one correspondence between the linear transformations $T : \mathbb{R}^n \to \mathbb{R}^m$ and the $m \times n$ matrices. It is easy to check that this correspondence is, in fact, an isomorphism between vector spaces. Thus for all practical purposes the space of linear maps from $\mathbb{R}^n$ to $\mathbb{R}^m$ is the same as the space of $m \times n$ matrices. In particular, the previous proposition can be used to define a norm on the vector space of linear transformations by setting $\|T\| = \|A\|$. Therefore, the topological concepts—open, closed, compact, and continuous—make sense for the vector space of linear maps from $\mathbb{R}^n$ to $\mathbb{R}^m$. Because all norms on $\mathbb{R}^{mn}$ are equivalent, these topological properties are independent of the norm used.

Since most of the time, m and n will be equal to d, for convenience the vector space of $d \times d$ matrices with real entries will be denoted by $\mathcal{M}_d(\mathbb{R})$. It is nothing more than $\mathbb{R}^{d^2}$ with the additional structure of matrix multiplication. Proposition 3.1 provides a useful norm on the vector space $\mathcal{M}_d(\mathbb{R})$ that is related to the matrix structure. Consequently, only norms given by *Proposition 3.1* with the additional restriction that $\|\cdot\|_a = \|\cdot\|_b$ will be used on $\mathcal{M}_d(\mathbb{R})$, so that $\|A\| = \sup\left\{\|Ax\|_a : \|\mathbf{x}\|_a = 1\right\}$. In this context, there is another important norm inequality, namely,

Remark *Let $\|\cdot\|_a$ be a norm on $\mathbb{R}^d$ and let A and B be $d \times d$ real matrices. Then,*

$$\|AB\| \le \|A\|\,\|B\|.$$

Proof. From $\|Ax\|_a \leq \|A\| \, \|x\|_a$, it follows that $\|AB\mathbf{x}\|_a \leq \|A\| \, \|B\| \, \|\mathbf{x}\|_a = \|A\| \, \|B\|$ when $\|\mathbf{x}\|_a = 1.$, Hence,

$$\|AB\| = \sup\left\{\|AB\mathbf{x}\|_a : \|\mathbf{x}\|_a = 1\right\} \leq \|A\| \, \|B\|. \; \square$$

Let $D = I \times \mathbb{R}^d$, where I is an interval in $\mathbb{R}$, and consider the differential equation $\dot{\mathbf{x}} = \mathbf{f}(t, \mathbf{x})$, where as usual $\mathbf{f} : D \to \mathbb{R}^d$ is continuous. Suppose that for each fixed t in I, the function $T(\mathbf{x}) = \mathbf{f}(t, \mathbf{x}) - \mathbf{f}(t, \mathbf{0})$ is linear in $\mathbf{x}$. Hence, for each $t \in I$, there exists a $d \times d$ matrix $A(t) = \left(a_{ij}(t)\right)$ such that $\mathbf{f}(t, \mathbf{x}) - \mathbf{f}(t, \mathbf{0}) = A(t)\mathbf{x}$. Since $a_{ij}(t)$ is the ith coordinate of $\mathbf{f}(t, \mathbf{e}_j) - \mathbf{f}(t, \mathbf{0})$, the function $a_{ij}(t)$ is continuous on I, and $A(t)$ is a continuous function from I into $\mathcal{M}_d(\mathbb{R})$. Let

$$\mathbf{f}(t, \mathbf{0}) = \mathbf{h}(t) = \begin{pmatrix} h_1(t) \\ h_2(t) \\ \cdots \\ h_d(t) \end{pmatrix}.$$

Clearly, $\mathbf{h}(t)$ is a continuous $\mathbb{R}^d$ valued function on I. The equation $\mathbf{f}(t, \mathbf{x}) - \mathbf{f}(t, \mathbf{0}) = A(t)\mathbf{x}$ can be written $\mathbf{f}(t, \mathbf{x}) = A(t)\mathbf{x} + \mathbf{h}(t)$ and the differential equation has the form

$$\dot{\mathbf{x}} = A(t)\mathbf{x} + \mathbf{h}(t) \tag{3.3}$$

or

$$\begin{pmatrix} \dot{x}_1 \\ \dot{x}_2 \\ \vdots \\ \dot{x}_d \end{pmatrix} = \begin{bmatrix} a_{11}(t) & a_{12}(t) & \cdots & a_{1d}(t) \\ a_{21}(t) & a_{22}(t) & \cdots & a_{2d}(t) \\ \vdots & \vdots & \vdots & \vdots \\ a_{d1}(t) & a_{d2}(t) & \vdots & a_{dd}(t) \end{bmatrix} \begin{pmatrix} x_1 \\ x_2 \\ \vdots \\ x_d \end{pmatrix} + \begin{pmatrix} h_1(t) \\ h_2(t) \\ \vdots \\ h_d(t). \end{pmatrix}$$

Equation *(3.3)* is called a *linear differential equation* or a *system of linear differential equations*. When it is known that $\mathbf{h}(t)$ is not identically $\mathbf{0}$ on I, the linear differential equation is said to be *nonhomogeneous*. When $\mathbf{h}(t) \equiv \mathbf{0}$ on I,

$$\dot{\mathbf{x}} = A(t)\mathbf{x} \tag{3.4}$$

is called a *homogeneous differential equation*. Throughout this section, it will be assumed that $A(t)$ and $\mathbf{h}(t)$ are continuous on an open interval I.

Theorem 3.3 *Let $A : I \to \mathcal{M}_d(\mathbb{R})$ and $\mathbf{h} : I \to \mathbb{R}^d$ be continuous on the open interval I. For any $(\tau, \boldsymbol{\xi}) \in D = I \times \mathbb{R}^d$, the initial-value problem*

$$\begin{aligned} \dot{\mathbf{x}} &= A(t)\mathbf{x}, \\ \boldsymbol{\xi} &= \mathbf{x}(\tau) \end{aligned}$$

has a unique solution and this solution can be continued to all of I.

Proof. Let K be any compact subset of D. Because K is closed and bounded, the t-coordinates for $(t, \mathbf{x}) \in K$ are a closed and bounded subset of I. In particular, there exist a, b in I, $a < b$, such that $a \leq t \leq b$ for all $(t, \mathbf{x}) \in K$. Because $A(t)$ is continuous, $|A(t)|$ is continuous and there exists a positive real number M satisfying $|A(t)| \leq M$ for $a \leq t \leq b$. Therefore, given $(t, \mathbf{x})$ and $(t, \mathbf{y}) \in K$

$$\big|A(t)\mathbf{x} + \mathbf{h}(t) - \big(A(t)\mathbf{y} + \mathbf{h}(t)\big)\big| \leq \big|A(t)\big|\,|\mathbf{x} - \mathbf{y}| \leq M|\mathbf{x} - \mathbf{y}|$$

and $A(t)\mathbf{x} + \mathbf{h}(t)$ is locally Lipschitz, which implies the uniqueness of solutions. The continuation part of the theorem follows immediately from *Theorem 2.10* because

$$\big|A(t)\mathbf{x} + \mathbf{h}(t)\big| \leq \big|A(t)\big|\,|\mathbf{x}| + \big|\mathbf{h}(t)\big|,$$

and $|A(t)|$ and $|\mathbf{h}(t)|$ are continuous on I. □

Let α and β be real numbers, let $\varphi(t)$ be a solution of $\dot{\mathbf{x}} = A(t)\mathbf{x} + \mathbf{h}(t)$, and let $\psi(t)$ be a solution of $\dot{\mathbf{x}} = A(t)\mathbf{x} + \mathbf{g}(t)$. Then clearly $\alpha\varphi + \beta\psi$ is a solution of $\dot{\mathbf{x}} = A(t)\mathbf{x} + \alpha\mathbf{h}(t) + \beta\mathbf{g}(t)$. This is called the *principle of superposition* and the next remark readily follows from it.

Remark *The solutions of* $\dot{\mathbf{x}} = A(t)\mathbf{x}$ *form a vector space* V *over* $\mathbb{R}$. *If* ψ_0 *is a particular solution of*

$$\dot{\mathbf{x}} = A(t)\mathbf{x} + \mathbf{h}(t), \tag{3.3}$$

then the set of all solutions of (3.3) is precisely

$$\{\varphi + \psi_0 : \varphi \in V\}.$$

Proof. When $\mathbf{h}(t)$ and $\mathbf{g}(t)$ are identically zero, the principle of superposition says that linear combinations of solutions of $\dot{\mathbf{x}} = A(t)\mathbf{x}$ are again solutions of the same equation. In particular, the set of all solutions of it form a vector space.

Let ψ be any solution of *(3.3)*. Then $\varphi = \psi - \psi_0$ is a solution of $\dot{\mathbf{x}} = A(t)\mathbf{x}$ and in V by superposition. Obviously, $\psi = \varphi + \psi_0$. Conversely, if φ is in V, then $\varphi + \psi_0$ is a solution of *(3.3)* again by superposition. □

The content of the previous remark is that one should first try to understand the vector space V of solutions of the homogeneous equations *(3.4)* before turning to the nonhomogeneous equation. Specifically, we want to calculate the dimension of V, the vector space of solutions to *(3.4)*, understand what constitutes a basis of solutions, and find a method for obtaining particular solutions of *(3.3)*. This program will provide a good theoretical picture of the solutions of *(3.3)*.

EXERCISES

1. Let $A(t)$ be a continuous matrix-valued function on the interval $0 < t < \infty$ and let $\mathbf{x}(t)$ be a nonzero solution of $\dot{\mathbf{x}} = A(t)\mathbf{x}$. Show that if $|A(t)|$ is bounded on the interval $1 \leq t < \infty$, then

$$\frac{\log |x(t)|}{t}$$

is also bounded on the interval $1 \leq t < \infty$.

2. Let $\mathbf{x}(t)$ be a solution of $\dot{\mathbf{x}} = A(t) + \mathbf{h}(t)$, where $A(t)$ and $\mathbf{h}(t)$ are continuous on an open interval $0 < t < \infty$. Prove that $|\mathbf{x}(t)|$ is bounded for $t \geq 1$, if both

$$\int_1^\infty |A(t)|dt < \infty \text{ and } \int_1^\infty |\mathbf{h}(t)|dt < \infty.$$

3. For each of the following, construct a sequence of $d \times d$ invertible (Invertible is defined on page 93 in the next section.) real matrices A_m having the specified properties:

   (a) As m goes to infinity, Det A_m goes to 0 and A_m converges to A with $|A| \neq 0$.

   (b) As m goes to infinity, Det A_m goes to 0 and $|A_m|$ goes to ∞.

   (c) As m goes to infinity, one of the rows of A_m goes to $\mathbf{0}$ and Det A_m goes to ∞.

   (d) As m goes to infinity, $|A_m^{-1}|$ goes to 0.

4. Let A_m be a sequence of invertible real $d \times d$ real matrices such that $|A_m|$ is a bounded sequence.

   (a) Show that if one row of A_m goes to $\mathbf{0}$ as m goes to infinity, then Det A_m goes to 0 as m goes to infinity.

   (b) Show that there exists a positive real number α such that $|A_m^{-1}| \geq \alpha$ for all m.

5. Let $A : I \to \mathcal{M}_d(\mathbb{R})$ and $B : I \to \mathcal{M}_d(\mathbb{R})$ be differentiable functions on the interval I, that is, every entry of $A(t)$ and of $B(t)$ is just a real-valued differentiable function on an interval. Prove that

$$\frac{dA(t)B(t)}{dt} = \dot{A}(t)B(t) + A(t)\dot{B}(t).$$

3.2 Fundamental Matrix Solutions

To study the vector space V of solutions of the differential equation

$$\dot{\mathbf{x}} = A(t)\mathbf{x}, \tag{3.4}$$

it is convenient to introduce the matrix differential equation

$$\dot{X} = A(t)X, \tag{3.5}$$

where X is a $d \times d$ variable matrix and $\dot{X}(t)$ is the $d \times d$ matrix obtained by differentiating every entry of a curve $X(t)$ in $\mathcal{M}_d(\mathbb{R})$. It is a system of d^2 linear differential equations.

Solutions of *(3.5)* have several elementary properties that are used repeatedly. They are recorded in the next remark and the proposition following it.

Remark *The following hold:*

(a) $X(t)$ is a solution of the matrix differential equation (3.5) if and only if every column of $X(t)$ is a solution of $\dot{\mathbf{x}} = A(t)\mathbf{x}$.

(b) The solutions of the matrix differential equation (3.5) are defined on I.

(c) If $X(t)$ is a solution of the matrix differential equation (3.5), then $X(t)\mathbf{v}$ is a solution of $\dot{\mathbf{x}} = A(t)\mathbf{x}$ for each $\mathbf{v} \in \mathbb{R}^d$.

(d) If $X(t)$ is a solution of the matrix differential equation (3.5), then $X(t)B$ is also a solution of the same matrix differential equation for each $B \in \mathcal{M}_d(\mathbb{R})$.

Proof. Part (a) follows from the observation that the jth column of $A(t)X$ is just $A(t)$ times the jth column of $X(t)$. Part (b) is an immediate consequence of part (a). Since $X(t)\mathbf{v}$ is a linear combination of the columns of $X(t)$, $X(t)\mathbf{v}$ is a linear combination of solutions of $\dot{\mathbf{x}} = A(t)\mathbf{x}$ by part (a). Thus $X(t)\mathbf{v}$ is a solution of $\dot{\mathbf{x}} = A(t)\mathbf{x}$ because its solutions form a vector space. Finally, part (d) follows form (a) and (c). $\square$

Proposition 3.4 *Let $X(t)$ be a solution of $\dot{X} = A(t)X$. The determinant of $X(t)$, written* $\mathrm{Det}\left[X(t)\right]$, *either vanishes identically on I or is never zero on I.*

Proof. It suffices to show that if $\mathrm{Det}\left[X(\tau)\right] = 0$ for some $\tau \in I$, then $\mathrm{Det}\left[X(t)\right] \equiv 0$ on I. Recall that for an $d \times d$ matrix B, $\mathrm{Det}\left[B\right] = 0$ if and only if there exists a nonzero vector $\mathbf{v}$ such that $B\mathbf{v} = \mathbf{0}$. So there exists $\mathbf{v} \neq \mathbf{0}$ such that $X(\tau)\mathbf{v} = \mathbf{0}$. Set $\varphi(t) = X(t)\mathbf{v}$. Then $\varphi(t)$ is a solution of $\dot{x} = A(t)\mathbf{x}$ satisfying $\varphi(\tau) = \mathbf{0}$. Since $\mathbf{x}(t) \equiv \mathbf{0}$ on I is also a solution of $\dot{\mathbf{x}} = A(t)\mathbf{x}$, by the uniqueness of solutions of *(3.4)*, $X(t)\mathbf{v} = \varphi(t) \equiv \mathbf{0}$ on I and $\mathrm{Det}\left[X(t)\right]$ vanishes on I. $\square$

A square matrix B is an *invertible matrix* or *nonsingular matrix* if there exists another matrix B^{-1} of the same size as B such that $BB^{-1} = B^{-1}B = \mathrm{I}$, the $d \times d$ identity matrix having ones on the diagonal and zeros off it. Recall that a matrix B is invertible if and only if $\operatorname{Det} B \neq 0$ if and only if $B\mathbf{x} \neq \mathbf{0}$ when $\mathbf{x} \neq \mathbf{0}$.

The interesting solutions of *(3.5)* are those for which $\operatorname{Det}\left[X(t)\right]$ is never zero on I, and hence $X(t)$ is invertible for all $t \in I$. This idea is worth naming. A *fundamental matrix solution* of *(3.5)* is a solution $X(t)$ such that $X(t)$ is nonsingular or invertible for all $t \in I$. It will be apparent after the next two theorems that $X(t)$ is a fundamental matrix precisely when the columns of $X(t)$ form a basis for V.

Theorem 3.5 *If $A : I \to \mathcal{M}_d(\mathbb{R})$ is continuous on the open interval I, then a fundamental matrix solution of $\dot{X} = A(t)X$ exists. If $X(t)$ is any fundamental matrix solution, then*

$$X(t)\big(X(\tau)\big)^{-1}\boldsymbol{\xi}$$

is the unique solution of the initial-value problem

$$\begin{aligned} \dot{\mathbf{x}} &= A(t)\mathbf{x}, \\ \boldsymbol{\xi} &= \mathbf{x}(\tau). \end{aligned}$$

Furthermore,

$$\mathbf{x}(t, \tau, \boldsymbol{\xi}) = X(t)\big(X(\tau)\big)^{-1}\boldsymbol{\xi}$$

and $D_f = I \times I \times \mathbb{R}^d$.

Proof. Let $\mathbf{v}_1, \ldots, \mathbf{v}_d$ be a basis for $\mathbb{R}^d$, let $\tau \in I$, and let $\boldsymbol{\varphi}_i(t)$ be the solution of the initial-value problem

$$\begin{aligned} \dot{\mathbf{x}} &= A(t)\mathbf{x} \\ \mathbf{v}_i &= \mathbf{x}(\tau), \end{aligned}$$

for $1 \leq i \leq d$. Let $X(t)$ be the matrix solution whose columns are $\boldsymbol{\varphi}_1(t), \ldots, \boldsymbol{\varphi}_d(t)$. Then clearly, $\operatorname{Det}\left[X(\tau)\right] \neq 0$, and by the previous lemma, $X(t)$ is a fundamental matrix solution.

Now, let $X(t)$ be any fundamental matrix solution. It follows from the Remark on page 92, that $\boldsymbol{\varphi}(t) = X(t)\big(X(\tau)\big)^{-1}\boldsymbol{\xi}$ is a solution of *(3.4)*. Moreover, it solves the required initial-value problem because $\boldsymbol{\varphi}(\tau) = X(\tau)\big(X(\tau)\big)^{-1}\boldsymbol{\xi} = \boldsymbol{\xi}$. Hence, $X(t)\big(X(\tau)\big)^{-1}\boldsymbol{\xi} = \mathbf{x}(t, \tau, \boldsymbol{\xi})$. Since t and τ are arbitrary real numbers in I and $\boldsymbol{\xi}$ is an arbitrary point in $\mathbb{R}^d$, the set $D_{\mathbf{f}}$ must equal $I \times I \times \mathbb{R}^d$ □

Theorem 3.6 *If $A : I \to \mathcal{M}_d(\mathbb{R})$ is continuous on the open interval I, then the vector space V of solutions of $\dot{\mathbf{x}} = A(t)\mathbf{x}$ is d-dimensional. Furthermore, the solutions $\boldsymbol{\varphi}_1, \ldots, \boldsymbol{\varphi}_d$ form a basis for V if and only if for some τ in I, the vectors $\boldsymbol{\varphi}_1(\tau), \ldots, \boldsymbol{\varphi}_n(\tau)$, form a basis for $\mathbb{R}^d$.*

Proof. Let $X(t)$ be a fundamental matrix solution and define a map $T : \mathbb{R}^d \to V$ by $T(\mathbf{v}) = X(t)\mathbf{v}$. Obviously, T is linear. From *Theorem 3.5*, it follows that T is onto. Suppose $T(\mathbf{v}) = X(t)\mathbf{v} \equiv 0$ on I. Since $X(t)$ has an inverse, $\mathbf{v} = \mathbf{0}$ and T is a vector space isomorphism. Therefore, the dimension of V is d.

For the second part, let $\varphi_1, \ldots, \varphi_d$ be solutions of *(3.4)*. Because the dimension of both V and $\mathbb{R}^d$ is d, it suffices to show that $\varphi_1, \ldots, \varphi_d$ are linearly dependent in V if and only if $\varphi_1(\tau), \ldots, \varphi_d(\tau)$ are linearly dependent in $\mathbb{R}^d$.

If there exist constants not all 0 so that

$$c_1\varphi_1(t) + \cdots + c_d\varphi_d(t) \equiv \mathbf{0},$$

on I, then, in particular,

$$c_1\varphi_1(\tau) + \cdots + c_d\varphi_d(\tau) = \mathbf{0}.$$

Conversely, suppose $\varphi_1(\tau), \ldots, \varphi_d(\tau)$ are linearly dependent, so the above equation at τ holds for some set of constants $c_1, \ldots, c_d$, which are not all zero. Then, by the uniqueness of solutions

$$\mathbf{0} \equiv c_1\varphi_1(t) + \cdots + c_d\varphi_n(t)$$

on I and the the solutions $\varphi_1, \ldots, \varphi_d$ are linearly dependent, completing the proof. □

Let $X(t)$ be any fundamental matrix solution of *(3.5)*, and let $s \in I$. Then $X(t,s) = X(t)\big(X(s)\big)^{-1}$ is another matrix solution, and $X(s,s) = \mathrm{I}$, the $d \times d$ identity matrix. Since $\mathrm{Det}\,\big[X(s,s)\big] = 1$, $X\big(t,s\big)$ is also a fundamental matrix solution for every value of s and is called the *principal matrix solution* of *(3.5)* at s.

The focus of the discussion has been on the connections between $\dot{\mathbf{x}} = A(t)\mathbf{x}$ and $\dot{X} = A(t)X$, but the latter can be examined in its own right. Because $\|AB\| \le \|A\|\,\|B\|$ and $A(t)$ is continuous, it follows that

$$\big|A(t)X - A(t)Y\big| \le \big|A(t)\big|\,\big|X - Y\big|,$$

and $A(t)X$ is locally Lipschitz. Now, *Theorem 1.23* applies and $X(t,\tau,B)$, the solution of *(3.5)* that equals $B \in \mathcal{M}_n(\mathbb{R})$ when $t = \tau$, is a continuous function on $I \times I \times \mathcal{M}_d(\mathbb{R})$. In fact, if $X(t)$ is any fundamental matrix solution,

$$\begin{aligned} X(t,\tau,B) &= X(t)\big(X(\tau)\big)^{-1}B \\ &= X(t,\tau)B. \end{aligned}$$

It follows that the principal matrix solution is given by

$$X(t,s) = X(t,s,\mathrm{I})$$

and that $X(t,s)$ is continuous on $I \times I$.

The next result establishes an important feature of linear systems. Namely, if a basis of solutions for the homogeneous problem is known, then the general solutions of the nonhomogeneous problem can be calculated. The formula depends on the continuity of the principal matrix solution $X(t,s)$on $I \times I$.

Theorem 3.7 (Variation of constants) *Let $A : I \to \mathcal{M}_d(\mathbb{R})$ and $\mathbf{h} : I \to \mathbb{R}^d$ be continuous on the open interval I. Then the unique solution of the initial-value problem $\dot{\mathbf{x}} = A(t)\mathbf{x} + \mathbf{h}(t)$, and $\mathbf{x}(\tau) = \boldsymbol{\xi}$ is given by*

$$\mathbf{x}(t,\tau,\boldsymbol{\xi}) = X(t,\tau)\boldsymbol{\xi} + \int_\tau^t X(t,s)\mathbf{h}(s)\,ds, \tag{3.6}$$

where $X(t,s)$ is the principal matrix solution of $\dot{X} = A(t)X$.

Proof. One can simply check that this formula works, but it is worth understanding its origin. Let $X(t)$ be a fundamental matrix solution. For any constant vector $\mathbf{v}$, the curve $X(t)\mathbf{v}$ is a solution of $\dot{\mathbf{x}} = A(t)\mathbf{x}$. Now "vary the constant" vector $\mathbf{v}$ and look for a solution of $\dot{\mathbf{x}} = A(t)\mathbf{x} + \mathbf{h}(t)$ of the form $\varphi(t) = X(t)\mathbf{v}(t)$. This amounts to requiring that

$$\dot{\varphi}(t) = \dot{X}(t)\mathbf{v}(t) + X(t)\dot{\mathbf{v}}(t) = A(t)X(t)\mathbf{v}(t) + \mathbf{h}(t).$$

Since $\dot{X}(t) = A(t)X(t)$, the above equation reduces to

$$\dot{\mathbf{v}}(t) = \big(X(t)\big)^{-1}\mathbf{h}(t)$$

or

$$\mathbf{v}(t) = C + \int_\tau^t \big(X(s)\big)^{-1}\mathbf{h}(s)\,ds.$$

Thus

$$\begin{aligned} \varphi(t) &= X(t)\left(C + \int_\tau^t \big(X(s)\big)^{-1}\mathbf{h}(s)\,ds\right) \\ &= X(t)C + \int_\tau^t X(t)\big(X(s)\big)^{-1}\mathbf{h}(s)\,ds. \end{aligned}$$

The initial condition forces

$$C = \big(X(\tau)\big)^{-1}\boldsymbol{\xi}$$

and produces the desired formula. □

It is worth noting that *(3.2)* is a special case of *(3.6)*. To see this, just observe that $1/\mu(t) = e^{g(t)}$ is a "fundamental matrix solution" of the scalar equation $\dot{x} = a(t)x$ and

$$X(t,\tau) = e^{\int_\tau^t a(s)\,ds}.$$

In many problems, a basis for the vector space of solutions of *(3.4)* cannot be readily determined and thus a fundamental matrix solution $X(t)$ is often not known. *Exercises 5 and 6* provide an intuitive sense of why finding fundamental matrix solutions is generally difficult. Sometimes, however, information regarding the behavior of solutions can be obtained without knowing a fundamental matrix solution. *Theorem 3.8* is one tool that can be used to glean information about the solutions of $\dot{\mathbf{x}} = A(t)\mathbf{x}$ and *Exercise 4* is an illustration of how it can be used.

Theorem 3.8 (Abel's formula) *Let $X(t)$ be a fundamental matrix solution of (3.5), then for any τ in I,*

$$\text{Det}\left[X(t)\right] = \text{Det}\left[X(\tau)\right] \exp\left(\int_{\tau}^{t} \text{Tr}\left[A(s)\right] ds\right) \tag{3.7}$$

where $\text{Tr}\left[A(s)\right] = \sum_{j=1}^{d} a_{jj}(s) =$ *trace of* $A(s)$.

Proof. It suffices to show that $\text{Det}\left[X(t)\right]$ is a solution of the scalar equation $\dot{y} = \text{Tr}\left[A(t)\right]y$. Let $\varphi_{ij}(t)$ denote the ijth entry of $X(t)$. From the definition of the determinant and the product rule, it follows that

$$\frac{d}{dt}\text{Det}\left[X(t)\right] = \sum_{i=1}^{d} \text{Det}\begin{bmatrix} \varphi_{11}(t) & \cdots & \varphi_{1d}(t) \\ \vdots & \ddots & \vdots \\ \dot{\varphi}_{i1}(t) & \cdots & \dot{\varphi}_{id}(t) \\ \vdots & \ddots & \vdots \\ \varphi_{d1}(t) & \cdots & \varphi_{dd}(t) \end{bmatrix}.$$

Using $\dot{\varphi}_{ij}(t) = \sum_{k=1}^{d} a_{ik}(t)\varphi_{kj}(t)$, the ith row of derivatives can be replaced by the linear sum of rows $\sum_{k=1}^{d} a_{ik}(t)\big(\varphi_{k1}(t), \ldots, \varphi_{kd}(t)\big)$.

Because the determinant is linear in each row,

$$\frac{d\text{Det}\left[X(t)\right]}{dt} = \sum_{i=1}^{d}\sum_{k=1}^{d} a_{ik}(t)\text{Det}\begin{bmatrix} \varphi_{11}(t) & \cdots & \varphi_{1d}(t) \\ \vdots & \ddots & \vdots \\ \varphi_{k1}(t) & \cdots & \varphi_{kd}(t) \\ \vdots & \ddots & \vdots \\ \varphi_{d1}(t) & \cdots & \varphi_{dd}(t) \end{bmatrix}.$$

(Here, as above, the row written in the middle denotes the generic ith row.)

When $k \neq i$, the above matrix will have two identical rows and the determinant will be 0. When $k = i$, this matrix will be $X(t)$ again. Therefore,

$$\frac{d\text{Det}\left[X(t)\right]}{dt} = \sum_{i=1}^{d} a_{ii}(t)\text{Det}\left[X(t)\right] = \text{Tr}\left[A(t)\right]\text{Det}\left[X(t)\right]$$

and the proof is finished. □

EXERCISES

1. Show that the definition of the principal matrix solution defined by the equation $X(t,s) = X(t)\big(X(s)\big)^{-1}$ is, in fact, independent of the fundamental matrix $X(t)$ and that

$$X(t,s)X(s,\tau) = X(t,\tau)$$

for $t, s, \tau \in I$.

2. Let $X(t)$ be a fundamental matrix solution of $\dot{X} = A(t)X$.

   (a) Show that $X(t)^{-1}$ is a fundamental matrix solution of $\dot{Y} = -YA(t)$.

   (b) Show that $X(t)^{\mathrm{t}}$ is a fundamental matrix solution of $\dot{Y} = YA(t)^{\mathrm{t}}$ where B^{t} denotes the usual transpose of a matrix B. (See page 137 for definition.)

3. Consider $\dot{\mathbf{x}} = A(t)\mathbf{x} + \mathbf{g}(t)$, where

$$A = \begin{bmatrix} 3 & 1 \\ 0 & 3 \end{bmatrix} \text{ and } \mathbf{g}(t) = \begin{pmatrix} \sin t \\ \cos t \end{pmatrix}.$$

Verify that

$$X(t) = \begin{bmatrix} e^{3t} & te^{3t} \\ 0 & e^{3t} \end{bmatrix}$$

is a fundamental matrix solution of $\dot{\mathbf{x}} = A\mathbf{x}$. Find a solution to the initial-value problem

$$\dot{\mathbf{x}} = A(t)\mathbf{x} = \mathbf{g}(t) \text{ and } \mathbf{x}(0) = \begin{pmatrix} 1 \\ -1 \end{pmatrix}.$$

4. Suppose $A : (0,\infty) \to \mathcal{M}_d(\mathbb{R})$ is continuous. Prove the following: If

$$\int_1^\infty \operatorname{Tr} A(t)dt = \infty$$

then there exists a solution $\mathbf{x}(t)$ of $\dot{\mathbf{x}} = A(t)\mathbf{x}$ such that $|\mathbf{x}(t)|$ is unbounded for $t \geq 1$.

5. Let $X(t)$ be a $d \times d$ matrix whose entries are differentiable functions on the interval $a < t < b$. Show that if $\operatorname{Det} X(t) \neq 0$ when $a < t < b$, then $X(t)$ is a fundamental matrix solution of some $\dot{X} = A(t)X$ on $a < t < b$.

6. Let $X(t)$ be a $d \times d$ matrix of rational functions of the real variable t. In other words, $x_{ij}(t) = p_{ij}(t)/q_{ij}(t)$ where both $p_{ij}(t)$ and $q_{ij}(t)$ are polynomials with real coefficients and $q_{ij}(t)$ is not the zero polynomial. Show that if $\operatorname{Det} X(\tau) \neq 0$ for some τ, then there exists a such that $\operatorname{Det} X(t) \neq 0$ for $t > a$ and a continuous matrix valued function $A(t)$ on $a < t$ such that $X(t)$ is a fundamental matrix solution of $\dot{X} = A(t)X$ on $a < t$.

3.3 Higher Order Linear Differential Equations

The linear differential equations first encountered are usually of the form

$$x^{(n)} + a_n(t)x^{(n-1)} + \cdots + a_1(t)x = g(t) \tag{3.8}$$

where $x^{(i)}$ denotes the ith derivative of y with respect to t. How does this equation fit into the general theory of linear equations developed in the previous section? To answer this question, assume that $a_1(t)$, ..., $a_n(t)$, and $g(t)$ are continuous on an open interval I and place *(3.8)* in standard form. That is, let

$$A(t) = \begin{bmatrix} 0 & 1 & 0 & \cdots & 0 \\ 0 & 0 & 1 & \cdots & 0 \\ \vdots & \vdots & \vdots & \ddots & \vdots \\ -a_1(t) & -a_2(t) & -a_3(t) & \cdots & -a_n(t) \end{bmatrix} \tag{3.9}$$

and

$$\mathbf{h}(t) = \begin{pmatrix} 0 \\ \vdots \\ 0 \\ g(t) \end{pmatrix}, \tag{3.10}$$

and consider $\dot{\mathbf{x}} = A(t)\mathbf{x} + \mathbf{h}(t)$, which is the same as the following linear system of differential equations:

$$\begin{aligned} \dot{x}_1 &= x_2 \\ \dot{x}_2 &= x_3 \\ \vdots \quad & \quad \vdots \\ \dot{x}_n &= -\big(a_1(t)x_1 + a_2(t)x_2 + \cdots + a_n(t)x_n\big) + g(t). \end{aligned}$$

It follows that $\varphi(t)$ is a solution of *(3.8)* if and only if

$$\begin{pmatrix} \varphi(t) \\ \varphi^{(1)}(t) \\ \vdots \\ \varphi^{(n-1)}(t) \end{pmatrix}$$

is a solution of $\dot{\mathbf{x}} = A(t)\mathbf{x} + \mathbf{h}(t)$ with $A(t)$ and $\mathbf{h}(t)$ as above. In other words, the solutions of *(3.8)* are precisely the first coordinates of the solutions of $\dot{\mathbf{x}} = A(t)\mathbf{x} + \mathbf{h}(t)$ with $A(t)$ and $\mathbf{h}(t)$ satisfying equations *(3.9)* and *(3.10)*. As a consequence of *Theorem 3.3*, we now have the following result:

Theorem 3.9 *Let $a_1(t)$, ..., $a_n(t)$, and $g(t)$ be continuous real-valued functions on an open interval I. Given τ in I and real numbers x_0, x_1, ..., x_{n-1}, there exists a unique solution of*

$$x^{(n)} + a_n(t)x^{(n-1)} + \cdots + a_1(t)x = g(t) \tag{3.8}$$

defined on I and satisfying $x(\tau) = x_0$, $x^{(1)}(\tau) = x_1$, ..., $x^{(n-1)}(\tau) = x_{n-1}$.

When $g(t)$ is identically zero on I, the equation is

$$x^{(n)} + a_n(t)x^{)(n-1)} + \cdots + a_1(t)x = 0, \tag{3.11}$$

and its solutions clearly form a vector space. The earlier results about dimension and basis apply to the vector-valued solution of $\dot{\mathbf{x}} = A(t)\mathbf{x}$ and not directly to the real-valued solutions of *(3.8)*. The Wronskian bridges this gap. Let φ_1, ..., φ_n be a set of $(n-1)$-times differentiable functions on I. The *Wronskian* is defined by

$$W(\varphi_1, \ldots, \varphi_n)(t) = \text{Det} \begin{bmatrix} \varphi_1(t) & \cdots & \varphi_n(t) \\ \varphi_1^{(1)}(t) & \cdots & \varphi_n^{(1)}(t) \\ \vdots & \ddots & \vdots \\ \varphi_1^{(n-1)}(t) & \cdots & \varphi_n^{(n-1)}(t) \end{bmatrix}.$$

Proposition 3.10 *If there exists $\tau \in I$ for which $W(\varphi_1, \ldots, \varphi_n)(\tau) \neq 0$, then φ_1, ..., φ_n are linearly independent in the vector space of real-valued functions on I.*

Proof. Let c_1, ..., c_n be real numbers such that the function $c_1\varphi_1(t) + \cdots + c_n\varphi_n(t) \equiv 0$ on I. It must be shown that $c_1 = c_2 = \cdots = c_n = 0$. The equation $c_1\varphi_1(t) + \cdots + c_n\varphi_n(t) \equiv 0$ can be differentiated $n-1$ times to get

$$c_1\varphi_1^{(i)}(t) + \cdots + c_n\varphi_n^{(i)}(t) \equiv 0$$

on I and $0 \leq i \leq n-1$. Substituting τ and writing down all n equations gives us the system

$$\begin{aligned} c_1\varphi_1(\tau) + \cdots + \quad c_n\varphi_n(\tau) \quad &= 0 \\ c_1\varphi_1^{(1)}(\tau) + \cdots + \quad c_n\varphi_n^{(1)}(\tau) \quad &= 0 \\ \vdots \qquad\qquad\qquad \vdots \quad &\;\; \vdots \\ c_1\varphi_1^{(n-1)}(\tau) + \cdots + c_n\varphi_n^{(n-1)}(\tau) &= 0. \end{aligned}$$

If the constants c_1, ..., c_n are thought of as unknowns, then the determinant of this system of equations is $W(\varphi_1, \ldots, \varphi_n)(\tau) \neq 0$ and , $c_1 = c_2 = \cdots = c_n = 0$. This proves φ_1, ..., φ_n are linearly independent. □

To see that the converse of this lemma is false, let

$$\varphi_1(t) = \begin{cases} 0, & t \leq 0 \\ t^2, & t > 0 \end{cases} \qquad \text{and} \qquad \varphi_2(t) = \begin{cases} t^2, & t < 0 \\ 0, & t \geq 0 \end{cases}$$

and calculate $W(\varphi_1, \varphi_2)$.

The relationship between the solutions of a higher order linear homogenous differential equation and the Wronskian can now be established.

Theorem 3.11 *The solutions of*

$$x^{(n)} + a_n(t)x^{(n-1)} + \cdots + a_1(t)x = 0 \tag{3.11}$$

form a vector space of dimension n. Moreover, if $\{\varphi_1, \ldots, \varphi_n\}$ is a set of n solutions of (3.11), then the following are equivalent:

(a) $W(\varphi_1, \ldots, \varphi_n)(\tau) \neq 0$ for some $\tau \in I$;

(b) $\{\varphi_1, \ldots, \varphi_n\}$ is a basis for the vector space of solutions of (3.11);

(c) $W(\varphi_1, \ldots, \varphi_n)(t) \neq 0$ for all $t \in I$.

Proof. Let

$$X(t) = \begin{bmatrix} \varphi_1(t) & \cdots & \varphi_n(t) \\ \vdots & \ddots & \vdots \\ \varphi_1^{(n-1)}(t) & \cdots & \varphi_n^{(n-1)}(t) \end{bmatrix}$$

be a fundamental matrix solution of the corresponding $\dot{X} = A(t)X$. Clearly, every solution of *(3.11)* is a linear combination of $\varphi_1, \ldots, \varphi_n$. Since on I

$$W(\varphi_1, \ldots, \varphi_n)(t) = \text{Det}\left[X(t)\right] \neq 0,$$

the solutions $\varphi_1, \ldots, \varphi_n$ are linearly independent. This proves the first part.

To prove the second part, we will show that (a) implies (b), then (b) implies (c), and finally (c) implies (a).

For the first step, assume that $W(\varphi_1, \ldots, \varphi_n)(\tau) \neq 0$. It follows from the previous lemma that $\varphi_1, \ldots, \varphi_n$ are linearly independent. Since it was just shown that the dimension of the vector space of solutions of *(3.11)* is n, it follows that $\{\varphi_1, \ldots, \varphi_n\}$ is a basis of it.

Next, assume $\{\varphi_1, \ldots, \varphi_n\}$ is a basis and consider the matrix solution

$$X(t) = \begin{bmatrix} \varphi_1(t) & \cdots & \varphi_n(t) \\ \varphi_1^{(1)}(t) & \cdots & \varphi_n^{(1)}(t) \\ \vdots & \ddots & \vdots \\ \varphi_1^{(n-1)}(t) & \cdots & \varphi_n^{(n-1)}(t) \end{bmatrix}.$$

If $X(t)$ is not a fundamental matrix solution, then as in the proof of *Proposition 3.4*, there exists $\mathbf{v} \neq \mathbf{0}$ such that

$$X(t)\mathbf{v} \equiv \mathbf{0},$$

on I and it follows that

$$v_1\varphi_1(t) + \cdots + v_n\varphi_n(t) \equiv 0,$$

on I, contradicting the linear independence of $\varphi_1, \ldots, \varphi_n$. Therefore, $X(t)$ is a fundamental matrix solution and

$$W(\varphi_1, \ldots, \varphi_n)(t) = \text{Det}\left[X(t)\right] \neq 0$$

for all $t \in I$. The last implication is trivial. □

Corollary *Let $A(t)$ be given by equation (3.9), and let $\varphi_1(t), \ldots, \varphi_n(t)$ be solutions of $\dot{\mathbf{x}} = A(t)\mathbf{x}$. Then $\varphi_1(t), \ldots, \varphi_n(t)$ is a basis for the vector space of solutions of $\dot{\mathbf{x}} = A(t)\mathbf{x}$ if and only if the first coordinate functions of $\varphi_1(t), \ldots, \varphi_n(t)$ are a basis for the vector space of solutions of $x^{(n)} + a_n(t)x^{)n-1)} + \cdots + a_1(t)x =$* 0.

EXERCISES

1. Consider $\dot{X} = A(t)X$ on the interval $0 < t < \infty$ where

$$A(t) = \begin{bmatrix} 0 & 1 & 0 \\ 0 & 0 & 1 \\ 6t^{-3} & -6t^{-2} & 3t^{-1} \end{bmatrix}$$

   (a) Show that

$$\begin{bmatrix} t^3 & t^2 & t \\ 3t^2 & 2t & 1 \\ 6t & 2 & 0 \end{bmatrix}$$

   is a fundamental matrix solution of $\dot{X} = A(t)X$.

   (b) Calculate $X(t, s)$.

   (c) Use $X(t, s)$ to solve the third-order initial-value problem

$$\frac{d^3y}{dt^3} + \frac{3}{t}\frac{d^2y}{dt^2} + \frac{6}{t^2}\frac{dy}{dt} + \frac{6}{t^3}y = 0$$

$$y(1) = 1,\ \dot{y}(1) = 2,\ \ddot{y}(1) = 3$$

2. Let $p(t)$, $q(t)$, and $g(t)$ be continuous real-valued functions on an open interval I and suppose $\varphi_1(t)$ and $\varphi_2(t)$ are linearly independent solutions of the second-order scalar equation

$$\ddot{y} + p(t)\dot{y} + q(t)y = 0.$$

   Derive an explicit formula for a particular solution of

$$\ddot{y} + p(t)\dot{y} + q(t)y = g(t).$$

3. Let $\varphi_1(t)$ and $\varphi_2(t)$ be linearly independent solutions of the second-order scalar equation $\ddot{y} + p(t)\dot{y} + q(t)y = 0$, where $p(t)$ and $q(t)$ are continuous functions on $\mathbb{R}$. Prove that the set of roots of φ_j,

$$\{t : \varphi_j(t) = 0\} \cap [-N, N],$$

   is a finite set for $N > 0$. Then, prove that between any two consecutive roots of φ_1 there is exactly one root of φ_2.

4. Let $\varphi_1, \ldots, \varphi_n$ be linearly independent solutions of an nth order homogenous linear differential equation. Find necessary and sufficient conditions for $W(\varphi_1, \ldots, \varphi_n)$ to be constant.

5. Consider $\ddot{y} + a(t)y = 0$ with $a(t)$ continuous on $0 < t < \infty$ and suppose $\varphi(t)$ and $\psi(t)$ are nonzero solutions that are not multiples of each other. Show that if
$$\lim_{t\to\infty} |\varphi(t)| + |\dot{\varphi}(t)| = 0$$
then
$$\lim_{t\to\infty} |\psi(t)| + |\dot{\psi}(t)| = \infty$$

3.4 Complex Linear Differential Equations

Let $\boldsymbol{\varphi} : I \to \mathbb{C}^d$, where, as usual, I is an open interval in $\mathbb{R}$ and $\mathbb{C}$ denotes the complex numbers. Thinking of $\boldsymbol{\varphi}(t)$ as a point in $\mathbb{R}^{2d}$ in the natural way, it can be written
$$\boldsymbol{\varphi}(t) = \big(u_1(t) + iv_1(t), \ldots, u_d(t) + iv_d(t)\big),$$
where $u_j(t)$ and $v_j(t)$ are real-valued functions on I and $i^2 = -1$. (We will be careful not to use i as an index when complex numbers are in play.) Then $\boldsymbol{\varphi}(t)$ is continuous, if the functions $u_j(t)$ and $v_j(t)$, $j = 1, \ldots, d$, are all continuous. In addition, $\boldsymbol{\varphi}(t)$ is differentiable on I, if the functions $u_j(t)$ and $v_j(t)$, $j = 1, \ldots, d$, are differentiable on I and
$$\dot{\boldsymbol{\varphi}}(t) = \big(\dot{u}_1(t) + i\,\dot{v}_1(t), \ldots, \dot{u}_d(t) + i\,\dot{v}_d(t)\big).$$

Since complex $m \times n$ matrices are points in $\mathbb{C}^{mn}$ written in a rectangular array instead of a column or row, the same comments apply to functions from I into $\mathcal{M}_d(\mathbb{C})$, the $d \times d$ complex matrices.

In this context, consider the complex linear differential equation
$$\dot{\mathbf{z}} = A(t)\mathbf{z} + \mathbf{h}(t), \tag{3.12}$$
where A and $\mathbf{h}$ are continuous functions on I with values in $\mathcal{M}_d(\mathbb{C})$ and $\mathbb{C}^d$, respectively. Since $\mathbb{R}^d$ is naturally contained in $\mathbb{C}^d$, the equation
$$\dot{\mathbf{x}} = A(t)\mathbf{x} + \mathbf{h}(t),$$
where A and $\mathbf{h}$ are continuous functions on I with values in $\mathcal{M}_d(\mathbb{R})$ and $\mathbb{R}^d$, respectively, can be considered as a special case of *(3.12)*.

In this section, the results of Sections 1 and 2 will be extended to *(3.12)*. Then it will follow that when the real equation $\dot{\mathbf{x}} = A(t)\mathbf{x} + \mathbf{h}(t)$ is viewed as being complex, the complex solutions satisfying real initial conditions have zero imaginary part. Thus the real solutions of the $\dot{\mathbf{x}} = A(t)\mathbf{x}+\mathbf{h}(t)$ are automatically recovered from the complex ones. Since the complex numbers are *algebraically closed*, that is, every complex polynomial has a root in the complex numbers, this opens the possibility of using this advantage of $\mathbb{C}$ over $\mathbb{R}$ to gain information about the solutions of $\dot{\mathbf{x}} = A(t)\mathbf{x} + \mathbf{h}(t)$. In Chapter 4, this point of view will be exploited when $A(t)$ is a constant function.

For the rest of this section, the function $A(t)$ will have its values in the a complex $d \times d$ matrices, $\mathcal{M}_d(\mathbb{C})$, and the function $\mathbf{h}(t)$ will have its values in $\mathbb{C}^d$. As usual, $A(t)$ and $h(t)$ depend continuously on $t \in I$, an open interval of $\mathbb{R}$. The entries of $A(t)$ will be written as

$$a_{jk}(t) = \alpha_{jk}(t) + i\beta_{jk}(t)$$

and the jth components of $h(t)$ as

$$h_j(t) = \gamma_j(t) + i\delta_j(t),$$

where $\alpha_{jk}(t)$, $\beta_{jk}(t)$, $\gamma_j(t)$, and $\delta_j(t)$ are continuous real-valued functions on I. Define a real $2d \times 2d$ matrix by

$$B(t) = \begin{bmatrix} \alpha_{11}(t) & -\beta_{11}(t) & \cdots & \alpha_{1d}(t) & -\beta_{1d}(t) \\ \beta_{11}(t) & \alpha_{11}(t) & \cdots & \beta_{1d}(t) & \alpha_{1d}(t) \\ \vdots & \vdots & \ddots & \vdots & \vdots \\ \alpha_{d1}(t) & -\beta_{d1}(t) & \cdots & \alpha_{dd}(t) & -\beta_{dd}(t) \\ \beta_{d1}(t) & \alpha_{d1}(t) & \cdots & \beta_{dd}(t) & \alpha_{dd}(t) \end{bmatrix}$$

and define $g : I \to \mathbb{R}^{2d}$ by

$$\mathbf{g}(t) = \begin{pmatrix} \gamma_1(t) \\ \delta_1(t) \\ \vdots \\ \gamma_d(t) \\ \delta_d(t) \end{pmatrix}.$$

Then $\boldsymbol{\varphi}(t) = \big(u_1(t) + iv_1(t), \ldots, u_d(t) + iv_d(t)\big)$ is a solution of the complex linear differential equation *(3.12)*, $\dot{\mathbf{z}} = A(t)\mathbf{z} + \mathbf{h}(t)$, if and only if

$$\dot{\varphi}_j(t) = \sum_{k=1}^{d} a_{jk}(t)\varphi_k(t) + h_j(t)$$

if and only if

$$\dot{u}_j(t) = \sum_{k=1}^{d} \alpha_{jk}(t)u_k(t) - \beta_{jk}(t)v_k(t) + \gamma_j(t)$$

and

$$\dot{v}_j(t) = \sum_{k=1}^{d} \beta_{jk}(t)u_k(t) + \alpha_{jk}(t)v_k(t) + \delta_j(t)$$

if and only if

$$\psi(t) = \big(u_1(t), v_1(t), u_2(t), v_2(t), \ldots, u_d(t), v_d(t)\big)$$

is a solution of the real differential equation $\dot{\mathbf{x}} = B(t)\mathbf{x}+\mathbf{g}(t)$. Now *Theorem 3.3* can be applied to prove its own complex analogue.

Theorem 3.12 *Let $A : I \to \mathcal{M}_d(\mathbb{C})$ and $\mathbf{h} : I \to \mathbb{C}^d$ be continuous functions on the open interval I. For any $(\tau, \boldsymbol{\zeta}) \in I \times \mathbb{C}^n$, the complex initial-value problem*

$$\begin{aligned} \dot{\mathbf{z}} &= A(t)\mathbf{z} + \mathbf{h}(t) \\ \boldsymbol{\zeta} &= \mathbf{z}(\tau) \end{aligned}$$

has a unique solution and this solution can be continued to all of I.

The proof of the results in Section 2 were based on the above theorem and linear algebra, and hence, they easily carry over to the complex setting.

Note that the principal of superposition holds as before. It follows that the solutions of $\dot{\mathbf{z}} = A(t)\mathbf{z}$ form a vector space W over $\mathbb{C}$ and if ψ_0 is a particular solution of $\dot{\mathbf{z}} = A(t)\mathbf{z} + \mathbf{h}(t)$ then the set of all solutions of this equation is precisely $\{\varphi + \psi_0 : \varphi \in W\}$.

As in Section 2, consider the matrix differential equation

$$\dot{Z} = A(t)Z. \tag{3.13}$$

The solutions of the above have the same basic properties as the solutions of $\dot{X} = A(t)X$ when $A(t) \in \mathcal{M}_d(\mathbb{R})$, in particular:

Lemma *Let $Z(t)$ be a solution of (3.13). Then* Det $\big[Z(t)\big]$ *either vanishes identically on I or is never zero on I.*

Proof. The same argument used to prove the corresponding result (*Proposition 3.4*) in the real case works here. □

Theorem 3.13 *Fundamental matrix solutions of $\dot{Z} = A(t)Z$ always exist. If $Z(t)$ is any fundamental matrix solution, then*

$$Z(t)\big(Z(\tau)\big)^{-1}\boldsymbol{\zeta}$$

is the unique solution of the initial-value problem

$$\begin{aligned} \dot{\mathbf{z}} &= A(t)\mathbf{z}, \\ \boldsymbol{\zeta} &= \mathbf{z}(\tau). \end{aligned}$$

Proof. Proceed as in the proof of *Theorem 3.5.* □

As before, $Z(t, \tau) = Z(t)Z(\tau)^{-1}$ is a principal matrix solution satisfying $Z(\tau, \tau) = \mathrm{I}$ and depends continuously on t and τ. Furthermore, the proof of *Theorem 3.6* applies here to prove:

Theorem 3.14 *The solutions of $\dot{\mathbf{z}} = A(t)\mathbf{z}$ form a d-dimensional vector space W over $\mathbb{C}$. Furthermore, the solutions $\varphi_1(t), \ldots, \varphi_d(t)$ form a basis of W if and only if for some τ in I, the vectors $\varphi_1(\tau), \ldots, \varphi_d(\tau)$ form a basis for $\mathbb{C}^d$.*

The integral of curves in $\mathbb{C}^d$, like $\varphi(t)$, are defined by

$$\int_\tau^t \varphi(s)\,ds =$$

$$\left(\int_\tau^t u_1(s)\,ds + i\int_\tau^t v_1(s)\,ds, \ldots, \int_\tau^t u_d(s)\,ds + i\int_\tau^t v_d(s)\,ds\right).$$

Recall that $e^{x+iy} = e^x(\cos y + i\sin y)$. By using the chain rule it is easy to verify that the solution of the complex scalar differential equation

$$\dot{z} = a(t)z$$

is still

$$z(t) = C\exp\left(\int_\tau^t a(s)\,ds\right).$$

The variation of constants formula and Abel's formula also hold in the complex case.

It is important to show that nothing has been lost by the generalization of real linear differential equations to complex linear differential equations. Specifically, when $A(t) \in \mathcal{M}_d(\mathbb{R})$ and $\mathbf{h}(t) \in \mathbb{R}^d$, then either the real linear differential equation $\dot{\mathbf{x}} = A(t)\mathbf{x} + \mathbf{h}(t)$ in $\mathbb{R}^d$ or the complex linear differential equation $\dot{\mathbf{z}} = A(t)\mathbf{z} + \mathbf{h}(t)$ in $\mathbb{C}^d$ can be considered. It needs to be demonstrated that the the solutions of $\dot{\mathbf{x}} = A(t)\mathbf{x} + \mathbf{h}(t)$ are an easily identified subset of the solutions of $\dot{\mathbf{z}} = A(t)\mathbf{z} + \mathbf{h}(t)$. The final result shows that the answer lies in the initial conditions.

Proposition 3.15 *Let $A : I \to \mathcal{M}_d(\mathbb{R})$ and $\mathbf{h} : I \to \mathbb{R}^d$ be continuous functions on the open interval I and let $\varphi(t)$ be a solution of the complex linear differential equation $\dot{\mathbf{z}} = A(t)\mathbf{z} + \mathbf{h}(t)$. Then the coordinates of $\varphi(t)$ all have identically zero imaginary parts and $\varphi(t)$ is a of solution of the real linear differential equation $\dot{\mathbf{x}} = A(t)\mathbf{x} + \mathbf{h}(t)$ if and only if for some initial time $\tau \in I$ the coordinates $\varphi(\tau)$ all have zero imaginary parts.*

Proof. Suppose $\varphi(\tau) = \boldsymbol{\xi} + i\mathbf{0}$ for some $\boldsymbol{\xi} \in \mathbb{R}^d$. Let $\mathbf{x}(t)$ be the unique solution to the real initial-value problem

$$\begin{aligned} \dot{\mathbf{x}} &= A(t)\mathbf{x} + \mathbf{h}(t) \\ \boldsymbol{\xi} &= \mathbf{x}(\tau). \end{aligned}$$

Let $\boldsymbol{\zeta} = \boldsymbol{\xi} + i\mathbf{0}$ and check that

$$\mathbf{z}(t) = \big(x_1(t) + 0i, \ldots, x_d(t) + 0i\big)$$

is a solution to the complex initial-value problem

$$\begin{aligned}\dot{\mathbf{z}} &= A(t)\mathbf{z} + \mathbf{h}(t) \\ \zeta &= \mathbf{z}(\tau).\end{aligned}$$

By uniqueness, $\varphi(t) = \mathbf{z}(t)$ and $\varphi(t)$ has the required properties. □

Corollary *Let $Z(t, \tau)$ be the principal matrix solution of the complex equation*

$$\dot{Z} = A(t)Z.$$

If $A(t)$ is in $\mathcal{M}_d(\mathbb{R})$ for all t in I, then $Z(t, \tau)$ is also the principal matrix solution of the real equation

$$\dot{X} = A(t)X.$$

Proof. The columns of $Z(t, \tau)$ satisfy real initial conditions at τ, and the proposition applies to each column. □

To see how the complex viewpoint can be a unifying idea, consider the parameterized second-order scalar differential equation

$$\ddot{x} - \mu x = 0$$

and try to explicitly express the dependence of the solutions on initial conditions and μ. From an elementary view-point one would usually obtain expressions like

$$\varphi(t) = \begin{cases} C_1 e^{\sqrt{\mu}t} + C_2 e^{-\sqrt{u}t} & \text{for } \mu > 0 \\ C_1 + C_2 t, & \text{for } \mu = 0 \\ C_1 \sin\sqrt{-\mu}t + C_2 \cos\sqrt{-\mu}t & \text{for } \mu < 0 \end{cases}$$

from which it is not immediately clear how they fit together to form the function $\mathbf{x}(t, \tau, \boldsymbol{\xi}, \mu)$.

Instead, write $\ddot{x} - \mu x = 0$ in the usual way as a parameterized system

$$\begin{aligned}\dot{x} &= y \\ \dot{y} &= \mu x,\end{aligned}$$

which is the same as

$$\begin{pmatrix} \dot{x} \\ \dot{y} \end{pmatrix} = \begin{bmatrix} 0 & 1 \\ \mu & 0 \end{bmatrix} \begin{pmatrix} x \\ y \end{pmatrix}.$$

Think of the function

$$A(t, \mu) = \begin{bmatrix} 0 & 1 \\ \mu & 0 \end{bmatrix}$$

as a 2×2 complex matrix that is a function of t and the parameter μ. In light of the above corollary, it is only necessary to find a fundamental matrix solution of the complex equation $\dot{Z} = A(t, \mu)Z$ and then compute $Z(t, \mu)\big(Z(\tau, \mu)\big)^{-1}\boldsymbol{\xi}$ to obtain an explicit formula for $\mathbf{x}(t, \tau, \boldsymbol{\xi}, \mu)$.

It is easy to check that when complex solutions are allowed,

$$Z(t, \mu) = \begin{bmatrix} e^{\sqrt{\mu}\,t} & e^{-\sqrt{\mu}\,t} \\ \sqrt{\mu}\,e^{\sqrt{\mu}\,t} & -\sqrt{\mu}\,e^{-\sqrt{\mu}\,t} \end{bmatrix}$$

is a fundamental matrix solution for all $\mu \neq 0$, not just $\mu > 0$. Straightforward calculation shows that for $\mu \neq 0$

$$\big(Z(t, \mu)\big)^{-1} = -\frac{1}{2\sqrt{\mu}} \begin{bmatrix} -\sqrt{\mu}\,e^{-\sqrt{\mu}\,t} & -e^{-\sqrt{\mu}\,t} \\ -\sqrt{\mu}\,e^{\sqrt{\mu}\,t} & e^{\sqrt{\mu}\,t} \end{bmatrix},$$

and

$$Z(t, \tau, \mu) = \begin{bmatrix} \dfrac{e^{\sqrt{\mu}\,(t-\tau)} + e^{-\sqrt{\mu}\,(t-\tau)}}{2} & \dfrac{e^{\sqrt{\mu}\,(t-\tau)} - e^{-\sqrt{\mu}\,(t-\tau)}}{2\sqrt{\mu}} \\ \sqrt{\mu}\dfrac{e^{\sqrt{\mu}\,(t-\tau)} - e^{-\sqrt{\mu}\,(t-\tau)}}{2} & \dfrac{e^{\sqrt{\mu}\,(t-\tau)} + e^{-\sqrt{\mu}\,(t-\tau)}}{2} \end{bmatrix}.$$

When $\mu = 0$, only the upper right corner of $Z(t, \tau, \mu)$ is not defined. Since $Z(t, \tau, \mu)$ is the global solution of $\dot{Z} = A(t, \mu)Z$ restricted to the initial matrix I, it is continuous in all variables. Taking the limit as μ goes to 0 shows that

$$Z(t, \tau, 0) = \begin{bmatrix} 1 & t-\tau \\ 0 & 1 \end{bmatrix}.$$

Letting $\boldsymbol{\xi} = \begin{pmatrix} \xi_0 \\ \xi_1 \end{pmatrix} \in \mathbb{R}^2$, produces

$$\mathbf{x}(t, \tau, \boldsymbol{\xi}, \mu) =$$

$$\begin{bmatrix} \dfrac{e^{\sqrt{\mu}\,(t-\tau)} + e^{-\sqrt{\mu}\,(t-\tau)}}{2} & \dfrac{e^{\sqrt{\mu}\,(t-\tau)} - e^{-\sqrt{\mu}\,(t-\tau)}}{2\sqrt{\mu}} \\ \sqrt{\mu}\dfrac{e^{\sqrt{\mu}\,(t-\tau)} - e^{-\sqrt{\mu}\,(t-\tau)}}{2} & \dfrac{e^{\sqrt{\mu}\,(t-\tau)} + e^{-\sqrt{\mu}\,(t-\tau)}}{2} \end{bmatrix} \begin{pmatrix} \xi_0 \\ \xi_1 \end{pmatrix}$$

for $\mu \neq 0$ and

$$\mathbf{x}(t, \tau, \boldsymbol{\xi}, 0) = \begin{bmatrix} 1 & t-\tau \\ 0 & 1 \end{bmatrix} \begin{pmatrix} \xi_0 \\ \xi_1 \end{pmatrix}.$$

Finally, the solution $\ddot{y} - \mu y = 0$ such that $y(\tau) = \xi_o$ and $\dot{y}(\tau) = \xi_1$, is given by

$$\xi_0 \left(\frac{e^{\sqrt{\mu}\,(t-\tau)} + e^{-\sqrt{\mu}\,(t-\tau)}}{2} \right) + \xi_1 \left(\frac{e^{\sqrt{\mu}\,(t-\tau)} - e^{-\sqrt{\mu}\,(t-\tau)}}{2\sqrt{\mu}} \right)$$

for $\mu \neq 0$ and the limiting formula when $\mu = 0$ is

$$\xi_0 + \xi_1(t-\tau).$$

When μ is positive, a cosh and a sinh term are present and when μ is negative, a cos and sin term appear. The parameterized complex linear differential equation $\dot{Z} = A(\mu)Z$ shows how they are naturally tied together, as solutions of $\ddot{y} - \mu y = 0$.

Although $\mathbf{x}(t, \tau, \boldsymbol{\xi}, \mu)$ is continuous at $\mu = 0$, there is also a dramatic change in the phase portrait as μ moves from negative-to-positive values. Namely, when $\mu < 0$ the origin is a saddle point and when $\mu > 0$ the origin is a center. Such changes are called *bifurcations*, and $\mu = 0$ is a bifurcation value of $\ddot{y} - \mu y = 0$.

EXERCISES

1. Let $\varphi : I \to \mathbb{C}$ and $\psi : I \to \mathbb{C}$ be differentiable. Show that the derivative of $\varphi\psi$ can be calculated using the usual formula $\dot{\varphi}\psi + \varphi\dot{\psi}$.

2. Let $a(t)$ and $h(t)$ be continuous complex-valued functions on some open interval I and consider the complex differential equation $\dot{z} = a(t)z + h(t)$. Derive a formula for $z(t, \tau, \zeta)$.

3. Let $A : I \to \mathcal{M}_d(\mathbb{C})$ and $\mathbf{h} : I \to \mathbb{C}^d$ be continuous on the open interval I. Prove the complex version of the variation of constants formula in this setting.

4. Let $A : I \to \mathcal{M}_d(\mathbb{C})$ be continuous on the open interval I. Prove the complex version of Abel's formula in this setting.

5. Let $A : I \to \mathcal{M}_d(\mathbb{C})$ be a continuous function on the open interval I such that $a_{jj}(t) = i\beta_{jj}(t)$. Show that if $Z(t)$ is a solution of the complex matrix differential equation $\dot{Z} = A(t)Z$, then $\operatorname{Det} Z(t)$ is bounded.

6. Let $A : I \to \mathcal{M}_d(\mathbb{R})$ be a continuous function on the open interval I. Show that $\mathbf{z}(t) = (u_1(t) + iv_1(t), \ldots, u_d(t) + iv_d(t))$, where $u_j(t)$ and $v(j(t)$ are differentiable real-valued functions on I for $1 \le j \le d$, is a solution of the complex differential equation $\dot{\mathbf{z}} = A(t)\mathbf{z}$ if and only if $\mathbf{u}(t) = (u_1(t), \ldots, u_d(t))$ and $\mathbf{v}(t) = (v_1(t), \ldots, v_d(t))$ are solutions of the real differential equation $\dot{\mathbf{x}} = A(t)\mathbf{x}$.

7. Let L be a given positive real number. Find all real values of μ for which $\ddot{x} - \mu x = 0$ has a non zero solution $\varphi(t)$ satisfying $\varphi(0) = 0 = \varphi(L)$. Problem like this arise naturally when solving partial differential equations with constant coefficients by separation of variables, but they are usually written in the form $\ddot{x} + \lambda x = 0$. See [4] or most any introduction to partial differential equations. These problems are called *eigenvalue problems* because differentiation is a linear operator on functions and they can be written as $D^2(\varphi) = \lambda\varphi$. An extensive treatment of such eigenvalue problems can be found in [7].

8. Find all real values of μ for which $\ddot{x} - \mu x = 0$ has a non zero solution $\varphi(t)$ satisfying $\dot{\varphi}(0) = 0 = \dot{\varphi}(L)$ with $L > 0$ as above.

Chapter 4

Constant Coefficients

Although the solutions of $\dot{\mathbf{x}} = A(t)\mathbf{x}$ possess important general properties such as being defined on the longest possible interval and forming a d-dimensional vector space, more explicit calculations of a fundamental matrix solution require significant restrictions on the matrix valued function $A(t)$. In fact, the best results are obtained when it is constant, that is, $A(t) \equiv A$ on $\mathbb{R}$. Thinking of $\dot{\mathbf{x}} = A(t)\mathbf{x}$ as a system of differential equations, the condition that $A(t)$ be constant means that the coefficients of x_1 to x_d in the system are all constant, hence the title of this chapter is *Constant Coefficients.* In this context, $A(t)$ can be written simply as A, and $\dot{\mathbf{x}} = A\mathbf{x}$ can be studied using finite-dimensional linear algebra.

The study of $\dot{\mathbf{x}} = A\mathbf{x}$ begins in a particularly nice environment. The vector field $\mathbf{f}(\mathbf{x}) = A\mathbf{x}$ is autonomous and locally Lipschitz on $\mathbb{R}^d$. The solutions of $\dot{\mathbf{x}} = A\mathbf{x}$ are defined on on $\mathbb{R}$ by *Theorem 3.3.* Consequently, $\mathbf{x}(t, \tau, \boldsymbol{\xi})$ is defined and continuous on $D_{\mathbf{f}} = \mathbb{R}^{d+2}$, and $\mathbf{x}(t, 0, \boldsymbol{\xi}) = \mathbf{x}(t, \boldsymbol{\xi})$ defines a flow on $\mathbb{R}^d$. These flows will play a substantive role in the remaining chapters.

The right-hand side of $\dot{\mathbf{x}} = A\mathbf{x}$ is a single linear transformation, namely, $T(\mathbf{x}) = A\mathbf{x}$ and the powerful theory of a single linear transformation from linear algebra can be applied to T. To carry out this program, the development of the necessary linear algebra and the theory of $\dot{\mathbf{x}} = A\mathbf{x}$ will be woven together. It is advantageous to focus on complex vector spaces and complex linear differential equations. Hence, the center of attention will be $\dot{\mathbf{z}} = A\mathbf{z}$ with $A \in \mathcal{M}_d(\mathbb{C})$. Specific results for the case of a real matrix A will be derived from the more general results for a complex matrix.

The first step is to show that the principal matrix solution of $\dot{\mathbf{z}} = A\mathbf{z}$ is a matrix analogue of e^{at}, the solution of the scalar equation $\dot{z} = az$. The importance and limitations of eigenvalues and eigenvectors then become apparent and lead naturally to studying generalized eigenvectors, eigenspaces, and canonical forms for a linear transformation. The resulting linear algebra provides a complete solution to $\dot{\mathbf{z}} = A\mathbf{z}$ that can be applied to the original linear differential equation $\dot{\mathbf{x}} = A\mathbf{x}$ with $A \in \mathcal{M}_d(\mathbb{R})$.

4.1 The Exponential of a Matrix

Let A be a $d \times d$ matrix. The differential equation $\dot{\mathbf{x}} = A\mathbf{x}$ is called *a linear differential equation with constant coefficients.* This is a special case of $\dot{\mathbf{x}} = A(t)\mathbf{x}$, where $A(t) = A$ for all $t \in I = \mathbb{R}$.

Because the general solution $\mathbf{x}(t, \boldsymbol{\xi})$ of $\dot{\mathbf{x}} = A\mathbf{x}$ is completely determined by a principal matrix solution of $\dot{X} = AX$, a natural objective is to determine the principal matrix solution of $\dot{X} = AX$ at $s = 0$. Since the function $\mathbf{f}(X) = AX$ can be thought of as vector field on the $d \times d$ invertible matrices, this amounts to studying a solution to an autonomous differential equation on the invertible matrices instead of $\mathbb{R}^d$. This perspective is particularly fruitful.

The important new ingredient for the constant coefficient problem $\dot{\mathbf{x}} = A\mathbf{x}$ is the eigenvalues of A and their generalization. They will play a very significant role in the study of $\dot{\mathbf{x}} = A\mathbf{x}$. Since the *eigenvalues* are roots of the polynomial equation $\operatorname{Det}[A - \lambda \mathrm{I}] = 0$, there will always be eigenvalues in $\mathbb{C}$ because $\mathbb{C}$ is algebraically closed, but there may not be any eigenvalues in $\mathbb{R}$. Consequently, the best strategy is to develop the theory for a complex $d \times d$ matrix A and treat a real $d \times d$ matrix, as a special case. It has already been shown (*Proposition 3.15*) that when A is real and the initial conditions are real, it does not matter whether $\dot{\mathbf{x}} = A\mathbf{x}$ is treated as a real or complex differential equation.

For the rest of this chapter, A will be a complex $d \times d$ matrix unless otherwise stated. When $d = 1$, the matrix differential equation $\dot{Z} = AZ$ is simply $\dot{z} = az$ and $e^{at} = e^{\alpha t}(\cos(\beta t) + i \sin(\beta t))$ for $a = \alpha + i\beta$ is the principal matrix solution at 0. By analogy, define e^{At} to be the principal matrix solution of $\dot{Z} = AZ$ at 0 . It follows immediately from the definition that

$$\frac{de^{At}}{dt} = Ae^{At}$$

and

$$e^{A0} = \mathrm{I}.$$

For fixed s, clearly, $e^{At}e^{As}$ is a matrix solution of $\dot{Z} = AZ$. Since $\dot{\mathbf{z}} = A\mathbf{z}$ is autonomous, $e^{A(t+s)}$ is also a matrix solution for fixed s. When $t = 0$, $e^{A0}\, e^{As} = \mathrm{I}\, e^{As} = e^{As}$ and $e^{A(0+s)} = e^{As}$. Thus by uniqueness of solutions

$$e^{A(t+s)} = e^{At}\, e^{As}$$

for all t and s in $\mathbb{R}$. It follows that

$$\left(e^{At}\right)^{-1} = e^{A(-t)} = e^{-At}$$

because $\mathrm{I} = e^{A0} = e^{A(t-t)} = e^{At}\, e^{A(-t)}$.

As before, set $|A| = \sup\left\{|A\mathbf{z}| : |\mathbf{z}| = 1\right\}$, where $|\mathbf{z}| = \sum_{i=1}^{d} |z_i|$ for $\mathbf{z} = (z_1, \cdots, z_d) \in \mathbb{C}^d$. This is the obvious extension of $|\cdot|$ from $\mathbb{R}^d$ to $\mathbb{C}^d$. These complex norms on $\mathbb{C}^d$ and $\mathcal{M}_d(\mathbb{C})$ satisfy the definition of a norm on page 7 with the scalar $\alpha \in \mathbb{R}$ replaced by $\alpha \in \mathbb{C}$. The norm of a complex matrix also has the same basic properties as in the real case. In particular,

$$\begin{aligned}|A\mathbf{z}| &\leq |A|\,|\mathbf{z}|,\\ |AB| &\leq |A|\,|B|,\end{aligned}$$

and hence

$$|A^n| \leq |A|^n.$$

Theorem 4.1 *Given $A \in \mathcal{M}_d(\mathbb{C})$, the sum*

$$\sum_{k=0}^{n} \frac{(At)^k}{k!}$$

converges uniformly on every closed interval of $\mathbb{R}$ to e^{At} as n goes to infinity and

$$e^{At} = \sum_{k=0}^{\infty} \frac{(At)^k}{k!}. \tag{4.1}$$

Proof. If $M < N$, then

$$\begin{aligned}\left|\sum_{k=0}^{N} \frac{(At)^k}{k!} - \sum_{k=0}^{M} \frac{(At)^k}{k!}\right| &= \left|\sum_{k=M+1}^{N} \frac{(At)^k}{k!}\right|\\ &\leq \sum_{k=M+1}^{N} \frac{|A|^k t^k}{k!}\end{aligned}$$

by the previous observation. Since $\sum_{k=0}^{n} \frac{|A|^k t^k}{k!}$ converges uniformly on closed intervals to $e^{|A|t}$, it follows that

$$\sum_{k=0}^{n} \frac{(At)^k}{k!}$$

is a Cauchy sequence of matrices and converges uniformly on closed intervals to some $d \times d$ matrix $\Gamma(t)$. (See *Exercise 9* on page 12.) Since the convergence is uniform on closed intervals, $\Gamma(t)$ is continuous and $\Gamma(0) = \mathrm{I}$. To complete the proof it suffices to show that

$$\Gamma(t) = \mathrm{I} + \int_0^t A\Gamma(s)\, ds.$$

Note that

$$\sum_{k=0}^{n} \frac{A^k t^k}{k!} = \mathrm{I} + \int_0^t A \sum_{k=0}^{n-1} \frac{A^k s^k}{k!}\, ds.$$

Consequently,

$$\begin{aligned}
\left|\Gamma(t) - \mathrm{I} - \int_0^t A\Gamma(s)\,ds\right| &= \\
\left|\Gamma(t) - \sum_{k=0}^{n} \frac{A^k t^k}{k!} + \sum_{k=0}^{n} \frac{A^k t^k}{k!} - \mathrm{I} - \int_0^t A\Gamma(s)\,ds\right| &\leq \\
\left|\Gamma(t) - \sum_{k=0}^{n} \frac{A^k t^k}{k!}\right| + \left|\int_0^t A\left(\sum_{k=0}^{n-1} \frac{A^k s^k}{k!}\right) - A\Gamma(s)\,ds\right| &\leq \\
\left|\Gamma(t) - \sum_{k=0}^{n} \frac{A^k t^k}{k!}\right| + \int_0^t |A| \left|\sum_{k=0}^{n-1} \frac{A^k s^k}{k!} - \Gamma(s)\right| ds, &
\end{aligned}$$

which clearly goes to 0 as n goes to ∞ to complete the proof. □

It is not true in general that $e^{A+B} = e^A e^B$, however, the following proposition shows that this formula is at least true when $AB = BA$.

Proposition 4.2 *Let A and B be $d \times d$ complex matrices. Then the following hold:*

(a) $Be^{At} = e^{At}B$ for all $t \in \mathbb{R}$ if and only if $AB = BA$; and

(b) $e^{(A+B)t} = e^{At}e^{Bt}$ for all $t \in \mathbb{R}$ if and only if $AB = BA$.

Proof. To start the proof of part (a), assume that $Be^{At} = e^{At}B$ for all t. Differentiating with respect to t yields

$$BAe^{At} = Ae^{At}B$$

and setting $t = 0$ implies that $BA = AB$. Now suppose $AB = BA$. Let C be an arbitrary $d \times d$ matrix and note that $C \to BC$ and $C \to CB$ are continuous maps of $\mathcal{M}_d(\mathbb{C})$ into itself. It follows from *Exercise 15* on page 12 that

$$\begin{aligned}
Be^{At} &= B \lim_{n\to\infty} \sum_{k=0}^{n} \frac{A^k t^k}{k!} \\
&= \lim_{n\to\infty} B \sum_{k=0}^{n} \frac{A^k t^k}{k!} \\
&= \lim_{n\to\infty} \left(\sum_{k=0}^{n} \frac{A^k t^k}{k!} B\right) \\
&= \left(\lim_{n\to\infty} \sum_{k=0}^{n} \frac{A^k t^k}{k!}\right) B \\
&= e^{At}B.
\end{aligned}$$

Turning to part (b), suppose $e^{(A+B)t} = e^{At}e^{Bt}$ for all t. Differentiate this expression with respect to t to obtain

$$(A+B)e^{(A+B)t} = Ae^{At}e^{Bt} + e^{At}Be^{Bt}$$

or

$$Ae^{At}e^{Bt} + Be^{At}e^{Bt} = Ae^{At}e^{Bt} + e^{At}Be^{Bt},$$

which simplifies to

$$Be^{At} = e^{At}B$$

because e^{Bt} is invertible. It follows from part (a) that $AB = BA$.

Finally, if $AB = BA$, then differentiating and using part (a) to simplify yields

$$\begin{aligned} \frac{d}{dt}e^{At}e^{Bt} &= Ae^{At}e^{Bt} + e^{At}Be^{Bt} \\ &= (A+B)e^{At}e^{Bt}. \end{aligned}$$

Therefore, $e^{At}e^{Bt}$ is a matrix solution of $\dot{Z} = (A+B)Z$, as is $e^{(A+B)t}$. Since $e^{A0}e^{B0} = \mathrm{I}$, by uniqueness of solutions $e^{At}e^{Bt} = e^{(A+B)t}$ as desired. □

Remark $\operatorname{Det}[e^A] = e^{\operatorname{Tr}[A]}$.

Proof. Apply Abel's formula. □

Let $GL_d(\mathbb{C})$ denote the set of $d \times d$ invertible matrices. It is an open subset of $\mathcal{M}_d(\mathbb{C})$ because

$$\begin{aligned} GL_d(\mathbb{C}) &= \{A \in \mathcal{M}_d(\mathbb{C}) : \operatorname{Det}[A] \neq 0\} \\ &= \operatorname{Det}^{-1}\big[\{z \in \mathbb{C} : z \neq 0\}\big], \end{aligned}$$

the Det : $\mathcal{M}_d(\mathbb{C}) \to \mathbb{C}$ is a continuous function, and $\{z \in \mathbb{C} : z \neq 0\}$ is an open set. Also note that $GL_d(\mathbb{C})$ is a group under matrix multiplication because matrix multiplication is associative, the product of two invertible matrices is invertible, and the inverse of an invertible matrix is invertible. $GL_d(\mathbb{C})$ is called the *general linear group for* $\mathbb{C}$. Of course the same remarks apply to $GL_d(\mathbb{R})$, the *general linear group for* $\mathbb{R}$, in $\mathcal{M}_d(\mathbb{R})$. These two groups are important examples of Lie groups.

A *one parameter subgroup* of $GL_d(\mathbb{C})$ or $GL_d(\mathbb{R})$ is a differentiable function $\varphi : \mathbb{R} \to GL_d(\cdot)$ such that $\varphi(s+t) = \varphi(s)\varphi(t)$. In other words, φ is a homomorphism of the additive group $\mathbb{R}$ into $GL_d(\mathbb{C})$ or $GL_d(\mathbb{R})$ that is also differentiable.

Let A be in $\mathcal{M}_d(\mathbb{C})$ or $\mathcal{M}_d(\mathbb{R})$ and set $\varphi(t) = e^{At}$. Then $\varphi(t+s) = e^{A(t+s)} = e^{At}e^{As} = \varphi(t)\varphi(s)$ and φ is a homomorphism of groups. Since e^{At} is a solution of a differential equation, $\varphi(t)$ is also differentiable and thus a one parameter subgroup. There are no others.

Proposition 4.3 *Let φ be a one parameter subgroup of $GL_d(\mathbb{R})$ or $GL_d(\mathbb{C})$. Then there exists A in $\mathcal{M}_d(\mathbb{R})$ or $\mathcal{M}_d(\mathbb{C})$ such that $\varphi(t) = e^{At}$.*

Proof. It must be shown that $\varphi(t)$ is a solution of $\dot{X} = AX$ for some A. Since $\varphi(0) = \mathrm{I}$ is the identity of $GL_d(\cdot)$, it will then follow that $\varphi(t) = e^{At}$. Set $A = \dot{\varphi}(0)$. Now

$$\begin{aligned}
\dot{\varphi}(t) &= \lim_{h\to 0} \frac{\varphi(h+t) - \varphi(t)}{h} \\
&= \lim_{h\to 0} \frac{\varphi(h)\varphi(t) - \varphi(t)}{h} \\
&= \left(\lim_{h\to 0} \frac{\varphi(h) - \mathrm{I}}{h}\right)\varphi(t) \\
&= \dot{\varphi}(0)\varphi(t) \\
&= A\varphi(t)
\end{aligned}$$

and $\varphi(t)$ is a solution of $\dot{X} = AX$. □

Thus the principal matrix solutions of $\dot{Z} = AZ$ at 0 are precisely the one parameter subgroups of $GL_d(\mathbb{C})$ and can be expressed as an exponential power series of a matrix. The next step is to develop the linear algebra needed to calculate e^{At}.

EXERCISES

1. Prove that if B is an invertible matrix, then
$$Be^{A}B^{-1} = e^{BAB^{-1}}.$$

2. Let A^{t} denotes the transpose of a matrix A as defined on page 137 and let $\overline{A}$ denote the matrix whose entries are the complex conjugates of the entries of A. Show that
$$\left(e^{A}\right)^{\mathrm{t}} = e^{A^{\mathrm{t}}}$$
and
$$\overline{e^{A}} = e^{\overline{A}}.$$

3. Find necessary and sufficient conditions for $\operatorname{Det} e^{A} = 1$ when A is real and when A is complex.

4. Calculate e^{At} for the following matrices:
$$\begin{bmatrix} 0 & 1 & 0 \\ 0 & 0 & 1 \\ 0 & 0 & 0 \end{bmatrix} \quad \begin{bmatrix} 2 & 0 & 0 \\ 0 & -3 & 0 \\ 0 & 0 & 1 \end{bmatrix} \quad \begin{bmatrix} -3 & 1 & 7 \\ 0 & 4 & -1 \\ 0 & 0 & 2 \end{bmatrix}$$
$$\begin{bmatrix} 1 & 1 & 1 \\ -1 & 1 & 0 \\ 0 & 0 & 1 \end{bmatrix} \quad \begin{bmatrix} 0 & -1 & 2 \\ 0 & 1 & 0 \\ 1 & 1 & -1 \end{bmatrix} \quad \begin{bmatrix} -2 & 1 & 0 \\ 0 & -2 & 1 \\ 0 & 0 & -2 \end{bmatrix}.$$

5. Let A be a real matrix. Show that the differential equation $\dot{\mathbf{x}} = A\mathbf{x}$ has the form $\dot{\mathbf{x}} = -\nabla F(\mathbf{x})$ where $F : \mathbb{R}^d \to \mathbb{R}$ has continuous first partial derivatives, called a *gradient system*, if and only if A is a symmetric matrix.

6. Let A be a real invertible $d \times d$ matrix and let $\mathbf{v}$ be a nonzero vector in $\mathbb{R}^d$. Show that $\dot{\mathbf{x}} = A\mathbf{x} + \mathbf{v}$ has a constant solution and use it to inductively determine the function $\mathbf{x}(t, \tau, \boldsymbol{\xi})$ for $\dot{\mathbf{x}} = A\mathbf{x} + t^k\mathbf{v}$.

4.2 Generalized Eigenspaces

Suppose $\mathbf{v} \in \mathbb{C}^d$ is an eigenvector of A for the eigenvalue λ of A. That is, $A\mathbf{v} = \lambda\mathbf{v}$ and $\mathbf{v} \neq \mathbf{0}$. Then $\mathbf{z}(t) = e^{\lambda t}\mathbf{v}$ is a solution to $\dot{\mathbf{z}} = A\mathbf{z}$ because $\dot{\mathbf{z}}(t) = \lambda e^{\lambda t}\mathbf{v} = e^{\lambda t}A\mathbf{v} = A(e^{\lambda t}\mathbf{v})$. Consequently, if $\mathbf{v}_1, \ldots, \mathbf{v}_n$ is a basis of $\mathbb{C}^n$ consisting of eigenvectors (for not necessarily distinct eigenvalues), then the matrix $Z(t)$ whose columns are $e^{\lambda_1 t}\mathbf{v}_1, \ldots, e^{\lambda_n t}\mathbf{v}_n$ is a fundamental matrix solution and $e^{At} = Z(t)Z(0)^{-1}$.

For example, if A is a real symmetric matrix, then there exists a basis of eigenvectors belonging to real eigenvalues. Another example is when a $d \times d$ complex matrix has d distinct eigenvalues. Most matrices, however, do not have a basis of eigenvectors and more sophisticated linear algebra is required to calculate e^{At}.

Let $T : \mathbb{C}^d \to \mathbb{C}^d$ be a linear transformation. The eigenvalue equation $T(\mathbf{v}) = \lambda\mathbf{v}$ can also be written as $(T - \lambda\mathrm{I})(\mathbf{v}) = \mathbf{0}$, which suggests that the solutions of $(T - \lambda\mathrm{I})^k(\mathbf{v}) = \mathbf{0}$ for $k > 1$ may also be relevant. With a slight abuse of notation we will use I to denote both the identity matrix with 1's on the diagonal and zeros elsewhere and the identity linear transformation $\mathrm{I}(\mathbf{v}) = \mathbf{v}$. Of course, the matrix of the linear transformation I with respect to any basis is the matrix I so they are almost synonymous. It will be clear from the context which meaning of I is in use.

A vector $\mathbf{v}$ is called a *generalized eigenvector of degree p* if for some $\lambda \in \mathbb{C}$

$$(T - \lambda\mathrm{I})^p(\mathbf{v}) = \mathbf{0}$$

and

$$(T - \lambda\mathrm{I})^{p-1}(\mathbf{v}) \neq \mathbf{0}.$$

Since the first equation can be written

$$(T - \lambda\mathrm{I})(T - \lambda\mathrm{I})^{p-1}(\mathbf{v}) = \mathbf{0},$$

λ must be an eigenvalue of T. Clearly, eigenvectors are generalized eigenvectors of degree 1. The next order of business is to study the generalized eigenvectors.

Proposition 4.4 *If $\mathbf{v}$ is a generalized eigenvector of degree p for the eigenvalue λ, then the vectors $\mathbf{v}$, $(T - \lambda\mathrm{I})(\mathbf{v}), \ldots, (T - \lambda\mathrm{I})^{p-1}(\mathbf{v})$ are linearly independent.*

Proof. Suppose not. Then it is easy to see that there exists k, $0 \leq k \leq p-2$ and $c_i \in \mathbb{C}$, $k+1 \leq j \leq p-1$ such that

$$(T - \lambda \mathrm{I})^k(\mathbf{v}) = \sum_{j=k+1}^{p-1} c_j (T - \lambda \mathrm{I})^j(\mathbf{v}).$$

Applying $(T - \lambda \mathrm{I})^{p-k-1}$ to the above equation produces the contradiction

$$0 \neq (T - \lambda \mathrm{I})^{p-1}(\mathbf{v}) = \sum_{j=k+1}^{p-1} c_j (T - \lambda \mathrm{I})^{p-k-1+j}(\mathbf{v}) = \mathbf{0}$$

because $j \geq k+1$ implies $p - k - 1 + j \geq p$. □

Let $\mathcal{N}(S) = \{\mathbf{v} : S(\mathbf{v}) = 0\}$ denote the *null space* or *kernel* of a linear transformation S and let λ be an eigenvalue of T. Clearly,

$$\{\mathbf{0}\} \subset \mathcal{N}(T - \lambda \mathrm{I}) \subset \mathcal{N}(T - \lambda \mathrm{I})^2 \subset \cdots .$$

Because each $\mathcal{N}(T - \lambda \mathrm{I})^k$ is a finite dimensional subspace of dimension at most d, there exists a smallest positive integer k such that $\mathcal{N}(T - \lambda \mathrm{I})^k = \mathcal{N}(T - \lambda \mathrm{I})^{k+1}$. Denote it by $r(\lambda)$ and let $M(\lambda) = \mathcal{N}(T - \lambda \mathrm{I})^{r(\lambda)}$, which is called the *generalized eigenspace for* λ.

Lemma *The following hold:*

(a) $\mathcal{N}(T - \lambda \mathrm{I})^k = M(\lambda)$ *for all* $k \geq r(\lambda)$*;*

(b) $M(\lambda) = \{\mathbf{v} : (T - \lambda \mathrm{I})^k(\mathbf{v}) = \mathbf{0}$ *for some* $k \geq 1\}$*;*

(c) $r(\lambda)$ *is the maximum degree of the generalized vectors for* λ*.*

(d) $\mathrm{Dim}\left[M(\lambda)\right] \geq r(\lambda)$*, where* $\mathrm{Dim}\left[\cdot\right]$ *denotes the dimension of a subspace.*

Proof. To prove (a), it suffices to show that $\mathcal{N}(T - \lambda \mathrm{I})^k = \mathcal{N}(T - \lambda \mathrm{I})^{k+1}$ implies $\mathcal{N}(T - \lambda \mathrm{I})^{k+1} = \mathcal{N}(T - \lambda \mathrm{I})^{k+2}$. Start by letting $\mathbf{v} \in \mathcal{N}(T - \lambda \mathrm{I})^{k+2}$ so $\mathbf{0} = (T - \lambda \mathrm{I})^{k+2}(\mathbf{v}) = (T - \lambda \mathrm{I})^{k+1}(T - \lambda \mathrm{I})(\mathbf{v})$. It follows that $(T - \lambda \mathrm{I})(\mathbf{v}) \in \mathcal{N}(T - \lambda \mathrm{I})^{k+1} = \mathcal{N}(T - \lambda \mathrm{I})^k$. Therefore, $(T - \lambda \mathrm{I})^{k+1}(\mathbf{v}) = \mathbf{0}$, and $\mathcal{N}(T - \lambda \mathrm{I})^{k+2} \subset \mathcal{N}(T - \lambda \mathrm{I})^{k+1}$. Since the other inclusion always holds, the proof of part (a) is finished.

Part (b) is an immediate consequence of (a).

By part (b,) every generalized eigenvector is in $M(\lambda)$ and of degree at most $r(\lambda)$. From the definition of $r(\lambda)$, there exists $\mathbf{v} \in M(\lambda)$, which is not in $\mathcal{N}(T - \lambda \mathrm{I})^{r(\lambda)-1}$. Such a $\mathbf{v}$ is clearly a generalized eigenvector of degree $r(\lambda)$, which finishes the proof of part (c).

Finally, by part (c) there exists a generalized eigenvector $\mathbf{v}$ of degree precisely $r(\lambda)$. Applying *Proposition 4.4* to $\mathbf{v}$ produces $r(\lambda)$ linearly independent vectors in $M(\lambda)$ and shows that $\mathrm{Dim}\left[M(\lambda)\right] \geq r(\lambda)$ □

Notice that if $r(\lambda) = 1$, then $M(\lambda) = \mathcal{N}(T - \lambda \mathrm{I})$ and every generalized eigenvector is an eigenvector. When $r(\lambda) = 1$, the eigenvalue λ is called a *simple eigenvalue.*

Given a linear transformation $T : V \to V$ of a vector space V into itself, a subspace W of V is said to be a T *invariant subspace* provided $T(\mathbf{v})$ is in W, whenever $\mathbf{v}$ is in W. When W is a T invariant subspace, restricting T to W produces a linear transformation of W into itself. One way to study a linear transformation of a vector space into itself is to examine its restrictions to special T invariant subspaces. The next result begins the process of constructing these special T invariant subspaces.

Proposition 4.5 *If λ is an eigenvalue of T, then $M(\lambda)$ and $R(\lambda)$, the range of $(T - \lambda \mathrm{I})^{r(\lambda)}$, are T invariant subspaces of $\mathbb{C}^d$ such that*

$$\mathbb{C}^d = M(\lambda) \oplus R(\lambda).$$

Proof. The T invariance of both $M(\lambda)$ and $R(\lambda)$ follows from $T(T - \lambda \mathrm{I})^k = (T - \lambda \mathrm{I})^k T$. Since $\mathrm{Dim}\left[R(\lambda)\right] + \mathrm{Dim}\left[M(\lambda)\right]$ = rank plus nullity of $(T - \lambda \mathrm{I})^{r(\lambda)}$, which equals d, it only remains to show that

$$M(\lambda) \cap R(\lambda) = \{\mathbf{0}\}.$$

Suppose $\mathbf{0} \neq \mathbf{v} = (T - \lambda \mathrm{I})^{r(\lambda)}(\mathbf{u}) \in R(\lambda)$ and $(T - \lambda \mathrm{I})^{r(\lambda)}(\mathbf{v}) = \mathbf{0}$. Then $\mathbf{u}$ is a generalized eigenvector of degree at least $r(\lambda) + 1$ which is impossible by the previous lemma. $\square$

Recall that the eigenvalues of T are the roots of the polynomial equation

$$\mathrm{Det}\,[T - \lambda \mathrm{I}] = \mathrm{Det}\,[A - \lambda \mathrm{I}] = 0,$$

where A is the matrix of T with respect to any basis of $\mathbb{C}^d$. Because $\mathbb{C}$ is algebraically closed

$$\mathrm{Det}\,[T - \lambda \mathrm{I}] = \pm(\lambda - \lambda_1)^{m_1} \cdots (\lambda - \lambda_\rho)^{m_\rho}$$

where $\lambda_j \neq \lambda_k$ for $j \neq k$, and m_j is called the *multiplicity* of λ_j. Of course, $d = \sum_{j=0}^{\rho} m_j$. The critical linear algebra theorem needed to study $\dot{\mathbf{z}} = Az$ is

Theorem 4.6 *Let $\lambda_1, \ldots, \lambda_\rho$ be the distinct eigenvalues with multiplicities $m_1, \ldots, m_\rho$ of the linear transformation $T : \mathbb{C}^d \to \mathbb{C}^d$, and let $M(\lambda_1), \ldots, M(\lambda_\rho)$ be the corresponding generalized eigenspaces. Then the dimension of $M(\lambda_j)$ is m_j and*

$$\mathbb{C}^d = M(\lambda_1) \oplus \cdots \oplus M(\lambda_\rho).$$

Proof. Fix j, $1 \leq j \leq \rho$. By the previous proposition, there is a basis $\mathbf{u}_1, \ldots, \mathbf{u}_d$ of $\mathbb{C}^d$ such that $\mathbf{u}_1, \ldots, \mathbf{u}_\sigma$ is a basis for $M(\lambda_j)$ and $\mathbf{u}_{\sigma+1}, \ldots, \mathbf{u}_d$ is a basis for $R(\lambda_j)$. Because $M(\lambda_j)$ and $R(\lambda_j)$ are T invariant, the matrix of T with respect to this basis has the form

$$\begin{bmatrix} A_1 & \mathrm{O} \\ \mathrm{O} & A_2 \end{bmatrix},$$

where A_1 is $\sigma \times \sigma$ and A_2 is $(d-\sigma) \times (d-\sigma)$. (The matrix of all zeros is being denoted by O to help distinguish it from the the zero vector $\mathbf{0}$ and the number 0.) It follows that

$$\text{Det}\,[A_1 - \lambda \text{I}]\,\text{Det}\,[A_2 - \lambda \text{I}] = \text{Det}\,[T - \lambda \text{I}] = \pm \prod_{k=1}^{\rho} (\lambda - \lambda_k)^{m_k}.$$

If $\lambda - \lambda_j$ divides $\text{Det}\,[A_2 - \lambda \text{I}]$, then $R(\lambda_j)$ must contain an eigenvector $\mathbf{v} \neq \mathbf{0}$ for λ_j. Since every eigenvector for λ_j lies in $M(\lambda_j)$, it follows that $\mathbf{0} \neq \mathbf{v} \in \mathbf{M}(\lambda_\mathbf{j}) \cap \mathbf{R}(\lambda_\mathbf{j})$, which is impossible by *Proposition 4.5*. Therefore, $\lambda - \lambda_j$ does not divide $\text{Det}\,[A_2 - \lambda \text{I}]$ and $(\lambda - \lambda_j)^{m_j}$ divides $\text{Det}\,[A_1 - \lambda \text{I}]$.

If $(\lambda - \lambda_k)$ divides $\text{Det}\,[A_1 - \lambda \text{I}]$ for some $k \neq j$, then $M(\lambda_j)$ contains an eigenvector $\mathbf{v}$ for λ_k. Because $\mathbf{v}$ would also be a generalized eigenvalue for λ_j of some degree p, the following contradiction occurs:

$$\begin{aligned}
\mathbf{0} &= (T - \lambda_j \text{I})^p(\mathbf{v}) \\
&= (T - \lambda_j \text{I})^{p-1}(T - \lambda_j \text{I})(\mathbf{v}) \\
&= (T - \lambda_j \text{I})^{p-1}(\lambda_k - \lambda_j)(\mathbf{v}) \\
&= (\lambda_k - \lambda_j)(T - \lambda_j \text{I})^{p-1}(\mathbf{v}) \\
&\neq \mathbf{0}.
\end{aligned}$$

Therefore, $\lambda - \lambda_k$ does not divide $\text{Det}\,[A_1 - \lambda \text{I}]$ and $\text{Det}\,[A_1 - \lambda \text{I}] = \pm(\lambda - \lambda_j)^{m_j}$. It follows that $m_j = \sigma$ and that the dimension of $M(\lambda_j)$ is m_j.

Because $m_1 + \cdots + m_\rho = d$, showing that $\mathbb{C}^d = M(\lambda_1) \oplus \cdots \oplus M(\lambda_\rho)$ is now reduced to showing that $\mathbf{v}_1 + \cdots + \mathbf{v}_\rho = \mathbf{0}$, $\mathbf{v}_j \in M(\lambda_j)$, implies $\mathbf{v}_j = 0$ for all j.

First, it must be shown that $M(\lambda_j) \cap M(\lambda_k) = \{\mathbf{0}\}$ when $j \neq k$. The restriction of T to $M(\lambda_j) \cap M(\lambda_k)$ is linear transformation of $M(\lambda_j) \cap M(\lambda_k)$ into itself because each $M(\lambda_i)$ is T invariant. If $M(\lambda_j) \cap M(\lambda_k) \neq \{\mathbf{0}\}$, then $M(\lambda_j) \cap M(\lambda_k)$ must contain an eigenvector for some eigenvalue μ of T restricted to $M(\lambda_j) \cap M(\lambda_k)$ because in complex vector spaces for every eigenvalue there is eigenvector. From the first part of the proof, it follows that $\lambda_j = \mu = \lambda_k$ and $j = k$. Thus $M(\lambda_j) \cap M(\lambda_k) = \{\mathbf{0}\}$ when $j \neq$ k.

Next, note that $M(\lambda_j)$ is also invariant under $T - \lambda_k \text{I}$. Consequently, for $\mathbf{0} \neq \mathbf{v} \in M(\lambda_j)$ and $k \neq j$,

$$\mathbf{0} \neq (T - \lambda_k \text{I})^{r(\lambda_k)}(\mathbf{v}) \in M(\lambda_j)$$

because $M(\lambda_j) \cap M(\lambda_k) = \{\mathbf{0}\}$.

Now applying

$$S = (T - \lambda_2 \text{I})^{r(\lambda_2)} \cdots (T - \lambda_\rho \text{I})^{r(\lambda_\rho)}$$

to $\mathbf{v}_1 + \cdots + \mathbf{v}_\rho = \mathbf{0}$ will force a contradiction unless $\mathbf{v}_1 = \mathbf{0}$. On the one hand, repeated applications of the preceding observation with $j = 1$ and $k = 2, \ldots, \rho$, to $\mathbf{v}_1$ implies that $S(\mathbf{v}_1) \neq \mathbf{0}$ when $\mathbf{v}_1 \neq \mathbf{0}$. On the other hand, clearly $S(\mathbf{v}_j) = \mathbf{0}$ for $j \geq 2$, and hence applying S to $\mathbf{v}_1 + \cdots + \mathbf{v}_\rho = \mathbf{0}$ implies that $S(\mathbf{v}_1) = \mathbf{0}$.

The result is the contradiction that $\mathbf{0} = S(\mathbf{v}_1) \neq \mathbf{0}$, unless $\mathbf{v}_1 = \mathbf{0}$. Therefore, $\mathbf{v}_1 = \mathbf{0}$, similarly $\mathbf{v}_j = \mathbf{0}$ for $2 \leq j \leq \rho$, and the proof is complete. □

The dimension of $M(\lambda_j)$ or m_j is called the *algebraic multiplicity of* λ_j because it is the multiplicity of the root λ_j of the polynomial Det $[T - \lambda \mathrm{I}]$. The dimension of the subspace $\mathcal{N}(T - \lambda_j \mathrm{I})$ is called the *geometric multiplicity of* λ_j. It follows from the construction of $M(\lambda_j)$ that the geometric multiplicity is less than or equal to the algebraic multiplicity. There are two extremes that can occur; the geometric multiplicity can equal the algebraic multiplicity or the geometric multiplicity can equal 1. The former occurs when the eigenvalue is simple or $r(\lambda_j) = 1$ and the latter occurs when $r(\lambda_j)$ equals the algebraic multiplicity.

Given $A \in \mathcal{M}_d(\mathbb{C})$, then $T(\mathbf{v}) = A\mathbf{v}$ defines a linear transformation $T : \mathbb{C}^d \to \mathbb{C}^d$ and associates with A generalized eigenspaces, $M(\lambda_1), \ldots, M(\lambda_\rho)$. In other words, by thinking of A as the matrix of a linear transformation T, the previous theorem applies to a matrix as well as a linear transformation and the eigenvalues are the roots of Det $[A - \lambda \mathrm{I}] = 0$. In this context a subspace W of $\mathbb{C}^d$ is A invariant if $A\mathbf{v}$ is in W whenever $\mathbf{v}$ is in W. The $M(\lambda_j)$ are A invariant subspaces.

The key idea that will be used in applying the previous theorem to $\dot{\mathbf{z}} = A\mathbf{z}$ is to reduce the calculation of $e^{At}\mathbf{v}$ to finite sums. For example, consider the very special case of a single eigenvalue λ with $\mathbb{C}^d = M(\lambda)$. Because $(A - \lambda \mathrm{I})$ and $\lambda \mathrm{I}$ commute $e^{At} = e^{\lambda \mathrm{I} t} e^{(A-\lambda \mathrm{I})t}$. Clearly, $e^{\lambda \mathrm{I} t} = e^{\lambda} \mathrm{I}$. Since $(A - \lambda \mathrm{I})^{r(\lambda)}\mathbf{v} = \mathbf{0}$, it follows that

$$e^{(A-\lambda \mathrm{I})t}\mathbf{v} = \lim_{n \to \infty} \left[\sum_{k=0}^{n} \frac{t^k}{k!}(A - \lambda \mathrm{I})^k \mathbf{v} \right] = \sum_{k=0}^{r(\lambda)-1} \frac{t^k}{k!}(A - \lambda \mathrm{I})^k \mathbf{v}$$

and

$$e^{At}\mathbf{v} = e^{\lambda t}\,\mathrm{I}\, e^{(A-\lambda)t}\mathbf{v} = e^{\lambda t} \sum_{k=0}^{r(\lambda)-1} \frac{t^k}{k!}(A - \lambda \mathrm{I})^k \mathbf{v}.$$

The next theorem uses this idea and *Theorem 4.6* to calculate $e^{AT}\boldsymbol{\zeta}$ more generally.

Theorem 4.7 *Let $\lambda_1, \ldots, \lambda_\rho$ be the distinct eigenvalues of the $d \times d$ complex matrix A and let $M(\lambda_1), \ldots, M(\lambda_\rho)$ be the corresponding generalized eigenspaces for A. If $\boldsymbol{\zeta} = \mathbf{v}_1 + \cdots + \mathbf{v}_\rho$ and $\mathbf{v}_j \in M(\lambda_j)$, then the solution of the initial-value problem*

$$\begin{aligned} \dot{\mathbf{z}} &= A\mathbf{z} \\ \boldsymbol{\zeta} &= \mathbf{z}(0) \end{aligned}$$

is given by

$$\mathbf{z}(t) = \sum_{j=1}^{\rho} e^{\lambda_j t} \left[\sum_{k=0}^{r(\lambda_j)-1} (A - \lambda_j \mathrm{I})^k \frac{t^k}{k!} \right] \mathbf{v}_j. \tag{4.2}$$

Proof. Since $\mathbf{z}(t) = e^{At}\boldsymbol{\zeta} = e^{At}\mathbf{v}_1 + \cdots + e^{At}\mathbf{v}_\rho$, it suffices to show that

$$e^{At}\mathbf{v}_j = e^{\lambda_j t}\left[\sum_{k=0}^{r(\lambda_j)-1}(A-\lambda_j \mathrm{I})^k \frac{t^k}{k!}\right]\mathbf{v}_j.$$

Because $(A - \lambda_j \mathrm{I})$ and $\lambda_j \mathrm{I}$ commute

$$e^{At} = e^{\lambda_j \mathrm{I} t} e^{(A-\lambda_j \mathrm{I})t} = e^{\lambda_j t} e^{(A-\lambda_j \mathrm{I})t}.$$

Finally, because $\mathbf{v}_j \in M(\lambda_j)$

$$\begin{aligned} e^{(A-\lambda_j \mathrm{I})t}\mathbf{v}_j &= \lim_{n\to\infty}\left[\sum_{k=0}^{n}(A-\lambda_j \mathrm{I})^k \frac{t^k}{k!}\mathbf{v}_j\right] \\ &= \sum_{k=0}^{r(\lambda_j)-1}(A-\lambda_j \mathrm{I})^k \frac{t^k}{k!}\mathbf{v}_j \end{aligned}$$

and when combined with the previous steps, gives the result. □

When A is a real matrix, the linear differential equation $\dot{\mathbf{x}} = A\mathbf{x}$ defines a flow on $\mathbb{R}^d$ given by $\mathbf{x}(t,\boldsymbol{\xi}) = e^{At}\boldsymbol{\xi}$. As a result of the previous theorem, $e^{At}\boldsymbol{\xi}$ can be calculated by treating $\boldsymbol{\xi}$ as an element of $\mathbb{C}^d$ and decomposing $\boldsymbol{\xi}$ as $\boldsymbol{\xi} = \mathbf{v}_1 + \cdots + \mathbf{v}_\rho$ with $\mathbf{v}_j \in M(\lambda_j)$. Even though some of the terms $e^{At}\mathbf{v}_j$ will lie outside of $\mathbb{R}^d$, their sum, $e^{At}\boldsymbol{\xi}$, will lie in $\mathbb{R}^d$ and describe an orbit of the flow.

Theorem 4.7 provides a way of calculating $e^{At}\boldsymbol{\zeta}$ but falls short of explicitly determining e^{At}. Calculating e^{At} requires a carefully constructed basis, $\mathbf{v}_1, \ldots, \mathbf{v}_d$ for which $e^{At}\mathbf{v}_j$ can be readily computed. The logical way to obtain such a basis is to construct natural basis for each $M(\lambda_j)$, and then apply *Theorem 4.6* and *Theorem 4.7*. This program will be carried out in Section 4.3.

EXERCISES

1. Show that subspaces of $\mathbb{R}^d$ and $\mathbb{C}^d$ are closed sets.

2. Let W be a subspace of $\mathbb{R}^d$ and let A be in $\mathcal{M}_d(\mathbb{R})$. Show that if W is A invariant ($\mathbf{v} \in W$ implies $A\mathbf{v} \in W$), then W is an invariant set for the flow $\mathbf{x}(t,\boldsymbol{\xi}) = e^{At}\boldsymbol{\xi}$.

3. Let $T : \mathbb{C}^3 \to \mathbb{C}^3$ be the linear transformation whose matrix with respect to the standard basis $\mathbf{e}_1, \mathbf{e}_2, \mathbf{e}_3$ is

$$\begin{bmatrix} 2 & 1 & 0 \\ 0 & 2 & 1 \\ 0 & 0 & 2 \end{bmatrix}.$$

Show that $\mathbf{e}_3$ is a generalized eigenvector of degree 3 for the eigenvalue 2. Calculate the vectors

$$\left(T^{-1}-(1/2)\mathrm{I}\right)\mathbf{e}_3,\ \left(T^{-1}-(1/2)\mathrm{I}\right)^2\mathbf{e}_3\ \left(T^{-1}-(1/2)\mathrm{I}\right)^3\mathbf{e}_3$$

and then calculate the matrix of T^{-1} with respect to the basis

$$\left(T^{-1}-(1/2)\mathrm{I}\right)^2\mathbf{e}_3,\ \left(T^{-1}-(1/2)\mathrm{I}\right)\mathbf{e}_3,\ \mathbf{e}_3.$$

4. Let $T:\mathbb{C}^d\to\mathbb{C}^d$ be an invertible linear transformation. Show that $\mathbf{v}$ is a generalized eigenvector of degree p for the eigenvalue λ of T if and only if $\mathbf{v}$ is a generalized eigenvector of degree p for the eigenvalue $1/\lambda$ of T^{-1}.

5. Show that a complex matrix A is *diagonalizable* (For some invertible matrix B, the matrix BAB^{-1} is a diagonal matrix.) if and only if every eigenvalue is simple.

6. Let A be a matrix in $\mathcal{M}_d(\mathbb{C})$, let $\mathbf{w}$ be a nonzero vector in $\mathbb{C}^d$, and let μ be a nonzero complex number that is not an eigenvalue of A. Determine the function $\mathbf{z}(t,\tau,\zeta,\mu)$ for $\dot{\mathbf{z}}=A\mathbf{z}+e^{\mu t}\mathbf{w}$.

7. Use the results in this section to show that $A^d=\mathrm{O}$ if $a_{jk}=0$ when $j\geq k$.

8. Let V be a subspace of $\mathbb{R}^d$ and set

$$W=\{\mathbf{z}\in\mathbb{C}^d:\mathbf{z}=\mathbf{u}+i\mathbf{v}\text{ for some }\mathbf{u}\text{ and}\mathbf{v}\in V\}.$$

Show that the dimensions of V and W are equal as subspaces of $\mathbb{R}^d$ and $\mathbb{C}^d$, respectively.

9. Let λ be a real eigenvalue of multiplicity m for a real $d\times d$ matrix A and set

$$M_{\mathbb{R}}(\lambda)=\left\{\mathbf{x}\in\mathbb{R}^d:(A-\lambda\mathrm{I})^k\mathbf{x}=\mathbf{0}\text{ for some }k>0\right\}.$$

Show that $M_{\mathbb{R}}(\lambda)$ is a subspace of $\mathbb{R}^d$ of dimension m and that

$$\mathbb{R}^d=M_{\mathbb{R}}(\lambda)\oplus R_{\mathbb{R}}(\lambda)$$

where

$$R_{\mathbb{R}}(\lambda)=\left\{\mathbf{u}\in\mathbb{R}^d:\mathbf{u}=(A-\lambda\mathrm{I})^{r(\lambda)}\mathbf{v}\text{ for some }\mathbf{v}\in\mathbb{R}^d\right\}.$$

4.3 Canonical Forms

Theorem 4.6 reduces the study a single linear transformation T to its behavior on the generalized eigenspaces $M(\lambda_1)$, ..., $M(\lambda_p)$. When T is restricted to $M(\lambda_j)$, its characteristic polynomial is $(\lambda - \lambda_j)^{m_j}$. Thus, to probe deeper into the structure of linear transformations, it is only necessary to investigate those linear transformations T with exactly one eigenvalue. This will involve a further decomposition of $M(\lambda_j)$ and will lead to a basis for which the matrix of T can be explicitly calculated. These special matrices of a linear transformation are known as canonical forms and fully describe the structure of the transformation. The exponential of a matrix A in canonical form is readily calculated, and thus, in principal, e^{At} can be calculated for any A.

Let W be an m-dimensional subspace of an n-dimensional vector space V. A set of vectors $\mathbf{v}_1, \ldots, \mathbf{v}_q \in V$ is said to be *linearly independent over* W provided that whenever

$$\alpha_1\mathbf{v}_1 + \cdots + \alpha_q\mathbf{v}_q \in W$$

then

$$\alpha_1 = \alpha_2 = \cdots = \alpha_q = 0.$$

Observe that a set of vectors is linearly independent if and only if it is linearly independent over the trivial subspace $W = \{\mathbf{0}\}$

Proposition 4.8 *Let W be an m-dimensional subspace of the n-dimensional vector space V, and let the vectors $\mathbf{v}_1, \ldots, \mathbf{v}_q$ be linearly independent over W. Then the following hold:*

(a) If $\mathbf{u}_1, \ldots, \mathbf{u}_r$ are linearly independent vectors in W, then

$$\mathbf{v}_1, \ldots, \mathbf{v}_q, \mathbf{u}_1, \ldots, \mathbf{u}_r$$

are linearly independent in V and $q \leq n - m$.

(b) If $q < n - m$, there exist vectors $\mathbf{v}_{q+1}, \ldots, \mathbf{v}_{n-m}$ such that $\mathbf{v}_1, \ldots, \mathbf{v}_{n-m}$ are linearly independent over W.

Proof. Exercise.

Fix λ_j, an eigenvalue of $T : \mathbb{C}^d \to \mathbb{C}^d$. Recall that $M(\lambda_j)$ is T-invariant, that is $T(M(\lambda_j)) \subset M(\lambda_j)$. The goal is to understand the action of T on $M(\lambda_j)$, in other words, study a linear transformation with one eigenvalue λ_j. For convenience, we simplify the notation by setting $\lambda = \lambda_j$, $r = r(\lambda_j)$, and $W_k = \mathcal{N}(T - \lambda\mathrm{I})^k$ for $k = 0, \ldots, r$. Note that $W_j \neq W_{j+1}$ for $j = 0, \ldots, r-1$ and

$$\mathbf{0} = W_0 \subset W_1 \subset W_2 \subset \ldots \subset W_{r-1} \subset W_r = M(\lambda).$$

Let $S = T - \lambda\mathrm{I}$ restricted to $M(\lambda) = W_r$, so $\mathcal{N}(S^k) = W_k$ and $T = S + \lambda\mathrm{I}$ on W_r. Note that $S(W_k) \subset W_{k-1}$. Set $p(k) = \mathrm{Dim}\, W_k - \mathrm{Dim}\, W_{k-1}$ for $k = 1, \ldots, r$.

Lemma *If $\mathbf{u}_1, \ldots, \mathbf{u}_q$ are in W_{k+1} and linearly independent over W_k, then* *(a)* $S(\mathbf{u}_1), \ldots, S(\mathbf{u}_q)$ *are in* W_k,

(b) $S(\mathbf{u}_1), \ldots, S(\mathbf{u}_q)$ *are linearly independent over* W_{k-1},

(c) $p(k+1) \leq p(k)$.

Proof. Obviously, $S(\mathbf{u}_j) \in W_k$ for $j = 1, \ldots q$. Suppose there exists $\alpha_j \in \mathbb{C}$, $1 \leq j \leq q$ such that

$$\alpha_1 S(\mathbf{u}_1) + \cdots + \alpha_q S(\mathbf{u}_q) \in W_{k-1}.$$

Because W_{k-1} is the null space of S^{k-1},

$$S^{k-1}\big(\alpha_1 S(\mathbf{u}_1) + \cdots + \alpha_q S(\mathbf{u}_q)\big) = \mathbf{0}$$

or

$$S^k(\alpha_1 \mathbf{u}_1 + \cdots + \alpha_q \mathbf{u}_q) = \mathbf{0}.$$

Thus $\alpha_1 \mathbf{u}_1 + \cdots + \alpha_q \mathbf{u}_q \in W_k$, the null space of S^k, and the α_j are all 0 by hypothesis. Since the maximum number of vectors in W_j that are linearly independent over W_{j-1} is precisely $p(j)$ by *Proposition 4.8*, it follows that $p(k+1) \leq p(k)$. □

The next step is to construct a special basis for W_r for which the matrix of T restricted to W_r has a particularly nice form. First, select a set of vectors

$$\mathcal{B}_r = \{\mathbf{v}_1, \ldots, \mathbf{v}_{p(r)}\}$$

in W_r that are linearly independent over W_{r-1}. By the above lemma, $S(\mathbf{v}_1)$, $\ldots$, $S(\mathbf{v}_{p(r)})$ are in W_{r-1} and are linearly independent over W_{r-2}. By part (b) of *Proposition 4.8*, there exist vectors $\mathbf{v}_{p(r)+1}, \ldots, \mathbf{v}_{p(r-1)}$ in W_{r-1} such that the set of vectors

$$\mathcal{B}_{r-1} = \{S(\mathbf{v}_1), \ldots, S(\mathbf{v}_{p(r)}), \mathbf{v}_{p(r)+1}, \ldots, \mathbf{v}_{p(r-1)}\}$$

is linearly independent over W_{r-2}. [It can happen that $p(r) = p(r-1)$ and no additional vectors are selected.] Now the lemma applies and

$$S^2(\mathbf{v}_1), \ldots, S^2(\mathbf{v}_{p(r)}), S(\mathbf{v}_{p(r)+1}), \ldots, S(\mathbf{v}_{p(r-1)})$$

are linearly independent over W_{r-2} and the process can be repeated to construct $\mathcal{B}_{r-2}$. After r steps the set of vectors

$$\mathcal{B}_1 = \{S^{r-1}(\mathbf{v}_1), \ldots, S^{r-2}(\mathbf{v}_{p(r)+1}), \ldots, \mathbf{v}_{p(1)}\}$$

is in W_1 and is linearly independent (over $\{\mathbf{0}\}$). At this point, the process terminates because another application of S sends all the vectors in $\mathcal{B}_1$ to $\mathbf{0}$. Thus sets of vectors $\mathcal{B}_k$ in W_k that are linearly independent over W_{k-1} have been systematically constructed for $1 \leq k \leq r$.

Set

$$\mathcal{B} = \bigcup_{j=1}^{r} \mathcal{B}_j$$

The total number of vectors in $\mathcal{B}$ is given by

$$\sum_{k=1}^{r} p(k) = \sum_{k=1}^{r} (\text{Dim}\,[W_k] - \text{Dim}\,[W_{k-1}]) = \text{Dim}\,[W_r].$$

Applying part (a) of *Proposition 4.8* to $\mathcal{B}_2$ in W_2 and $\mathcal{B}_1$ in W_1, which is a subspace of W_2, shows that $\mathcal{B}_1 \cup \mathcal{B}_2$ is a linearly independent set of vectors in W_2. Repeated applications of the same argument using *Proposition 4.8*, shows that $\mathcal{B}$ is a linearly independent set of vectors in W_r. Since the number of these vectors is $\text{Dim}\, W_r$, the set of vectors $\mathcal{B}$ is a basis for $W_r = M(\lambda)$.

To calculate the matrix of T restricted to W_r with respect to the basis $\mathcal{B}$, it is necessary to organize $\mathcal{B}$ in a different way. You can thinks of the construction of $\mathcal{B}$ as an array of increasingly longer rows of vectors $\mathcal{B}_j$ in W_j, and then shift your point of view to the columns. For example, the first column would be

$$\begin{array}{c} \mathbf{v}_1 \\ S(\mathbf{v}_1) \\ \vdots \\ S^{r-1}(\mathbf{v}_1). \end{array}$$

Before describing the entire reorganization of $\mathcal{B}$, it is worthwhile to use the first column to show why this is a fruitful point of view.

Let V_1 denote the span of $\mathbf{v}_1, S(\mathbf{v}_1), \ldots, S^{r-1}(\mathbf{v}_1)$. Because $\mathcal{B}$ is a set of linearly independent vectors, the subset of vectors

$$\{S^{r-1}(\mathbf{v}_1), S^{r-2}(\mathbf{v}_1), \ldots S^2(\mathbf{v}_1), S(\mathbf{v}_1), \mathbf{v}_1\}$$

is also linearly independent and forms a basis for V_1 and $\text{Dim}\,[V_1] = r$. (Although the basis vectors could be written in any order, the order used is the more usual choice.) The subspace V_1 is S invariant. In fact, S maps the first vector to zero, and the others map to the preceding vector in the basis. Thus the $r \times r$ matrix of S restricted to V_1 has the simple form

$$\begin{bmatrix} 0 & 1 & 0 & \cdots & 0 & 0 \\ 0 & 0 & 1 & \cdots & 0 & 0 \\ \vdots & \vdots & \vdots & \ddots & \vdots & \vdots \\ 0 & 0 & 0 & \cdots & 0 & 1 \\ 0 & 0 & 0 & \cdots & 0 & 0 \end{bmatrix}.$$

Since T restricted to W_r is $\lambda\text{I} + S$, the subspace V_1 is also T invariant and the matrix of T restricted to V_1 is

$$\begin{bmatrix} \lambda & 1 & 0 & \cdots & 0 & 0 \\ 0 & \lambda & 1 & \cdots & 0 & 0 \\ \vdots & \vdots & \vdots & \ddots & \vdots & \vdots \\ 0 & 0 & 0 & \cdots & \lambda & 1 \\ 0 & 0 & 0 & \cdots & 0 & \lambda \end{bmatrix}. \tag{4.3}$$

Consequently, with enough patience to carefully determine the basis $\mathcal{B}$ of W_r and then construct the decomposition of W_r into $p(1)$ subspaces V_k, the matrix of T restricted to $W_r = M(\lambda)$ will have the form of a block diagonal matrix made up of matrices of the form shown above in *(4.3)*.

The process used to construct the basis $\mathcal{B}$ for $W_r = M(\lambda)$ produced a special set of vectors, $\mathbf{v}_1, \ldots, \mathbf{v}_{p(1)}$. Following the above construction of V_1, these vectors can be used to construct a decomposition of $M(\lambda)$ into smaller invariant subspaces. For $k = 1, \ldots, p(1)$, which is the number of columns when we think of $\mathcal{B}$ as an array of rows $\mathcal{B}_j$, set

$$V_k = \operatorname{Span}\left\{S^j(\mathbf{v}_k) : j \geq 0\right\}.$$

If $p(q-1) < k \leq p(q)$, then $S^{q-1}(\mathbf{v}_k), \ldots, S(\mathbf{v}_k), \mathbf{v}_k$ are nonzero vectors in $\mathcal{B}$ and form a basis for V_k. Clearly, V_k is S, and thus, T invariant because T equals $\lambda\mathrm{I} + S$ on $W_r = M(\lambda)$. Since each vector in $\mathcal{B}$ is in exactly one V_k, it follows that

$$M(\lambda) = V_1 \oplus V_2 \oplus \cdots \oplus V_{p(1)}.$$

Furthermore, this decomposition of $M(\lambda)$ has the following properties

(a) The number of subspaces V_k or $p(1)$ is precisely the dimension of $\mathcal{N}(T - \lambda\mathrm{I})$ or the geometric multiplicity.

(b) The dimension of V_1 is $r = r(\lambda)$.

(c) The dimension of V_k is at most $r(\lambda)$ for all k.

(d) With respect to the basis $\{S^{q-1}(\mathbf{v}_k), \ldots, S(\mathbf{v}_k), \mathbf{v}_k\}$ of V_k the matrix of S restricted to V_k is

$$\begin{bmatrix} 0 & 1 & 0 & \cdots & 0 & 0 \\ 0 & 0 & 1 & \cdots & 0 & 0 \\ \vdots & \vdots & \vdots & \ddots & \vdots & \vdots \\ 0 & 0 & 0 & \cdots & 0 & 1 \\ 0 & 0 & 0 & \cdots & 0 & 0 \end{bmatrix}.$$

(e) With respect to the basis $\{S^{q-1}(\mathbf{v}_k), \ldots, S(\mathbf{v}_k), \mathbf{v}_k\}$ of V_k the matrix of $T = S + \lambda\mathrm{I}$ restricted V_k is

$$\hat{A}(k) = \begin{bmatrix} \lambda & 1 & 0 & \cdots & 0 & 0 \\ 0 & \lambda & 1 & \cdots & 0 & 0 \\ \vdots & \vdots & \vdots & \ddots & \vdots & \vdots \\ 0 & 0 & 0 & \cdots & \lambda & 1 \\ 0 & 0 & 0 & \cdots & 0 & \lambda \end{bmatrix}.$$

(f) There exist a basis for $M(\lambda)$ such that the matrix of T with respect to this basis has the block diagonal form

$$\text{Diag}\left[\hat{A}(1), \ldots, \hat{A}(p(1))\right]$$

and each $\hat{A}(j)$ has the form specified in part (e).

This completes the analysis of the structure of T on a single $M(\lambda)$. The next theorem extends these ideas to $\mathbb{C}^d$ using all the eigenvalues and *Theorem 4.6*. Consequently, the above construction will have to be indexed by eigenvalue as well as by the $V_1, \ldots, V_{p(1)}$.

Theorem 4.9 (Jordan Canonical Form) *If $T : \mathbb{C}^d \to C^d$ is a linear transformation with distinct eigenvalues $\lambda_1, \ldots, \lambda_\rho$, then there exist a basis for $\mathbb{C}^d$ such that the matrix of T with respect to this basis has the block diagonal form*

$$\text{Diag}\left[\hat{A}(1,1), \ldots, \hat{A}(1,d(1)), \hat{A}(2,1), \ldots, \hat{A}(2,d(2)), \ldots, \hat{A}(\rho, d(\rho))\right]$$

where $d(j) = \text{Dim}\left[\mathcal{N}(T - \lambda_j \mathrm{I})\right]$, the geometric multiplicity of λ_j, and

$$\hat{A}(j,k) = \begin{bmatrix} \lambda_j & 1 & 0 & \cdots & 0 & 0 \\ 0 & \lambda_j & 1 & \cdots & 0 & 0 \\ \vdots & \vdots & \vdots & \ddots & \vdots & \vdots \\ 0 & 0 & 0 & \cdots & \lambda_j & 1 \\ 0 & 0 & 0 & \cdots & 0 & \lambda_j \end{bmatrix}.$$

Furthermore, $\hat{A}(j,k)$ is at most $r(\lambda_j) \times r(\lambda_j)$, and $\hat{A}(j,1)$ is $r(\lambda_j) \times r(\lambda_j)$.

Proof. For each eigenvalue λ_j, construct a basis for $M(\lambda_j)$ of the form $\mathcal{B}$. Let $\mathcal{B}(j,k) = \{S^{q-1}(\mathbf{v}_k), \ldots, S(\mathbf{v}_k), \mathbf{v}_k\}$ be the basis of the $V_{(j,k)}$ subspace of $M(\lambda_j)$ as constructed in the preceding discussion. By *Theorem 4.6*, the amalgamation of the bases $\mathcal{B}(j,k)$s is a basis for $\mathbb{C}^d$. The matrix of T with respect to this basis must be a block diagonal matrix because each $V_{(j,k)}$ is T invariant. The rest follows from the analysis of T on $M(\lambda)$ immediately preceding the statement of the theorem. □

The next theorem shows how e^{At} can be computed from the Jordan canonical form of A.

Theorem 4.10 *Let A be a complex $d \times d$ matrix having distinct eigenvalues $\lambda_1, \ldots, \lambda_\rho$. Then there exists an invertible matrix B and a matrix $\hat{A}$ such that $e^{At} = Be^{\hat{A}t}B^{-1}$ and $e^{\hat{A}t}$ has the block diagonal form*

$$\text{Diag}\left[e^{\hat{A}(1,1)t}, \ldots, e^{\hat{A}(1,d(1))t}, \ldots, e^{\hat{A}(\rho,d(\rho))t}\right]$$

where $d(j) = \text{Dim}\left[\mathcal{N}(A - \lambda_j \mathrm{I})\right]$ *and*

$$e^{\hat{A}(j,k)t} = e^{\lambda_j t} \begin{bmatrix} 1 & t & t^2/2 & \cdots & t^{q-1}/(q-1)! \\ 0 & 1 & t & \cdots & t^{q-2}/(q-2)! \\ \vdots & \vdots & \vdots & \ddots & \vdots \\ 0 & 0 & 0 & \cdots & t \\ 0 & 0 & 0 & \cdots & 1 \end{bmatrix}. \tag{4.4}$$

Furthermore, $\hat{A}(j,k)$ *is at most* $r(\lambda_j) \times r(\lambda_j)$*, and* $\hat{A}(j,1)$ *is* $r(\lambda_j) \times r(\lambda_j)$.

Proof. Let T be the linear transformation given by $T(\mathbf{z}) = A\mathbf{z}$, so A is the matrix of T with respect to the usual basis for $\mathbb{C}^d$. By the previous theorem, there exists another basis such that the matrix of T with respect to this new basis has the form

$$\hat{A} = \text{Diag}\left[\hat{A}(1,1), \ldots, \hat{A}(1,d(1)), \ldots \hat{A}(\rho, d(\rho))\right].$$

Since A and $\hat{A}$ are matrices for the same linear transformation, there exists a $d \times d$ invertible matrix B such that $A = B\hat{A}B^{-1}$. It follows from the continuity of matrix multiplication and *Theorem 4.1* that

$$e^{At} = Be^{\hat{A}t}B^{-1}$$

and that

$$e^{\hat{A}t} = \text{Diag}\left[e^{\hat{A}(1,1)t}, \ldots, e^{\hat{A}(\rho,d(\rho))t}\right].$$

It remains to compute

$$e^{\hat{A}(j,k)t} = e^{\lambda_j t} e^{(\hat{A}(j,k) - \lambda_j \mathrm{I})t}$$

using $T = \lambda_j \mathrm{I} + S$ as before. Because

$$\hat{A}(j,k) - \lambda_j \mathrm{I} = \begin{bmatrix} 0 & 1 & 0 & \cdots & 0 & 0 \\ 0 & 0 & 1 & \cdots & 0 & 0 \\ \vdots & \vdots & \vdots & \ddots & \vdots & \vdots \\ 0 & 0 & 0 & \cdots & 0 & 1 \\ 0 & 0 & 0 & \cdots & 0 & 0 \end{bmatrix},$$

it follows that

$$\left(\hat{A}(j,k) - \lambda_j \mathrm{I}\right)^2 = \begin{bmatrix} 0 & 0 & 1 & 0 & \cdots & 0 & 0 \\ 0 & 0 & 0 & 1 & \cdots & 0 & 0 \\ \vdots & \vdots & \vdots & \vdots & \ddots & \vdots & \vdots \\ 0 & 0 & 0 & 0 & \cdots & 0 & 1 \\ 0 & 0 & 0 & 0 & \cdots & 0 & 0 \\ 0 & 0 & 0 & 0 & \cdots & 0 & 0 \end{bmatrix},$$

and so forth. Since $\left(\hat{A}(j,k) - \lambda_j \mathrm{I}\right)^q = 0$ (assuming $\hat{A}(j,k)$ is $q \times q$), $e^{(\hat{A}(j,k)-\lambda_j \mathrm{I})t}$ is a finite calculation and has the required form. □

Theorem 4.10 has several immediate corollaries that are worth noting.

Corollary *Let $\lambda_1, \ldots, \lambda_\rho$ be the distinct eigenvalues with multiplicities $m_1, \ldots, m_\rho$ of $A \in \mathcal{M}_d(\mathbb{C})$. Then e^{λ_j}, $j = 1, \ldots, \rho$ is an eigenvalue of e^A with multiplicity*

$$\sum \left\{ m_k : e^{\lambda_k} = e^{\lambda_j} \right\}.$$

Proof. Let $\hat{A}$ be the Jordan canonical form of A. Since $\hat{A}$ and $e^{\hat{A}}$ are upper triangular, the eigenvalues and their multiplicities can be read off the diagonal of $\hat{A}$ and $e^{\hat{A}}$. In particular, the eigenvalues of $e^{\hat{A}}$ are exponentials of the eigenvalues of $\hat{A}$, but they are not necessarily distinct. Hence, the multiplicities are given by $\sum \left\{ m_k ; e^{\lambda_k} = e^{\lambda_j} \right\}$.

Since there exists an invertible matrix B such that $e^{At} = B e^{\hat{A}t} B^{-1}$ by *Theorem 4.10*, the matrices e^A and $e^{\hat{A}}$ have the same eigenvalues with the same multiplicities. □

Corollary *The identity matrix equals e^A if and only if*

$$A = B \mathrm{Diag}\left[2\pi k_1 i, \ldots, 2\pi k_d i\right] B^{-1}$$

for $k_j \in \mathbb{Z}$ and $B \in GL_d(\mathbb{C})$.

Proof. Suppose $e^A = \mathrm{I}$, the $d \times d$ identity matrix. Then $e^{\hat{A}} = \mathrm{I}$, where $\hat{A}$ is the Jordan canonical form of A. It follows from *Theorem 4.10* that $\hat{A}$ must be a diagonal matrix. The previous Corollary implies that the eigenvalues of A are all of the form $2\pi k i$, $k \in \mathbb{Z}$. Therefore,

$$\hat{A} = \mathrm{Diag}\left[2\pi k_1 i, \ldots, 2\pi k_n i\right]$$

with $k_j \in \mathbb{Z}$. □

It is also possible to derive a *real canonical form* for real matrices using *Theorem 4.9*. The argument relies on the fact that complex eigenvalues occur in conjugate pairs when the matrix is real. Like the Jordan form, the real canonical form is a block diagonal with the blocks determined by the eigenvalues. We will derive the form of the blocks without formally stating the result.

Any vector $\mathbf{w} \in \mathbb{C}^d$ can be written as $\mathbf{w} = \mathbf{u} + i\mathbf{v}$, where $\mathbf{u}$, $\mathbf{v} \in \mathbb{R}^d$. Set $\overline{\mathbf{w}} = \mathbf{u} - i\mathbf{v}$. It is easy to see that given $\mathbf{u}_1, \ldots, \mathbf{u}_k \in \mathbb{R}^d$, they are linearly independent over $\mathbb{R}$ or span $\mathbb{R}^d$ if and only if they are also linearly independent over $\mathbb{C}$ or span$\mathbb{C}^d$ when viewed as vectors in $\mathbb{C}^d$.

Let $T : \mathbb{R}^d \to \mathbb{R}^d$ be a linear transformation and extend T to $\mathbb{C}^d$ by setting $T(\mathbf{w}) = T(\mathbf{u} + i\mathbf{v}) = T(\mathbf{u}) + iT(\mathbf{v})$. Clearly, this extension is linear,

$$T(\overline{\mathbf{w}}) = \overline{T(\mathbf{w})},$$

and $T(\mathbf{w}) = \mathbf{0}$ if and only if $T(\mathbf{u}) = \mathbf{0} = T(\mathbf{v})$.

Let λ be a real eigenvalue for T and verify for $\mathbf{w} = \mathbf{u}+i\mathbf{v}$ that $\mathbf{w} \in \mathcal{N}(T-\lambda\mathrm{I})^k$ if and only if $\mathbf{u}, \mathbf{v} \in \mathcal{N}(T - \lambda\mathrm{I})^k$. It follows that the subspace

$$\{\mathbf{u} \in \mathbb{R}^n : (T - \lambda\mathrm{I})^k(\mathbf{u}) = \mathbf{0}\}$$

of $\mathbb{R}^d$ has the same dimension as $\mathcal{N}(T - \lambda\mathrm{I})^k$ in $\mathbb{C}^d$. Therefore, the basis $\mathcal{B}$ for $M(\lambda)$ can be constructed using vectors from $\mathbb{R}^d$, and there are invariant subspaces of $\mathbb{R}^d$ and bases for which the matrix of T has the same form as in the Jordan canonical form for T extended to $\mathbb{C}^d$.

Next, consider a complex eigenvalue $\lambda = \alpha + i\beta$, $\beta \neq 0$. Then $\overline{\lambda} = \alpha - i\beta$ is also an eigenvalue of T. The strategy is to construct a suitable basis of real vectors for the invariant subspace $M(\lambda) \oplus M(\overline{\lambda})$, so the size of the diagonal block for λ will be twice the multiplicity of λ. It is easy to see that

$$M(\overline{\lambda}) = \{\overline{\mathbf{w}} : \mathbf{w} \in M(\lambda)\}$$

and the conjugate $\overline{\mathcal{B}}$ of the basis $\mathcal{B}$ for $M(\lambda)$ is the same kind of a basis for $M(\overline{\lambda})$.

Fix $\mathbf{v}_k$ in the basis $\mathcal{B}$; this determines an invariant subspace V_k of $M(\lambda)$ with basis $\{\mathbf{w}_1 = S^{q-1}(\mathbf{v}_k), \ldots, S(\mathbf{v}_k), \mathbf{w}_q = \mathbf{v}_k\}$ where q is the degree of the generalized eigenvector $\mathbf{v}_k$ and $S = T - \lambda\mathrm{I}$. It follows from the Jordan canonical form that

$$T(\mathbf{w}_k) = \lambda\mathbf{w}_k + \mathbf{w}_{k-1}$$

for $1 < k \leq q$ and

$$T(\mathbf{w}_1) = \lambda\mathbf{w}_1.$$

Set $V_k' = \{\mathbf{v} + \overline{\mathbf{w}} : \mathbf{v}, \mathbf{w} \in V_k\}$. Clearly V_k' is invariant and

$$M(\lambda) \oplus M(\overline{\lambda}) = V_1' \oplus \cdots \oplus V_{p(1)}'.$$

Next, set

$$\mathbf{u}_j = (\mathbf{w}_j + \overline{\mathbf{w}}_j)/2 \in \mathbb{R}^d$$

and

$$\mathbf{u}_j' = (\mathbf{w}_j - \overline{\mathbf{w}}_j)/2i \in \mathbb{R}^d.$$

It is easy to see that $\{\mathbf{u}_1, \mathbf{u}_1', \ldots, \mathbf{u}_q, \mathbf{u}_q'\}$ is a basis for V_k'.

The last step is to calculate the matrix of T on V_k' with respect to this basis. This is done by verifying that

$$\begin{aligned}
T(\mathbf{u}_1) &= \alpha\mathbf{u}_1 - \beta\mathbf{u}_1' \\
T(\mathbf{u}_1') &= \beta\mathbf{u}_1 + \alpha\mathbf{u}_1' \\
T(\mathbf{u}_k) &= \alpha\mathbf{u}_k - \beta\mathbf{u}_k' + \mathbf{u}_{k-1} \\
T(\mathbf{u}_k') &= \beta\mathbf{u}_k + \alpha\mathbf{u}_k' + \mathbf{u}_{k-1}'
\end{aligned}$$

from which it follows that the matrix has the form

$$\begin{bmatrix} \alpha & \beta & 1 & 0 & \cdots & 0 & 0 & 0 & 0 \\ -\beta & \alpha & 0 & 1 & \cdots & 0 & 0 & 0 & 0 \\ 0 & 0 & \alpha & \beta & \cdots & 0 & 0 & 0 & 0 \\ 0 & 0 & -\beta & \alpha & \cdots & 0 & 0 & 0 & 0 \\ \vdots & \vdots & \vdots & \vdots & \ddots & \vdots & \vdots & \vdots & \vdots \\ 0 & 0 & 0 & 0 & \cdots & \alpha & \beta & 1 & 0 \\ 0 & 0 & 0 & 0 & \cdots & -\beta & \alpha & 0 & 1 \\ 0 & 0 & 0 & 0 & \cdots & 0 & 0 & \alpha & \beta \\ 0 & 0 & 0 & 0 & \cdots & 0 & 0 & -\beta & \alpha \end{bmatrix}. \tag{4.5}$$

Therefore, each pair of conjugate eigenvalues contributes block diagonal matrices of the above form to the real canonical form.

EXERCISES

1. Prove *Proposition 4.8.*

2. Construct a Jordan canonical form $\hat{A}$ for a 10×10 matrix A whose only eigenvalue is -2 with geometric multiplicity 5 and with $r(-2) = 4$. Then calculate $e^{\hat{A}t}$.

3. Determine all the possible Jordan canonical forms for a 6×6 matrix with $\text{Det}\,[A - \lambda \text{I}] = (\lambda - 3)^2(\lambda - 5)^4$.

4. Find the principal matrix solution of $\dot{\mathbf{x}} = A\mathbf{x}$ at 0 when

$$A = \begin{bmatrix} 1 & 3 & 3 \\ 3 & 1 & 3 \\ -3 & -3 & 15 \end{bmatrix}.$$

5. Let A be a real invertible matrix. Show that if A has a pure imaginary eigenvalue, then $\dot{\mathbf{x}} = A\mathbf{x}$ has infinitely many periodic solutions.

6. Let $\{\lambda_1, \ldots, \lambda_\rho\}$ be the distinct eigenvalues of A and form the polynomial

$$p(\lambda) = \prod_{j=1}^{\rho} (\lambda - \lambda_j)^{r(\lambda_j)}.$$

   Show that

$$p(A) = \prod_{j=1}^{\rho} (A - \lambda_j \text{I})^{r(\lambda_j)} = 0.$$

   Prove that if the degree of a polynomial $q(\lambda)$ is less than that of $p(\lambda)$, then $q(A) \neq 0$. [The polynomial $p(\lambda)$ is called the minimum polynomial for A.]

7. Let μ be an eigenvalue of a matrix e^A. Show that μ is a simple eigenvalue of e^A if and only if λ is a simple eigenvalue of A when $e^\lambda = \mu$.

8. Let R be the usual $d \times d$ matrix with 1 in each $j, j+1$ entry and 0 elsewhere, and let

$$q(z) = \sum_{j=0}^{N} c_j z^j$$

be a complex polynomial of degree $N > d$. Show that

$$q(\lambda \mathrm{I} + R) = \sum_{k=0}^{d} \frac{q^{(k)}(\lambda)}{k!} R^k$$

where $q^{(k)}(\lambda)$ denotes the usual kth derivative of $q(\lambda)$. Hint: Apply the binomial theorem to $(\lambda \mathrm{I} + R)^m$ and carefully interchange the order of summation.

9. Let

$$\sum_{j=1}^{\infty} c_j z^j$$

be a complex power series with radius of convergence $r > 0$. Use the previous exercise to show that the power series of matrices

$$\sum_{j=0}^{\infty} c_j (\lambda \mathrm{I} + R)^j$$

converges when $\lambda < r$. Now prove that

$$\sum_{j=0}^{\infty} c_j A^j$$

converges when A is a $d \times d$ matrix whose eigenvalues have absolute value less than r.

10. If the absolute value of every eigenvalue of A is less than one, show that

$$\sum_{k=o}^{\infty} A^k = (\mathrm{I} - A)^{-1}$$

4.4 Higher Order Equations

The general theory developed for $\dot{\mathbf{z}} = A\mathbf{z}$ can be applied to the familiar nth-order linear differentiable equation with constant coefficients, namely,

$$x^{(n)} + a_n x^{(n-1)} + \cdots + a_2 x^{(1)} + a_1 x = 0.$$

The theoretical analysis of this higher order homogenous differential equation, which arises frequently in physics and engineering, completely explains the routine process for finding its general solution that is taught in most elementary courses on differential equations.

It can be best analyzed as a complex linear system

$$\dot{\mathbf{z}} = A\mathbf{z}$$

where A is the $d \times d$ matrix

$$A = \begin{bmatrix} 0 & 1 & 0 & \cdots & 0 & 0 \\ 0 & 0 & 1 & \cdots & 0 & 0 \\ \vdots & \vdots & \vdots & \ddots & \vdots & \vdots \\ 0 & 0 & 0 & \cdots & 1 & 0 \\ 0 & 0 & 0 & \cdots & 0 & 1 \\ -a_1 & -a_2 & -a_3 & \cdots & -a_{n-1} & -a_n \end{bmatrix}$$

The heart of the analysis is showing that the geometric multiplicity of every eigenvalue is one.

Proposition 4.11 *The eigenvalues of A are the roots of*

$$\lambda^n + a_n \lambda^{n-1} + \cdots + a_2 \lambda + a_1 = 0.$$

If λ_j is an eigenvalue of multiplicity m_j, then $\mathrm{Dim}\,[\mathcal{N}(A - \lambda_j \mathrm{I})] = d(j) = 1$. *Furthermore, $M(\lambda_j) = V_{j,1}$, and $r(\lambda_j) = m_j$.*

Proof. For the first part, it suffices to show that

$$\mathrm{Det}\,[A - \lambda \mathrm{I}] = (-1)^n \big(\lambda^n + a_n \lambda^{n-1} + \cdots + a_2 \lambda + a_1\big).$$

The proof proceeds by induction starting with $n = 2$. In this case,

$$\mathrm{Det} \begin{bmatrix} -\lambda & 1 \\ -a_1 & (-a_2 - \lambda) \end{bmatrix} = \lambda(a_2 + \lambda) + a_1 = \lambda^2 + a_2\lambda + a_1$$

Assume the formula is true for $n - 1$ and expand $\mathrm{Det}\,[A - \lambda \mathrm{I}]$ on the first column as follows:

$$\mathrm{Det}\,[A - \lambda \mathrm{I}] = \ \mathrm{Det} \begin{bmatrix} -\lambda & 1 & 0 & \cdots & 0 & 0 \\ 0 & -\lambda & 1 & \cdots & 0 & 0 \\ \vdots & \vdots & \vdots & \ddots & \vdots & \vdots \\ 0 & 0 & 0 & \cdots & -\lambda & 1 \\ -a_1 & -a_2 & -a_3 & \cdots & -a_{n-1} & (-a_n - \lambda) \end{bmatrix}$$

$$= (-1)^{1+1}(-\lambda) \quad \text{Det} \begin{bmatrix} -\lambda & 1 & 0 & \cdots & 0 & 0 \\ 0 & -\lambda & 1 & \cdots & 0 & 0 \\ \vdots & \vdots & \vdots & \ddots & \vdots & \vdots \\ 0 & 0 & 0 & \cdots & -\lambda & 1 \\ -a_2 & -a_3 & -a_4 & \cdots & -a_{n-1} & (-a_n - \lambda) \end{bmatrix} +$$

$$(-1)^{n+1}(-a_1) \quad \text{Det} \begin{bmatrix} 1 & 0 & 0 & \cdots & 0 & 0 \\ -\lambda & 1 & 0 & \cdots & 0 & 0 \\ 0 & -\lambda & 1 & \cdots & 0 & 0 \\ \vdots & \vdots & \vdots & \ddots & \vdots & \vdots \\ 0 & 0 & 0 & \cdots & 1 & 0 \\ 0 & 0 & 0 & \cdots & -\lambda & 1 \end{bmatrix}$$

$$\begin{aligned} &= (-\lambda)(-1)^{n-1}\left(\lambda^{n-1} + a_n\lambda^{n-2} + \cdots + a_3\lambda + a_2\right) + (-1)^n a_1 \\ &= (-1)^n\left(\lambda^n + a_n\lambda^{n-1} + \cdots + a_3\lambda^2 + a_2 + a_1\right). \end{aligned}$$

Consequently, the eigenvalues are the roots of

$$\lambda^n + a_{n-1}\lambda^{n-1} + \cdots + a_2\lambda + a_1 = 0.$$

Suppose λ_j is an eigenvalue of multiplicity m_j. Clearly, the rank of

$$(A - \lambda_j \mathrm{I}) = \begin{bmatrix} -\lambda_j & 1 & 0 & \cdots & 0 & 0 \\ 0 & -\lambda_j & 1 & \cdots & 0 & 0 \\ \vdots & \vdots & \vdots & \ddots & \vdots & \vdots \\ 0 & 0 & 0 & \cdots & -\lambda_j & 1 \\ -a_1 & -a_2 & -a_3 & \cdots & -a_{n-1} & (-\lambda_j - a_n) \end{bmatrix}$$

is at least $n-1$ and the nullity is at least 1 because λ_j is an eigenvalue. Therefore, the nullity or dimension of $\mathcal{N}(A - \lambda_j \mathrm{I}) = d(j) = 1$ because the rank plus nullity must equal n. It follows that $M(\lambda_j) = V_{j,1}$. Finally, m_j equals the dimension of $M(\lambda_j)$ and $r(\lambda_j)$ equals the dimension $V_{j,1}$, so they are equal. This completes the proof. □

Suppose $\lambda_1, \ldots, \lambda_\rho$ are the distinct eigenvalues of A, that is, the roots of $\lambda^n + a_n\lambda^{n-1} + \cdots + a_2\lambda + a_1 = 0$ with multiplicities $m_1, \ldots, m_\rho$. It follows from *Theorem 4.9* and *Proposition 4.11* that there exists a basis of $\mathbb{C}^n$ of the form

$$\mathbf{u}_{(1,1)}, \ldots, \mathbf{u}_{(1,n_1)}, \ldots, \mathbf{u}_{(\rho,1)}, \ldots, \mathbf{u}_{(\rho,n_\rho)}$$

such that $\mathbf{u}_{(j,1)}, \ldots, \mathbf{u}_{(j,m_j)}$ is a basis for $M(\lambda_j)$ and $(A - \lambda_j \mathrm{I})\mathbf{u}_{(j,k)} = \mathbf{u}_{(j,k-1)}$ for $k > 1$ and 0 for $k = 1$.

Then $e^{At}\mathbf{u}_{(j,k)}$, $1 \le j \le p$, $1 \le k \le m_j$, form a basis for the solutions of $\dot{\mathbf{z}} = A\mathbf{z}$, and by direct calculation using the formula $e^{At} = e^{\lambda \mathrm{I} t}e^{(A-\lambda \mathrm{I})t}$

$$e^{At}\mathbf{u}_{(j,k)} = e^{\lambda_j t}\left(\frac{t^k}{k!}\mathbf{u}_{(j,1)} + \frac{t^{k-1}}{(k-1)!}\mathbf{u}_{(j,2)} + \cdots + \mathbf{u}_{(j,k)}\right).$$

The corollary to *Theorem 3.11* guarantees that the first coordinate functions of $e^{At}\mathbf{u}_{(j,k)}$, $1 \le j \le p$, $1 \le k \le m_j$, form a basis for the space of solutions of

$$x^{(n)} + a_n x^{(n-1)} + \cdots + a_1 x = 0.$$

(Some of these solutions may be complex, which is acceptable since the equation makes sense with the a_j real or complex.)

Let $\gamma_{(j,k)}$ denote the first coordinate of the vector $\mathbf{u}_{(j,k)}$ in $\mathbb{C}^d$. Using the formula for $e^{At}\mathbf{u}_{(j,k)}$, the first coordinate functions of $e^{At}\mathbf{u}_{(j,k)}$, $1 \le k \le m_j$ are

$$\begin{aligned}
&\gamma_{(j,1)} e^{\lambda_j t}\\
&\gamma_{(j,1)} t e^{\lambda_j t} + \gamma_{(j,2)} e^{\lambda_j t}\\
&\gamma_{(j,1)} \frac{t^2}{2} e^{\lambda_j t} + \gamma_{(j,2)} t e^{\lambda_j t} + \gamma_{(j,3)} e^{\lambda_j t}\\
&\vdots\\
&\gamma_{(j,1)} \frac{t^{(m_j-1)}}{(m_j-1)!} e^{\lambda_j t} + \cdots + \gamma_{(j,m_j)} e^{\lambda_j t}.
\end{aligned}$$

Although the above functions are a basis for the solutions of

$$x^{(n)} + a_n x^{(n-1)} + \cdots + a_2 x^{(1)} + a_1 x = 0,$$

the formulas are unnecessarily cumbersome and can be greatly simplified by applying the next lemma in the vector space of complex valued functions defined on $\mathbb{R}$.

Lemma *Let V be a vector space and let $\mathbf{v}_1, \ldots, \mathbf{v}_m \in V$. If the vectors*

$$\begin{array}{lllllllll}
\alpha_1\mathbf{v}_1 & & & & & & & & \\
\alpha_1\mathbf{v}_2 & + & \alpha_2\mathbf{v}_1 & & & & & & \\
\alpha_1\mathbf{v}_3 & + & \alpha_2\mathbf{v}_2 & + & \alpha_3\mathbf{v}_1 & & & & \\
 & \vdots & & \vdots & & & & & \\
\alpha_1\mathbf{v}_m & + & \alpha_2\mathbf{v}_{m-1} & + & \alpha_3\mathbf{v}_{m-2} & + & \cdots & + & \alpha_m\mathbf{v}_1
\end{array}$$

are linearly independent, then the vectors $\mathbf{v}_1, \ldots, \mathbf{v}_m$ are linearly independent and span the same subspace as the original set of vectors.

Proof. Exercise.

This lemma applies to the functions

$$e^{\lambda_j t}, te^{\lambda_j t}, t^2 e^{\lambda_j t}, \ldots, t^{(m_j-1)} e^{\lambda_j t}$$

in the vector space of continuous complex valued functions defined on $\mathbb{R}$. Consequently, a basis for the complex vector space of solutions is given by

$$\{t^k e^{\lambda_j t} : 0 \le k < m_j \text{ and } j = 1, \ldots, \rho\}.$$

Thus far, everything works for a_j real or complex. Now, assume the a_j are all real. If $\lambda_j = \alpha_j + i\beta_j$ is an eigenvalue and $\beta_j \neq 0$, then $\overline{\lambda_j} = \alpha_j - i\beta_j$ is also an eigenvalue, and a basis for the space of real solutions is obtained by replacing $t^k e^{\lambda_j t}$ and $t^k e^{\overline{\lambda_j} t}$ with

$$\left(\frac{1}{2}\right)\left(t^k e^{\lambda_j t} + t^k e^{\overline{\lambda_j} t}\right) = t^k e^{\alpha_j t} \cos \beta_j t$$

and

$$\left(\frac{1}{2i}\right)\left(t^k e^{\lambda_j t} - t^k e^{\overline{\lambda_j} t}\right) = t^k e^{\alpha_j t} \sin \beta_j t.$$

Thus from the general theory for $\dot{\mathbf{z}} = A\mathbf{z}$, we have rigorously established the familiar but usually not justified basis of solutions for

$$x^{(n)} + a_n x^{(n-1)} + \cdots + a_2 x^{(1)} + a_1 x = 0.$$

EXERCISES

1. Prove the lemma on page 134.

2. Show that $x^{(n)} - x = 0$ has a periodic solution if and only if n is a multiple of 4.

3. Prove that the solutions of $x^{(n)} + a_n x^{(n-1)} + \cdots + a_2 x^{(1)} + a_1 x = 0$ are all bounded if and only if the real parts of the roots of $\lambda^n + a_n \lambda^{n-1} + \cdots + a_2 \lambda + a_1 = 0$ are less than or equal to zero and any pure imaginary roots have multiplicity 1.

4. Show that there exists a $2n$th-order linear differentiable equation with constant coefficients such that every nonzero solution is periodic.

5. Consider again the eigenvalue problem $\ddot{x} - \mu x = 0$ and $\varphi(0) = 0 = \varphi(L)$ with $L > 0$ but now allow μ to be a complex number. (See *Exercise 7* on page 108.) Show that μ must be real when there exists a non zero solution.

4.5 The Range of the Exponential Map

Given $A \in \mathcal{M}_d(\mathbb{C})$, the curve e^{At} in $GL_d(\mathbb{C})$ has played a central role in the preceding sections, but A can also be thought of as a variable. From this point of view, $A \to e^A$ is a map of $\mathcal{M}_d(\mathbb{C})$ into $GL_d(\mathbb{C})$ and it is natural to ask when can the equation $e^A = B$ be solved for A given B. The answer will be useful in the last section of chapter 5.

Theorem 4.12 *Let B be an $d \times d$ complex matrix. If* $\operatorname{Det} B \neq 0$*, then there exists a complex matrix A such that $e^A = B$.*

Proof. Since $Pe^AP^{-1} = e^{PAP^{-1}}$ for any invertible $d \times d$ matrix P, it can be assumed without loss of generality that B is in Jordan canonical form, that is,

$$\begin{aligned} B &= \operatorname{Diag}(B_1, \ldots, B_p), \\ B_j &= \lambda_j I + R_j, \end{aligned}$$

and

$$R_j = \begin{bmatrix} 0 & 1 & 0 & \cdots & 0 \\ 0 & 0 & 1 & \cdots & 0 \\ 0 & 0 & 0 & \cdots & 0 \\ \vdots & \vdots & \vdots & \cdots & \vdots \\ 0 & 0 & 0 & \cdots & 1 \\ 0 & 0 & 0 & \cdots & 0 \end{bmatrix}.$$

Clearly, it suffices to find A_j such that $e^{A_j} = B_j$. Dropping the subscripts for convenience, $B = \lambda I + R$ and noting that $\lambda \neq 0$ because it is an eigenvalue of the invertible matrix B, it must be shown that there exists A such that $e^A = B$.

Set

$$S = -\sum_{j=1}^{d} \frac{(-R)^j}{j\lambda^j}$$

and

$$A = (\log \lambda)I + S,$$

where $\log \lambda$ is any complex number such that $e^{\log \lambda} = \lambda$. It follows that $e^A = \lambda e^S$, and hence to complete the proof, it suffices to show that

$$e^S = \mathrm{I} + \frac{1}{\lambda} R.$$

Consider the following polynomial in the complex variable z:

$$\sum_{k=0}^{N} \frac{1}{k!} \left(-\sum_{j=1}^{N} \frac{(-z)^j}{j} \right)^k = \sum_{m=0}^{N} c_m z^m + \sum_{m=N+1}^{N^2} d_m(N) z^m.$$

Observe (by staring hard enough at the formula) that c_m is independent of N for $N > m$. It can be shown that the left-hand side of the above converges uniformly to $e^{\log(1+z)} = 1 + z$ as $N \to \infty$ on $|z| \leq r$, $0 < r < 1$. It follows that $\sum_{m=0}^{\infty} c_m z^m$ is the Taylor series for $1 + z$, and hence, $c_0 = c_1 = 1$ and $c_m = 0$ for $m \geq 2$. In particular, for $N \geq 2$, we have

$$\sum_{k=0}^{N} \frac{1}{k!} \left(-\sum_{j=1}^{N} \frac{(-z)^j}{j} \right)^k = 1 + z + \sum_{m=N+1}^{N^2} d_m(N) z^m. \tag{4.6}$$

Since any polynomial formula in z holds when z is replaced by a complex $d \times d$ matrix, replacing z by $(1/\lambda)R$ when $N > d$ yields

$$\sum_{k=0}^{N} \frac{1}{k!} S^k = I + \frac{1}{\lambda} R.$$

Hence, $e^S = I + \left(\frac{1}{\lambda}\right) R$ and the proof is complete. □

The situation with $GL_d(\mathbb{R})$ and $\mathcal{M}_d(\mathbb{R})$ is very different. Since $\operatorname{Det}[e^A] = e^{\operatorname{Tr}[A]}$ by Abel's formula, $\operatorname{Det}[e^A]$ is positive when A is a real matrix. Hence, the map $A \to e^A$ cannot map $\mathcal{M}_d(\mathbb{R})$ onto $GL_d(\mathbb{R})$ because e^A cannot have a negative determinant.

A *subgroup of the general linear group* $GL_d(\mathbb{C})$ is a subset of $GL_d(\mathbb{C})$ which is closed under matrix multiplication and taking inverses. Clearly, $GL_d(\mathbb{R})$ is a subgroup of $GL_d(\mathbb{C})$. There are quite a few other mathematically important subgroups of $GL_d(\mathbb{C})$ that have been studied extensively and are part of the theory of Lie groups. One of them, the orthogonal group, will be examined here and used later. For an introduction to the very large subject of Lie groups see [33].

The *transpose of a matrix* A is the matrix A^t whose $j\,k$ entry is the $k\,j$ entry of A. In other words, the entries above the main diagonal are switched with the entries below it. The usefulness of the transpose stems from its relationship with the dot product.

Remark *If A is a real $d \times d$ matrix, then*

$$A\mathbf{x} \cdot \mathbf{y} = \mathbf{x} \cdot A^t\mathbf{y}$$

for every $\mathbf{x}$ *and* $\mathbf{y}$ *in* $\mathbb{R}^d$.

Proof. Just calculate both sides to get

$$A\mathbf{x} \cdot \mathbf{y} = \sum_{j=1}^{d} \sum_{k=1}^{d} a_{jk} x_k y_j$$

and

$$\mathbf{x} \cdot A^t\mathbf{y} = \sum_{k=1}^{d} \sum_{j=1}^{d} a_{jk} y_j x_k.$$

Since the order of summation can be interchanged here, the formula holds. □

A matrix $A \in GL_d(\mathbb{R})$ is *orthogonal* if $A^{-1} = A^t$ and $T(\mathbf{x}) = A\mathbf{x}$ is an *orthogonal transformation* when A is orthogonal. The elementary properties of orthogonal matrices are contained in the following proposition:

Proposition 4.13 *If A and B are orthogonal matrices, then the following hold:*

(a) $A\mathbf{x} \cdot A\mathbf{y} = \mathbf{x} \cdot \mathbf{y}$ *for every* $\mathbf{x}$ *and* $\mathbf{y}$ *in* $\mathbb{R}^d$.

(b) $A^{-1} = A^{\mathrm{t}}$ *is orthogonal.*

(c) AB *is orthogonal.*

(d) $\mathrm{Det}\,[A] = 1$.

Proof. Exercise.

Let $O_d(\mathbb{R})$ denote the set of $d \times d$ real matrices that are orthogonal. Part (a) of the above proposition implies that orthogonal transformations preserve Euclidean distance and angles. It follows from (b) and (c) in the above proposition that $O_d(\mathbb{R})$ is subgroup of $GL_d(\mathbb{R})$. The orthogonal transformations along with the translations $(\mathbf{x} \to \mathbf{x} + \mathbf{w})$ are the rigid motions for the Euclidean geometry of $\mathbb{R}^d$.

The defining condition $AA^{\mathrm{t}} = \mathrm{I}$ for a matrix to be orthogonal can be written as d^2 algebraic equations. Specifically, a real matrix $A = (a_{jk})$ is orthogonal if and only if it satisfies the d^2 equations

$$\sum_{m=1}^{d} a_{jm} a_{km} = \delta_{jk}$$

where as usual $\delta_{jk} = 1$ if and only if $j = k$ and 0 otherwise.

When $j = k$, these equations require every row to have a Euclidean norm of 1. When $j \neq k$, they require every pair of rows to be orthogonal. Since A^{t} is orthogonal when A is orthogonal, the columns of an orthogonal matrix also have Euclidean norm 1 and are orthogonal to each other. Because orthogonal matrices are invertible, the columns are basis. Conversely, given a basis of orthogonal vectors of Euclidean norm 1, called an *orthonormal basis*, placing the coordinates of these vectors in rows or columns will form an orthogonal matrix.

Proposition 4.14 *The subgroup $O_d(\mathbb{R})$ is a compact set.*

Proof. By the Heine-Borel theorem (*Theorem 1.1*) it suffices to show that $O_d(\mathbb{R})$ is both closed and bounded in $\mathcal{M}_d(\mathbb{R}) = \mathbb{R}^{d^2}$. Let $A = (a_{jk})$ be an arbitrary $d \times d$ matrix. The criteria that A is orthogonal if and only if it satisfies the equations

$$\sum_{m=1}^{d} a_{jm} a_{km} = \delta_{jk}$$

will be used to complete the proof.

Individually, each of these conditions defines a closed set of matrices because they are of the form $f^{-1}(C)$ where $f : \mathcal{M}_d(\mathbb{R}) \to \mathbb{R}$ is a continuous function and

C is a closed subset of $\mathbb{R}$. (See *Exercise 14* on page 12.) Since the intersection of a finite number of closed sets is closed, $O_d(\mathbb{R})$ is closed.

When $j = k$, the the condition $\sum_{m=1}^{d} a_{jm}a_{km} = \delta_{jk}$ becomes

$$\sum_{m=1}^{d} a_{jm}^2 = 1$$

and implies that $-1 \le a_{jm} \le 1$ for $m = 1, \ldots, d$ and $j = 1, \ldots, d$. Hence, $O_d(\mathbb{R})$ is bounded. □

A matrix A is said to be *skew symmetric*, if $A^{\mathrm{t}} = -A$. If A is skew symmetric, then

$$\left(e^{At}\right)^{-1} = e^{-At} = e^{A^{\mathrm{t}}t} = \left(e^{At}\right)^{\mathrm{t}}$$

and e^{At} is orthogonal for every real t. By part (a) of *Proposition 4.13*

$$\left\|e^{At}\boldsymbol{\xi}\right\|^2 = \|\boldsymbol{\xi}\|^2$$

and $\|\mathbf{x}\|^2$ is an integral for $\dot{\mathbf{x}} = A\mathbf{x}$. Thus when A is skew symmetric each solution of $\dot{\mathbf{x}} = A\mathbf{x}$ lies on a sphere, $\{\mathbf{x} : \|\mathbf{x}\| = r\}$.

The goal is to understand the structure of orthogonal matrices and to show that if B is orthogonal with $\mathrm{Det}\,[B] = 1$, then $B = e^A$ for a skew symmetric matrix A. Once again the matrix B must first be studied as a complex matrix so that each eigenvalue has an eigenvector. This requires the complex analogue of the dot product and leads naturally to unitary matrices.

For $\mathbf{z}, \mathbf{w} \in \mathbb{C}^d$ set

$$\mathbf{z} \cdot \mathbf{w} = \sum_{j=1}^{d} z_j \overline{w}_j. \tag{4.7}$$

Note that $\mathbf{w} \cdot \mathbf{z} = \overline{\mathbf{z} \cdot \mathbf{w}}$. The use of the conjugate in this formula ensures that

$$\|\mathbf{z}\|^2 = \mathbf{z} \cdot \mathbf{z} = \sum_{j=1}^{d} |z_i|^2.$$

and that $\|\mathbf{z}\|^2 = \mathbf{z} \cdot \mathbf{z}$ defines a norm on $\mathbb{C}^d$. Note, however, the conjugate does affect how scalars move in and out of the dot product, specifically,

$$\alpha(\mathbf{z} \cdot \mathbf{w}) = (\alpha\mathbf{z}) \cdot \mathbf{w} = \mathbf{z} \cdot (\overline{\alpha}\mathbf{w}).$$

Following the definition for $\mathbb{R}^d$, two vectors $\mathbf{z}$ and $\mathbf{w}$ in $\mathbb{C}^d$ are *orthogonal vectors* if $\mathbf{z} \cdot \mathbf{w} = 0$. If E is any subset of $\mathbb{C}^d$ or $\mathbb{R}^d$, set $E^{\perp} = \{\mathbf{z} : \mathbf{z} \cdot \mathbf{w} = 0 \text{ for all } \mathbf{w} \in E\}$. It is easy to verify that $E^{\perp}$ is always a subspace. The following lemma will be useful in the next proposition.

Lemma *If V is a subspace of $\mathbb{R}^d$ or $\mathbb{C}^d$, then*

(a) V has an orthonormal basis and

(b) $V \oplus V^{\perp}$ equals $\mathbb{R}^d$ or $\mathbb{C}^d$, respectively.

Proof. For (a), let $\mathbf{u}_1, \ldots, \mathbf{u}_q$ be a basis for V. Define $\mathbf{v}_j$ inductively by $\mathbf{v}_1 = \mathbf{u}_1$ and

$$\mathbf{v}_k = \mathbf{u}_k - \sum_{j=1}^{k-1} \frac{\mathbf{u}_k \cdot \mathbf{v}_j}{\mathbf{v}_j \cdot \mathbf{v}_j} \mathbf{v}_j$$

and check that $\mathbf{v}_1, \ldots, \mathbf{v}_q$ is an orthogonal basis. (This called the *Gram-Schmidt orthogonalization process*). Simply multiply $\mathbf{v}_j$ by $1/\|\mathbf{v}_j\|$ to make the basis orthonormal.

For (b,) it suffices to show that any $\mathbf{u}$ can be written as $\mathbf{u} = \mathbf{u}_1 + \mathbf{u}_2$ where $\mathbf{u}_1 \in V$ and $\mathbf{u}_2 \in V^{\perp}$ because obviously $V \cap V^{\perp} = \mathbf{0}$. Let $\mathbf{v}_1, \ldots, \mathbf{v}_q$ be an orthonomral basis of V. Set

$$\mathbf{u}_1 = \sum_{j=1}^{q} (\mathbf{u} \cdot \mathbf{v}_j)\mathbf{v}_j$$

and $\mathbf{u}_2 = \mathbf{u} - \mathbf{u}_1$ and check that $\mathbf{u}_1 \in V$ and $\mathbf{u}_2 \in V^{\perp}$. □

Given $A \in \mathcal{M}_d(\mathbb{C})$, there exist another matrix A^* such that for $\mathbf{z}$ and $\mathbf{w}$ in $\mathbb{C}^d$

$$(A\mathbf{z}) \cdot \mathbf{w} = \mathbf{z} \cdot (A^*\mathbf{w}),$$

but now it is necessary to conjugate each element of the matrix A and take the transpose, that is, the *conjugate transpose* of A is defined by

$$A^* = \overline{A}^{\mathrm{t}}$$

and satisfies $(A\mathbf{z}) \cdot \mathbf{w} = \mathbf{z} \cdot (A^*)\mathbf{w}$ for $\mathbf{z}$ and $\mathbf{w}$ in $\mathbb{C}^d$. (Just calculate both sides and interchange the order of summation.) For a real matrix A, of course, $A^* = A^{\mathrm{t}}$.

The analogue of the orthogonal matrix for $\mathbb{C}^d$ is the *unitary matrix* defined by $A^{-1} = A^*$. It follows that $U\mathbf{u} \cdot U\mathbf{v} = \mathbf{u} \cdot \mathbf{v}$ holds for a unitary matrix U. A real orthogonal matrix is unitary when viewed as an element of $GL_d(\mathbb{C})$. The following proposition about unitary matrices will apply to real orthogonal matrices and be used to prove that the real canonical form of an orthogonal matrix is particularly simple.

Proposition 4.15 *If U is a unitary matrix in $GL_d(\mathbb{C})$, then each eigenvalue of U is simple and has absolute value 1. In addition, eigenvectors for distinct eigenvalues are orthogonal.*

Proof. Let λ_1 be an arbitrary eigenvalue of U and let $\mathbf{v}_1$ be an eigenvector for λ_1 such that $\|\mathbf{v}_1\|^2 = \mathbf{v}_1 \cdot \mathbf{v}_1 = 1$. Then,

$$|\lambda_1|^2 = \lambda_1\overline{\lambda_1}(\mathbf{v}_1 \cdot \mathbf{v}_1) = (\lambda_1\mathbf{v}_1) \cdot (\lambda_1\mathbf{v}_1) = (U\mathbf{v}_1) \cdot (U\mathbf{v}_1) = \mathbf{v}_1 \cdot \mathbf{v}_1 = 1$$

and the absolute value of λ_1 is 1.

Since $U^*(U\mathbf{v}_1) = \mathbf{v}_1$, it follows that $\lambda_1(U^*\mathbf{v}_1) = \mathbf{v}_1$ or $U^*\mathbf{v}_1 = (1/\lambda_1)\mathbf{v}_1 = \overline{\lambda}_1\mathbf{v}_1$. (Note $|\lambda_1| = 1$, implies $1/\lambda_1 = \overline{\lambda}_1$.) Hence, a λ_1 eigenvector for U is automatically a $\overline{\lambda}_1$ eigenvector for U^*.

Now let $\mathbf{v}_2$ be an eigenvector for another eigenvalue $\lambda_2 \neq \lambda_1$. Then,

$$\lambda_1(\mathbf{v}_1 \cdot \mathbf{v}_2) = (\lambda_1\mathbf{v}_1) \cdot \mathbf{v}_2 = (U\mathbf{v}_1) \cdot \mathbf{v}_2 = \mathbf{v}_1 \cdot (U^*\mathbf{v}_2) = \mathbf{v}_1 \cdot (\overline{\lambda_2}\mathbf{v}_2) = \lambda_2(\mathbf{v}_1 \cdot \mathbf{v}_2)$$

and $\mathbf{v}_1 \cdot \mathbf{v}_2 = 0$ because $\lambda_2 \neq \lambda_1$. Therefore, eigenvectors for distinct eigenvectors are orthogonal in $\mathbb{C}^d$.

It remains to show that λ_1 is simple or its geometric multiplicity equals its algebraic multiplicity. Let $V = \mathcal{N}(U - \lambda_1\mathrm{I})$. Then by the lemma $\mathbb{C}^d = V \oplus V^\perp$. The first step is to show that $V^\perp$ is U invariant, that is, $(U\mathbf{z})$ is in $V^\perp$ when $\mathbf{z} \in V^\perp$. If $\mathbf{z}$ is an element of $V^\perp$ and $\mathbf{w}$ is an arbitrary element of V, then

$$(U\mathbf{z}) \cdot \mathbf{w} = \mathbf{z} \cdot (U^*\mathbf{w}) = \mathbf{z} \cdot (\overline{\lambda}_1\mathbf{w}) = 0,$$

and $V^\perp$ is U invariant. Clearly, V is also U invariant.

There exists an orthonormal basis $\mathbf{v}_1, \ldots, \mathbf{v}_d$ for $\mathbb{C}^d$ such $\mathbf{v}_1, \cdots, \mathbf{v}_q$ is an orthonormal basis for V and $\mathbf{v}_{q+1}, \ldots, \mathbf{v}_d$ is an orthonormal basis for $V^\perp$. Because the subspaces V and $V^\perp$ are U invariant, the matrix of the linear transformation $T(\mathbf{z}) = U\mathbf{z}$ with respect to the basis $\mathbf{v}_1, \ldots, \mathbf{v}_d$ has the block form

$$\begin{bmatrix} \lambda_1\mathrm{I} & \mathrm{O} \\ \mathrm{O} & C \end{bmatrix}$$

and

$$\mathrm{Det}\,[U - \lambda\mathrm{I}] = \mathrm{Det}\,[T - \lambda\mathrm{I}] = (\lambda - \lambda_1)^q\mathrm{Det}\,[C - \lambda\mathrm{I}].$$

If q is less than m_1, the algebraic multiplicity of λ_1, then $(\lambda - \lambda_1)$ divides $\mathrm{Det}\,[C - \lambda\mathrm{I}]$ and there is an eigenvector for λ_1 in $V^\perp$,which is impossible because all the λ_1 eigenvectors are in $V = \mathcal{N}(U - \lambda_1\mathrm{I}) = \mathcal{N}(T - \lambda_1\mathrm{I})$. Therefore, the geometric multiplicity of λ_1 or the dimension of $\mathcal{N}(U - \lambda_1\mathrm{I})$ equals m_1, the algebraic multiplicity of λ_1, and λ_1 is simple. Since λ_1 was an arbitrary eigenvalue, the proof is complete. □

Corollary *If U is a unitary matrix, then the Jordan canonical form for U is a diagonal matrix.*

Since an orthogonal matrix is also a unitary matrix, we can apply the preceding proposition to orthogonal matrices in $\mathbb{C}^d$ and then extract information about their structure within $\mathbb{R}^d$.

Proposition 4.16 *If B is an orthogonal matrix, then there exists an orthogonal matrix C such that the orthogonal matrix $\hat{B} = C^{\mathrm{t}}BC$ is a block diagonal matrix of the form*

$$\mathrm{Diag}\left(\mathrm{I}_{m_1}, -\mathrm{I}_{m_2}, \hat{B}_1, \ldots, \hat{B}_p\right)$$

where m_1 and m_2 are the algebraic multiplicities of 1 and -1 (they can be 0), I_{m_j} is an $m_j \times m_j$ identity matrix, and each $\hat{B}_k$ has the form

$$\begin{bmatrix} \alpha & \beta \\ -\beta & \alpha \end{bmatrix}$$

with $\alpha^2 + \beta^2 = 1$.

Proof. It follows from the previous proposition that $\mathcal{N}(A - \lambda_j \mathrm{I}) = M(\lambda_j)$ and that an amalgamation of orthonormal bases for each $\mathcal{N}(A - \lambda_j \mathrm{I})$ will produce an orthonormal basis for $\mathbb{C}^d$ by *Theorem 4.6.* Because the eigenvalues all have absolute value 1, there are only two possible real eigenvalues, 1 and -1. Denote their multiplicities by m_1 and m_2.

The ideas used to construct the real canonical forms at the end of Section 4.3 can be applied to B. In particular, there exists orthonormal bases of real vectors for $\mathcal{N}(B \pm \mathrm{I}) = M(\pm 1)$, and there will be m_1 and m_2 vectors, respectively, in each of these bases. They will produce the first two blocks in $\hat{B}$. Of course, if m_1 and/or m_2 is 0, then I_{m_1} and/or I_{m_2} will not appear in the canonical form for B.

Since B is real, its complex eigenvalues come in conjugate pairs. Choose the bases for complex eigenvalues so that when $\mathbf{v}_1, \ldots, \mathbf{v}_{m_j}$ is the selected basis for $\mathcal{N}(B - \lambda_j)$, then $\overline{\mathbf{v}}_1, \ldots, \overline{\mathbf{v}}_{m_j}$ is the selected basis for $\mathcal{N}(B - \overline{\lambda}_j)$. As in the construction of the real canonical form, a real basis is obtained by replacing $\mathbf{v}_j$ and $\overline{\mathbf{v}}_j$ with $(\mathbf{v}_j + \overline{\mathbf{v}}_j)/2$ and $(\mathbf{v}_j - \overline{\mathbf{v}}_j)/(2i)$. A routine calculation shows that these new vectors are orthogonal and have Euclidean norm 1. Thus an orthonormal basis for $\mathbb{R}^d$ can be constructed from the orthonormal basis of eigenvectors for $\mathbb{C}^d$ given by the previous proposition. Denote this orthonormal basis by $\mathbf{u}_1, \ldots, \mathbf{u}_d$.

It is easy to see that the matrix $\hat{B}$ for $T(\mathbf{x}) = B\mathbf{x}$ with respect to the basis $\mathbf{u}_1, \ldots, \mathbf{u}_d$ has the required form if the basis elements are numbered carefully. If the coordinates of the $\mathbf{u}_j$ vectors are used to form the columns of a matrix C, then C changes $\mathbf{u}_1, \ldots, \mathbf{u}_d$ coordinates into standard coordinates and $\hat{B} = C^{-1}BC$. Finally, because $\mathbf{u}_1, \ldots, \mathbf{u}_d$ is an orthonormal basis, C is an orthogonal matrix and the proof is finished. □

The stage is now set to prove the following theorem about the range of the exponential map.

Theorem 4.17 *If B is an orthogonal matrix and* Det $[B] = 1$*, then there exists a skew symmetric matrix A such that $B = e^A$.*

Proof. Let $\hat{B} = C^{\mathrm{t}}BC$ be given by *Theorem 4.16.* It suffices to find a skew symmetric matrix $\hat{A}$ such that $e^{\hat{A}} = \hat{B}$ because then

$$B = C\hat{B}C^{\mathrm{t}} = Ce^{\hat{A}}C^{\mathrm{t}} = e^{C\hat{A}C^{\mathrm{t}}}$$

and

$$(C\hat{A}C^{\mathrm{t}})^{\mathrm{t}} = C\hat{A}^{\mathrm{t}}C^{\mathrm{t}} = C(-\hat{A})C^{\mathrm{t}} = -C\hat{A}C^{\mathrm{t}}$$

Proposition 4.16 reduces the proof to finding a solution for each of the three types of diagonal blocks that form $\hat{B}$. This is trivial for I_{m_1}. Because the $\mathrm{Det}[B] = 1$, the multiplicity m_2 of -1 must be even. Hence, $-\mathrm{I}_{m_2}$ can be thought of as a block diagonal matrix of $m_2/2$ more 2×2 matrices of the same form as $\hat{B}_j$ with $\alpha = -1$ and $\beta = 0$. Because $\alpha^2 + \beta^2 = 1$, each $\hat{B}_j$ has the form

$$\begin{bmatrix} \cos\theta & \sin\theta \\ -\sin\theta & \cos\theta \end{bmatrix}.$$

The following lemma completes the proof. □

Lemma *If*

$$\Theta = \begin{bmatrix} 0 & \theta \\ -\theta & 0 \end{bmatrix}$$

for a real number θ, *then*

$$e^{\Theta} = \begin{bmatrix} \cos\theta & \sin\theta \\ -\sin\theta & \cos\theta \end{bmatrix}.$$

Proof. A straightforward calculation shows that

$$\sum_{k=0}^{2n} \frac{\Theta^k}{k!} = \begin{bmatrix} \sum_{k=0}^{n} \frac{(-1)^{k-1}\theta^{2k}}{(2k)!} & \sum_{k=1}^{n} \frac{(-1)^{k-1}\theta^{2k-1}}{(2k-1)!} \\ -\sum_{k=1}^{n} \frac{(-1)^{k-1}\theta^{2k-1}}{(2k-1)!} & \sum_{k=0}^{n} \frac{(-1)^{k-1}\theta^{2k}}{(2k)!} \end{bmatrix}.$$

For odd terms the summation index for the lower left-hand entry and the upper right-hand entry increases to $n+1$. In either case, the entries are the familiar Maclaurin series for cosine and sine functions. □

Note that the exponential of both

$$A = \begin{bmatrix} 0 & \pi \\ -\pi & 0 \end{bmatrix} \text{ and } \pi i\mathrm{I} = \begin{bmatrix} \pi i & 0 \\ 0 & \pi i \end{bmatrix}$$

is $-\mathrm{I}_2$. Their one parameter subgroups, however, behave very differently. Specifically, e^{At} is in $O_2(\mathbb{R})$ for all t and $e^{\pi i \mathrm{I} t}$ is only in $O_2(\mathbb{R})$ when t is an integer.

It is now possible to characterize the range of the map $A \to e^A$ on $\mathcal{M}_d(\mathbb{R})$. Since this result is not of any particular importance in the remaining chapters, the proof is just sketched.

Theorem 4.18 *Let* B *be a* $d \times d$ *real matrix. There exist a real matrix* A *such that* $e^A = B$ *if and only if there exists a real invertible matrix* C *such that* $C^2 = B$.

Proof. If there exist a real matrix A such that $e^A = B$, then it is an exercise to show that there exists a real invertible matrix C such that $C^2 = B$.

For the other implication it suffices to prove the result for C^2 where C is one of the standard blocks in the real canonical form (page 128) of a matrix. The proof makes extensive use of the ideas in the proof of *Theorem 4.12*

When C has the form of the matrix *(4.3)*, it can be written as $\lambda \mathrm{I} + R$ as in the proof of *Theorem 4.12*. There are two cases to consider - $\lambda > 0$ and $\lambda < 0$. For $\lambda > 0$, choose the usual positive real $\log \lambda$. Then $A = (\log \lambda)\mathrm{I} + S$ is real, $e^A = C$ and $e^{2A} = C^2$.

For $\lambda < 0$, use $\log \lambda = \log(|\lambda|) \pm \pi i$ with $\log(|\lambda|) > 0$. These choices produce conjugate matrices A and $\overline{A}$ such that $e^A = e^{\overline{A}} = C$ and $A\overline{A} = \overline{A}A$ is real. Therefore, $e^{A+\overline{A}} = C^2$ and $A + \overline{A}$ is a real matrix.

In the remaining case, C has the form of the matrix *(4.5)*. Let R' be the matrix obtained by replacing the 2×2 diagonal blocks in *(4.5)* with the 2×2 zero matrix and replacing the 2×2 off diagonal identity matrices with

$$\begin{bmatrix} \alpha & -\beta \\ \beta & \alpha \end{bmatrix}.$$

By using the previous lemma, it is easy to show that there exist a real matrix D such that

$$e^D = \begin{bmatrix} \alpha & \beta \\ -\beta & \alpha \end{bmatrix}.$$

Let D' be the block diagonal matrix $\mathrm{Diag}\,[D, D, \ldots, D]$ of the same size as R' and verify that R' and D' commute. Set

$$S' = -\sum_{j=1}^{d} (-R')^j$$

and

$$A = D' + S'.$$

To complete the proof, calculate e^S using equation *(4.6)* and show that $e^A = C$.
□

EXERCISES

1. Let $\mathbf{f} : GL_d(\mathbb{R}) \to \mathcal{M}_d(\mathbb{R})$ be a vector field on $GL_d(\mathbb{R})$,an open subset of $\mathcal{M}_d(\mathbb{R}) = \mathbb{R}^{d^2}$. The vector field $\mathbf{f}$ is *right invariant* if for all $B \in GL_d(\mathbb{R})$the vector fields $\mathbf{f}(XB)$ and $\mathbf{f}(X)B$ are identical on $GL_d(\mathbb{R})$. Show that $\mathbf{f}$ is right invariant if and only if $\mathbf{f}(X) = AX$. Determine the function $X(t, \Xi)$ for the vector field $\mathbf{f}(X) = AX$ and show that it defines a flow on $GL_d(\mathbb{R})$. (The function $X(t, \Xi)$ is the matrix version of $\mathbf{x}(t, \boldsymbol{\xi})$ and denotes the solution of $\dot{X} = AX$ that passes through Ξ when $t = 0$.)

2. Let A be a real matrix and let $X(t, \Xi)$ be the flow on $GL_d(\mathbb{R})$ determined by $\dot{X} = AX$. Show that $\overline{\mathcal{O}}(\mathrm{I})$ is an abelian subgroup of $GL_d(\mathbb{R})$.

3. Provide a proof for *Proposition 4.13* on page 138.

4. Show that the orbit closures are compact for the flow on $GL_d(\mathbb{R})$ determined by the matrix differential equation $\dot{X} = AX$ when A is a real skew symmetric matrix.

5. Show that there exists a real matrix B such that

$$e^B = \begin{bmatrix} \alpha & -\beta \\ \beta & \alpha \end{bmatrix}$$

   where $\alpha, \beta \in \mathbb{R}$ and $\beta \neq 0$.

6. Show by example that the map $A \to e^A$ is not one-to-one even for real matrices.

7. Show that there is no real matrix A such that

$$e^A = \begin{bmatrix} -1 & 0 \\ 0 & -3 \end{bmatrix}.$$

8. Define the *special linear group* by $SL_d(\mathbb{R}) = \{A \in GL_d(\mathbb{R}) : \operatorname{Det} A = 1\}$ and prove the following:

   (a) $SL_d(\mathbb{R})$ is a subgroup of $GL_d(\mathbb{R})$.

   (b) $SL_d(\mathbb{R})$ is not compact.

   (c) e^{At} is in $SL_d(\mathbb{R})$ for all real t if and only if $\operatorname{Tr} A = 0$.

   (d) There exists $B \in SL_d(\mathbb{R})$ such that $B \neq e^A$ for all $A \in \mathcal{M}_d(\mathbb{R})$.

9. Show that $SL_d(\mathbb{R})$ is an invariant set for the flow on $GL_d(\mathbb{R})$ determined by $\dot{X} = AX$ if $A^d = O$.

10. Let A and B be in $\mathcal{M}_d(\mathbb{C})$. Show the following:

    (a) If $A + A^* = O$, then e^{At} is unitary for all $t \in \mathbb{R}$.

    (b) If B is unitary, then $B = e^A$ for some A satisfying $A + A^* = O$

11. Let B be an invertible $d \times d$ matrix. Show that there exists $C \in GL_d(\mathbb{R})$ such that $C^2 = B$ if and only if for every integer $p \geq 2$ there exists $C \in GL_d(\mathbb{R})$ such that $C^p = B$

Chapter 5

Stability

The defining question for the subject of dynamical systems is to determine the behavior of solutions over infinite time intervals. Stability is one class of concepts and results that fits this mold. The question of whether or not a solution is stable is intuitively a fundamental question.

The idea of a stable solution is that a slight deviation in the initial position should not produce a solution that differs wildly from the stable one. For example, trying to place the bob of a pendulum exactly in the rest position by hand is unlikely to be successful. But the pendulum will oscillate only a small amount around the rest position if the placement is close because the rest position is a stable solution. In other words, an initial position near the rest position will produce a solution that stays near it forever.

Continuity in initial conditions guarantees only that two solutions starting sufficiently near each other will remain close over a preselected bounded time interval. But how close the initial positions must be at the start can go to zero extremely rapidly as the prescribed time interval grows. The result is a sensitive dependence on the initial conditions that can cause solutions starting close together to differ significantly from each other after sufficient time has evolved. Stability is the opposite kind of behavior and a typical stability theorem is akin to continuity in initial conditions over an infinite interval.

Fixed points of autonomous differential equations provide the most tractable place to begin a study of stability. Section 1 examines two types of stability for fixed points that correspond roughly to the behavior of the resting solutions for undamped and damped pendulums. In Section 2, the stability of the origin in a linear system with constant coefficients is characterized in terms of the eigenvalues. This is the first step toward studying the behavior of solutions near a fixed point by linearization in later chapters. Section 3 is devoted to the stability of more general solutions with an emphasis on linear systems. The stability of linear systems with periodic coefficients is the topic of the final section.

5.1 Stability at Fixed Points

Throughout this section only autonomous differential equations of the form $\dot{\mathbf{x}} = \mathbf{f}(\mathbf{x})$ will be considered and the standing hypotheses that the function $\mathbf{f}(\mathbf{x})$ be locally Lipschitz on Ω, an open subset of $\mathbb{R}^d$, will be in force. The locally Lipschitz hypothesis will not be repeated in the statement of the results to help keep the statement of the hypotheses as simple as possible. So it will be understood throughout the section that solutions are unique and continuity in initial conditions holds. In this context, it is always convenient to use the simpler notation of $\mathbf{x}(t, \boldsymbol{\xi})$ for $\mathbf{x}(t, 0, \boldsymbol{\xi})$.

Also in this section, $\mathbf{p}$ will consistently denote a fixed point of $\dot{\mathbf{x}} = \mathbf{f}(\mathbf{x})$, so that, $\mathbf{f}(\mathbf{p}) = \mathbf{0}$ and $\mathbf{x}(t, \mathbf{p}) = \mathbf{p}$ is a solution defined on $\mathbb{R}$.

The general question is If a solution starts near $\mathbf{p}$, under what conditions will it stay near $\mathbf{p}$ or approach $\mathbf{p}$ as time increases? This is of interest because the initial conditions may come from physical measurements that by their nature have limited accuracy, and it is desirable to know whether or not slight deviations will significantly change the qualitative behavior of the solution. In other words, how much qualitative information is lost when there are small measurement errors in the initial conditions.

The fixed point $\mathbf{p}$ is said to be *positively stable* provided the following conditions hold:

(a) There exists $r > 0$ such that when $|\boldsymbol{\xi} - \mathbf{p}| < r$ the solution $\mathbf{x}(t, \boldsymbol{\xi})$ is defined for all $t \geq 0$. (This condition is automatically satisfied for flows.)

(b) Given $\varepsilon > 0$ there exists $\delta > 0$ such that $|\mathbf{x}(t, \boldsymbol{\xi}) - \mathbf{p}| < \varepsilon$ for all $t \geq 0$ when $|\boldsymbol{\xi} - \mathbf{p}| < \delta$.

The fixed point $\mathbf{p}$ is said to be *positively asymptotically stable*, when in addition to being positively stable, $\mathbf{p}$ has the following additional property:

(c) There exists $\gamma > 0$ such that $\lim_{t \to \infty} \mathbf{x}(t, \boldsymbol{\xi}) = \mathbf{p}$ whenever $|\boldsymbol{\xi} - \mathbf{p}| < \gamma$.

Obviously, there are parallel concepts of *negative stability* and *negative asymptotic stability*. There are also a number of other variations of stability, and the terminology becomes rather wordy.

Examples of stability occur readily in dimension one. With $\Omega = \mathbb{R}$ and $\dot{x} = x(x-1)$ the set of fixed points is $\{0, 1\}$. Since $x(x-1) < 0$ precisely when $0 < x < 1$, it is obvious from the phase portrait (*Figure 5.1*) that 0 is positively asymptotically stable and 1 is negatively asymptotically stable.

For the familiar linear planar system,

$$\begin{aligned} \dot{x}_1 &= -x_2 \\ \dot{x}_2 &= x_1 \end{aligned}$$

on $\mathbb{R}^2$, the origin is a positively and negatively stable fixed point that is neither positively asymptotically stable nor negatively asymptotically stable. (See

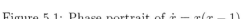

Figure 5.1: Phase portrait of $\dot{x} = x(x-1)$.

Figure 2.2 on page 63.) Of course, the origin is neither positively stable nor negatively stable for the system

$$\begin{aligned}\dot{x}_1 &= x_2\\ \dot{x}_2 &= x_1,\end{aligned}$$

whose phase portrait is shown in *Figure 2.3* on 64.

For the above examples, we can actually solve the equations, but is there a way to determine whether or not a fixed point is positively stable without knowing the solutions of the differential equation? This is precisely what Lyapunov's method accomplishes. The only catch is that to apply this method, one must select a test function and doing so may or may not be easy.

Suppose $V : \Omega_0 \to \mathbb{R}$ has continuous first partial derivatives on Ω_0, an open subset of Ω, and suppose $\mathbf{x}(t)$ is a solution of $\dot{\mathbf{x}} = \mathbf{f}(\mathbf{x})$. When $\mathbf{x}(t)$ is in Ω_0, the derivative of $V(\mathbf{x}(t))$ with respect to t can be calculated using the chain rule and yields

$$\frac{dV(\mathbf{x}(t))}{dt} = \nabla V(\mathbf{x}(t)) \cdot \dot{\mathbf{x}}(t) = \nabla V(\mathbf{x}(t)) \cdot \mathbf{f}(\mathbf{x}(t)).$$

Thus it can be determined whether or not V is increasing or decreasing along the solution through $\mathbf{x}$ by simply calculating

$$\dot{V}(\mathbf{x}) = \nabla V(\mathbf{x}) \cdot \mathbf{f}(\mathbf{x}). \tag{5.1}$$

Note that V is an integral precisely when $\dot{V} \equiv 0$ and V is neither increasing nor decreasing along any solution.

The idea of Lyapunov's method and related results is to impose conditions on the functions V and $\dot{V}$, which both imply stability and are easy to verify, and then for a given $\dot{\mathbf{x}} = \mathbf{f}(\mathbf{x})$ and fixed point $\mathbf{p}$, to find a V that satisfies these conditions. The functions V will be called a *Lyapunov function.*

Let Ω_0 be an open neighborhood of $\mathbf{p}$. A function $V : \Omega_0 \to \mathbb{R}$ is said to be *positive definite at* $\mathbf{p}$, if

(a) $V(\mathbf{x}) \geq 0$ for all $\mathbf{x} \in \Omega_0$ and

(b) $V(\mathbf{x}) = 0$ if and only if $\mathbf{x} = \mathbf{p}$.

Of course, *negative definite* is defined in the obvious way.

Theorem 5.1 (Lyapunov's method) *Let $V : \Omega_0 \to \mathbb{R}$ be a function with continuous first partial derivatives on Ω_0, a neighborhood of the fixed-point $\mathbf{p}$ of $\dot{\mathbf{x}} = \mathbf{f}(\mathbf{x})$. If V is positive definite at $\mathbf{p}$ and $\dot{V} \leq 0$ on Ω_0, then $\mathbf{p}$ is a positively stable fixed point. If, in addition, $\dot{V}$ is negative definite at $\mathbf{p}$, then $\mathbf{p}$ is a positively asymptotically stable fixed point.*

Proof. Let r be a positive real number such that $\{\mathbf{x} : |\mathbf{x} - \mathbf{p}| \le r\} \subset \Omega_0$. For $0 < \varepsilon \le r$, set $k(\varepsilon) = \min\{V(\mathbf{x}) : |\mathbf{x} - \mathbf{p}| = \varepsilon\}$. The hypothesis that V is positive definite implies that $k(\varepsilon) > 0$ and the continuity of $V(\mathbf{x})$ implies that there exists $\delta(\varepsilon) < \varepsilon$ such that $V(\mathbf{x}) < k(\varepsilon)$ when $|\mathbf{x} - \mathbf{p}| < \delta(\varepsilon)$. It suffices to restrict attention to ε satisfying $0 < \varepsilon \le r$.

Consider a solution $\mathbf{x}(t, \boldsymbol{\xi})$ of $\dot{\mathbf{x}} = \mathbf{f}(\mathbf{x})$ defined on an open interval I such that $|\boldsymbol{\xi} - \mathbf{p}| < \delta(\varepsilon)$. Suppose that $|\mathbf{x}(t', \boldsymbol{\xi}) - \mathbf{p}| \ge \varepsilon$ for some t'. Set

$$\tau = \sup\{t : |\mathbf{x}(s, \boldsymbol{\xi}) - \mathbf{p}| < \varepsilon \text{ for } 0 \le s \le t\}.$$

It follow that τ has the following properties:

(a) $|\mathbf{x}(\tau, \boldsymbol{\xi}) - \mathbf{p}| = \varepsilon$,

(b) $|\mathbf{x}(t, \boldsymbol{\xi}) - \mathbf{p}| < \varepsilon$ for $0 \le t < \tau$,

(c) $V(\mathbf{x}(\tau, \boldsymbol{\xi})) \ge k(\varepsilon)$.

Because $\dot{V}(\mathbf{x}) \le 0$, it also follows that

$$V(\mathbf{x}(\tau, \boldsymbol{\xi})) = V(\boldsymbol{\xi}) + \int_0^{\tau} \dot{V}(\mathbf{x}(s, \boldsymbol{\xi}))ds \le V(\boldsymbol{\xi}) < k(\varepsilon)$$

and there is a contradiction. Therefore, $|\mathbf{x}(t, \boldsymbol{\xi}) - \mathbf{p}| < \varepsilon$ for positive $t \in I$. By *Theorem 1.19*, I must contain the positive reals. And then $|\mathbf{x}(t, \boldsymbol{\xi}) - \mathbf{p}| < \varepsilon$ for all $t \ge 0$ when $|\boldsymbol{\xi} - \mathbf{p}| < \delta(\varepsilon)$ proving that $\mathbf{p}$ is positively stable with $W = \{\mathbf{x} : |\mathbf{x} - \mathbf{p}| < \delta(r)\}$.

Now suppose that $\dot{V}(\mathbf{x})$ is also negative definite at $\mathbf{p}$ and set $\gamma = \delta(r)$. To complete the proof it will be first shown that it suffices to prove that if $\boldsymbol{\xi}$ is any point satisfying $|\boldsymbol{\xi} - \mathbf{p}| < \gamma$, then for every $\varepsilon < r$ there exists τ_ε such that $|\mathbf{x}(\tau_\varepsilon, \boldsymbol{\xi}) - \mathbf{p}| < \delta(\varepsilon)$. By uniqueness, $\mathbf{x}(s + \tau_\varepsilon, \boldsymbol{\xi}) = \mathbf{x}(s, \mathbf{x}(\tau_\varepsilon, \boldsymbol{\xi}))$ and the first part of the proof can be applied to the right side of this equation to guarantee that $|\mathbf{x}(t, \boldsymbol{\xi}) - \mathbf{p}| < \varepsilon$ for $t = s + \tau_\varepsilon > \tau_\varepsilon$ when $|\mathbf{x}(\tau_\varepsilon, \boldsymbol{\xi}) - \mathbf{p}| < \delta(\varepsilon)$.

Proceeding by contradiction, assume that for some $\boldsymbol{\xi} \in W$ and $\varepsilon > 0$, $|\mathbf{x}(t, \boldsymbol{\xi}) - \mathbf{p}| \ge \delta(\varepsilon)$ for all $t \ge 0$. Set $\alpha = \max\{\dot{V}(\mathbf{x}) : \delta(\varepsilon) \le |\mathbf{x} - \mathbf{p}| \le r\}$, which is negative because $\dot{V}$ is negative definite at $\mathbf{p}$. Since

$$\delta(\varepsilon) \le |\mathbf{x}(t, \boldsymbol{\xi}) - \mathbf{p}| \le r$$

for all $t \ge 0$, it follows that

$$0 \le V(\mathbf{x}(t, \boldsymbol{\xi})) = V(\boldsymbol{\xi}) + \int_0^t \dot{V}(\mathbf{x}(s, \boldsymbol{\xi}))ds \le V(\boldsymbol{\xi}) + \alpha t,$$

which is the desired contradiction because the last term on the right goes to negative infinity as t goes to positive infinity. □

Many examples can be handled by this theorem. For example, anything of the form

$$\begin{aligned} \dot{x} &= -x^{2k-1} + \alpha x y^2 \\ \dot{y} &= -y^{2m-1} - \beta x^2 y, \end{aligned}$$

where k and m are positive integers and α and β are positive real numbers. Just set $V(x,y) = x^2/\alpha + y^2/\beta$, which is positive definite at $(0,0)$, and note that $\dot{V}(x,y) = -2(x^{2k} + y^{2m})$ which is negative definite at $(0,0)$. It follows that $(0,0)$ is positively asymptotically stable.

A variation of the technique in the previous theorem can also be used to show that a fixed point is not positively stable.

Theorem 5.2 *Let $V : \Omega_0 \to \mathbb{R}$ be a function with continuous first partial derivatives on Ω_0, a neighborhood of the fixed point $\mathbf{p}$ of $\dot{\mathbf{x}} = \mathbf{f}(\mathbf{x})$. If there exists an open set U contained in Ω_0 such that :*

(a) $\mathbf{p} \in \partial U$, the boundary of U,

(b) Both $V(\mathbf{x})$ and $\dot{V}(\mathbf{x})$ are positive on U,

(c) $V(\mathbf{x}) = 0$ on $\partial U \cap \Omega_0$,

then $\mathbf{p}$ is not positively stable.

Proof. Assume $\mathbf{p}$ is positively stable and choose r such that $\{\mathbf{x} : |\mathbf{x} - \mathbf{p}| < r\} \subset \Omega_0$ and such that $\mathbf{x}(t, \boldsymbol{\xi})$ is defined for all $t \geq 0$ when $|\boldsymbol{\xi} - \mathbf{p}| < r$. Because $\mathbf{p} \in \partial U$, there exists a maximal solution $\varphi(t)$ such that $\varphi(0) \in U \cap \Omega_0$ and $|\varphi(t) - \mathbf{p}| < r$ for all $t \geq 0$. Set

$$\tau = \sup\{t : \varphi(s) \in U \text{ for } 0 \leq s \leq t\}.$$

If $\tau \neq \infty$, then clearly $\varphi(\tau) \in \partial U \cap \Omega_0$ and $V(\varphi(\tau)) = 0$. But

$$V(\varphi(\tau)) = V(\varphi(0)) + \int_0^\tau \dot{V}(\varphi(s))ds > V(\varphi(0)) > 0$$

because V and $\dot{V}$ are positive on $U \cap \Omega_0$. Hence, $\tau = \infty$, and it follows that $\varphi(t) \in U$ for all $t \geq 0$ and $V(\varphi(t)) > V(\varphi(0))$ for all $t \geq 0$.

Now, set

$$\Gamma = \{\mathbf{x} : |\mathbf{x} - \mathbf{p}| \leq r, \ \mathbf{x} \in \overline{U}, \text{ and } V(\mathbf{x}) \geq V(\varphi(0))\}.$$

Clearly Γ is closed and bounded and hence compact. Note that $\varphi(t) \in \Gamma$ for all $t \geq 0$. Also $V(\mathbf{x}) = 0$ on ∂U and $V(\mathbf{x}) \geq V(\varphi(0)) > 0$ for $\mathbf{x} \in \Gamma$ imply that $\Gamma \cap \partial U = \phi$. Hence, Γ must be contained in U and

$$\min\{\dot{V}(\mathbf{x}) : \mathbf{x} \in \Gamma\} = \alpha > 0$$

by *Theorem 1.3.*

Since $\varphi(t) \in \Gamma$ for all $t \geq 0$,

$$V(\varphi(t)) = V(\varphi(0)) + \int_0^t \dot{V}(\varphi(s))ds \geq V(\varphi(0)) + \alpha t.$$

It follows that V is unbounded and continuous on the compact set Γ, which is impossible. Therefore, $\mathbf{p}$ is not positively stable. □

For example, consider the planar system of differential equations

$$\begin{aligned}\dot{x} &= yh_1(x,y)\\ \dot{y} &= xh_2(x,y),\end{aligned}$$

where h_1 and h_2 have continuous first partial derivatives on $\mathbb{R}^2$ and are positive at $(0,0)$. The previous theorem can be used to show that $(0,0)$ is not positively stable.

Choose r so that $h_j(x,y) > 0$, $j = 1,2$, when $|x|+|y| < r$, and set $\Omega_0 = \{(x,y) : |x|+|y| < r\}$. Define $V : \Omega_0 \to \mathbb{R}$ by $V(x,y) = xy$ and let U be the first quadrant. Then $\dot{V}(x,y) = y^2h_1(x,y) + x^2h_2(x,y) > 0$ on $U \cap \Omega_0$, and $V(x,y) = 0$ on $\partial U \cap \Omega_0$ because it is contained in coordinate axes. Now *Theorem 5.2* applies and the origin is not positively stable.

These two initial theorems can also be applied to *gradient system*, that is, $\dot{\mathbf{x}} = -\nabla F(\mathbf{x})$, where $F : \Omega \to \mathbb{R}$ has continuous second partial derivatives. Note that the critical points (the points were the first partial derivatives vanish) of F are precisely the fixed points of $\dot{\mathbf{x}} = -\nabla F(\mathbf{x})$. In fact, critical points and fixed points are often synonymous in differential equation literature.

Proposition 5.3 *Let $F : \Omega \to \mathbb{R}$ have continuous second partial derivatives on Ω and consider $\dot{\mathbf{x}} = -\nabla F(\mathbf{x})$. Let $\mathbf{p}$ be an isolated critical point of F. Then $\mathbf{p}$ is a positively asymptotically stable, negatively asymptotically stable, neither positively nor negatively stable according as $\mathbf{p}$ is a local minimum, local maximum, neither.*

Proof. First suppose $\mathbf{p}$ is a local minimum. There exists a neighborhood Ω_0 of $\mathbf{p}$ such that $F(\mathbf{p}) < F(\mathbf{x})$ for all $\mathbf{x} \in \Omega_0$ and $\mathbf{x} \neq \mathbf{p}$. Set $V(\mathbf{x}) = F(\mathbf{x}) - F(\mathbf{p})$ and note that $\dot{V}(\mathbf{x}) = -\|\nabla F(\mathbf{x})\|^2$. So *Theorem 5.1* applies.

When $\mathbf{p}$ is a local maximum, let $V(\mathbf{x}) = F(\mathbf{p}) - F(\mathbf{x})$ and use the analogue of *Theorem 5.1* for negative asymptotic stability.

Suppose $\mathbf{p}$ is neither a local minimum nor a local maximum and let $U = \{\mathbf{x} : F(\mathbf{x}) < F(\mathbf{p})\}$. Clearly, U is open and $\mathbf{p} \in \partial U$ by continuity and the assumption that $\mathbf{p}$ is not a local minimum. Let $V(\mathbf{x}) = F(\mathbf{p}) - F(\mathbf{x})$. Then $\dot{V}(\mathbf{x}) = \|\nabla F(\mathbf{x})\|^2$ is positive definite in some neighborhood Ω_0 of $\mathbf{p}$ because $\mathbf{p}$ is an isolated critical point of F. Hence, V and $\dot{V}$ are positive on $U \cap \Omega_0$ and $\mathbf{p}$ is not positively stable by *Theorem 5.2*. Apply the analogue of *Theorem 5.2* for negative stability to complete the proof. □

When $\mathbf{p}$ is a positively stable fixed point, a solution $\mathbf{x}(t, \boldsymbol{\xi})$ is defined for all $t \geq 0$ and $|\mathbf{x}(t,\boldsymbol{\xi}) - \mathbf{p}| < \varepsilon$ when $|\boldsymbol{\xi} - \mathbf{p}| < \delta$. Since the set $\{\mathbf{x} : |\mathbf{x} - \mathbf{p}| \leq \varepsilon\}$ is compact, $\mathbf{x}(t_m, \boldsymbol{\xi})$ must have a convergent subsequence whenever $|\boldsymbol{\xi} - \mathbf{p}| < \delta$ and $t_m \to \infty$ as $m \to \infty$. Describing such limit points is a way of better understanding how solutions behave as time goes to infinity, and introducing these ideas here fits into the study of stability.

The omega limit set of a point will be defined as the intersection of the closure of increasingly far out pieces of the positive orbit. Recall from *Exercise 7* on page 11 that the closure of a set U is denoted by $\overline{U}$ and is the smallest closed set containing U. If the maximal solution $\mathbf{x}(t, \boldsymbol{\xi})$ of $\dot{\mathbf{x}} = \mathbf{f}(\mathbf{x})$ is defined for all $t \geq 0$, then the *omega limit set of* $\boldsymbol{\xi}$ is the subset of Ω defined by

$$\omega(\boldsymbol{\xi}) = \bigcap_{m=0}^{\infty} \left(\overline{\{\mathbf{x}(t, \boldsymbol{\xi}) : t \geq m\}} \cap \Omega \right).$$

The *alpha limit set* $\alpha(\boldsymbol{\xi})$ is defined similarly when $\mathbf{x}(t, \boldsymbol{\xi})$ is defined for all $t \leq 0$. For a flow, $\omega(\boldsymbol{\xi})$ and $\alpha(\boldsymbol{\xi})$ are defined for all $\boldsymbol{\xi}$ in the space but may be empty and are very much part of the study of flows.

The condition, in addition to being positively stable, required for a critical point to be positively asymptotically stable, namely,

(c) There exists $\gamma > 0$ such that $\lim_{t \to 0} \mathbf{x}(t, \boldsymbol{\xi}) = \mathbf{p}$ whenever $|\boldsymbol{\xi} - \mathbf{p}| < \gamma$

can now be stated as

(c) There exists $\gamma > 0$ such that $\omega(\boldsymbol{\xi}) = \{\mathbf{p}\}$ whenever $|\boldsymbol{\xi} - \mathbf{p}| < \gamma$.

The next few propositions establish the basic properties of the omega limit set for future use.

Proposition 5.4 *A point* $\mathbf{q}$ *is in* $\omega(\boldsymbol{\xi})$ *if and only if there exists a sequence of real numbers* $\{t_k\}$ *such that* $\lim_{k \to \infty} t_k = \infty$ *and* $\lim_{k \to \infty} \mathbf{x}(t_k, \boldsymbol{\xi}) = \mathbf{q} \in \Omega$.

Proof. Let $\mathbf{q} \in \omega(\boldsymbol{\xi})$. For each integer k, there exists $t_k \geq k$ such that $|\mathbf{x}(t_k, \boldsymbol{\xi}) - \mathbf{q}| < 1/k$ because

$$\mathbf{q} \in \overline{\{\mathbf{x}(t, \boldsymbol{\xi}) : t \geq k\}} \cap \Omega.$$

Clearly, $\lim_{k \to \infty} t_k = \infty$ and $\lim_{k \to \infty} \mathbf{x}(t_k, \boldsymbol{\xi}) = \mathbf{q} \in \Omega$. So the first half of the proof is done.

For the converse, suppose $\lim_{k \to \infty} t_k = \infty$ and $\lim_{k \to \infty} \mathbf{x}(t_k, \boldsymbol{\xi}) = \mathbf{q} \in \Omega$. Given a positive integer m, there exists a positive integer K such that $t_k \geq m$ for all $k \geq K$. Hence,

$$\mathbf{q} = \lim_{k \to \infty} \mathbf{x}(t_k, \boldsymbol{\xi}) \in \overline{\{\mathbf{x}(t, \boldsymbol{\xi}) : t \geq m\}} \cap \Omega.$$

Since the above holds for all $m \geq 0$, it follows that $\mathbf{q} \in \omega(\boldsymbol{\xi})$. □

Proposition 5.5 *If* $\mathbf{q} \in \omega(\boldsymbol{\xi})$, *then the following hold:*

(a) Every point on the solution $\mathbf{x}(t, \mathbf{q})$ *is also in* $\omega(\boldsymbol{\xi})$.

(b) For every point on the solution $\mathbf{x}(t, \boldsymbol{\xi})$, *the point* $\mathbf{q}$ *is in* $\omega(\mathbf{x}(t, \boldsymbol{\xi}))$ *and* $\omega(\boldsymbol{\xi}) = \omega(\mathbf{x}(t, \boldsymbol{\xi}))$.

Proof. From the previous proposition, $\mathbf{q} = \lim_{k\to\infty} \mathbf{x}(t_k, \boldsymbol{\xi})$ for some sequence t_k such that $\lim_{k\to\infty} t_k = \infty$. This set up will be used in the proof of both parts.

For the first part, let τ be in the domain of $\mathbf{x}(t, \mathbf{q})$. Then $\lim_{k\to\infty} \tau + t_k = \infty$ and by uniqueness of solutions

$$\mathbf{x}(\tau + t_k, \boldsymbol{\xi}) = \mathbf{x}(\tau, \mathbf{x}(t_k, \boldsymbol{\xi})).$$

Continuity in initial conditions implies that

$$\begin{aligned}\mathbf{x}(\tau, \mathbf{q}) &= \lim_{k\to\infty} \mathbf{x}(\tau, \mathbf{x}(t_k, \boldsymbol{\xi})) \\ &= \lim_{k\to\infty} \mathbf{x}(\tau + t_k, \boldsymbol{\xi}) \in \omega(\boldsymbol{\xi})\end{aligned}$$

and the proof of the first part is finished.

For the second part, let τ be in the domain of $\mathbf{x}(t, \boldsymbol{\xi})$. Then $t_k - \tau \to \infty$ and

$$\mathbf{x}(t_k - \tau, \mathbf{x}(\tau, \boldsymbol{\xi})) = \mathbf{x}(t_k, \boldsymbol{\xi}) \to \mathbf{q}$$

as $k \to \infty$. Hence, $\mathbf{q} \in \omega(\mathbf{x}(\tau, \boldsymbol{\xi}))$ and $\omega(\boldsymbol{\xi}) \subset \omega(\mathbf{x}(\tau, \boldsymbol{\xi}))$, since $\mathbf{q}$ was an arbitrary point in $\omega(\boldsymbol{\xi})$. Applying this to $\omega(\mathbf{x}(\tau, \boldsymbol{\xi}))$ and $-\tau$, yields

$$\omega(\mathbf{x}(\tau, \boldsymbol{\xi})) \subset \omega(\mathbf{x}(-\tau, \mathbf{x}(\tau, \boldsymbol{\xi}))) = \omega(\boldsymbol{\xi})$$

and completes the proof. □

Proposition 5.6 *If there exists a compact set $K \subset \Omega$ such that*

$$\{\mathbf{x}(t, \boldsymbol{\xi}) : t \geq 0\} \subset K,$$

then the following hold:

(a) $\omega(\boldsymbol{\xi}) \neq \phi$,

(b) $\omega(\boldsymbol{\xi})$ *is compact,*

(c) $\mathbf{x}(t, \mathbf{q})$ *is defined on* $\mathbb{R}$ *and contained in* $\omega(\boldsymbol{\xi})$ *when* $\mathbf{q} \in \omega(\boldsymbol{\xi})$,

(d) $\omega(\boldsymbol{\xi})$ *is connected.*

Proof. For part (a), note that the sequence $\mathbf{x}(k, \boldsymbol{\xi})$, k a positive integer, is contained in K and therefore must have a convergent subsequence because K is compact. Thus $\omega(\boldsymbol{\xi}) \neq \phi$.

Because

$$\overline{\{\mathbf{x}(t, \boldsymbol{\xi}) : t \geq 0\}} \subset K \subset \Omega$$

it follows that $\omega(\boldsymbol{\xi})$ is closed and bounded, proving (b).

Part (c) follows from part (a) of the previous proposition and *Theorem 1.19.*

The proof of (d) is left as an exercise. □

Corollary *If* $\mathbf{p}$ *is a positively stable critical point, then there exists* $r > 0$ *such that when* $\|\boldsymbol{\xi} - \mathbf{p}\| < r$ *the omega limit set,* $\omega(\boldsymbol{\xi})$*, is a nonempty compact connected set of solutions defined on* $\mathbb{R}$ *.*

Proof. Because $\mathbf{p}$ is positively stable, there exists $\delta > 0$ such that $\|\mathbf{x}(t, \boldsymbol{\xi}) - \mathbf{p}\| < 1$ for all $t \geq 0$. Apply the proposition to the compact set $\{\mathbf{x} : \|\mathbf{x} - \mathbf{p}\| \leq 1\}$. □

Theorem 5.7 (LaSalle) *Given* $\dot{\mathbf{x}} = \mathbf{f}(\mathbf{x})$ *on* Ω*, suppose* $V : \Omega_0 \to \mathbb{R}$ *has continuous first partial derivatives on* Ω_0 *an open subset of* Ω*. If for some* $r \in \mathbb{R}$ *the set* $K = \{\mathbf{x} : V(\mathbf{x}) \leq r\}$ *is compact and* $\dot{V}(\mathbf{x}) \leq 0$ *on* K*, then* $V(\boldsymbol{\xi}) < r$ *implies the following:*

(a) $\mathbf{x}(t, \boldsymbol{\xi})$ *is defined for all* $t \geq 0$*,*

(b) V *is constant on* $\omega(\boldsymbol{\xi})$*,*

(c) $\dot{V}(\mathbf{x}) = 0$ *on* $\omega(\boldsymbol{\xi})$*,*

(d) $\omega(\boldsymbol{\xi}) \subset M = \{\mathbf{q} \in K : \dot{V}(x(t, \mathbf{q})) = 0 \text{ for all } t \in \mathbb{R}\}$.

Proof. Suppose $V(\boldsymbol{\xi}) < r$. If $V(\mathbf{x}(t, \boldsymbol{\xi})) < r$ for the positive t in the domain of $\mathbf{x}(t, \boldsymbol{\xi})$, then all positive t are in the domain of $\mathbf{x}(t, \boldsymbol{\xi})$ by *Theorem 1.19* because K is compact. If this is not the case, $\tau = \sup\{t : V(\mathbf{x}(s, \boldsymbol{\xi})) < r \text{ for } 0 \leq s \leq t\}$ is in the domain of $\mathbf{x}(t, \boldsymbol{\xi})$. As before

$$r = V(\mathbf{x}(\tau, \boldsymbol{\xi})) = V(\boldsymbol{\xi}) + \int_0^\tau \dot{V}(\mathbf{x}(s, \boldsymbol{\xi}))ds \leq V(\boldsymbol{\xi}) < r,$$

which is impossible. Therefore, the solution $\mathbf{x}(t, \boldsymbol{\xi})$ is defined for all $t \geq 0$ and remains in K whenever $V(\boldsymbol{\xi}) < r$. This proves part (a).

Next note that

$$\lim_{t \to \infty} V(\mathbf{x}(t, \boldsymbol{\xi})) = c > -\infty$$

because $V(\mathbf{x}(t, \boldsymbol{\xi}))$ is nonincreasing and V is bounded below on K because K is compact. If $\mathbf{q} \in \omega(\boldsymbol{\xi})$, then

$$V(\mathbf{q}) = \lim_{k \to \infty} V(\mathbf{x}(t_k, \boldsymbol{\xi})) = c$$

because V is continuous and by *Proposition 5.4* there exists a sequence of real numbers $\{t_k\}$ such that $\lim_{k \to \infty} t_k = \infty$ and $\lim_{k \to \infty} \mathbf{x}(t_k, \boldsymbol{\xi}) = \mathbf{q}$. Part (b) is now established and part (c) is an immediate consequences of part (b). Finally, part (d) follows from part (c) and *Proposition 5.6.* □

For an example, consider $\ddot{x} + x^2\dot{x} + h(x) = 0$ where $h(x)$ has a continuous derivative, $h(0) = 0$, and $xh(x) > 0$ for $x \neq 0$. LaSalle's theorem will be used to show that there exits $\Omega_0 \subset \mathbb{R}^2$ with the following property: If $x(t)$ is a solution such that $(x(0), \dot{x}(0)) \in \Omega_0$, then

$$\lim_{t \to \infty} |x(t)| + |\dot{x}(t)| = 0,$$

in other words, the solution damps out as time increases.

Because $xh(x) > 0$ for $x \neq 0$, the function

$$H(x) = \int_0^x h(w)dw$$

is decreasing for $x < 0$ and increasing for $x > 0$. Moreover, $0 = H(0) < H(x)$ for $x \neq 0$.

The corresponding system, which is obviously locally Lipschitz, is

$$\begin{aligned} \dot{x} &= y \\ \dot{y} &= -x^2 y - h(x). \end{aligned}$$

The only critical point is $(0,0)$.

Set

$$V(x,y) = H(x) + \frac{y^2}{2}.$$

Check that $\dot{V}(x,y) = -x^2y^2 \leq 0$ and that $V(x,y)$ is positive definite at $(0,0)$. Hence the critical point $(0,0)$ is positively stable by Lyapunov's method, (*Theorem 5.1*).

Let r be the minimum of $H(-1)$ and $H(1)$. Then there exit $\alpha < 0 < \beta$ satisfying $H(\alpha) = r = H(\beta)$. It follows that $K = \{(x,y) : V(x,y) \leq r\}$ consists of a simple closed curve and its interior. Hence, K is a compact set. Now LaSalle's theorem can be applied.

If $\boldsymbol{\varphi}(\tau) = (a,0)$ with $a \neq 0$ for some solution $\boldsymbol{\varphi}(t)$, then $\boldsymbol{\varphi}(t)$ must cross the x-axis because $\dot{\boldsymbol{\varphi}}(\tau) = (0, h(a))$ and $h(a) \neq 0$. Similarly, if $\boldsymbol{\varphi}(\tau) = (0,b)$ with $b \neq 0$. Therefore, $M = \{(0,0)\}$ and $\omega(\boldsymbol{\xi}) = \{(0,0)\}$ for all $\boldsymbol{\xi} \in K$.

The set $\Omega_0 = \{(x,y) : V(x,y) \leq r\}$ is open and contains $(0,0)$. It follows that $(0,0)$ is positively asymptotically stable and $\lim_{t\to\infty} |\boldsymbol{\varphi}(t)| = 0$, which is the desired conclusion.

The omega and alpha limit sets will reappear in Chapter 6 and play an important role there in the study of planar dynamics.

EXERCISES

1. Prove that a positively asymptotically stable fixed point cannot be negatively stable.

2. Show that the real-valued function

$$f(x) = \begin{cases} x^2 \sin(1/x) & \text{if } x \neq 0 \\ 0 & \text{if } x = 0 \end{cases}$$

   is continuously differentiable. Find the positively and negatively asymptotically stable fixed points of $\dot{x} = f(x)$.

3. Let $\mathbf{p}$ be a positively asymptotically stable fixed point of $\dot{\mathbf{x}} = \mathbf{f}(\mathbf{x})$ and let $E(\mathbf{x})$ be an integral for $\dot{\mathbf{x}} = \mathbf{f}(\mathbf{x})$. Show that there exists a neighborhood of $\mathbf{p}$ on which $E(\mathbf{x})$ is constant.

4. Given $\dot{\mathbf{x}} = \mathbf{f}(\mathbf{x})$ on Ω an open subset of $\mathbb{R}^d$, suppose there exists a compact set K contained in Ω and a point $\boldsymbol{\xi}$ such that $\mathbf{x}(t, \boldsymbol{\xi}) \in K$ for all $t \geq 0$. Prove that $\omega(\boldsymbol{\xi})$ is connected, that is, show that it is impossible to find disjoint open sets U_1, and U_2 such that $\omega(\boldsymbol{\xi}) \subset U_1 \cup U_2$ and $\omega(\boldsymbol{\xi}) \cap U_j \neq \phi$ for $j = 1, 2$.

5. Consider the following system of differential equations on $\mathbb{R}^2$:

$$\begin{aligned} \dot{x} &= y - \alpha x y^2 \\ \dot{y} &= -x - \beta y \end{aligned}$$

where α and β are positive real constants. Show that the origin is positively asymptotically stable and determine the omega limit set for every point in the plane.

5.2 Stability and Constant Coefficients

An important technique for studying the behavior of solutions near a fixed point is to linearize the system, that is, to use the linear system $\dot{\mathbf{x}} = A\mathbf{x}$, where A is is the Jacobian matrix of $\mathbf{f}(\mathbf{x})$ at the fixed point $\mathbf{p}$, to obtain information about the behavior of solutions of $\dot{\mathbf{x}} = \mathbf{f}(\mathbf{x})$ near the fixed point $\mathbf{p}$. So, naturally it is important first to understand completely the stability of $\dot{\mathbf{x}} = A\mathbf{x}$ at the origin. The two theorems in this section address this problem using the eigenvalues. The first theorem is a general result about complex matrices, and the second one applies the first theorem to characterize the stability of the origin for $\dot{\mathbf{x}} = A\mathbf{x}$, when A is a real matrix.

For real and complex $d \times d$ matrices the notation is exactly the same as in Chapter 4. An eigenvalue λ is said to be *simple* if every generalized eigenvector is an eigenvector or equivalently $r(\lambda) = 1$. In particular, if λ is a simple eigenvalue, then $A\mathbf{v} = \lambda\mathbf{v}$ for all $\mathbf{v} \in M(\lambda)$.

Theorem 5.8 *Let $\lambda_1, \ldots, \lambda_\rho$ be the distinct eigenvalues of a complex $d \times d$ matrix A. The following hold:*

(a) There exists a positive constant K such that $|e^{At}| \leq K$ for all $t \geq 0$ if and only if $Re\lambda_j \leq 0$ for $j = 1, \ldots \rho$ and λ_j is a simple eigenvalue when $Re\lambda_j = 0$. ($Re\lambda$ denotes the real part of a complex number λ.)

(b) There exist positive constants K and α such that $|e^{At}| \leq Ke^{-\alpha t}$ for all $t \geq 0$ if and only if $Re\lambda_j < 0$ for $j = 1, \ldots, \rho$.

Proof. Let $\boldsymbol{\zeta} \in \mathbb{C}^d$ and let $\mathbf{z}(t)$ be the solution $e^{At}\boldsymbol{\zeta}$ of $\dot{\mathbf{z}} = A\mathbf{z}$. Write $\lambda_j = \alpha_j + i\beta_j$ where α_j and β_j are real numbers. Using *Theorem 4.6*, $\boldsymbol{\zeta}$ can be written as $\boldsymbol{\zeta} = \mathbf{v}_1 + \ldots + \mathbf{v}_\rho$ with $\mathbf{v}_j \in M(\lambda_j)$, the generalized eigenspace for λ_j in $\mathbb{C}^d$. By *Theorem 4.7*,

$$\mathbf{z}(t) = \sum_{j=1}^{\rho} e^{\lambda_j t} \sum_{k=0}^{r(\lambda_j)-1} (A - \lambda_j I)^k \frac{t^k}{k!}\mathbf{v}_j$$

and

$$\begin{aligned} |\mathbf{z}(t)| &\le \sum_{j=1}^{\rho} e^{\lambda_j t} \sum_{k=0}^{r(\lambda_j)-1} |A - \lambda_j I|^k \frac{|t^k|}{k!}|\mathbf{v}_j| \\ &\le \sum_{j=1}^{\rho} e^{\alpha_j t} \sum_{k=0}^{r(\lambda_j)-1} M\frac{|t^k|}{k!}|\mathbf{v}_j|, \end{aligned}$$

where $M = \max\{|A - \lambda_j I|^k : 1 \le j \le \rho \text{ and } 0 \le k \le r(\lambda_j) - 1\}$.

It is easy to check that $\|\mathbf{w}\|_a = |\mathbf{w}_1| + \ldots + |\mathbf{w}_\rho|$ defines a norm on $\mathbb{C}^d$ when $\mathbf{w} = \mathbf{w}_1 + \ldots + \mathbf{w}_\rho$ and $\mathbf{w}_j \in M(\lambda)$ because $\mathbb{C}^d = M(\lambda_1) \oplus \ldots \oplus M(\lambda_\rho)$. Since all norms are equivalent (a norm on $\mathbb{C}^d$ automatically defines a norm on $\mathbb{R}^{2d}$), there exists a positive constant C such that $\|\mathbf{w}\|_a \le C|\mathbf{w}|$. In particular, $|\mathbf{v}_j| \le \|\boldsymbol{\zeta}\|_a \le C|\boldsymbol{\zeta}|$ in the estimate of $|\mathbf{z}(t)|$. Thus

$$|\mathbf{z}(t)| \le \sum_{j=1}^{\rho} e^{\alpha_j t} \sum_{k=0}^{r(\lambda_j)-1} \frac{|t^k|}{k!} MC|\boldsymbol{\zeta}| \le g(t)MC|\boldsymbol{\zeta}|,$$

where

$$g(t) = \sum_{j=1}^{\rho} e^{\alpha_j t} \sum_{k=0}^{r(\lambda_j)-1} \frac{|t^k|}{k!}.$$

This estimate can be used in both parts of the theorem to prove that the required property of $|e^{At}|$ follows from the conditions on the eigenvalues because

$$|e^{At}| = \sup\{|e^{At}\boldsymbol{\zeta}| : |\boldsymbol{\zeta}| = 1\} \le g(t)MC.$$

First, suppose that $Re\lambda_j \le 0$ for all j and the eigenvalue λ_j is simple when $Re\lambda_j = 0$, so $r(\lambda_j) = 1$ when $Re\lambda_j = 0$. If $\alpha_j < 0$, then $e^{\alpha_j t}t^k$ is bounded for $t \ge 0$, and

$$e^{\alpha_j t} \sum_{k=0}^{r(\lambda_j)-1} \frac{t^k}{k!}$$

is also bounded for $t \ge 0$. If $\alpha_j = 0$, then because $r(\lambda_j) = 1$ it follows that

$$e^{\alpha_j t} \sum_{k=0}^{r(\lambda_j)-1} \frac{t^k}{k!} = 1.$$

Therefore, $g(t)$ is bounded above by say K' for $t \geq 0$ and

$$|e^{At}| \leq K'MC = K,$$

if $Re\lambda_j \leq 0$ for $j = 1, \ldots, \rho$ and λ_j is a simple eigenvalue when $Re\lambda_j = 0$.

Next, suppose that $Re\lambda_j < 0$ for all j. Choose $\alpha > 0$ such that $\alpha_j < -\alpha$ for all j. Since

$$\lim_{t\to\infty} \frac{e^{\alpha_j t}t^k}{e^{-\alpha t}} = 0,$$

it follows that

$$\lim_{t\to\infty} \frac{g(t)}{e^{-\alpha t}} = 0.$$

Hence, $|g(t)/e^{-\alpha t}| \leq K'$ for $t \geq 0$ and some K' or $|g(t)| \leq K'e^{-\alpha t}$ for $t \geq 0$. Therefore,

$$|e^{At}| \leq MCK'e^{-\alpha t} = Ke^{-\alpha t}.$$

for all $t \geq 0$, if $Re\lambda_j < 0$ for $j = 1, \ldots, \rho$.

To prove the reverse implication for part (a), it suffices to show that if some eigenvalue, λ_j, does not satisfy the required conditions, then $|e^{At}| \leq K$ does not hold for $t \geq 0$. First, suppose $\alpha_j > 0$ and let $\mathbf{v} \in \mathbb{C}^d$ be an eigenvector for λ_j with $|\mathbf{v}| = 1$. Then

$$|e^{At}| \geq |e^{At}\mathbf{v}| = e^{\alpha_j t} \to \infty$$

as $t \to \infty$.

Second, suppose $\alpha_j = 0$ and $r(\lambda_j) > 1$. Then there exists a generalized eigenvector $\mathbf{v}$ of degree 2 with $|\mathbf{v}| = 1$, and

$$|e^{At}| \geq |e^{At}\mathbf{v}| = |e^{\lambda_j t}(\mathbf{v} + t(A - \lambda_j I)\mathbf{v})| = |\mathbf{v} + t(A - \lambda_j I)\mathbf{v}| \to \infty$$

as $t \to \infty$ because $(A - \lambda_j I)\mathbf{v} \neq 0$. This completes part (a).

Finally, to complete the proof of (b), it remains to show that if $\alpha_j = 0$, then $|e^{At}| \leq Ke^{-\alpha t}$ cannot hold for any positive constants K and α. Let $\mathbf{v}$ be an eigenvector for $\lambda_j = i\beta_j$ with $|\mathbf{v}| = 1$. Then for all $t \geq 0$,

$$|e^{At}| \geq |e^{At}\mathbf{v}| = |e^{\lambda_j t}\mathbf{v}| = |\mathbf{v}| = 1$$

and $|e^{At}|$ does not decay exponentially. This completes the proof of the theorem. □

What is needed next is to have the same result for a real $d \times d$ matrix A and the usual norm for $\mathcal{M}_d(\mathbb{R})$. This turns out to be an immediate corollary of the above theorem because the restriction of the norm used on the matrices in the proof of *Theorem 5.8* to real matrices, namely,

$$\|A\|_c = \sup\{|A\mathbf{z}| : \mathbf{z} \in \mathbb{C}^d \text{ and } |\mathbf{z}| = 1\},$$

defines a norm on the real $d \times d$ matrices. Obviously,

$$|A| = \sup\{|A\mathbf{x}| : \mathbf{x} \in \mathbb{R}^d \text{ and } |\mathbf{x}| = 1\} \leq \|A\|_c.$$

Since any two norms on $\mathbb{R}^{d^2}$ are equivalent, there exists a constant L such that $\|A\|_c \leq L|A|$.

For a real $d \times d$ matrix A, it follows that $|e^{At}| \leq K_1$ for $t \geq 0$ if and only if $\|e^{At}\|_c \leq K_2$ for $t \geq 0$. By *Theorem 5.8*, this is equivalent to $Re\lambda_j \leq 0$ for $j = 1, \ldots, \rho$ and λ_j is a simple eigenvalue when $Re\lambda_j = 0$. Similarly, $|e^{At}| \leq K_1 e^{-\alpha t}$ for $t \geq 0$ if and only if $\|e^{At}\|_c \leq K_2 e^{-\alpha t}$ for $t \geq 0$ if and only if $Re\lambda_j < 0$ for $j = 1, \ldots, \rho$. This proves the following corollary:

Corollary *Let $\lambda_1, \ldots, \lambda_\rho$ be the distinct eigenvalues (real and complex) of a real $d \times d$ matrix A. The following hold:*

(a) *There exists a positive constant K such that $|e^{At}| \leq K$ for all $t \geq 0$ if and only if $Re\lambda_j \leq 0$ for $j = 1, \ldots, \rho$ and λ_j is a simple eigenvalue when $Re\lambda_j = 0$.*

(b) *There exist positive constants K and α such that $|e^{At}| \leq Ke^{-\alpha t}$ for all $t \geq 0$ if and only if $Re\lambda_j < 0$ for $j = 1, \ldots, \rho$.*

Not surprisingly, the eigenvalue conditions in *Theorem 5.8* provide necessary and sufficient conditions for the origin to be positively stable or positively asymptotically stable.

Theorem 5.9 *Let $\lambda_1, \ldots, \lambda_\rho$ be the distinct eigenvalues (real and complex) of a real $d \times d$ matrix A and consider $\dot{\mathbf{x}} = A\mathbf{x}$.*

(a) *The fixed point $\mathbf{0}$ of $\dot{\mathbf{x}} = A\mathbf{x}$ is positively stable if and only if $Re\lambda_j \leq 0$ for $j = 1, \ldots, \rho$ and λ_j is a simple eigenvalue when $Re\lambda_j = 0$.*

(b) *The fixed point $\mathbf{0}$ of $\dot{\mathbf{x}} = A\mathbf{x}$ is positively asymptotically stable if and only if $Re\lambda_j < 0$ for $j = 1, \ldots, \rho$.*

Proof. Suppose $\mathbf{0}$ is positively stable; so there exists $\delta > 0$ such that $|\mathbf{v}| < \delta$ implies that $|\mathbf{x}(t, \mathbf{v})| < 1$ for all $t \geq 0$. Since $e^{At}\mathbf{v} = \mathbf{x}(t, \mathbf{v})$, it follows that

$$\begin{aligned} |e^{At}| &= \sup\{|e^{At}\mathbf{v}| : |\mathbf{v}| = 1\} \\ &= \frac{1}{\delta} \sup\{|e^{At}\mathbf{v}| : |\mathbf{v}| = \delta\} \\ &\leq \frac{1}{\delta}. \end{aligned}$$

It now follows from the above corollary that $Re\lambda_j \leq 0$ for $j = 1, \ldots, \rho$ and λ_j is a simple eigenvalue when $Re\lambda_j = 0$.

The converse is even easier. Given $\varepsilon > 0$, let $\delta = \varepsilon / K$ where K is given by the corollary. It then follows that

$$|\mathbf{x}(t, \mathbf{v})| = |e^{At}\mathbf{v}| \leq K|\mathbf{v}| < \varepsilon$$

and part (a) is done.

Turning to (b), suppose $\mathbf{0}$ is positively asymptotically stable. By part (a), it is known that $Re\lambda_j \leq 0$ for $j = 1, \ldots, \rho$, and by definition there exists γ such that

$$\lim_{t\to\infty} |\mathbf{x}(t, \mathbf{v})| = \lim_{t\to\infty} |e^{At}\mathbf{v}| = 0$$

whenever $|\mathbf{v}| < \gamma$. It follows that $|e^{At}\mathbf{v}| \to 0$ as $t \to \infty$ for all $\mathbf{v} \in \mathbb{R}^d$. Since

$$|e^{At}| = \max\{|e^{At}\mathbf{e}_j| : 1 \leq j \leq d\}$$

by *Theorem 3.2*, it follows that

$$\lim_{t\to\infty} |e^{At}| = 0$$

and

$$\lim_{t\to\infty} \|e^{At}\|_c = 0.$$

If $Re\lambda_j = 0$ and $\mathbf{v} \in \mathbb{C}^d$ is an eigenvector for λ_j with $|\mathbf{v}| = 1$, then

$$\|e^{At}\|_c \geq |e^{At}\mathbf{v}| = |e^{i\beta_j t}\mathbf{v}| = |\mathbf{v}| = 1,$$

a contradiction.

The converse and last step in the proof is an obvious consequence of part (b) of the corollary. □

When A is a 2×2 real matrix, it is easy to exhibit all the possible phase portraits for $\dot{\mathbf{x}} = A\mathbf{x}$ and describe in more detail the qualitative behavior of solutions near $\mathbf{0}$. It is instructive to examine this familiar example as part of the general theory of linear differential equations. These phase portraits will also provide useful mental images in subsequent chapters. In what follows λ_1 and λ_2 will denote the eigenvalues of the matrix

$$A = \begin{bmatrix} a & b \\ c & d \end{bmatrix} \neq \mathbf{0}$$

with real entries. In this case, the eigenvalues are the roots of

$$\lambda^2 - (a + d)\lambda + (ad - bc) = \lambda^2 - \mathrm{Tr}\,[A]\lambda + \mathrm{Det}\,[A] = 0,$$

where as usual $\mathrm{Tr}\,[A]$ and $\mathrm{Det}\,[A]$ denote the trace and determinant of A. The discussion will be divided into three cases—distinct real eigenvalues, repeated real eigenvalues, and complex conjugate eigenvalues.

First, assume that λ_1 and λ_2 are real and $\lambda_1 < \lambda_2$. Let $\mathbf{v}_1$ and $\mathbf{v}_2$ be eigenvectors for λ_1 and λ_2, respectively, and let L_1 and L_2 be the lines through the origin in the directions of the vectors $\mathbf{v}_1$ and $\mathbf{v}_2$, respectively. Since $\mathbf{v}_1$ and $\mathbf{v}_2$ are linearly independent and both $e^{\lambda_1 t}\mathbf{v}_1$ and $e^{\lambda_2 t}\mathbf{v}_2$ are solutions, every solution has the form

$$\mathbf{x}(t) = c_1 e^{\lambda_1 t}\mathbf{v}_1 + c_2 e^{\lambda_2 t}\mathbf{v}_2.$$

The next remark shows that every solution like this has an asymptotic direction as t goes to plus or minus infinity.

Remark *If* $\mathbf{x}(t) = c_1e^{\lambda_1 t}\mathbf{v}_1 + c_2e^{\lambda_2 t}\mathbf{v}_2$ *is any solution of* $\dot{\mathbf{x}} = A\mathbf{x}$ *with both* c_1 *and* c_2 *nonzero, then*

$$\lim_{t\to\infty} \frac{\dot{\mathbf{x}}(t)}{\|\dot{\mathbf{x}}(t)\|} = \frac{\operatorname{sgn}(c_2\lambda_2)\mathbf{v}_2}{\|\mathbf{v}_2\|}$$

and

$$\lim_{t\to-\infty} \frac{\dot{\mathbf{x}}(t)}{\|\dot{\mathbf{x}}(t)\|} = \frac{\operatorname{sgn}(c_1\lambda_1)\mathbf{v}_1}{\|\mathbf{v}_1\|},$$

where sgn(c) *denotes the sign of* c.

Proof. Factoring $e^{\lambda_2 t}$ out of both the numerator and denominator yields

$$\frac{\dot{\mathbf{x}}(t)}{\|\dot{\mathbf{x}}(t)\|} = \frac{c_1\lambda_1 e^{(\lambda_1-\lambda_2)t}\mathbf{v}_1 + c_2\lambda_2\mathbf{v}_2}{\|c_1\lambda_1 e^{(\lambda_1-\lambda_2)t}\mathbf{v}_1 + c_2\lambda_2\mathbf{v}_2\|}$$

from which it follows that

$$\lim_{t\to\infty} \frac{\dot{\mathbf{x}}(t)}{\|\dot{\mathbf{x}}(t)\|} = \frac{c_2\lambda_2\mathbf{v}_2}{\|c_2\lambda_2\mathbf{v}_2\|} = \frac{\operatorname{sgn}(c_2\lambda_2)\mathbf{v}_2}{\|\mathbf{v}_2\|}.$$

the second limit follows in the same way. □

There are now several subcases to consider.

(i) If $\lambda_1 < \lambda_2 < 0$, then $\mathbf{0}$ is clearly positively asymptotically stable and is called a *stable node*. The lines L_1 and L_2 each decompose into three orbits—the origin, $e^{\lambda_j t}\mathbf{v}_j$, and $-e^{\lambda_j t}\mathbf{v}_j$. The latter two approach the origin as t goes to infinity. From the above remark, it follows that all the other solutions approach the origin tangentially to L_2 as t goes to infinity. *Figure 5.2* is a phase portrait of a stable node with L_1 and L_2 perpendicular to each other, making it easier to show the asymptotic behavior of the other orbits. In general, however, L_1 and L_2 are not perpendicular.

(ii) When $0 < \lambda_1 < \lambda_2$, the origin is called an *unstable node*. The phase portrait of an unstable node looks like a stable node with the arrows reversed and the subscripts on L_1 and L_2 switched.

(iii) When $\lambda_1 < 0 < \lambda_2$, it is obvious that $\lim_{t\to\pm\infty} |\mathbf{x}(t)| = \infty$, if both c_1 and c_2 are not equal to zero and the phase portrait is the familiar hyperbolic phase portrait or saddle in *Figure 2.3* on page 64. Again the lines L_1 and L_2 are usually not perpendicular.

(iv) If $\lambda_1 < \lambda_2 = 0$, then the entire line L_2 consists of fixed points, while the line L_1 has the same orbit structure as in (i), namely, the origin, $e^{\lambda_1 t}\mathbf{v}_1$, and $-e^{\lambda_1 t}\mathbf{v}_1$. Moreover every solution has the form $\mathbf{x}(t) = c_1e^{\lambda_1 t}\mathbf{v}_1 + c_2\mathbf{v}_2$. which lies on a line parallel to L_1 and approaches $c_2\mathbf{v}_2$ as $t \to \infty$. In this case, the origin and every point on L_2 is a positively stable fixed point but not positively asymptotically stable. This is illustrated in *Figure 5.3* with L_2 represented by the line of dots for fixed points.

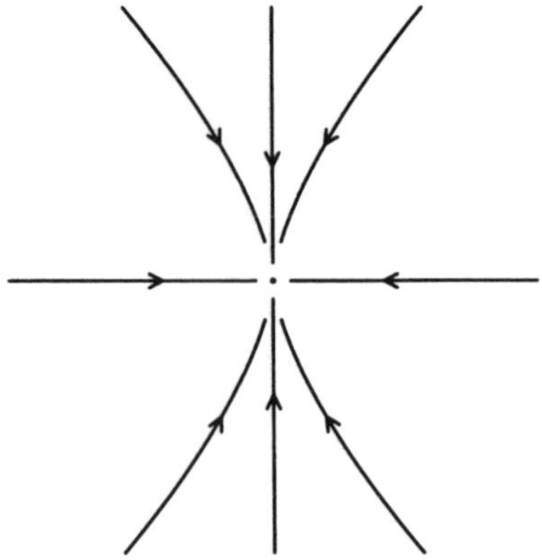

Figure 5.2: Phase portrait of $\dot{x} = -1.5x$ and $\dot{y} = -y$.

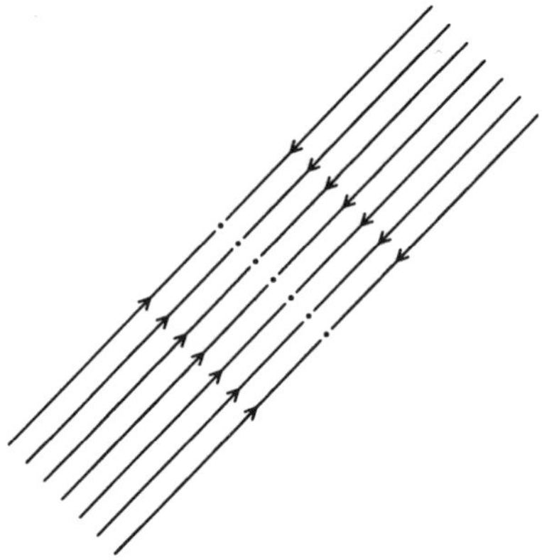

Figure 5.3: Phase portrait of $\dot{x} = x + y$ and $\dot{y} = x + y$.

(v) The phase portrait when $0 = \lambda_1 < \lambda_2$ looks like the phase portrait in *Figure 5.3* with the arrows reversed and the subscripts on L_1 and L_2 switched.

The second major case is $\lambda = \lambda_1 = \lambda_2 \in \mathbb{R}$. Here there need not exist a basis of eigenvectors. If, in fact, there is a basis of eigenvectors, then every nonzero $\mathbf{v} \in \mathbb{R}^2$ is an eigenvector for λ and

$$A = \begin{bmatrix} \lambda & 0 \\ 0 & \lambda \end{bmatrix}$$

Consequently, $\mathbf{x}(t) = e^{\lambda t}\mathbf{v}$ is a solution for every $\mathbf{v} \in \mathbb{R}^2$. This produces two simple phase portraits, and every ray emanating from the origin is an orbit.

(i) If $\lambda < 0$ and there exists a basis of eigenvectors, then every solution moves toward the origin as t goes to infinity. Once again the origin is positively asymptotically stable. It has a similar but simpler phase portrait than when

$\lambda_1 < \lambda_2 < 0$ as shown in *Figure 5.4*.

(ii) If $0 < \lambda$ and there exists a basis of eigenvectors, the phase portrait is the same as in the previous case with the arrows reversed.

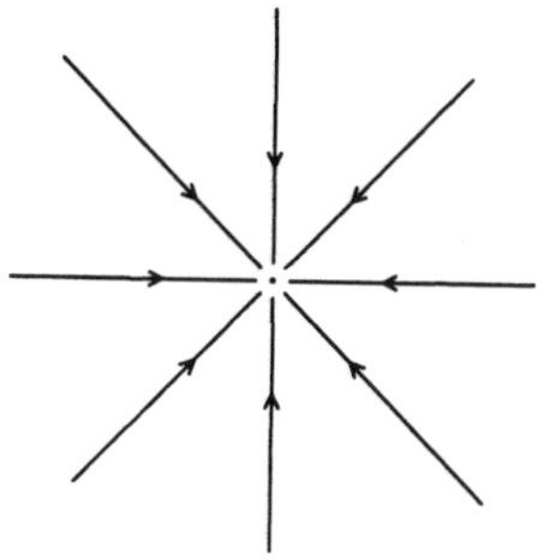

Figure 5.4: Phase portrait of an improper stable node.

The two previous cases are called *improper stable node and improper unstable node*, respectively. The word *improper* is used because the slightest change in A can (and usually does) produce two distinct eigenvalues and destroy this type phase portrait. In contrast, observe that when λ_1 and λ_2 are nonzero real numbers and $\lambda_1 < \lambda_2$, slight changes in A will not change the relationship between the eigenvalues or alter their signs. Consequently, the phase portraits will also have the same general appearance and from the qualitative perspective be identical.

Next, assume there is not a basis of eigenvectors. As usual let $\mathbf{v}_1$ be an eigenvector for λ and let L_1 be a line through the origin in the direction of the vector $\mathbf{v}_1$, but now let $\mathbf{v}_2$ be a vector perpendicular to $\mathbf{v}_1$. Shifting to $\mathbb{C}^2$ for a minute, the only possibility is that $M(\lambda) = \mathbb{C}^2$ and $r(\lambda) = 2$. Hence $(A - \lambda I)^2\mathbf{v}_2 = \mathbf{0}$ and $(A - \lambda I)\mathbf{v}_2$ is an eigenvector. So it can be assumed that $(A - \lambda I)\mathbf{v}_2 = \mathbf{v}_1$ and that $\|\mathbf{v}_1\| = 1$ by adjusting the length of $\mathbf{v}_2$. Clearly $\mathbf{v}_1$ and $\mathbf{v}_2$ are linearly independent and form a basis for $\mathbb{R}^2$.

If $\mathbf{x}(t)$ is any solution of $\dot{\mathbf{x}} = A\mathbf{x}$ and $\mathbf{x}(0) = \mathbf{v} = c_1\mathbf{v}_1 + c_2\mathbf{v}_2$, then

$$\begin{aligned}\mathbf{x}(t) &= e^{At}\mathbf{v} \\ &= e^{\lambda t}[\mathbf{v} + t(A - \lambda I)\mathbf{v}] \\ &= e^{\lambda t}[c_1\mathbf{v}_1 + c_2\mathbf{v}_2 + tc_2\mathbf{v}_1] \\ &= e^{\lambda t}[(c_1 + tc_2)\mathbf{v}_1 + c_2\mathbf{v}_2].\end{aligned}$$

An easy calculation shows that

$$\lim_{t\to\pm\infty} \frac{\dot{\mathbf{x}}(t)}{\|\dot{\mathbf{x}}(t)\|} = \operatorname{sgn}(c_2)\mathbf{v}_1.$$

Moreover, the function $e^{\lambda t}(c_1 + tc_2)$, which describes the behavior of $\mathbf{x}(t)$ in the $\mathbf{v}_1$ direction, has a unique extreme point at $t = -(1/\lambda) - (c_1/c_2)$. The final two subcases for a single real eigenvalue can now be described.

(iii) If $\lambda < 0$, the origin is a positively asymptotically stable fixed point and is again called a improper stable node. *Figure 5.5* shows the phase portrait a typical improper stable node without a basis of eigenvectors.

(iv) When the arrows are reversed in the previous phase portrait, we have another improper unstable node that occurs when $0 < \lambda$.

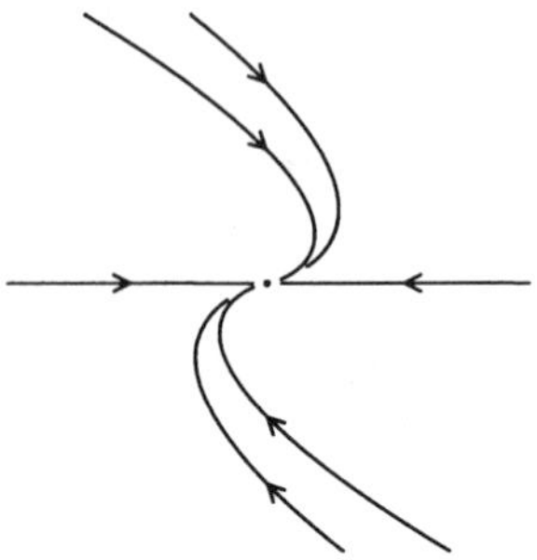

Figure 5.5: Phase portrait of $\dot{x} = -x + y$ and $\dot{y} = -y$.

The final major case occurs when there is a non-real eigenvalue. Because A is real, non-real eigenvalues must occur in conjugate pairs, $\lambda_2 = \overline{\lambda}_1$, and neither eigenvalue is real. If $\mathbf{v}_1$ is a complex eigenvector for λ_1, then $\mathbf{v}_2 = \overline{\mathbf{v}}_1$ is a complex eigenvector for λ_2. Obviously, $\lambda_1 \neq \lambda_2$, the vectors $\mathbf{v}_1$ and $\mathbf{v}_2$ are a basis for $\mathbb{C}^2$, and $r(\lambda_1) = 1 = r(\lambda_2)$. Let $\mathbf{v} \in \mathbb{R}^2 \subset \mathbb{C}^2$ and write $\mathbf{v} = c_1\mathbf{v}_1 + c_2\mathbf{v}_2$. Then by taking the conjugate, $\mathbf{v} = \overline{\mathbf{v}} = \overline{c}_1\mathbf{v}_2 + \overline{c}_2\mathbf{v}_1$ and $\overline{c}_1 = c_2$. Writing $c_1 = ae^{i\delta}$, $\lambda_1 = \alpha + i\beta$, and $\mathbf{v}_1 = \boldsymbol{\mu} + i\boldsymbol{\nu}$ where a, α, β and δ are real and $\boldsymbol{\mu}$ and $\boldsymbol{\nu}$ are in $\mathbb{R}^2$, produces

$$\begin{aligned} \mathbf{x}(t, \mathbf{v}) &= e^{\lambda_1 t} c_1 \mathbf{v}_1 + e^{\lambda_2 t} \overline{c}_1 \mathbf{v}_2 \\ &= e^{(\alpha + i\beta)t} a e^{i\delta} (\boldsymbol{\mu} + i\boldsymbol{\nu}) + e^{(\alpha - i\beta)t} a e^{-i\delta} (\boldsymbol{\mu} - i\boldsymbol{\nu}) \\ &= 2ae^{\alpha t} [\cos(\beta t + \delta)\boldsymbol{\mu} - \sin(\beta t + \delta)\boldsymbol{\nu}]. \end{aligned}$$

(i) If $\alpha = 0$, every solution is periodic of period $2\pi/\beta$. The orbits are all ellipses centered at $\mathbf{0}$ and the origin is called a *center*. *Figure 2.2* on page 63 is an example of a center with circular orbits.

(ii) If $\alpha < 0$, then the origin is positively asymptotically stable and solutions spiral toward $(0, 0)$. This is called a *stable focus* and is illustrated in *Figure 5.6.*

(iii) If $\alpha > 0$, the origin is negatively asymptotically stable and is called an *unstable focus.* The phase portrait looks like a stable focus with the arrows reversed. Note that for both stable and unstable foci the direction of the spiral is determined by the sign of β.

All the cases just described for 2×2 linear systems can be neatly summarized in a single picture. The eigenvalues are the roots of the equation $\lambda^2 - \operatorname{Tr} A\lambda + \operatorname{Det} A = 0$ and by the quadratic formula equal

$$\frac{\operatorname{Tr}[A] \pm \sqrt{\operatorname{Tr}[A]^2 - 4\operatorname{Det}[A]}}{2}.$$

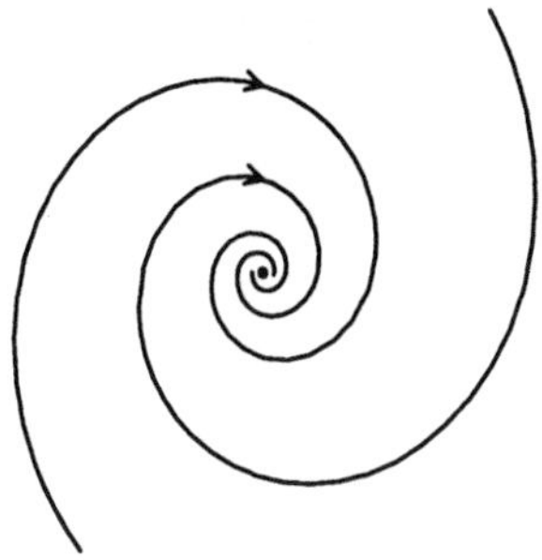

Figure 5.6: Phase portrait of $\dot{x} = -x/4 + y$ and $\dot{y} = -x - y/4$.

Set $\Delta A = \text{Tr}\,[A]^2 - 4\text{Det}\,[A]$, which is called the *discriminant.* For 2×2 real matrices, the trace, the determinant, and the discriminant completely determine the nature of the eigenvalues as described in the next remark.

Remark *The following statements are valid for a 2×2 real matrix A:*

(a) 0 is an eigenvalue if and only if $\text{Det}\,[A] = 0$.

(b) The eigenvalues are equal if and only if $\Delta A = 0$.

(c) The eigenvalues are pure imaginary numbers if and only if $\text{Det}\,[A] > 0$ *and* $\text{Tr}\,[A] = 0$.

(d) The eigenvalues are real if and only if $\Delta A \geq 0$.

(e) The eigenvalues are real and of opposite sign if and only if $\text{Det}\,[A] < 0$.

Using $\text{Tr}\,[A]$ and $\text{Det}\,[A]$ as the axes, the graph of $\Delta A = 0$ is a parabola. All the different phase portraits just described for the 2×2 linear system with constant coefficients, $\dot{\mathbf{x}} = A\mathbf{x}$, correspond to regions and curves in this figure. In particular, it is evident that centers and improper nodes are very fragile phenomena because slight changes in A can produce very different phase portraits. Proper nodes and foci, however, persist under slight perturbations of the matrix A.

On the one hand, *Figures 5.2, 5.4, 5.5, and 5.6* illustrate the four different ways that the orbits can approach the origin when it is a positively asymptotically stable fixed point of $\dot{\mathbf{x}} = A\mathbf{x}$. These different types of behaviors are completely determined by the eigenvalues. On the other hand, these figures are all the same in the sense that in all of them $\omega(\boldsymbol{\xi}) = \{(0,0)\}$ and $\alpha(\boldsymbol{\xi}) = \phi$ for every $\boldsymbol{\xi} \in \mathbb{R}^2$, $\boldsymbol{\xi} \neq \mathbf{0}$. For higher dimensions, the analysis of the fine distinctions in these figures quickly becomes impossible and cruder classifications are necessary. We will eventually make precise the idea that these four phase portraits are all the same, but distinct from the hyperbolic phase portrait that occurs when $\lambda_1 < 0 < \lambda_2$.

EXERCISES

1. Consider $\dot{\mathbf{x}} = A\mathbf{x}$ for a 3×3 real matrix A. Suppose that $\text{Det}\,[A - \lambda \mathrm{I}] = (\lambda^2 + 1)(\lambda + 1)$. Let $\boldsymbol{\xi}$ be an arbitrary point in $\mathbb{R}^3$. Show that $\omega(\boldsymbol{\xi})$ is either the origin or a periodic orbit.

2. Construct real matrices A and B such that:

   (a) $\text{Det}\,[A - \lambda \mathrm{I}] = \text{Det}\,[B - \lambda \mathrm{I}]$.

   (b) The origin is a positively stable fixed point for $\dot{\mathbf{x}} = A\mathbf{x}$.

   (c) The origin is not a positively stable fixed point for $\dot{\mathbf{x}} = B\mathbf{x}$.

3. Construct an invertible real invertible matrix A having the following properties:

   (a) A has an eigenvalue of multiplicity 4 for which the maximal order of a generalized eigenvector is 3.

   (b) $\text{Det}\, e^{At} = e^{-4t}$.

   (c) The origin is a positively stable but not positively asymptotically stable fixed point for $\dot{\mathbf{x}} = A\mathbf{x}$.

4. Let A be a real matrix and prove that the following are equivalent:

   (a) The origin is a positively and negatively stable fixed point of $\dot{\mathbf{x}} = A\mathbf{x}$.

   (b) There exist an invertible matrix B such that BAB^{-1} is a skew symmetric matrix.

   (c) There exist an invertible matrix B such that $Be^{At}B^{-1}$ an orthogonal matrix for all $t \in \mathbb{R}$.

5. Consider $\dot{\mathbf{x}} = A\mathbf{x}$ for a real $d \times d$ matrix. Show that the origin is neither positively nor negatively stable when $\lambda^k - \alpha$ divides $\text{Det}\,[A - \lambda \mathrm{I}]$ for some $k \geq 3$ and $\alpha \neq 0$.

6. Verify the remark on page 165.

7. Graph the $\Delta A = 0$ in a plane with axes $\text{Tr}\, A$ and $\text{Det}\, A$ and label where the different types of phase portraits occur for a 2×2 real matrix.

8. Show that when $(0, 0)$ is a center for $\dot{\mathbf{x}} = A\mathbf{x}$ with $A \in \mathcal{M}_2(\mathbb{R})$, arbitrarily small perturbations of the matrix can produce a positively or negatively asymptotically stable phase portrait.

5.3 Stability and General Linear Systems

The theme of the previous two sections was to study the qualitative behavior of solutions near a fixed point or constant solution. In this section, the qualitative behavior near more general solutions is studied. Instead of a constant solution, consider a solution $\varphi(t)$, which has some significance for a particular problem and examine the qualitative behavior of nearby solutions. Assume the open interval on which the solution $\varphi(t)$ is defined contains $[\beta, \infty)$ for some real β. The natural question to ask is Do solutions starting near φ stay near it or approach it asymptotically? In this context, the two fundamental types of stability near a fixed point each split into two different concepts. Moreover, the distinctions between these four types of stability are more subtle and less easy to picture.

Consider the standard nonautonomous differential equation $\dot{\mathbf{x}} = \mathbf{f}(t, \mathbf{x})$ with $\mathbf{f}(t, \mathbf{x})$ continuous on an open subset $D \subset \mathbb{R}^{d+1}$. To ensure uniqueness of solutions and continuity in initial conditions, it will be assumed that $\mathbf{f}(t, \mathbf{x})$ is locally Lipschitz on D throughout this section.

Let $\varphi(t)$ be a solution of $\dot{\mathbf{x}} = \mathbf{f}(t, \mathbf{x})$ defined on an open interval containing $[\beta, \infty)$. Then for any closed interval $[\beta, T]$ solutions starting near enough to $\varphi(\beta)$ will stay near $\varphi(t)$ throughout $[\beta, T]$. Specifically, given $\varepsilon > 0$ there exists $\delta > 0$ such that $|\mathbf{x}(t, \beta, \boldsymbol{\xi}) - \varphi(t)| < \varepsilon$ for $\beta \leq t \leq T$ when $|\boldsymbol{\xi} - \varphi(\beta)| < \delta$. This follows easily from the continuity of $\mathbf{x}(t, \tau, \boldsymbol{\xi})$. The point of stability is to study situations in which similar concrete results hold for the behavior of solutions on infinite intervals.

To start, simply turn the behavior on $[\beta, T]$ into a definition on $[\beta, \infty)$. The solution $\varphi(t)$ is *Lyapunov stable* if given $\varepsilon > 0$ there exists $\delta > 0$ such that when $|\boldsymbol{\xi} - \varphi(\beta)| < \delta$, the solution $\mathbf{x}(t, \beta, \boldsymbol{\xi})$ is defined on $[\beta, \infty)$ and $|\mathbf{x}(t, \beta, \boldsymbol{\xi}) - \varphi(t)| < \varepsilon$ for $\beta \leq t < \infty$. The first thing to do is show that the starting time β is not important in the definition.

Proposition 5.10 *Let $\varphi(t)$ be a solution of $\dot{\mathbf{x}} = \mathbf{f}(t, \mathbf{x})$ defined on (α, ∞) and suppose $\alpha < \beta < \gamma$. Then $\varphi(t)$ is Lyapunov stable on $[\beta, \infty)$ if and only if $\varphi(t)$ is Lyapunov stable on $[\gamma, \infty)$.*

Proof. Suppose $\varphi(t)$ is Lyapunov stable on $[\gamma, \infty)$. Given $\varepsilon > 0$, let $\delta(\gamma)$ be determined by the Lyapunov stability on $[\gamma, \infty)$. By continuity in initial conditions there exists $\delta(\beta) > 0$ such that $|\mathbf{x}(t, \beta, \boldsymbol{\xi}) - \varphi(t)| < \min\{\varepsilon, \delta(\gamma)\}$, whenever $|\boldsymbol{\xi} - \varphi(\beta)| < \delta(\beta)$ and $\beta \leq t \leq \gamma$. It follows that if $|\boldsymbol{\xi} - \varphi(\beta)| < \delta(\beta)$, then $|\mathbf{x}(\gamma, \beta, \boldsymbol{\xi}) - \varphi(\gamma)| < \delta(\gamma)$. Consequently, $\mathbf{x}(t, \gamma, \mathbf{x}(\gamma, \beta, \boldsymbol{\xi}))$ is defined on $[\gamma, \infty)$ and

$$|\mathbf{x}(t, \gamma, \mathbf{x}(\gamma, \beta, \boldsymbol{\xi})) - \varphi(t)| < \varepsilon,$$

for $t \geq \gamma$. Therefore, φ is Lyapunov stable on $[\beta, \infty)$ because

$$\mathbf{x}(t, \gamma, \mathbf{x}(\gamma, \beta, \boldsymbol{\xi})) = \mathbf{x}(t, \beta, \boldsymbol{\xi}).$$

The same kind of argument proves the converse. $\square$

It is clear from the previous argument that when φ is Lyapunov stable, $\delta(\beta)$ and $\delta(\gamma)$ need not be equal for a given ε. This leads to the concept of uniform stability. Specifically, $\varphi(t)$ is *uniformly stable* if given $\varepsilon > 0$ there exists $\delta > 0$ such that when $|\boldsymbol{\xi} - \varphi(\tau)| < \delta$ for any $\tau \geq \beta$, $\mathbf{x}(t, \tau, \boldsymbol{\xi})$ is defined on $[\tau, \infty)$ and $|\mathbf{x}(t, \tau, \boldsymbol{\xi}) - \varphi(t)| < \varepsilon$ for $\tau \leq t < \infty$. Again the starting time β is not important in the definition.

Proposition 5.11 *Suppose $\varphi(t)$ be a solution of $\dot{\mathbf{x}} = \mathbf{f}(t, \mathbf{x})$ defined on (α, ∞) and suppose $\alpha < \beta < \gamma$. Then $\varphi(t)$ is uniformly stable on $[\beta, \infty)$ if and only if $\varphi(t)$ is uniformly stable on $[\gamma, \infty)$.*

Proof. Exercise. □

For fixed points of an autonomous differential equations, Lyapunov and uniform stability simplify to positive stability as the next proposition shows.

Proposition 5.12 *Suppose $\mathbf{p}$ is a fixed point of the autonomous differential equation $\dot{\mathbf{x}} = \mathbf{f}(\mathbf{x})$ and $\varphi(t) = \mathbf{p}$ for all t. Then the following are equivalent:*

(a) $\mathbf{p}$ is positively stable.

(b) φ is Lyapunov stable.

(c) φ is uniformly stable.

Proof. First, observe that the definition of positively stable for $\mathbf{p}$ is the same as the definition of $\varphi(t)$ being Lyapunov stable on $[0, \infty)$. Thus (a) and (b) are equivalent. Clearly, (c) implies (a). So it remains to show that in this setting (b) implies (c).

Suppose φ is Lyapunov stable and let $\beta = 0$. Given $\varepsilon > 0$, let $\delta > 0$ be determined by the Lyapunov stability of φ so

$$|\mathbf{x}(t, 0, \boldsymbol{\xi}) - \varphi(t)| = |\mathbf{x}(t, 0, \boldsymbol{\xi}) - \mathbf{p}| < \varepsilon$$

when $|\boldsymbol{\xi} - \mathbf{p}| < \delta$ and $t \geq 0$. Note that $\mathbf{x}(t - \tau, 0, \boldsymbol{\xi})$ is a solution because the differential equation is autonomous and that $\mathbf{x}(t, \tau, \boldsymbol{\xi}) = \mathbf{x}(t - \tau, 0, \boldsymbol{\xi})$ because these two solutions agree at $t = \tau$. For $\tau > 0$ and $|\boldsymbol{\xi} - \mathbf{p}| < \delta$, it follows that

$$|\mathbf{x}(t, \tau, \boldsymbol{\xi}) - \varphi(t)| = |\mathbf{x}(t - \tau, 0, \boldsymbol{\xi}) - \mathbf{p}| < \varepsilon$$

when $t - \tau \geq 0$ or $t \geq \tau$. Hence, (c) is holds and the proof is done. □

The Lyapunov and uniform stable behavior of a trajectory can be visualized more easily in the nonautonomous case when $\varphi(t) \equiv \mathbf{0}$. It turns out that this can always be arranged with a change of variables. Given φ a solution of $\dot{\mathbf{x}} = \mathbf{f}(t, \mathbf{x})$, consider

$$\dot{\mathbf{y}} = \mathbf{f}(t, \mathbf{y} + \varphi(t)) - \mathbf{f}(t, \varphi(t)) = \mathbf{f}(t, \mathbf{y} + \varphi(t)) - \dot{\varphi}(t) = \mathbf{g}(t, \mathbf{y}).$$

It is easy to check that $\mathbf{y}(t)$ is a solution of $\dot{\mathbf{y}} = \mathbf{g}(t, \mathbf{y})$ if and only if $\mathbf{x}(t) = \mathbf{y}(t) + \boldsymbol{\varphi}(t)$ is a solution of $\dot{\mathbf{x}} = \mathbf{f}(t, \mathbf{x})$. Hence, $\boldsymbol{\psi}(t) \equiv \mathbf{0}$ on the domain of $\boldsymbol{\varphi}$ is a solution of $\dot{\mathbf{y}} = \mathbf{g}(t, \mathbf{y})$. Since $|\mathbf{x}(t) - \boldsymbol{\varphi}(t)| = |\mathbf{y}(t) - \mathbf{0}| = |\mathbf{y}(t) - \boldsymbol{\psi}(t)|$, the solutions $\boldsymbol{\varphi}$ and $\boldsymbol{\psi}$ will have the same stability properties.

The stability notions discussed in this section are arguably most applicable to the general linear equation $\dot{\mathbf{x}} = A(t)\mathbf{x}$ where $A(t)$ is continuous on (α, ∞). Obviously, $\mathbf{x}(t, 0, \mathbf{0}) \equiv \mathbf{0}$ is a solution. If $\boldsymbol{\varphi}(t)$ is any solution of $\dot{\mathbf{x}} = A(t)\mathbf{x}$, then it is defined on (α, ∞) and

$$\begin{aligned} \dot{\mathbf{y}} &= \mathbf{f}(t, \mathbf{y} + \boldsymbol{\varphi}(t)) - \mathbf{f}(t, \boldsymbol{\varphi}(t)) \\ &= A(t)\big(\mathbf{y} + \boldsymbol{\varphi}(t)\big) - A(t)\boldsymbol{\varphi} \\ &= A(t)\mathbf{y}. \end{aligned}$$

Therefore, by the argument in the previous paragraph every solution of $\dot{\mathbf{x}} = A(t)\mathbf{x}$ has the same stability properties as the zero solution. For example, one solution of $\dot{\mathbf{x}} = A(t)\mathbf{x}$ is Lyapunov stable if and only if all solutions of it are Lyapunov stable. We will study the stability of the solution $\boldsymbol{\varphi}(t) = \mathbf{0}$ for all t, but speak about the stability of the linear system $\dot{\mathbf{x}} = A(t)\mathbf{x}$.

Theorem 5.13 *Let $X(t)$ be a fundamental matrix solution for $\dot{\mathbf{x}} = A(t)\mathbf{x}$ where $A(t)$ is continuous on (α, ∞). Then the following are equivalent:*

(a) $\dot{\mathbf{x}} = A(t)\mathbf{x}$ is Lyapunov stable,

(b) for all $\beta > \alpha$ there exist K such that $|X(t)| \leq K$ for $t \geq \beta$,

(c) for some $\beta > \alpha$ there exist K such that $|X(t)| \leq K$ for $t \geq \beta$.

Proof. Assume the solution $\boldsymbol{\varphi}(t) \equiv \mathbf{0}$ is Lyapunov stable and let $\beta > \alpha$. Recall that $\mathbf{x}(t, \beta, \boldsymbol{\xi}) = X(t)X(\beta)^{-1}\boldsymbol{\xi}$. There exists a $\delta > 0$ such that $|X(t)X(\beta)^{-1}\boldsymbol{\xi}| \leq 1$ when $|\boldsymbol{\xi}| \leq \delta$ and $t \geq \beta$. By the definition of the matrix norm,

$$\begin{aligned} |X(t)X(\beta)^{-1}| &= \sup\{|X(t)X(\beta)^{-1}\mathbf{v}| : |\mathbf{v}| = 1\} \\ &= (1/\delta) \sup\{|X(t)X(\beta)^{-1}\mathbf{v}| : |\mathbf{v}| = \delta\} \leq (1/\delta) \end{aligned}$$

It follows that $|X(t)X(\beta)^{-1}| \leq (1/\delta)$ for $t \geq \beta$ and then

$$|X(t)| = |X(t)X(\beta)^{-1}X(\beta)| \leq |X(t)X(\beta)^{-1}|\,|X(\beta)| \leq (1/\delta)|X(\beta)|$$

for $t \geq \beta$. Thus, (a) implies (b).

Obviously, (b) implies (c). So it remains to assume K exists as prescribed for some β and prove that $\dot{\mathbf{x}} = A(t)\mathbf{x}$ is Lyapunov stable. Let $L = |X(\beta)^{-1}|$. Given $\varepsilon > 0$, set $\delta = \varepsilon/KL$. If $|\boldsymbol{\xi}| < \delta$, then

$$|\mathbf{x}(t, \beta, \boldsymbol{\xi})| = |X(t)X(\beta)^{-1}\boldsymbol{\xi}| \leq |X(t)||X(\beta)^{-1}||\boldsymbol{\xi}| < KL\delta = \varepsilon,$$

to complete the proof. □

Theorem 5.14 *Let* $\dot{\mathbf{x}} = A(t)\mathbf{x}$ *be a linear differential equation such that* $A(t)$ *is continuous on* (α, ∞). *Then the following are equivalent:*

(a) The linear differential equation $\dot{\mathbf{x}} = A(t)\mathbf{x}$ *is uniformly stable.*

(b) For every $\beta > \alpha$ *there exist* K *such that* $|X(t,\tau)| \le K$, *whenever* $\beta \le \tau \le t$.

(c) For some $\beta > \alpha$ *there exist* K *such that* $|X(t,\tau)| \le K$, *whenever* $\beta \le \tau \le t$.

Proof. Assume that $\dot{\mathbf{x}} = A(t)\mathbf{x}$ is uniformly stable, and let $\beta > \alpha$. Recall that $X(t,\tau) = X(t)X(\tau)^{-1}$. By the uniform stability, there exists $\delta > 0$ such that $|\mathbf{x}(t,\tau,\boldsymbol{\xi})| = |X(t)X(\tau)^{-1}\boldsymbol{\xi}| < 1$, whenever $\beta \le \tau \le t$ and $|\boldsymbol{\xi}| \le \delta$. As in the previous proof

$$|X(t)X(\tau)^{-1}| = \sup(1/\delta)\{|X(t)X(\tau)^{-1}\mathbf{v}| : |\mathbf{v}| = \delta\} \le \frac{1}{\delta}$$

for $\beta \le \tau \le t$. Thus (a) implies (b).

It remains to prove that (c) implies (a), since (b) implies (c) is again trivial. To this end, assume that for some $\beta > \alpha$, we have $|X(t,\tau)| \le K$, whenever $\beta \le \tau \le t$. Given $\varepsilon > 0$, set $\delta = \varepsilon/K$. It follows that

$$|\mathbf{x}(t,\tau,\boldsymbol{\xi})| = |X(t,\tau)\boldsymbol{\xi}| \le |X(t,\tau)|\,|\boldsymbol{\xi}| < \varepsilon$$

when $\beta \le \tau \le t$ and $|\boldsymbol{\xi}| < \delta$. □

There are also two concepts of asymptotic stability for solutions. The weaker concept will be called Lyapunov asymptotically stable, although some authors simply call it asymptotically stable. Specifically, the solution $\varphi(t)$ defined on an open interval containing $[\beta, \infty)$ is said to be *Lyapunov asymptotically stable* if it is Lyapunov stable and there exits $\kappa > 0$ such that

$$\lim_{t\to\infty} |\mathbf{x}(t,\beta,\boldsymbol{\xi}) - \varphi(t)| = 0,$$

whenever $|\boldsymbol{\xi} - \varphi(\beta)| < \kappa$. Using continuity in initial conditions as in the proof of *Theorem 5.10*, it can be shown that the definition is independent of the starting time β. The proof is left as an exercise.

Clearly κ depends on β and can go to zero as β goes to infinity. Moreover, the rate at which $|\mathbf{x}(t,\beta,\boldsymbol{\xi}) - \varphi(t)|$ goes to zero can also depend on the starting time β. Uniformly asymptotically stable, which is defined in the next paragraph, will require a uniform κ and a uniform rate of convergence for all β as well the uniform stability of $\varphi(t)$. Not surprisingly it is also the most complex of the four notions of stability for solutions.

Let $\varphi(t)$ be a solution of $\dot{\mathbf{x}} = \mathbf{f}(t,\mathbf{x})$ defined on (α, ∞). The solution $\varphi(t)$ is *uniformly asymptotically stable* if the following conditions are satisfied for some $\beta > \alpha$:

(a) $\varphi(t)$ is uniformly stable,

(b) there exists $\kappa > 0$ such that

$$\lim_{t\to\infty} |\mathbf{x}(t,\tau,\boldsymbol{\xi}) - \varphi(t)| = 0$$

whenever $|\boldsymbol{\xi} - \varphi(\tau)| < \kappa$ and $\tau \geq \beta$,

(c) given $\eta > 0$, there exists $T(\eta) > 0$ such that

$$|\mathbf{x}(t,\tau,\boldsymbol{\xi}) - \varphi(t)| < \eta$$

whenever $|\boldsymbol{\xi} - \varphi(\tau)| < \kappa$ and $t > T(\eta) + \tau$ for all $\tau \geq \beta$.

Condition (c) in the above definition is the condition that guarantees a uniform rate of convergence to zero of $|\mathbf{x}(t,\tau,\boldsymbol{\xi}) - \varphi(t)|$. Also notice that uniformly asymptotically stable implies both uniformly stable and Lyapunov asymptotically stable and they both imply Lyapunov stable. Uniformly asymptotically stable and Lyapunov asymptotically stable are not equivalent; an example can be found in [15] on page 88. Again it is an exercise to show that the definition is independent of the starting time β.

Proposition 5.15 *Suppose* $\mathbf{p}$ *is a fixed point of the autonomous differential equation* $\dot{\mathbf{x}} = \mathbf{f}(\mathbf{x})$ *and* $\varphi(t) = \mathbf{p}$ *for all* t*. Then the following are equivalent:*

(a) $\mathbf{p}$ *is positively asymptotically stable.*

(b) φ *is Lyapunov asymptotically stable.*

(c) φ *is uniformly asymptotically stable.*

Proof. Clearly (a) implies (b) and (c) implies (a). It remains to prove in this context that Lyapunov asymptotic stability implies uniform asymptotic stability.

Assume φ Lyapunov asymptotically stable. Then φ is Lyapunov stable and by *Proposition 5.12* it is uniformly stable. From the definition of Lyapunov asymptotic stability, there exists κ such that

$$\lim_{s\to\infty} |\mathbf{x}(s,0,\boldsymbol{\xi}) - \mathbf{p}| = 0,$$

whenever $|\boldsymbol{\xi} - \mathbf{p}| < \kappa$. Because the differential equation is autonomous, it follows that

$$\lim_{t\to\infty} |\mathbf{x}(t,\tau,\boldsymbol{\xi}) - \mathbf{p}| = \lim_{s\to\infty} |\mathbf{x}(s+\tau,\tau,\boldsymbol{\xi}) - \mathbf{p}| = \lim_{s\to\infty} |\mathbf{x}(s,0,\boldsymbol{\xi}) - \mathbf{p})| = 0.$$

Thus the first two conditions in the definition of uniform asymptotic stability have been established.

To establish the third condition let $\eta > 0$ and note that by uniform stability there exists $\delta > 0$ such that if $|\boldsymbol{\xi} - \mathbf{p}| < \delta$ and $\tau \geq 0$, then

$$|\mathbf{x}(t,\tau,\boldsymbol{\xi}) - \mathbf{p}| < \eta$$

for $t \geq \tau$. Finishing the proof requires a slightly delicate compactness argument on the set

$$\{\boldsymbol{\xi} : |\boldsymbol{\xi} - \mathbf{p}| \leq \kappa/2\}.$$

For each $\boldsymbol{\xi}$ satisfying $|\boldsymbol{\xi} - \mathbf{p}| \leq \kappa/2$, there exits $T(\boldsymbol{\xi}) > 0$ such that

$$|\mathbf{x}(T(\boldsymbol{\xi}), 0, \boldsymbol{\xi}) - \mathbf{p}| < \delta$$

because $\lim_{t\to\infty} |\mathbf{x}(t, 0, \boldsymbol{\xi}) - \mathbf{p}| = 0$. Then by continuity in initial conditions there exists $\kappa(\boldsymbol{\xi}) > 0$ such that $|\mathbf{x}(T(\boldsymbol{\xi}), 0, \mathbf{q}) - \mathbf{p}| < \delta$ whenever $|\mathbf{q} - \boldsymbol{\xi}| < \kappa(\boldsymbol{\xi})$. The choice of δ implies that

$$|\mathbf{x}(t, T(\boldsymbol{\xi}), \mathbf{x}(T(\boldsymbol{\xi}), 0, \mathbf{q})) - \mathbf{p}| < \eta$$

for $t \geq T(\boldsymbol{\xi})$ when $|\mathbf{q} - \boldsymbol{\xi}| < \kappa(\boldsymbol{\xi})$. Because

$$\mathbf{x}(t, 0, \mathbf{q}) = \mathbf{x}(t, T(\boldsymbol{\xi}), \mathbf{x}(T(\boldsymbol{\xi}), 0, \mathbf{q})),$$

it follows that

$$|\mathbf{x}(t, 0, \mathbf{q}) - \mathbf{p}| < \eta$$

for $t \geq T(\boldsymbol{\xi})$ whenever $|\mathbf{q} - \boldsymbol{\xi}| < \kappa(\boldsymbol{\xi})$.

By compactness, there exists a finite number of points $\boldsymbol{\xi}_1, \boldsymbol{\xi}_2, \ldots, \boldsymbol{\xi}_m$ such that

$$\{\boldsymbol{\xi} : |\boldsymbol{\xi} - \mathbf{p}| \leq \kappa/2\} \subset \bigcup_{j=1}^{m} \{\boldsymbol{\xi} : |\boldsymbol{\xi} - \boldsymbol{\xi}_j| < \kappa(\boldsymbol{\xi}_j)\}.$$

Set

$$T(\eta) = \max\{T(\boldsymbol{\xi}_j) : j = 1, \ldots, m\}.$$

Given $\boldsymbol{\xi}$ such that $|\boldsymbol{\xi} - \mathbf{p}| < \kappa/2$, there is a $j = 1, \ldots, m$ such that $|\boldsymbol{\xi} - \boldsymbol{\xi}_j| < \kappa(\boldsymbol{\xi}_j)$ and then

$$|\mathbf{x}(t, 0, \boldsymbol{\xi}) - \mathbf{p}| < \eta$$

for $t \geq T(\eta) \geq T(\boldsymbol{\xi}_j)$. The final step makes critical use of the hypothesis that the differential equation is autonomous. Given $\tau > 0$ and $|\boldsymbol{\xi} - \mathbf{p}| < \kappa/2$,

$$|\mathbf{x}(t, \tau, \boldsymbol{\xi}) - \mathbf{p}| = |\mathbf{x}(t - \tau, 0, \boldsymbol{\xi}) - \mathbf{p}| < \eta$$

for $t - \tau \geq T(\eta)$ or $t \geq \tau + T(\eta)$. □

For general linear systems, there are also simple characterizations of Lyapunov asymptotic stability and uniform asymptotic stability in terms of fundamental solutions similar to those for Lyapunov and uniform stability. These characterizations are the content of the next two theorems.

Theorem 5.16 *Let $X(t)$ be a fundamental matrix solution for $\dot{\mathbf{x}} = A(t)\mathbf{x}$ where $A(t)$ is continuous on (α, ∞). Then $\dot{\mathbf{x}} = A(t)\mathbf{x}$ is Lyapunov asymptotically stable if and only if $\lim_{t\to\infty} |X(t)| = 0$.*

Proof. First, suppose $\dot{\mathbf{x}} = A(t)\mathbf{x}$ is Lyapunov asymptotically stable on $[\beta, \infty)$ and let $X(t)$ be a fundamental matrix solution. Recall that for a, a real number, $\mathbf{x}(t, \beta, a\boldsymbol{\xi}) = a\mathbf{x}(t, \beta, \boldsymbol{\xi})$ because the solutions form a vector space. Thus the Lyapunov asymptotic stability implies that $\lim_{t\to\infty} |\mathbf{x}(t, \beta, \boldsymbol{\xi})| = 0$ for all $\boldsymbol{\xi}$. Therefore, $\lim_{t\to\infty} |X(t)| = 0$ because $X(t)\mathbf{e}_j = \mathbf{x}(t, \beta, X(\beta)\mathbf{e}_j)$ and

$$|X(t)| = \max\left\{|X(t)\mathbf{e}_j| : 1 \le j \le d\right\}.$$

by *Proposition 3.2.*

For the converse, observe that $\lim_{t\to\infty} |X(t)| = 0$ implies that $|X(t)|$ is bounded on $[\beta, \infty)$, and hence $\dot{\mathbf{x}} = A(t)\mathbf{x}$ is Lyapunov stable. Since $\mathbf{x}(t, \beta, \boldsymbol{\xi})$ has the form $X(t)\mathbf{v}$ for some vector $\mathbf{v}$, it follows that $\lim_{t\to\infty} |\mathbf{x}(t, \beta, \boldsymbol{\xi})| = 0$ for every $\boldsymbol{\xi}$. □

The characterization of uniform asymptotic stability in terms of the principal matrix solution $X(t, \tau)$ is more surprising.

Theorem 5.17 *Let $\dot{\mathbf{x}} = A(t)\mathbf{x}$ be a linear differential equation such that $A(t)$ is continuous on (α, ∞). Then $\dot{\mathbf{x}} = A(t)\mathbf{x}$ is uniformly asymptotically stable if and only if there exist constants β, K and θ such that*

$$|X(t, \tau)| \le Ke^{-\theta(t-\tau)},$$

whenever $\beta \le \tau \le t$.

Proof. Suppose $\dot{\mathbf{x}} = A(t)\mathbf{x}$ is uniformly asymptotically stable with β as the starting time. Let κ be given by the definition of uniformly asymptotic stability. Choose $\eta < \kappa$ and let $T(\eta)$ also be given by the definition of uniformly asymptotic stability. Since in this context $\varphi(t) \equiv \mathbf{0}$ for all $t > \alpha$ and $\mathbf{x}(t, \tau, \boldsymbol{\xi}) = X(t, \tau)\boldsymbol{\xi}$, it follows that $|X(t, \tau)\boldsymbol{\xi}| < \eta$ when $|\boldsymbol{\xi}| < \kappa$ and $t > \tau + T(\eta)$.

As in the proof of *Theorem 5.13* it follows that $|X(t, \tau)| \le \eta/\kappa < 1$ when $t > \tau + T(\eta)$. Using $X(t, \sigma)X(\sigma, \tau) = X(t, \tau)$, the following trick brings $(\eta/\kappa)^k$ into the argument:

$$\begin{aligned} |X(\tau + kT(\eta), \tau)| &= \left| \prod_{j=0}^{k-1} X(\tau + (k-j)T(\eta), \tau + (k-j-1)T(\eta)) \right| \\ &\le \prod_{j=0}^{k-1} |X(\tau + (k-j)T(\eta), \tau + (k-j-1)T(\eta))| \\ &\le (\eta/\kappa)^k. \end{aligned}$$

Since $\dot{\mathbf{x}} = A(t)\mathbf{x}$ is also uniformly stable, there exists M such that $|X(t, \sigma)| \le M$ when $t \ge \sigma \ge \beta$ by *Theorem 5.14*. Set

$$\theta = -\frac{\log(\eta/\kappa)}{T(\eta)} > 0$$

and

$$K = Me^{\theta T(\eta)}.$$

Given $t \geq \tau$, there exists an integer k such that $\tau + kT(\eta) \leq t < \tau + (k+1)T(\eta)$ and

$$\begin{aligned}
\left|X(t,\tau)\right| &= \left|X(t, \tau + kT(\eta))X(\tau + kT(\eta), \tau)\right| \\
&\leq \left|X(t, \tau + kT(\eta))\right| \left|X(\tau + kT(\eta), \tau)\right| \\
&\leq M(\eta/\kappa)^k \\
&= Me^{-\theta k T(\eta)} \\
&= Ke^{-\theta(k+1)T(\eta)} \\
&\leq Ke^{-\theta(t-\tau)}
\end{aligned}$$

as required to complete the first half of the proof.

Now suppose there exist constants K and θ such that

$$|X(t,\tau)| \leq Ke^{-\theta(t-\tau)}$$

whenever $\beta \leq \tau \leq t$. Clearly, $\dot{\mathbf{x}} = A(t)\mathbf{x}$ is uniformly stable by *Theorem 5.14* because $e^{-\theta(t-\tau)} \leq 1$ for $\beta \leq \tau \leq t$ and the second condition in the definition of uniform asymptotically stable holds with $\kappa = 1$. It remains to establish the third condition.

It suffices to consider $0 < \eta < K$. Set

$$T(\eta) = \frac{-\ln(\eta/K)}{\theta}.$$

Then given $|\boldsymbol{\xi}| < 1$ and $\tau \geq \beta$,

$$\begin{aligned}
|\mathbf{x}(t,\tau,\boldsymbol{\xi})| &= |X(t,\tau)\boldsymbol{\xi}| \\
&< Ke^{-\theta(t-\tau)} \\
&\leq Ke^{-\theta T(\eta)} \\
&= \eta.
\end{aligned}$$

for $t \geq \tau + T(\eta)$. This completes the proof. $\square$

If the linear differential equation $\dot{\mathbf{x}} = A(t)\mathbf{x}$ has some stability property, then it is natural to ask when a perturbed system of the form $\dot{\mathbf{x}} = [A(t) + B(t)]\mathbf{x}$ will have the same stability property. This section will close with an example of such a result. It represents a class of problems, and more examples of results of this type can be found in [15].

The proof of this perturbation theorem requires the following generalization of the variations of constants formula, which will also be useful in later chapters.

Proposition 5.18 *Consider* $\dot{\mathbf{x}} = A(t)\mathbf{x} + \mathbf{g}(t,\mathbf{x})$, *where* $A(t)$ *is continuous on an open interval* I *and* $\mathbf{g}(t,\mathbf{x})$ *is continuous on a domain* $D \subset \mathbb{R}^{d+1}$ *such that*

$t \in I$, whenever $(t, \mathbf{x}) \in D$. Let $X(t)$ be a fundamental matrix solution of $\dot{\mathbf{x}} = A(t)\mathbf{x}$. If $\mathbf{x}(t)$ is a solution of $\dot{\mathbf{x}} = A(t)\mathbf{x} + \mathbf{g}(t, \mathbf{x})$ and t and τ are in its domain, then $\mathbf{x}(t)$ is a solution of the integral equation

$$\mathbf{x}(t) = X(t, \tau)\mathbf{x}(\tau) + \int_\tau^t X(t, s)g(s, \mathbf{x}(s))ds.$$

Proof. Because $X(t)^{-1}$ is a matrix solution of $\dot{Y} = -YA(t)$ (*Exercise 2* on page 97),

$$\begin{aligned} \frac{d\left[X(t)^{-1}\mathbf{x}(t)\right]}{dt} &= -X(t)^{-1}A(t)\mathbf{x}(t) + X(t)^{-1}[A(t)\mathbf{x}(t) + \mathbf{g}(t, \mathbf{x}(t))] \\ &= X(t)^{-1}\mathbf{g}(t, \mathbf{x}(t)). \end{aligned}$$

Integrating from τ to t produces

$$X(t)^{-1}\mathbf{x}(t) - X(\tau)^{-1}\mathbf{x}(\tau) = \int_\tau^t X(s)^{-1}\mathbf{g}(t, \mathbf{x}(s))ds$$

and then multiplying by $X(t)$ essentially finishes it. □

Theorem 5.19 *Let $A(t)$ and $B(t)$ be two continuous functions from (α, ∞) to $\mathcal{M}_d(\mathbb{R})$. If $\dot{\mathbf{x}} = A(t)\mathbf{x}$ is uniformly stable and $\int_\beta^\infty |B(t)|dt < \infty$, then $\dot{\mathbf{x}} = [A(t) + B(t)]\mathbf{x}$ is also uniformly stable.*

Proof. Let $X(t)$ be a fundamental solution of $\dot{\mathbf{x}} = A(t)\mathbf{x}$ and let $\mathbf{x}(t)$ be a solution of $\dot{\mathbf{x}} = A(t)\mathbf{x} + B(t)\mathbf{x}$. By the *Proposition 5.18*,

$$\mathbf{x}(t) = X(t, \tau)\mathbf{x}(\tau) + \int_\tau^t X(t, s)B(s)x(s)ds.$$

By *Theorem 5.14* there exits K such that $|X(t, \tau)| \leq K$ when $\beta \leq \tau \leq t$. Thus

$$|\mathbf{x}(t)| \leq K|\mathbf{x}(\tau)| + \int_\tau^t K|B(s)|\,|\mathbf{x}(s)|ds.$$

The corollary to Gronwall's inequality (*Theorem 2.9*) can now be applied and yields

$$|\mathbf{x}(t)| \leq K|\mathbf{x}(\tau)| \exp\left(K \int_\tau^t |B(s)|ds\right)$$

for $\tau \leq t$. Since $|B(s)| \geq 0$,

$$|\mathbf{x}(t)| \leq K|\mathbf{x}(\tau)| \exp\left(K \int_\beta^\infty |B(s)|ds\right).$$

Given $\varepsilon > 0$, set

$$\delta = \frac{\varepsilon}{K} \exp\left(-K \int_\tau^\infty |B(s)|ds\right).$$

It suffices to check that $|\mathbf{x}(t)| < \varepsilon$ for $\beta \le \tau \le t$ when $|\mathbf{x}(\tau)| < \delta$. This is an immediate consequence of the above inequality. □

Corollary *Let $\lambda_1, \dots, \lambda_\rho$ be the distinct eigenvalues (real and complex) of a real $d \times d$ matrix A, and let $B : (\alpha, \infty) \to \mathcal{M}_d(\mathbb{R})$ be a continuous function. If $Re\lambda_j \le 0$ for $j = 1, \dots, \rho$ and λ_j is a simple eigenvalue when $Re\lambda_j = 0$ and if $\int_\beta^\infty |B(t)|dt < \infty$, then $\dot{\mathbf{x}} = [A + \mu B(t)]\mathbf{x}$ is uniformly stable for all real μ.*

Proof. By *Theorem 5.9* the origin is a positively stable fixed point of $\dot{\mathbf{x}} = A\mathbf{x}$, and by *Theorem 5.12*, the solution $\varphi(t) \equiv 0$ is a uniformly stable solution of $\dot{\mathbf{x}} = A\mathbf{x}$. Now the theorem applies for all μ. □

EXERCISES

1. Prove *Proposition 5.11.*

2. Suppose $X(t)$ is a fundamental matrix solution of $\dot{X} = A(t)X$ on the interval $a < t < \infty$, and suppose the entries of $X(t)$ are rational functions of the real variable t. In other words, $x_{ij}(t) = p_{ij}(t)/q_{ij}(t)$ where both $p_{ij}(t)$ and $q_{ij}(t)$ are polynomials with real coefficients and $q_{ij}(t)$ is not the zero polynomial.

   (a) Prove that $\dot{\mathbf{x}} = A(t)\mathbf{x}$ is Lyapunov stable if and only if the degree of p_{ij} is less than or equal to the degree of q_{ij} for all i and j.

   (b) Prove that $\dot{\mathbf{x}} = A(t)\mathbf{x}$ is Lyapunov asymptotically stable if and only if the degree of p_{ij} is less than the degree of q_{ij} for all i and j.

3. Suppose $\varphi(t)$ is a solution of $\dot{\mathbf{x}} = \mathbf{f}(t, \mathbf{x})$ defined on (α, ∞) and suppose $\alpha < \beta < \gamma$.

   (a) Show that $\varphi(t)$ is Lyapunov asymptotically stable on $[\beta, \infty)$ if and only if $\varphi(t)$ is Lyapunov asymptotically stable on $[\gamma, \infty)$.

   (b) Show that $\varphi(t)$ is uniformly asymptotically stable on $[\beta, \infty)$ if and only if $\varphi(t)$ is uniformly asymptotically stable on $[\gamma, \infty)$.

4. Define a real-valued function $a(t)$ on $\mathbb{R}$ by setting

$$a(t) = -\sin t$$

   when $t < 2\pi$ and for $m = 2, 3, \dots$ setting

$$a(t) = (-1)^m \sin(t/m)$$

   when $m(m-1) \le t \le m(m+1)$. Show that the scalar differential equation $\dot{x} = a(t)x$ is Lyapunov stable but not uniformly stable.

5. Give an example of a differential equation that is both uniformly stable and Lyapunov asymptotically stable but is not uniformly asymptotically stable. Hint: You need look no further than the scalar equation $\dot{x} = a(t)x$.

6. Suppose $\mathbf{p}$ is a periodic point for the locally Lipschitz autonomous differential equation $\dot{\mathbf{x}} = \mathbf{f}(\mathbf{x})$. Prove the following:
   (a) The solution $\mathbf{x}(t, 0, \mathbf{p})$ is Lyapunov stable if and only if it is uniformly stable.
   (b) The solution $\mathbf{x}(t, 0, \mathbf{p})$ is Lyapunov asymptotically stable if and only if it is uniformly asymptotically stable.
7. Prove the converse of *Proposition 5.18.*

5.4 Linear Systems with Periodic Coefficients

Another class of differential equations for which interesting results about the long-term behavior of solutions can be obtained are those linear equations $\dot{\mathbf{x}} = A(t)\mathbf{x}$ with a periodic coefficient matrix. In this case, there is a constant matrix associated with $\dot{\mathbf{x}} = A(t)\mathbf{x}$, and its eigenvalues determine the stability of the system. The first step in studying linear differential equations $\dot{\mathbf{x}} = A(t)\mathbf{x}$ where $A(t+T) = A(t)$ depends critically on *Theorem 4.12.*

Theorem 5.20 (Floquet) *If $A : \mathbb{R} \to \mathcal{M}_d(\mathbb{R})$ is continuous and periodic with period $T > 0$, then every fundamental matrix solution of $\dot{\mathbf{x}} = A(t)\mathbf{x}$ has the form*

$$X(t) = P(t)e^{Bt},$$

where $B \in \mathcal{M}_d(\mathbb{C})$ and $P : \mathbb{R} \to GL_d(\mathbb{R})$ is continuous on $\mathbb{R}$ and periodic with period T.

Proof. Let $X(t)$ be a fundamental matrix solution and observe that

$$\dot{X}(t+T) = A(t+T)X(t+T) = A(t)X(t+T).$$

In other words, $X(t+T)$ is another fundamental matrix solution, and hence, there exists a invertible real matrix C such that $X(t+T) = X(t)C$. By *Theorem 4.12*, there exists $B \in \mathcal{M}_d(\mathbb{C})$ such that $e^{BT} = C$. Set $P(t) = X(t)e^{-Bt}$, so $X(t) = P(t)e^{Bt}$. Clearly, $P(t)$ is continuous and invertible. Finally,

$$\begin{aligned} P(t+T) &= X(t+T)e^{-B(t+T)} \\ &= X(t)Ce^{-BT}e^{-Bt} \\ &= X(t)e^{BT}e^{-BT}e^{-Bt} \\ &= P(t), \end{aligned}$$

proving that $P(t)$ is periodic. □

A matrix C such that $X(t+T) = X(t)C$ for some fundamental matrix solution $X(t)$ of $\dot{\mathbf{x}} = A(t)\mathbf{x}$ is called a *monodromy matrix.* Substituting $t = 0$ in the formula for the monodromy matrix, implies that $C = X(0)^{-1}X(T)$. When $X(t) = X(t, 0)$, the principal matrix solution at 0, then by definition $X(0) = \mathrm{I}$ and $X(T)$ is a monodromy matrix.

Proposition 5.21 *Let C and D be monodromy matrices for $\dot{\mathbf{x}} = A(t)\mathbf{x}$ with $A(t)$ continuous and periodic. Then there exists an invertioble matrix E such that $D = E^{-1}DE$. In particular, C and D have the same characteristic polynomial and have the same eigenvalues with the same multiplicities.*

Proof. There exist fundamental matrix solutions $X(t)$ and $Y(t)$ of $\dot{\mathbf{x}} = A(t)\mathbf{x}$ for which $X(t+T) = X(t)C$ and $Y(t+T) = Y(t)D$. There also exists a nonsingular matrix E such that $Y(t) = X(t)E$. Now $Y(t+T) = X(t+T)E = X(t)CE = Y(t)E^{-1}CE$ and $D = E^{-1}CE$. □

From this proposition, it follows that the eigenvalues of a monodromy matrix are invariants of the equation $\dot{\mathbf{x}} = A(t)\mathbf{x}$. The eigenvalues of a monodromy matrix are called the *characteristic multipliers* of $\dot{\mathbf{x}} = A(t)\mathbf{x}$, and a complex number λ such that $e^{\lambda T}$ is a characteristic multiplier is called a *characteristic exponent.* The characteristic exponents are not uniquely determined because $e^{\lambda T} = e^{\lambda T + 2\pi i}$ and both λ and $\lambda + 2\pi i/T$ are characteristic exponents for the same characteristic multiplier.

Proposition 5.22 *Let $A : \mathbb{R} \to \mathcal{M}_d(\mathbb{R})$ be continuous and periodic with period $T > 0$. There exists a nonzero (possibly complex) solution of $\dot{\mathbf{x}} = A(t)\mathbf{x}$ of the form $e^{\lambda t}\mathbf{p}(t)$, where $\mathbf{p} : \mathbb{R} \to \mathbb{C}^d$ is a periodic function of period T if and only if λ is a characteristic exponent for $\dot{\mathbf{x}} = A(t)\mathbf{x}$. Furthermore, if λ is a positive real characteristic exponent, then there exists a nonzero real solution of the form $e^{\lambda t}\mathbf{p}(t)$ where $\mathbf{p}(t)$ is periodic of period T.*

Proof. Suppose $e^{\lambda t}\mathbf{p}(t)$ is a nonzero solution such that $\mathbf{p}(t+T) = \mathbf{p}(t)$. There must exist a nonzero vector $\mathbf{v} \in \mathbb{C}^d$ such that $e^{\lambda t}\mathbf{p}(t) = P(t)e^{Bt}\mathbf{v}$, where $P(t)e^{Bt}$ is a fundamental matrix solution in the form given by Floquet's theorem (*Theorem 5.20*). It follows that

$$\begin{aligned} P(t)e^{B(t+T)}\mathbf{v} &= P(t+T)e^{B(t+T)}\mathbf{v} \\ &= e^{\lambda(t+T)}\mathbf{p}(t+T) \\ &= e^{\lambda T}e^{\lambda t}\mathbf{p}(t) \\ &= e^{\lambda T}P(t)e^{Bt}\mathbf{v}. \end{aligned}$$

The equality of the first and last terms in the above calculation can be written as

$$P(t)e^{Bt}\left[e^{BT} - e^{\lambda T}\mathrm{I}\right]\mathbf{v} = 0.$$

Since $P(t)e^{Bt}$ is invertible, $\left[e^{BT} - e^{\lambda T}\mathrm{I}\right]\mathbf{v} = 0$ and $e^{\lambda T}$ is an eigenvalue of the monodromy matrix e^{BT}, or equivalently λ is a characteristic exponent.

For the converse, let λ be a characteristic exponent for $\dot{\mathbf{x}} = A(t)\mathbf{x}$, and let C be a monodromy matrix. Then, $e^{\lambda T}$ is an eigenvalue of $C = e^{BT}$ and without loss of generality, λ is an eigenvalue of B. Let $\mathbf{v} \in \mathbb{C}^d$ be an eigenvector for B belonging to λ and set $\mathbf{p}(t) = P(t)\mathbf{v}$ where as above $X(t) = P(t)e^{Bt}$. The periodic function $\mathbf{p}(t)$ is not identically zero because $P(t) \in GL_d(\mathbb{R})$ and $\mathbf{v} \neq \mathbf{0}$.

Since $e^{Bt}\mathbf{v} = e^{\lambda t}\mathbf{v}$ for all t, the solution $P(t)e^{Bt}\mathbf{v} = P(t)e^{\lambda t}\mathbf{v} = e^{\lambda t}P(t)v = e^{\lambda t}\mathbf{p}(t)$ has the required form.

Finally, if λ is a positive real characteristic exponent, then the real and imaginary parts of the nonzero solution of the form $\mathbf{p}(t)e^{\lambda t}$ have the required form and are solutions because $A(t)$ is real. One of them must be nonzero. □

Corollary *The number 1 (alternatively -1) is a characteristic multiplier of $\dot{\mathbf{x}} = A(t)\mathbf{x}$ if and only if $\dot{\mathbf{x}} = A(t)\mathbf{x}$ has a nonzero real periodic solution of period T (alternatively $2T$).*

Proof. If 1 is a characteristic multiplier, then 0 is a characteristic exponent and $\dot{\mathbf{x}} = A(t)\mathbf{x}$ has a solution of the form $e^{0t}\mathbf{p}(t) = \mathbf{p}(t)$ where $\mathbf{p}(t)$ is periodic and nonzero. Since $A(t)$ is real, both the real and complex parts of $\mathbf{p}(t)$ are solutions and clearly one of them is nonzero.

Conversely, if there exists a nonzero periodic solution of period T, then there is a nonzero solution of the form $e^{0t}\mathbf{p}(t)$ such that $\mathbf{p}(t+T) = \mathbf{p}(t)$ and 0 is a characteristic exponent.

The proof of the alternative statement is left as an exercise. □

Proposition 5.23 *If $\mu_j = e^{\lambda_j T}$, $j = 1, \ldots, \rho$ are the distinct characteristic multipliers with multiplicities $m_1, \ldots, m_\rho$ of $\dot{\mathbf{x}} = A(t)\mathbf{x}$, then*

$$\prod_{j=1}^{\rho} \mu_j^{m_j} = \exp\left(\int_0^T \operatorname{Tr} A(s)\, ds\right)$$

and

$$\sum_{j=1}^{\rho} m_j \lambda_j = \frac{1}{T}\int_0^T \operatorname{Tr} A(s)\, ds \qquad (\operatorname{mod} 2\pi i/T).$$

Proof. Let $X(t)$ be a principal matrix solution at 0, that is, $X(0) = \mathrm{I}$, and let C be the monodromy matrix for $X(t)$. Thus, $X(T) = X(0)C = C$. Recall Abel's formula (*Theorem 3.8* on page 96) that

$$\operatorname{Det} X(t) = \operatorname{Det} X(\tau)\exp\left(\int_\tau^t \operatorname{Tr} A(s)\, ds\right).$$

The first formula now follows easily because

$$\begin{aligned}
\prod_{j=1}^{\rho} \mu_j^{m_j} &= \operatorname{Det} C \\
&= \operatorname{Det} X(T) \\
&= \operatorname{Det} X(0)\exp\left(\int_0^T \operatorname{Tr} A(s)\, ds\right) \\
&= \exp\left(\int_0^T \operatorname{Tr} A(s)\, ds\right).
\end{aligned}$$

Since we also have

$$\prod_{j=1}^{\rho} \mu_j^{m_j} = \exp\left(\sum_{j=1}^{\rho} m_j \lambda_j T\right),$$

it follows from the above that mod$2\pi i$

$$T\sum_{j=1}^{\rho} m_j\lambda_j = \left(\int_0^T \operatorname{Tr} A(s)\, ds\right)$$

and the second formula is established. □

Theorem 5.24 *If $A : \mathbb{R} \to \mathcal{M}_d(\mathbb{R})$ is continuous and periodic with period $T > 0$, then the following are equivalent:*

(a) $\dot{\mathbf{x}} = A(t)\mathbf{x}$ is uniformly stable;

(b) $\dot{\mathbf{x}} = A(t)\mathbf{x}$ is Lyapunov stable;

(c) The absolute value of every characteristic multiplier is less than or equal to one and those of absolute value 1 are simple eigenvalues of the monodromy matrices for $\dot{\mathbf{x}} = A(t)\mathbf{x}$.

Proof. It is only necessary to prove (b) implies (c) and (c) implies (a), because (a) always implies (b).

Start by assuming $\dot{\mathbf{x}} = A(t)\mathbf{x}$ is Lyapunov stable and as usual write $X(t) = P(t)e^{Bt}$. It follows from *Theorem 5.13*, that $|X(t)|$ is bounded for $t \geq 0$. Because $P(t)$ is continuous, periodic and invertible, its inverse is also continuous and periodic. Consequently, $|P(t)^{-1}|$ is bounded for all t, and hence so is $|e^{Bt}|$ for $t \geq 0$. Now *Theorem 5.8* implies that the eigenvalue $\lambda_1, \ldots, \lambda_k$ of B have nonpositive real part and those with zero real part are simple. Moreover, $Re(\lambda_i) \leq 0$ implies $|e^{\lambda_i T}| \leq 1$. If every eigenvalue of B with zero real part is simple, then every eigenvalue of absolute value one for e^{Bt} is a simple eigenvalue of e^{BT}. (See *Exercise 7* on page 131.)

Now assume that (c) holds or equivalently the real parts of the characteristic exponents are less than or equal to zero and the pure imaginary characteristic exponents are simple. Then *Theorem 5.8* implies that $|e^{Bt}| \leq K$ for $t \geq 0$. It follows that

$$\begin{aligned}
|X(t)X(s)^{-1}| &= |P(t)e^{Bt}e^{-Bs}P(s)^{-1}| \\
&\leq |P(t)|\,|e^{B(t-s)}|\,|P(s)^{-1}| \\
&\leq |P(t)|K|P(s)^{-1}|
\end{aligned}$$

for $t \geq s \geq 0$. Since both $|P(t)|$ and $|P(t)^{-1}|$ are bounded functions of t, *Theorem 5.14* applies to complete the proof. □

Theorem 5.25 *If $A : \mathbb{R} \to \mathcal{M}_d(\mathbb{R})$ is continuous and periodic with period $T > 0$, then the following are equivalent:*

(a) $\dot{\mathbf{x}} = A(t)\mathbf{x}$ *is uniformly asymptotically stable;*

(b) $\dot{\mathbf{x}} = A(t)\mathbf{x}$ *is Lyapunov asymptotically stable; and*

(c) The absolute value of every characteristic multiplier is less than 1.

Proof. The proof is similar to the proof of the previous theorem. Assume $\dot{\mathbf{x}} = A(t)\mathbf{x}$ is Lyapunov asymptotically stable. By the previous theorem, the absolute values of the characteristic multipliers are at most one. Suppose $|\mu_j| = |e^{\lambda_j T}| = 1$, and hence, $\lambda_j = \alpha i$, $\alpha \in \mathbb{R}$. Then there exists a solution of the form $e^{\lambda_j t}\mathbf{p}(t)$, such that $\mathbf{p}(t+T) = \mathbf{p}(t)$ by *Proposition 5.22*. Clearly, $|e^{\lambda_j t}\mathbf{p}(t)| = |\mathbf{p}(t)|$ does not go to 0 as $t \to \infty$. This contradicts the asymptotic stability of $\dot{\mathbf{x}} = A(t)\mathbf{x}$ even if $e^{\lambda_j t}\mathbf{p}(t)$ is complex because both the real and the imaginary parts are solutions. Hence, $|\mu_j| < 1$ for all j and (b) implies (c).

If (c) holds, then *Theorem 5.8* can be applied via the characteristic exponents to conclude that there exist positive constants K and α such that $|e^{Bt}| \leq Ke^{-\alpha t}$ for $t \geq 0$. Set $L_1 = \sup\{|P(t)| : 0 \leq t \leq T\}$ and $L_2 = \sup\{|P(t)^{-1}| : 0 \leq t \leq T\}$. It follows that

$$\big|X(t,s)\big| = \big|X(t)X(s)^{-1}\big| \leq \big|P(t)\big|\,\big|e^{B(t-s)}\big|\,\big|P(s)^{-1}\big| \leq L_1 L_2 K e^{-\alpha(t-s)}$$

for $t \geq s \geq 0$. Therefore, $\dot{\mathbf{x}} = A(t)\mathbf{x}$ is uniformly asymptotically stable by *Theorem 5.17*. □

To illustrate the previous ideas, this section concludes with a brief analysis of the characteristic multipliers and stability of the equation

$$\ddot{x} + \big(\eta + \theta(t)\big)x = 0, \qquad \theta(t+\pi) = \theta(t),$$

which is known as Hill's equation. This is equivalent to the following system with periodic coefficients

$$\begin{pmatrix} \dot{x} \\ \dot{y} \end{pmatrix} = \begin{bmatrix} 0 & 1 \\ -\eta - \theta(t) & 0 \end{bmatrix} \begin{pmatrix} x \\ y \end{pmatrix}.$$

In this system, η should be be viewed as a parameter.

Let $X(t,\tau,\eta)$ be the principal fundamental matrix at τ for the parameter value η. It is a real matrix because $X(\tau,\tau,\eta) = \mathrm{I}$. Recall that $X(t,\tau,\eta)$ depends continuously on the parameter η as well as on t and τ (*Theorem 1.25*). A monodromy matrix C is given by

$$X(\pi,0,\eta) = X(0,0,\eta)C = \mathrm{I}C = C$$

and is simply $X(\pi,0,\eta)$.

Denote the characteristic multipliers by μ_1 and μ_2, suppressing their dependence on η. Since

$$\operatorname{Det}\big[X(t,0,\eta)\big] = \operatorname{Det}\big[X(0,0,\eta)\big]\exp\left(\int_0^t \operatorname{Tr} A(s)\,ds\right) = \exp\left(\int_0^t 0\,ds\right),$$

it follows that Det $\left[X(t,0,\eta)\right] = 1$ for all t. Now *Proposition 5.23* implies that $\mu_1\mu_2 = 1$ for all η. Hence, $|\mu_1| < 1$ if and only if $|\mu_2| > 1$ and $|\mu_2| < 1$ if and only if $|\mu_1| > 1$. Therefore Hill's equation is never Lyapunov asymptotically stable by *Theorem 5.25*. It can, however, be stable.

The characteristic multipliers are the roots of

$$\text{Det}\left[X(\pi,0,\eta) - \lambda \text{I}\right] = 0$$

which for a 2×2 matrix equals

$$\lambda^2 - \text{Tr}\left[X(\pi,0,\eta)\right]\lambda + \text{Det}\left[X(\pi,0,\eta)\right] = 0.$$

or

$$\lambda^2 - \text{Tr}\left[X(\pi,0,\eta)\right]\lambda + 1 = 0$$

because Det $\left[X(t,0,\mu)\right] \equiv 1$. Therefore, the behavior of Hill's equation is governed by the continuous function Tr $\left[X(\pi,0,\eta)\right]$. In particular, the characteristic multipliers are given by

$$\frac{\text{Tr}\left[X(\pi,0,\eta)\right] \pm \sqrt{\text{Tr}\left[X(\pi,0,\eta)\right]^2 - 4}}{2}.$$

For example, if Tr $[X(\pi,0,\eta)] > 2$, then one of the characteristic multipliers, say μ_1, is greater than 1. By *Theorem 5.22* there exists a nonzero real solution of the form $\mathbf{p}(t)e^{\lambda_1 t}$ where $\lambda_1 > 0$. Consequently, Hill's equation has an unbounded solution. This may appear to be somewhat circular since it requires the calculation of $X(\pi,0,\eta)$, but Eulers method (*Theorem 1.16*) can be used to determine if Tr $[X(\pi,0,\eta)] > 2$. Eulers method cannot, however, be used to determine whether or not Tr $[X(\pi,0,\eta)] = 2$. Additional consequences of the above formula for the characteristic multipliers of Hill's equation are included in the exercises. An extensive analysis of Hill's equation can be found in [15] beginning on page 121.

EXERCISES

1. Show that if the characteristic multipliers are all simple eigenvalues of the monodromy matrix, then there is basis of complex solutions of the form $\mathbf{p}(t)e^{\lambda_j t}$.

2. Verify the following basic facts about Hill's equation:

   (a) If Tr $[X(\pi,0,\eta)] = \pm 2$, then the characteristic multipliers are both equal to 1 or both -1.

   (b) If $\left|\text{Tr}\,[X(\pi,0,\eta)]\right| < 2$, then the characteristic multipliers are distinct complex conjugates of modulus 1.

(c) If $\left|\text{Tr}\left[X(\pi, 0, \eta)\right]\right| > 2$, then the characteristic multipliers are real and one has absolute value greater than 1.

3. Prove the following about Hill's equation:

   (a) If $\text{Tr}\left[X(\pi, 0, \eta)\right] = \pm 2$, then there exists a periodic solution, and the equation is stable if and only if there are linearly independent periodic solutions of period π or 2π.

   (b) If $\left|\text{Tr}\left[X(\pi, 0, \eta)\right]\right| < 2$, the equation is uniformly stable, there are independent complex solutions of the form $e^{\beta t i} p(t)$, where $\beta \in \mathbb{R}$ and $p(t + \pi) = p(t)$, and every solution is bounded for all time.

   (c) If $\left|\text{Tr}\left[X(\pi, 0, \eta)\right]\right| > 2$, then there exists a solution that goes to ∞ as $t \to \infty$.

4. Prove the following variant of *Theorem 5.20*: If $A : \mathbb{R} \to \mathcal{M}_d(\mathbb{R})$ is continuous and periodic with period $T > 0$, then every fundamental matrix solution of $\dot{\mathbf{x}} = A(t)\mathbf{x}$ has the form

$$X(t) = P(t)e^{Bt}$$

where $B \in \mathcal{M}_d(\mathbb{R})$ and $P : \mathbb{R} \to GL_d(\mathbb{R})$ is continuous on $\mathbb{R}$ and periodic with period $2T$. Hint: Use *Theorem 4.18*.

Chapter 6

The Poincaré Return Map

A solution of an autonomous differential equation that wanders off to infinity as time goes to infinity is dynamically rather uninteresting. There is little else to say about its evolution as time increases. The dynamics of a solution, however, becomes more interesting when the omega limit set is nonempty. What is the structure of the omega limit set of a solution and how does the solution approach the omega limit set as time evolves are fundamental dynamical questions. The more complicated the omega limit set is the more the challenging these questions are.

When there is a moving point (not a fixed point) in the omega limit set of a solution, the solution must get very close to this moving point and by continuity in initial conditions follow it for a long time. But then it must find a way to swing back to get even closer to the same moving point in its omega limit set and repeat the process. This process is particularly interesting when a solution starting at a point comes back close to itself in some manner. This phenomenon is generally known as recurrence.

Local sections and return maps provide a means for keeping track of how a solution comes back closer and closer to a nonfixed point. The existence and properties of local sections and return maps are established in the first section. These ideas are applied in the second section to study planar dynamics. The results about dynamics in the plane are quite elegant and are generally known as Poincaré -Bendixson theory. The final section is devoted to showing through examples that the dynamics of higher dimensions is far more complicated than what occurs on the plane.

6.1 Local Sections

Let $\dot{\mathbf{x}} = \mathbf{f}(\mathbf{x})$ be an autonomous differential equation with $\mathbf{f}$ locally Lipschitz on Ω an open subset of $\mathbb{R}^d$. This will be the context for the entire section. The hypothesis that $\mathbf{f}(\mathbf{x})$ is at least locally Lipschitz is absolutely essential because continuity in initial conditions (*Theorem 1.23*) plays a critical role in all the proofs. The simpler notation $\mathbf{x}(t, \boldsymbol{\xi})$ for $\mathbf{x}(t, 0, \boldsymbol{\xi})$ will also be used in this chapter.

Consider $\mathbf{p} \in \Omega$ such that $\mathbf{f}(\mathbf{p}) \neq \mathbf{0}$, so $\mathbf{p}$ is not a fixed point of the differential equation. The first goal is to obtain a more refined understanding of the local behavior of solutions near $\mathbf{p}$. Specifically, we want to show that near a nonfixed point the solutions look like a bundle of wires passing through a wall without any nasty kinks or short circuits. The first task is to put this idea into a mathematical context.

Set $V = \{\mathbf{x} : \mathbf{x} \cdot \mathbf{f}(\mathbf{p}) = 0\}$. It is an $(d-1)$-dimensional subspace of $\mathbb{R}^d$. Let $T : \mathbb{R}^{d-1} \to V$ be any convenient linear transformation of $\mathbb{R}^{d-1}$ onto V. Of course, such a T is automatically one-to-one. Define $\boldsymbol{\psi} : \mathbb{R}^{d-1} \to \mathbb{R}^d$ by $\boldsymbol{\psi}(\mathbf{v}) = T(\mathbf{v}) + \mathbf{p}$. The image of $\boldsymbol{\psi}$ is the hyperplane through $\mathbf{p}$ and perpendicular to $\mathbf{f}(\mathbf{p})$. It is the wall.

Because $\mathbf{f}(\boldsymbol{\psi}(\mathbf{v})) \cdot \mathbf{f}(\mathbf{p})$ is continuous in $\mathbf{v}$ and $\mathbf{f}(\boldsymbol{\psi}(\mathbf{0})) \cdot \mathbf{f}(\mathbf{p}) = \|\mathbf{f}(\mathbf{p})\|^2 \neq 0$, there exist $b > 0$ such that $\mathbf{f}(\boldsymbol{\psi}(\mathbf{v})) \cdot \mathbf{f}(\mathbf{p}) \neq 0$ when $\|\mathbf{v}\| \leq b$. The wires will be the solutions through the points $\boldsymbol{\psi}(\mathbf{v})$ with $\|\mathbf{v}\| \leq r$ for some positive r.

The hyperplane $V + \mathbf{p}$ is also described by the equation $(\mathbf{x} - \mathbf{p}) \cdot \mathbf{f}(\mathbf{p}) = 0$ and divides $\mathbb{R}^d$ into two half-spaces described by the inequalities $(\mathbf{x} - \mathbf{p}) \cdot \mathbf{f}(\mathbf{p}) > 0$ and $(\mathbf{x} - \mathbf{p}) \cdot \mathbf{f}(\mathbf{p}) < 0$. The idea of the local section is to find a small disk in the hyperplane so that over a short time interval, solutions must cross it from the negative side to the positive side . The main result is the following:

Theorem 6.1 *If $\mathbf{p}$ is not a fixed point of $\dot{\mathbf{x}} = \mathbf{f}(\mathbf{x})$ and $\boldsymbol{\psi}$ is constructed as above, then there exist $r > 0$ and $\eta > 0$ such that the following hold:*

(a) The map

$$\mathbf{h} : [-\eta, \eta] \times \{\mathbf{v} : \|\mathbf{v}\| \leq r\} \to \Omega$$

defined by $\mathbf{h}(t, \mathbf{v}) = \mathbf{x}(t, \boldsymbol{\psi}(\mathbf{v}))$is continuous and one-to-one.

(b) The expression $(\mathbf{x}(t, \mathbf{v}) - \mathbf{p}) \cdot \mathbf{f}(\mathbf{p})$ for $\|\mathbf{v}\| \leq r$ is positive or negative according as $0 < t \leq \eta$ or $-\eta \leq t < 0$.

(c) The set

$$\mathbf{h}\big((-\eta, \eta) \times \{\mathbf{v} : \|\mathbf{v} - \mathbf{u}\| < r'\}\big)$$

is an open set whenever

$$\{\mathbf{v} : \|\mathbf{v} - \mathbf{u}\| < r'\} \subset \{\mathbf{v} : \|\mathbf{v}\| < r\}$$

for some $\mathbf{u}$.

Proof. There exist positive numbers σ and β such that $\mathbf{x}(t, \boldsymbol{\xi})$ is defined whenever $|\boldsymbol{\xi} - \mathbf{p}| < \beta$ and $-\sigma < t < \sigma$. Choose b' satisfying $0 < b' < b$ and such that $\|\mathbf{v}\| \leq b'$ implies that $|\boldsymbol{\psi}(\mathbf{v}) - \mathbf{p}| < \beta$. It is now clear that for $0 < \eta < \sigma$ and $0 < r < b'$, the map $\mathbf{h}$ is defined and continuous on $[-\eta, \eta] \times \{\mathbf{v} : \|\mathbf{v}\| \leq r\}$. To make the proof easier to follow, it will be broken down into four pieces, which will be called claims, and each claim will have its own proof.

Claim 1 *There exist* η_1 *and* r_1 *such that* $\mathbf{h}(t, \mathbf{v})$ *is one-to-one on the set*

$$[-\eta, \eta] \times \{\mathbf{v} : \|\mathbf{v}\| \leq r\}$$

when $0 < \eta \leq \eta_1$ *and* $0 < r \leq r_1$.

Proof of Claim 1. Assume that for every $\eta < \sigma$ and $r < b'$ the function $\mathbf{h}$ is not one-to-one on $[-\eta, \eta] \times \{\mathbf{v} : \|\mathbf{v}\| \leq r\}$ and derive a contradiction. In this case, there would exist two sequences of vectors $\mathbf{v}_m$ and $\mathbf{w}_m$ tending to $\mathbf{0}$ in $\mathbb{R}^{d-1}$ and two sequences of real numbers s_m and t_m tending to 0 in $\mathbb{R}$ such that $(s_m, \mathbf{v}_m) \neq (t_m, \mathbf{w}_m)$ and $\mathbf{h}(s_m, \mathbf{v}_m) = \mathbf{h}(t_m, \mathbf{w}_m)$ or

$$\mathbf{x}(s_m, \boldsymbol{\psi}(\mathbf{v}_m)) = \mathbf{x}(t_m, \boldsymbol{\psi}(\mathbf{w}_m)).$$

It follows by uniqueness of solutions that

$$\mathbf{x}(t, \boldsymbol{\psi}(\mathbf{v}_m)) = \mathbf{x}(t + t_m - s_m, \boldsymbol{\psi}(\mathbf{w}_m)).$$

Thus $\mathbf{x}(0, \boldsymbol{\psi}(\mathbf{v}_m)) = \boldsymbol{\psi}(\mathbf{v}_m)$ and $\mathbf{x}(s_m - t_m, \boldsymbol{\psi}(\mathbf{v}_m)) = \boldsymbol{\psi}(\mathbf{w}_m)$. If $s_m = t_m$, then $\boldsymbol{\psi}(\mathbf{v}_m) = \boldsymbol{\psi}(\mathbf{w}_m)$ and $\mathbf{v}_m = \mathbf{w}_m$, which is contrary to our assumption that $(s_m, \mathbf{v}_m) \neq (t_m, \mathbf{x}_m)$. Therefore, we can assume that $s_m \neq t_m$ and without loss of generality $s_m > t_m$.

Now, write $\mathbf{x}(t, \boldsymbol{\psi}(\mathbf{v}_m))$ in terms of coordinate functions as

$$\mathbf{x}(t, \boldsymbol{\psi}(\mathbf{v}_m)) = \big(x_1(t, \boldsymbol{\psi}(\mathbf{v}_m)), \ldots, x_d(t, \boldsymbol{\psi}(\mathbf{v}_m))\big)$$

and apply the mean-value theorem to each $x_i(t, \boldsymbol{\psi}(\mathbf{v}_m))$ on $[0, s_m - t_m]$ to obtain d sequences of real numbers $\tau(m, j)$ satisfying $0 < \tau(m, j) < s_m - t_m$ and

$$\frac{x_j(s_m - t_m, \boldsymbol{\psi}(\mathbf{v}_m)) - x_j(0, \boldsymbol{\psi}(\mathbf{v}_m))}{s_m - t_m} = \dot{x}_j(\tau(m, j), \boldsymbol{\psi}(\mathbf{v}_m)).$$

Writing $\mathbf{f} = (f_1, \ldots, f_d)$ and using $\dot{x}_j(t, \boldsymbol{\psi}(\mathbf{v}_m)) = f_j\big(\mathbf{x}(t, \boldsymbol{\psi}(\mathbf{v}_m)\big)$, the above equations, $j = 1, \ldots, d$ can be rewritten as the single equation

$$\frac{\boldsymbol{\psi}(\mathbf{w}_m) - \boldsymbol{\psi}(\mathbf{v}_m)}{s_m - t_m} = \big(f_1(\mathbf{x}(\tau(m, 1), \boldsymbol{\psi}(\mathbf{v}_m))), \ldots, f_d(\mathbf{x}(\tau(m, d), \boldsymbol{\psi}(\mathbf{v}_m)))\big).$$

To finish the proof that $\mathbf{h}$ is one-to-one for some choice of η and r, it suffices to derive a contradiction by letting m go to infinity. Clearly, $\tau(m, j) \to 0$ for $j = 1, \ldots, d$ as $m \to \infty$ from which it follows that

$$\lim_{m \to \infty} \big(f_1(\mathbf{x}(\tau(m, 1), \boldsymbol{\psi}(\mathbf{v}_m))), \ldots, f_d(\mathbf{x}(\tau(m, d), \boldsymbol{\psi}(\mathbf{v}_m)))\big) = \mathbf{f}(\mathbf{p}).$$

Therefore, the left-hand side is also convergent. The limit, however, must be in the subspace V because

$$\frac{\boldsymbol{\psi}(\mathbf{w}_m) - \boldsymbol{\psi}(\mathbf{v}_m)}{s_m - t_m} = \frac{T(\mathbf{w}_m) - T(\mathbf{v}_m)}{s_m - t_m} \in V.$$

This is a contradiction because $\mathbf{f}(\mathbf{p})$ is not in V. Hence, there exist η_1 and r_1 such that $\mathbf{h}(t, \mathbf{v})$ is one-to-one on $[\eta_1, \eta_1] \times \{\mathbf{v} : \|\mathbf{v}\| \le r_1\}$. Obviously, $\mathbf{h}(t, \mathbf{v})$ is still one-to-one when $0 < \eta \le \eta_1$ and $0 < r \le r_1$. □

Claim 2 *There exists η_2 such that $0 < \eta_2 \le \eta_1$ and $(\mathbf{x}(t, \mathbf{p}) - \mathbf{p}) \cdot \mathbf{f}(\mathbf{p})$ is positive or negative according as $0 < t \le \eta_2$ or $-\eta_2 \le t < 0$. Furthermore, there exists r_2 with $0 < r_2 \le r_1$ and such that $(\mathbf{x}(\eta_2, \boldsymbol{\psi}(\mathbf{v})) - \mathbf{p}) \cdot \mathbf{f}(\mathbf{p}) > 0$ and $(\mathbf{x}(-\eta_2, \boldsymbol{\psi}(\mathbf{v})) - \mathbf{p}) \cdot \mathbf{f}(\mathbf{p}) < 0$ when $\|\mathbf{v}\| < r_2$.*

Proof of Claim 2. Observe that

$$\lim_{t \to 0} \frac{\mathbf{x}(t, \mathbf{p}) - \mathbf{p}}{t} = \dot{\mathbf{x}}(0, \mathbf{p}) = \mathbf{f}(\mathbf{p}),$$

and hence

$$\lim_{t \to 0} \frac{(\mathbf{x}(t, \mathbf{p}) - \mathbf{p}) \cdot \mathbf{f}(\mathbf{p})}{t} = \|\mathbf{f}(\mathbf{p})\|^2.$$

By the definition of a limit, there exists $\eta_2 > 0$ such that

$$\frac{(\mathbf{x}(t, \mathbf{p}) - \mathbf{p}) \cdot \mathbf{f}(\mathbf{p})}{t} > \frac{\|\mathbf{f}(\mathbf{p})\|^2}{2}$$

when $|t| \le \eta_2$. It follows that

$$(\mathbf{x}(t, \mathbf{p}) - \mathbf{p}) \cdot \mathbf{f}(\mathbf{p}) > \frac{t\|\mathbf{f}(\mathbf{p})\|^2}{2} > 0,$$

when $t > 0$ and

$$(\mathbf{x}(t, \mathbf{p}) - \mathbf{p}) \cdot \mathbf{f}(\mathbf{p}) < \frac{t\|\mathbf{f}(\mathbf{p})\|^2}{2} < 0,$$

when $t < 0$.

The existence of the required r_2 follows from the first part of the claim by continuity in initial conditions. □

Claim 3 *There exists r_3 with $0 < r_3 < r_2$ such that for $\|\mathbf{v}\| \le r_3$ the expression $(\mathbf{x}(t, \mathbf{v}) - \mathbf{p}) \cdot \mathbf{f}(\mathbf{p})$ is positive or negative according as $0 < t \le \eta_2$ or $-\eta_2 \le t < 0$.*

Proof of Claim 3. Assume that an r_3 meeting the required conditions for this η_2 does not exist. Then, by the second part of Claim 2 and the intermediate-value theorem, there must exist sequences $t_m \ne 0$ with $|t_m| \le \eta_2$ and $\mathbf{v}_m \in \mathbb{R}^{d-1}$ such that $\mathbf{v}_m \to \mathbf{0}$ and $(\mathbf{x}(t_m, \boldsymbol{\psi}(\mathbf{v}_m)) - \mathbf{p}) \cdot \mathbf{f}(\mathbf{p}) = 0$. It follows from the first part of Claim 2 and continuity in initial conditions that $t_m \to 0$.

Since $(\mathbf{x}(t_m, \boldsymbol{\psi}(\mathbf{v}_m)) - \mathbf{p}) \cdot \mathbf{f}(\mathbf{p}) = 0$, the point $\mathbf{x}(t_m, \boldsymbol{\psi}(\mathbf{v}_m))$ is in $V + \mathbf{p}$ and equals $\boldsymbol{\psi}(\mathbf{w}_m) = \mathbf{x}(0, \boldsymbol{\psi}(\mathbf{w}_m))$. If $\|\mathbf{w}_m\| \le r_2$, then $\mathbf{h}(t_m, \mathbf{v}_m) = \mathbf{h}(0, \mathbf{w}_m)$ and $t_m = 0$ and $\mathbf{v}_m = \mathbf{w}_m$ by Claim 1. Therefore, $\|\mathbf{w}_m\| \ge r_2$ and $\|\mathbf{v}_m - \mathbf{w}_m\| \ge (r_2/2)$ for large m. This produces a contradiction.

On the one hand,

$$\lim_{m \to \infty} \boldsymbol{\psi}(\mathbf{w}_m) = \lim_{m \to \infty} \mathbf{x}(t_m, \boldsymbol{\psi}(\mathbf{v}_m)) = \mathbf{p}$$

by continuity in initial conditions. Since ψ^{-1} is obviously continuous, it follows that $\lim_{m\to\infty} \mathbf{w}_m = \mathbf{0}$ and $\lim_{m\to\infty} \|\mathbf{w}_m - \mathbf{v}_m\| = 0$. On the other hand, $\lim_{m\to\infty} \|\mathbf{w}_m - \mathbf{v}_m\| \geq (r_2/2) > 0$. Thus there exists an η_3 meeting the required conditions for $r_3 = r_2/2$. □

Claim 4 *If $\eta \leq \eta_2$ and $r \leq r_3$, then the set*

$$\mathbf{h}((\eta, \eta) \times \{\mathbf{v} : \|\mathbf{v} - \mathbf{u}\| < r'\})$$

is an open set whenever

$$\{\mathbf{v} : \|\mathbf{v} - \mathbf{u}\| < r'\} \subset \{\mathbf{v} : \|\mathbf{v}\| < r\}$$

for some $\mathbf{u}$.

Proof of Claim 4. If the set $\mathbf{h}((\eta, \eta) \times \{\mathbf{v} : \|\mathbf{v} - \mathbf{u}\| < r'\})$ is not open, then for some point $\mathbf{x}(\rho, \psi(\mathbf{w}))$ with $|\rho| < \eta$ and $\|\mathbf{w} - \mathbf{u}\| < r'$ there exists a sequence of points $\mathbf{p}_m \in \mathbb{R}^d$ that are not in $\mathbf{h}((-\eta, \eta) \times \{\mathbf{v} : \|\mathbf{v} - \mathbf{u}\| < r'\})$ but converge to $\mathbf{x}(\rho, \psi(\mathbf{v}))$.

Claim 3 implies that $\mathbf{x}(\rho, \psi(\mathbf{w}))$ and $\mathbf{x}(-\rho, \psi(\mathbf{w}))$ are on opposite sides of $V + \mathbf{p}$ and that $\mathbf{x}(t, \psi(\mathbf{w}))$ lies on $V + \mathbf{p}$ if and only if $t = 0$. Since $\mathbf{x}(t, \mathbf{x}(\rho, \psi(\mathbf{w}))) = \mathbf{x}(t + \rho, \psi(\mathbf{w}))$, the solution $\mathbf{x}(t, \mathbf{x}(\rho, \psi(\mathbf{w})))$ crosses $V + \mathbf{p}$ for t between -2ρ and 0 precisely when $t = -\rho$ and at no other time.

By continuity in initial conditions, the solutions $\mathbf{x}(t, \mathbf{p}_m)$ must also cross $V + \mathbf{p}$ for large m at some t between -2ρ and 0. (This could be either $-2\rho < t < 0$ or $0 < t < -2\rho$ according as $\rho > 0$ or $\rho < 0$.) Thus there exists a sequence of vectors $\mathbf{v}_m$ in $\mathbb{R}^{d-1}$ and sequence of real numbers τ_m between -2ρ and 0 such that $\mathbf{x}(\tau_m, \mathbf{p}_m) = \psi(\mathbf{v}_m)$. By taking a subsequence, we can assume that τ_m converges to τ, which is also between -2ρ and 0.

Using continuity initial conditions again,

$$\lim_{m\to\infty} \psi(\mathbf{v}_m) = \lim_{m\to\infty} \mathbf{x}(\tau_m, \mathbf{p}_m) = \mathbf{x}(\tau, \mathbf{x}(\rho, \psi(\mathbf{w}))).$$

Consequently, $\mathbf{x}(\tau, \mathbf{x}(\rho, \psi(\mathbf{w})))$ is in $V + \mathbf{p}$ because $V + \mathbf{p}$ is closed and the sequence $\psi(\mathbf{v}_m)$ is in V. It follows that $\tau = -\rho$ and, consequently, that

$$\lim_{m\to\infty} \psi(\mathbf{v}_m) = \mathbf{x}(-\rho, \mathbf{x}(\rho, \psi(\mathbf{w}))) = \psi(\mathbf{w}).$$

Therefore, $\|\mathbf{v}_m - \mathbf{u}\| < r'$ and $|\tau_m||leq\eta$ for large m and $\mathbf{p}_m$ is in $\mathbf{h}((-\eta, \eta) \times \{\mathbf{v} : \|\mathbf{v} - \mathbf{u}\| < r'\})$ for large m. This contradicts the basic property of the sequence $\mathbf{p}_m$ and completes the proof of the last claim. □

Since $\eta_2 \leq \eta_1$ and $r_3 \leq r_2 \leq r_1$, the three conclusions of the theorem hold for any $\eta \leq \eta_2$ and $r \leq r_3$ by Claims 1, 2, and 4. This completes the proof of the theorem. □

When the conclusion of *Theorem 6.1* holds for η and r, the set

$$L = \{\psi(\mathbf{v}) : \|\mathbf{v}\| \leq r\} \tag{6.1}$$

is called a *local section or transversal at* $\mathbf{p}$ *of length* 2η and

$$\Gamma = \mathbf{h}([-\eta, \eta] \times \{\mathbf{v} : \|\mathbf{v}\| \leq r\}) \tag{6.2}$$

is called a *flow box or closed path cylinder of length* 2η. Notice the length of a local section is not unique; a smaller positive length will work just as well. Rather the length is just twice some positive number for which the three conclusions of *Theorem 6.1* hold. The expression "L is a local section" simply means there is some unspecified length that works.

Figure 6.1 is an actual flow box Γ for the linear system $\dot{x} = y$ and $\dot{y} = x$ at the point $(2, 2)$ with $\eta = 0.75$. The slightly darker line across the middle is a local section L and the dot at its mid-point marks the point $(2, 2)$. The local section L divides Γ into two pieces and the longer narrower piece in the upper right is the positive side of L. The other piece on the lower left is the negative side of L. The only way a solution can cross L is to enter Γ at the lower left, cross L, and then exit at the upper right before it can reenter the negative side. This is the critical property of local sections in all dimensions and why they are useful.

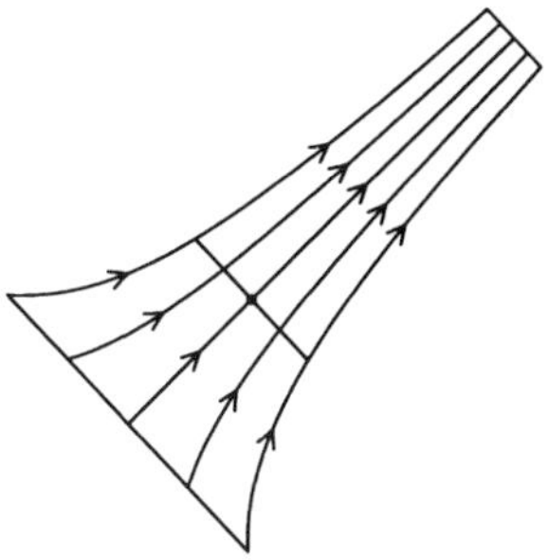

Figure 6.1: Sample flow box or path cylinder.

The flow box can be described more easily using the local section, namely,

$$\Gamma = \{\mathbf{x}(t, \mathbf{q}) : |t| \leq \eta \text{ and } \mathbf{q} \in L\}. \tag{6.3}$$

Note that Γ is compact because it is the continuous image of a compact set. It is convenient to let

$$L^o = \psi(\{\mathbf{v} : \|\mathbf{v}\| < r\}) \tag{6.4}$$

and then

$$\Gamma^o = \{\mathbf{x}(t, \mathbf{q}) : |t| < \eta \text{ and } \mathbf{q} \in L^o\} \tag{6.5}$$

is an open set of $\mathbb{R}^d$ by part (c) of *Theorem 6.1*. The next proposition gives some additional properties of local sections.

Proposition 6.2 *If* L *is a local section of length* 2η *for* $\dot{\mathbf{x}} = \mathbf{f}(\mathbf{x})$ *and* Γ *is its flow box, then the following hold for any solution* $\varphi(t)$*:*

(a) *If* $\varphi(\sigma) \in \Gamma$ *for some* σ, *then* $\varphi(\tau) \in L$ *for exactly one* τ *such that* $|\tau - \sigma| \leq \eta$

(b) *If* $\varphi(\tau_1)$ *and* $\varphi(\tau_2)$ *are in* L *and* $\tau_1 \neq \tau_2$, *then* $|\tau_1 - \tau_2| > 2\eta$.

(c) *The set* $\{t : \varphi(t) \in L\}$ *is a discrete subset of* $\mathbb{R}$.

Proof. (a) If $\varphi(\sigma) \in \Gamma$, then $\varphi(\sigma) = \mathbf{x}(s, \mathbf{q})$ for some $\mathbf{q} \in L$ and $|s| \leq \eta$. Hence, $\varphi(t) = \mathbf{x}(t + s - \sigma, \mathbf{q})$ and $\varphi(\sigma - s) = \mathbf{q}$. Let $\tau = \sigma - s$.

If $\varphi(\tau')$ is also in L and $|\tau' - \sigma| \leq \eta$, then $\varphi(\sigma) = \mathbf{x}(\sigma - \tau', \varphi(\tau'))$ or $\mathbf{h}(s, \mathbf{q}) = \mathbf{h}(\sigma - \tau, \varphi(\tau'))$. Because $\mathbf{h}$ is one-to-one it follows that $\varphi(\tau') = \mathbf{q}$ and $s = \sigma - \tau'$. Hence, $\tau = \tau'$.

(b) Assume $\tau_1 < \tau_2$ and $\tau_2 - \tau_1 \leq 2\eta$. Hence, $0 < (\tau_2 - \tau_1)/2 = s \leq \eta$ and $\tau_1 + s = \tau_2 - s$. Then

$$\varphi(\tau_1 + s) \in \{\mathbf{x}(t, \mathbf{q}) : 0 < t \leq \eta \text{ and } \mathbf{q} \in L\} = P \subset \Gamma$$

and

$$\varphi(\tau_2 - s) \in \{\mathbf{x}(t, \mathbf{q}) : 0 < -t \leq \eta \text{ and } \mathbf{q} \in L\} = N \subset \Gamma.$$

Consequently, $\varphi(\tau_1 + s) = \varphi(\tau_2 - s) \in P \cap N$. This is impossible because $P \cap N = \phi$ by part (b) of *Theorem 6.1* .

(c) This follows trivially from part (b). □

Corollary *Let* L *be a local section for* $\dot{\mathbf{x}} = \mathbf{f}(\mathbf{x})$. *If* $\omega(\boldsymbol{\xi}) \cap L^o \neq \phi$, *then there exists a sequence of times* t_m *such that* $\mathbf{x}(t_m, \boldsymbol{\xi}) \in L^o$ *and* $t_m \to \infty$ *as* $m \to \infty$.

Proof. Exercise.

Given L, a local section for a flow box Γ of length 2η for $\dot{\mathbf{x}} = \mathbf{f}(\mathbf{x})$, define $\lambda : \Gamma \to L$ and $\kappa : \Gamma \to [-\eta, \eta]$ by $\lambda(\mathbf{x}(t, \mathbf{q})) = \mathbf{q}$ and $\kappa(\mathbf{x}(t, \mathbf{q})) = t$, where as usual $\mathbf{q} \in L$ and $|t| \leq \eta$. These function are well defined because $\mathbf{h}$ is one-to-one.

Proposition 6.3 *If* L *is a local section for a flow box* Γ *of length* 2η *for* $\dot{\mathbf{x}} = \mathbf{f}(\mathbf{x})$, *then the functions* $\lambda : \Gamma \to L$ *and* $\kappa : \Gamma \to [-\eta, \eta]$ *are continuous.*

Proof. It suffices to show that if a sequence of points $\mathbf{x}(t_m, \mathbf{q}_m)$ in Γ with $|t_m| \leq \eta$ and $\mathbf{q}_m \in L$ converges to $\mathbf{x}(t, \mathbf{q})$ with $|t| \leq \eta$ and $\mathbf{q} \in L$, then $(t_m, \mathbf{q}_m)$ converges to $(t, \mathbf{q})$ because this is equivalent to $\kappa(\mathbf{x}(t_m, \mathbf{q}_m))$ converging to $\kappa(\mathbf{x}(t, \mathbf{q}))$ and $\lambda(\mathbf{x}(t_m, \mathbf{q}_m))$ converging to $\lambda(\mathbf{x}(t, \mathbf{q}))$.

Since Γ is compact, it can be assumed without loss of generality that $(t_m, \mathbf{q}_m)$ converges to $(t', \mathbf{q}')$. Then, by continuity in initial conditions, $x(t_m, \mathbf{q}_m)$ converges to $\mathbf{x}(t', \mathbf{q}')$ and $\mathbf{x}(t', \mathbf{q}') = \mathbf{x}(t, \mathbf{q})$. Because $\mathbf{h}$ is one-to-one, it follows that $(t', \mathbf{q}') = (t, \mathbf{q})$ to complete the proof. □

The Poincaré return map can now be constructed for a local section $L = \boldsymbol{\psi}(\{\mathbf{v} : \|\mathbf{v}\| \leq r\})$ of length 2η on a subset of $\{\mathbf{v} : \|\mathbf{v}\| \leq r\}$. First, define a function $\rho : L \to (0, \infty]$ by

$$\rho(\mathbf{v}) = \min\{t : t > 0 \text{ and } \mathbf{x}(t, \boldsymbol{\psi}(\mathbf{v})) \in L\}$$

with the convention that $\rho(\mathbf{v}) = \infty$ when $\{t : t > 0 \text{ and } \mathbf{x}(t, \boldsymbol{\psi}(\mathbf{v})) \in L\} = \phi$. The function $\rho(\mathbf{v})$ is the *first return time* of $\boldsymbol{\psi}(\mathbf{v})$ to L. Then the *Poincaré return map* is defined by

$$R(\mathbf{v}) = \boldsymbol{\psi}^{-1}(\mathbf{x}(\rho(\mathbf{v}), \boldsymbol{\psi}(\mathbf{v})))$$

when $\rho(\mathbf{v}) \neq \infty$. The function $R(\mathbf{v})$ maps points in the disk $\{\mathbf{v} : \|\mathbf{v}\| \leq r\}$ such that $\rho(\mathbf{v}) \neq \infty$ to points in the same disk.

Clearly, the Poincaré return map is continuous when the first return time is continuous, and the two functions are closely related. Because L is a hyperdisk and has an edge, the first return time need not be continuous in general. Specifically, ρ is not likely to be continuous at $\mathbf{v}$ when $\mathbf{x}(\rho(\mathbf{v}), \boldsymbol{\psi}(\mathbf{v}))$ is on the edge of L or $\|R(\mathbf{v})\| = r$. Staying away from the edge of L is, however, all that is needed to ensure that both the first return time, $\rho(\mathbf{v})$, and the Poincaré return map, $R(\mathbf{v})$, are continuous functions.

Theorem 6.4 *The following hold for the set*

$$W = \{\mathbf{v} : \boldsymbol{\psi}(\mathbf{v}) \in L^o,\ \rho(\mathbf{v}) \neq \infty, \text{ and } \mathbf{x}(\rho(\mathbf{v}), \boldsymbol{\psi}(\mathbf{v})) \in L^o\}$$

in $\mathbb{R}^{d-1}$*:*

(a) The first return time, $\rho(\mathbf{v})$*, is continuous on* W*.*

(b) The set W *is an open set.*

(c) The Poincaré return map, $R(\mathbf{v})$*, is continuous on* W*.*

Proof. (a) Consider a specific point $\mathbf{w}$ in W. It follows that $\mathbf{x}(\rho(\mathbf{w}), \boldsymbol{\psi}(\mathbf{w}))$ is in Γ^o as defined by equation *(6.5)*. Because Γ^o is open, by continuity in initial conditions, there exists $\delta > 0$ such that $\mathbf{x}(\rho(\mathbf{w}), \boldsymbol{\psi}(\mathbf{v})) \in \Gamma^o$, when $\|\mathbf{v} - \mathbf{w}\| < \delta$. Consequently, if $|\mathbf{v} - \mathbf{w}| < \delta$, then $\mathbf{x}(\tau, \boldsymbol{\psi}(\mathbf{v})) \in L^o$ for a unique τ satisfying $0 < \rho(\mathbf{w}) - \eta \leq \tau \leq \rho(\mathbf{w}) + \eta$ by part (a) of *Proposition 6.2*. Hence, $\rho(\mathbf{v}) \leq \rho(\mathbf{w}) + \eta$ when $\|\mathbf{v} - \mathbf{w}\| < \delta$.

Suppose ρ is not continuous at $\mathbf{w}$. Then there exists a sequence $\mathbf{v}_m$ converging to $\mathbf{w}$ such that $\rho(\mathbf{v}_m)$ does not converge to $\rho(\mathbf{w})$. Without loss of generality, it can be assumed that $\|\mathbf{v}_m - \mathbf{w}\| < \delta$, and hence $\rho(\mathbf{v}_m) \leq \rho(\mathbf{w}) + \eta$. Observe that $\rho(\mathbf{v}) > 2\eta$ in general . Thus it can also be assumed that $\rho(\mathbf{v}_m)$ converges to $t' \neq \rho(\mathbf{w})$ such that $2\eta \leq t' \leq \rho(\mathbf{w}) + \eta$. It follows that $\mathbf{x}(\rho(t_m), \boldsymbol{\psi}(\mathbf{v}_m))$ converges to $\mathbf{x}(t', \boldsymbol{\psi}(\mathbf{w}))$ and $\mathbf{x}(t', \boldsymbol{\psi}(\mathbf{w}))$ is in L.

There are now two possibilities for t' and both of them will lead to contradictions. If $\rho(\mathbf{w}) - \eta \leq t' \leq \rho(\mathbf{w}) + \eta$, then $0 < |t' - \rho(\mathbf{w})| \leq \eta$ which is contrary to *Proposition 6.2* unless $t' = \rho(\mathbf{w})$. If $2\eta \leq t' < \rho(\mathbf{w}) - \eta$, then there is an earlier crossing of L by $\boldsymbol{\psi}(\mathbf{w})$ and $\min\{t : t > 0 \text{ and } \mathbf{x}(t, 0, \boldsymbol{\psi}(\mathbf{w})) \in L\} \leq t' < \rho(\mathbf{w})$ contradicting the definition of $\rho(\mathbf{w})$. The two contradictions imply that ρ is continuous at $\mathbf{w}$.

(b) As in part (a) start with $\mathbf{w}$ in W. By using continuity in initial conditions and part (a), there exists $\delta > 0$ such that $\mathbf{x}(\rho(\mathbf{w}), \boldsymbol{\psi}(\mathbf{v})) \in \Gamma^o$ and $|\rho(\mathbf{v}) - \rho(\mathbf{w})| <$

η when $\|\mathbf{v}-\mathbf{w}\|<\delta$. As before, if $\|\mathbf{v}-\mathbf{w}\|<\delta$, then $\mathbf{x}(t,\psi(\mathbf{v}))\in L^o$ for a unique τ satisfying $\rho(\mathbf{w})+\eta\geq\tau\geq\rho(\mathbf{w})-\eta>0$. But now it is also true that $\rho(\mathbf{w})+\eta\geq\rho(\mathbf{v})\geq\rho(\mathbf{w})-\eta>0$ when $|\mathbf{v}-\mathbf{w}|<\delta$. Therefore, $\tau=\rho(\mathbf{v})$ and $\mathbf{v}$ is in W when $\|\mathbf{v}-\mathbf{w}\|<\delta$.

(c) The function $R(\mathbf{v})=\psi^{-1}(\mathbf{x}(\rho(\mathbf{v}),\psi(\mathbf{v})))$ is continuous on W because $\psi(\mathbf{v})$, $\psi^{-1}(\mathbf{x})$, and $\mathbf{x}(t,\boldsymbol{\xi})$ are continuous functions on their domains and $\rho(\mathbf{v})$ is continuous on W. □

In general, the set on which $R(\cdot)$ is defined can be empty or very complicated. However, $R(\mathbf{w})=\mathbf{w}$ implies that $\psi(\mathbf{w})$ is a periodic point for $\dot{\mathbf{x}}=\mathbf{f}(\mathbf{x})$. This observation can be used to study the behavior of solutions near periodic points or to prove the existence of periodic points.

The condition that $R(\mathbf{w})=\mathbf{w}$ is an example of a fixed-point condition, and existence theorems for fixed points can sometimes be used to prove the existence of periodic points. For example, the Brouwer fixed-point theorem guarantees that any continuous map of a closed ball into itself has a fixed point, that is, $\mathbf{g}:\{\mathbf{v}:\|\mathbf{v}-\mathbf{u}\|\leq r'\}\to\{\mathbf{v}:\|\mathbf{v}-\mathbf{u}\|\leq r'\}$ implies $\mathbf{g}(\mathbf{w})=\mathbf{w}$ for some $\mathbf{w}$ satisfying $\|\mathbf{w}-\mathbf{u}\|\leq r'$. (See [24] or [34] for more information about the Brouwer fixed-point theorem.) Consequently, if some closed ball in the domain of the Poincaré return map is mapped into itself by the Poincaré return map, then the Poincaré return map has a fixed point and the the original autonomous differential equation has a periodic point transversal to the local section.

EXERCISES

1. Show that $\mathbf{h}^{-1}$ is a continuous map from Γ to $[-\eta,\eta]\times\{\mathbf{v}:\|\mathbf{v}\|\leq r\}$ when η and r satisfy *Theorem 6.1*

2. Let L be a local section for $\dot{\mathbf{x}}=\mathbf{f}(\mathbf{x})$. Show that if $\{\mathbf{x}(t,\boldsymbol{\xi}):t\geq 0\}\cap L$ contains infinitely many points, then $\omega(\boldsymbol{\xi})\cap L\neq\phi$.

3. Suppose that $\dot{\mathbf{x}}=\mathbf{f}(\mathbf{x})$ defines a flow on Ω and that $\mathbf{p}\in\Omega$ is neither a fixed point nor a periodic point. Prove that if $\eta>0$, then there exists $r>0$ such that the map $\mathbf{h}:[-\eta,\eta]\times\{\mathbf{v}:\|\mathbf{v}\|\leq r\}\to\Omega$ defined by $\mathbf{h}(t,\mathbf{v})=\mathbf{x}(t,\psi(\mathbf{v}))$ satisfies conclusions (a) and (c) of *Theorem 6.1.* Hint: While obtaining r such that (a) holds also obtain an $\eta_1\leq\eta$ such that (c) holds and then use it to prove that (c) holds for η.

4. Prove the corollary on page 190.

5. With W defined as in *Theorem 6.4* for the Poincaré return map R, show that $R(W)$ is an open set and R^{-1} is defined and continuous on $R(W)$.

6.2 Planar Dynamics

The material in this section is frequently referred to as Poincaré -Bendixson theory and has its origins in the joint work of Poincaré and Bendixson at the beginning of the twentieth century. Although not precisely defined, Poincaré -Bendixson theory generally refers to the limitations of orbit behavior for autonomous differential equations in the plane and the more complete descriptions of omega limit sets that are possible because of these limitations.

The principal reason that orbit behavior in the plane is more limited than in higher dimensions is the Jordan curve theorem. Local sections will be the primary tool used to exploit the Jordan curve theorem. To be more specific, in the plane a local section is a segment of straight line, and a Jordan curve can often be constructed from a piece of a solution and a piece of local section. Then the Jordan curve theorem can be applied to extract dynamical information.

A *Jordan curve* or *simple closed curve* is the image of a continuous map $\gamma : [0,1] \to \mathbb{R}^2$ for which $\gamma(s) = \gamma(t)$ and $t \neq s$ if and only if $\{s,t\} = \{0,1\}$. Although it is intuitively clear that a Jordan curve divides the plane into two parts, writing down a formal proof requires sophisticated algebraic topology. The Jordan curve theorem is stated below and will be used at several key junctions in this section. A proof of the Jordan curve theorem can be found in [34].

Theorem 6.5 (Jordan curve theorem) *If J is a Jordan curve, then*

$$\mathbb{R}^2 - J = \{\mathbf{x} \in \mathbb{R}^2 : \mathbf{x} \notin J\} = S_E \cup S_I,$$

where S_E and S_I are disjoint open arc-wise connected sets with S_I, the interior, bounded and S_E, the exterior, unbounded. Moreover, $\overline{S}_E = S_E \cup J$ and $\overline{S}_I = S_I \cup J$.

Corollary *If $\mathbf{p}$ and $\mathbf{q}$ are distinct points in $\mathbb{R}^2$ not lying on a Jordan curve J and there is a curve $\gamma : [0,1] \to \mathbb{R}^2$ joining them ($\gamma(0) = \mathbf{p}$ and $\gamma(1) = \mathbf{q}$) such that if $\gamma(t) \notin J$ for $0 \leq t \leq 1$, then $\mathbf{p}$ and $\mathbf{q}$ are both in S_E or they are both in S_I.*

Throughout this section Ω will be an open subset of R^2 and $\dot{\mathbf{x}} = \mathbf{f}(\mathbf{x})$ a locally Lipschitz autonomous differential equation defined on Ω. The fundamental question is how solutions evolve as time goes to infinity, or more specifically how to characterize the omega limit sets as completely as possible.

To construct a Jordan curve using an orbit and local section, start with a point $\mathbf{p}$ that is not a fixed point of $\dot{\mathbf{x}} = \mathbf{f}(\mathbf{x})$. By *Theorem 6.1* there exists a local section L at $\mathbf{p}$. Using the notation from the previous section $L = \{\boldsymbol{\psi}(\mathbf{v}) : \|\mathbf{v}\| \leq r\}$, where $\boldsymbol{\psi}(\mathbf{v}) = T(\mathbf{v}) + \mathbf{p}$ and $T : \mathbb{R}^{d-1} \to \mathbb{R}^d$ is a linear isomorphism of $\mathbb{R}^{d-1}$ onto the subspace $V = \{\mathbf{x} \in \mathbb{R}^d : \mathbf{x} \cdot \mathbf{f}(\mathbf{p}) = 0\}$ of $\mathbb{R}^d$.

Since d now equals 2, V is a line through the origin, $V + \mathbf{p}$ is a parallel line through $\mathbf{p}$, and $\{\mathbf{v} : \|\mathbf{v}\| \leq r\}$ is the closed interval $[-r, r]$ in $\mathbb{R}$. By replacing T with rT, it can and will be assumed that $r = 1$. Then L is the line segment joining $\psi(-1)$ and $\psi(1)$ and is a copy of $[-1, 1]$ in $\mathbb{R}^2$ with $\mathbf{p}$ at its midpoint. If $\boldsymbol{\xi}_1$ and $\boldsymbol{\xi}_2$ are points in L, then the line segment joining $\boldsymbol{\xi}_1$ and $\boldsymbol{\xi}_2$ is like an interval and will be denoted by $[\boldsymbol{\xi}_1, \boldsymbol{\xi}_2]_L$. This set is just the set of all convex combinations $\{\alpha\boldsymbol{\xi}_1 + (1-\alpha)\boldsymbol{\xi}_2 : 0 \leq \alpha \leq 1\}$ of $\boldsymbol{\xi}_1$ and $\boldsymbol{\xi}_2$. Similarly, $(\boldsymbol{\xi}_1, \boldsymbol{\xi}_2)_L$ will denote the line segment joining the points $\boldsymbol{\xi}_1$ and $\boldsymbol{\xi}_2$ in L, without the endpoints $\boldsymbol{\xi}_1$ and $\boldsymbol{\xi}_2$.

Recall that the solution $\mathbf{x}(t, \boldsymbol{\xi})$ crosses a local section at discrete times (*Proposition 6.2*), and hence it is possible to talk about consecutive crossings. Specifically, $\mathbf{x}(t_1, \boldsymbol{\xi})$ and $\mathbf{x}(t_2, \boldsymbol{\xi})$ are *consecutive crossings of* L by the solution $\mathbf{x}(t, \boldsymbol{\xi})$ if both $\mathbf{x}(t_1, \boldsymbol{\xi})$ and $\mathbf{x}(t_2, \boldsymbol{\xi})$ are points in L and if

$$\{\mathbf{x}(t, \boldsymbol{\xi}) : t_1 < t < t_2\} \cap L = \phi.$$

The following remark is obvious.

Remark *If* $\mathbf{x}(t_1, \boldsymbol{\xi})$ *and* $\mathbf{x}(t_2, \boldsymbol{\xi})$ *are consecutive crossings of a local section* L*, then*

$$J = \{\mathbf{x}(t, \boldsymbol{\xi}) : t_1 \leq t \leq t_2\} \cup [\mathbf{x}(t_1, \boldsymbol{\xi}), \mathbf{x}(t_2, \boldsymbol{\xi})]_L$$

is a Jordan curve.

If $\boldsymbol{\xi}$ is periodic, then it is possible that $\mathbf{x}(t_1, \boldsymbol{\xi}) = \mathbf{x}(t_2, \boldsymbol{\xi})$ and the set $(\mathbf{x}(t_1, \boldsymbol{\xi}), \mathbf{x}(t_2, \boldsymbol{\xi}))_L$ is empty. When the consecutive crossings are distinct points, the Jordan curve J constructed as above has critical properties that are essential in the proofs of the main theorems in this section. These properties are the content of the next several propositions.

Proposition 6.6 *If* $\boldsymbol{\xi}_1 = \mathbf{x}(t_1, \boldsymbol{\xi})$ *and* $\boldsymbol{\xi}_2 = \mathbf{x}(t_2, \boldsymbol{\xi})$ *with* $t_1 < t_2$ *are distinct consecutive crossings of a local section* L *of length* 2η *at* $\mathbf{p}$ *by the solution* $\mathbf{x}(t, \boldsymbol{\xi})$*, then the Jordan curve*

$$J = \{\mathbf{x}(t, \boldsymbol{\xi}) : t_1 \leq t \leq t_2\} \cup [\boldsymbol{\xi}_1, \boldsymbol{\xi}_2]_L$$

has the property that one of the sets

$$\{\mathbf{x}(t, \mathbf{q}) : 0 < t < \eta \text{ and } \mathbf{q} \in (\boldsymbol{\xi}_1, \boldsymbol{\xi}_2)_L\}$$

and

$$\{\mathbf{x}(t, \mathbf{q}) : -\eta < t < 0 \text{ and } \mathbf{q} \in (\boldsymbol{\xi}_1, \boldsymbol{\xi}_2)_L\}$$

is a subset of S_E *and the other is a subset of* S_I.

Proof. It follows from the corollary to the Jordan curve theorem and *Theorem 6.1* that each of the sets

$$\{\mathbf{x}(t, \mathbf{q}) : 0 < t < \eta \text{ and } \mathbf{q} \in (\boldsymbol{\xi}_1, \boldsymbol{\xi}_2)_L\}$$

and

$$\{\mathbf{x}(t, \mathbf{q}) : -\eta < t < 0 \text{ and } \mathbf{q} \in (\boldsymbol{\xi}_1, \boldsymbol{\xi}_2)_L\}$$

lies entirely in S_E or S_I. If they both lie in say S_E, then by *Theorem 6.1* the set

$$\{\mathbf{x}(t,\mathbf{q}) : -\eta < t < \eta \text{ and } \mathbf{q} \in (\boldsymbol{\xi}_1, \boldsymbol{\xi}_2)_L\}$$

is an open set containing $(\boldsymbol{\xi}_1, \boldsymbol{\xi}_2)_L$ that does not intersect S_I. Consequently,

$$(\boldsymbol{\xi}_1, \boldsymbol{\xi}_2)_L \cap \overline{S_I} = \phi.$$

But the Jordan curve theorem requires that $\overline{S_I} = S_I \cup J$. This contradiction completes the proof. □

One way of viewing the previous proposition is that $[\boldsymbol{\xi}_1, \boldsymbol{\xi}_2]_L$ is a gate between S_E and S_I that solutions can only cross in one direction. This idea will be made precise in the *Proposition 6.7* and will be critical in what follows. For simplicity whichever of the two sets, S_E and S_I, contain

$$\{\mathbf{x}(t,\mathbf{q}) : 0 < t < \eta \text{ and } \mathbf{q} \in (\boldsymbol{\xi}_1, \boldsymbol{\xi}_2)_L\}$$

will be called the *positive side of J* and the other one the *negative side.* If the positive and negative sides are denoted by S_P and S_N, respectively, then the Jordan curve theorem implies that S_P and S_N are are disjoint open arc-wise connected sets such that $\mathbb{R}^2 - J = S_P \cup S_N$, $\overline{S}_P = S_P \cup J$, and $\overline{S}_N = S_N \cup J$.

Proposition 6.7 *If $\boldsymbol{\xi}_1 = \mathbf{x}(t_1, \boldsymbol{\xi})$ and $\boldsymbol{\xi}_2 = \mathbf{x}(t_2, \boldsymbol{\xi})$ with $t_1 < t_2$ are distinct consecutive crossings of a local section L of length 2η at $\mathbf{p}$ by the solution $\mathbf{x}(t, \boldsymbol{\xi})$, then the Jordan curve*

$$J = \{\mathbf{x}(t,\boldsymbol{\xi}) : t_1 \le t \le t_2\} \cup [\boldsymbol{\xi}_1, \boldsymbol{\xi}_2]_L$$

has the following properties:

(a) If $\mathbf{q}$ is on the positive side of J and there is no t such that $\mathbf{q} = \mathbf{x}(t, \boldsymbol{\xi})$, then $\mathbf{x}(t, \mathbf{q})$ is also in the positive side of J for $t > 0$ in its domain.

(b) If $\mathbf{q}$ is on the negative side of J and there is no t such that $\mathbf{q} = \mathbf{x}(t, \boldsymbol{\xi})$, then $\mathbf{x}(t, \mathbf{q})$ is also in the negative side of J for $t < 0$ in its domain.

Proof. The proofs of the two parts are essentially the same and the proof of the second part will be left to the reader. To prove the first part, suppose there is a $t > 0$ such that $\mathbf{x}(t, \mathbf{q}) \in S_N$ and set $\tau = \inf\{t > 0 : \mathbf{x}(t, \mathbf{q}) \in S_N\}$. Clearly, $\tau > 0$ and $\mathbf{x}(\tau, \mathbf{q}) \in J$ because S_P and S_N are open sets. Since $\mathbf{q}$ does not lie on the solution $\mathbf{x}(t, \boldsymbol{\xi})$, neither does $\mathbf{x}(\tau, \mathbf{q})$. Thus $\mathbf{x}(\tau, \mathbf{q})$ lies in $(\boldsymbol{\xi}_1, \boldsymbol{\xi}_2)_L$.

Choose δ such that $0 < \delta < \min\{\eta, \tau\}$. *Proposition 6.6* implies that $\mathbf{x}(\tau - \delta, \mathbf{q}) \in S_N$, and hence that $\tau \neq \inf\{t > 0 : \mathbf{x}(t, \mathbf{q}) \in S_N\}$. This contradiction completes the proof. □

Not surprisingly, periodic orbits cannot cross a local section in the plane at more than one point This fact requires a proof and is not true in higher dimensions.

Proposition 6.8 *If* $\boldsymbol{\xi}_1 = \mathbf{x}(t_1, \boldsymbol{\xi})$ *and* $\boldsymbol{\xi}_2 = \mathbf{x}(t_2, \boldsymbol{\xi})$ *with* $t_1 < t_2$ *are distinct consecutive crossings of a local section* L *of length* 2η *at* $\mathbf{p}$ *by the solution* $\mathbf{x}(t, \boldsymbol{\xi})$*, then* $\boldsymbol{\xi}$ *is not a periodic point.*

Proof. Suppose $\boldsymbol{\xi}$ is periodic. It is an exercise to show that $\mathbf{x}(t_2 + \eta, \boldsymbol{\xi})$ lies in S_P and that $\mathbf{x}(t_1 - \eta, \boldsymbol{\xi})$ lies in S_N. Consequently, $\mathbf{x}(t, \boldsymbol{\xi}) \in S_N$ for some $t > t_2 + \eta$ and it is possible to set $\tau = \inf\{t > t_2 + \eta : \mathbf{x}(t, \boldsymbol{\xi}) \in S_N\}$. Clearly, $\tau > t_2 + \eta$ and $\mathbf{x}(\tau, \boldsymbol{\xi}) \in J$ because S_P and S_N are open sets. The argument in the previous proof also works here to show that $\mathbf{x}(\tau, \boldsymbol{\xi}) \notin [\mathbf{x}(t_1, \boldsymbol{\xi}), \mathbf{x}(t_2, \boldsymbol{\xi})]_L$.

The only other possibility is that $\mathbf{x}(\tau, \boldsymbol{\xi}) = \mathbf{x}(t_3, \boldsymbol{\xi})$ for some t_3 such that $t_1 \le t_3 \le t_2$. Set $\boldsymbol{\xi}' = \mathbf{x}(t_2 + \eta, \boldsymbol{\xi})$. Then,

$$\mathbf{x}(t, \boldsymbol{\xi}') = \begin{cases} \mathbf{x}(t + t_2 + \eta, \boldsymbol{\xi}) & \text{for} \quad 0 \le t \le \tau - t_2 - \eta \\ \mathbf{x}(t - \tau + t_2 + \eta, \mathbf{x}(t_3, \boldsymbol{\xi})) & \text{for} \quad \tau - t_2 - \eta \le t \le \tau - t_3. \end{cases}$$

Observe that

$$\mathbf{x}(\tau - t_3, \boldsymbol{\xi}') = \mathbf{x}(-t_3 + t_2 + \eta, \mathbf{x}(t_3, \boldsymbol{\xi})) = \mathbf{x}(t_2 + \eta, \boldsymbol{\xi}) = \boldsymbol{\xi}',$$

and the point $\mathbf{x}(t_1 - \eta, \boldsymbol{\xi}) \in S_N$ on the orbit of $\boldsymbol{\xi}$ does not lie on the solution $\mathbf{x}(t, \boldsymbol{\xi}')$ because the $\boldsymbol{\xi}'$ is periodic and its orbit lies in $S_P \cup J$. This is a contradiction because $\boldsymbol{\xi}'$ lies on the orbit of $\boldsymbol{\xi}$, and by *Proposition 2.2* the orbits of $\boldsymbol{\xi}$ and $\boldsymbol{\xi}'$ must be identical. □

It is now possible to drop the requirement that "there is no t such that $\mathbf{q} = \mathbf{x}(t, \boldsymbol{\xi})$" from *Theorem 6.7* and obtain the simpler result.

Corollary *If* $\boldsymbol{\xi}_1 = \mathbf{x}(t_1, \boldsymbol{\xi})$ *and* $\boldsymbol{\xi}_2 = \mathbf{x}(t_2, \boldsymbol{\xi})$ *with* $t_1 < t_2$ *are distinct consecutive crossings of a local section* L *of length* 2η *at* $\mathbf{p}$ *by the solution* $\mathbf{x}(t, \boldsymbol{\xi})$*, then* $\mathbf{x}(t, \boldsymbol{\xi})$ *is in* S_P *for* $t > t_2$ *and in* S_N *for* $t < t_1$. .

Proof. Exercise.

Two substantive results are now within easy reach.

Theorem 6.9 *If* $\boldsymbol{\xi} \in \omega(\boldsymbol{\xi})$*, then either* $\boldsymbol{\xi}$ *is a fixed point or a periodic point.*

Proof. It suffices to assume that $\boldsymbol{\xi}$ is neither a fixed point nor a periodic point and derive a contradiction. If $\boldsymbol{\xi}$ is not a fixed point, then by *Theorem 6.1* there exists a local section L such that $\boldsymbol{\xi} \in L$. Because $\boldsymbol{\xi} \in \omega(\boldsymbol{\xi})$, the corollary on page 190 guarantees that there exists a next crossing $\mathbf{p} = \mathbf{x}(t_1, \boldsymbol{\xi}) \in L$ with $t_1 > 0$ and $\mathbf{x}(t, \boldsymbol{\xi}) \notin L$ for $0 < t < t_1$. Note that $\mathbf{p} \neq \boldsymbol{\xi}$ because we are assuming that $\boldsymbol{\xi}$ is not periodic.

The corollary on page 196 can now be applied to the Jordan curve

$$J = \{\mathbf{x}(t, \boldsymbol{\xi}) : 0 \le t \le t_1\} \cup [\boldsymbol{\xi}, \mathbf{p}]_L.$$

In particular, the point $\mathbf{q} = \mathbf{x}(-1, \boldsymbol{\xi})$ is in S_N and $\mathbf{x}(t, \boldsymbol{\xi})$ is in S_P for all $t > t_1$. This set up is illustrated in *Figure 6.2* with S_N as the interior of J. In this figure solutions, can only cross the segment of L between $\boldsymbol{\xi}$ and $\mathbf{p}$ from the interior to the exterior.

Because S_N is open and $\mathbf{q} \in \omega(\boldsymbol{\xi})$ by *Proposition 5.5*, there must exist $t_2 > t_1$ such that $\mathbf{x}(t_2, \boldsymbol{\xi})$ is in S_N. But this is impossible because $\mathbf{x}(t, \boldsymbol{\xi})$ is in S_P for all $t > t_1$. □

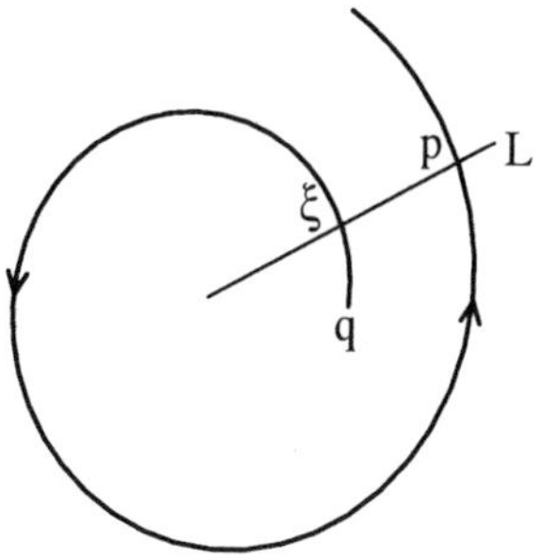

Figure 6.2: The picture proof of *Theorem 6.9*.

When a solution $\mathbf{x}(t, \boldsymbol{\xi})$ is defined for all $t \geq 0$, the *positive semiorbit of* $\boldsymbol{\xi}$ is defined by $\mathcal{O}^+(\boldsymbol{\xi}) = \{\mathbf{x}(t, \boldsymbol{\xi}) : t \geq 0\}$. The closures of semiorbits will play a large role in the rest of this section and will be denoted as follows:

$$\overline{\mathcal{O}}^+(\boldsymbol{\xi}) = \overline{\{\mathbf{x}(t, \boldsymbol{\xi}) : t \geq 0\}} \tag{6.6}$$

Naturally, if the solution is defined on $\mathbb{R}$, the *orbit* of $\boldsymbol{\xi}$ is defined by $\mathcal{O}(\boldsymbol{\xi}) = \{\mathbf{x}(t, \boldsymbol{\xi}) : t \in \mathbb{R}\}$ and the orbit closure denoted in like manner. They will be important in the next section.

Theorem 6.10 *If* $\overline{\mathcal{O}}^+(\boldsymbol{\xi})$ *is a compact set contained in* Ω *and there are no fixed points in* $\overline{\mathcal{O}}^+(\boldsymbol{\xi})$, *then* $\omega(\boldsymbol{\xi})$ *is not empty and contains only periodic orbits.*

Proof. If $\boldsymbol{\xi}$ is periodic, then $\omega(\boldsymbol{\xi}) = \mathcal{O}(\boldsymbol{\xi})$ and the conclusion holds. So it can be assumed that $\boldsymbol{\xi}$ is not periodic. From *Proposition 5.6*, it follows that $\omega(\boldsymbol{\xi})$ is a nonempty compact set and that $\mathbf{x}(t, \mathbf{q})$ is defined on $\mathbb{R}$ when $\mathbf{q} \in \omega(\boldsymbol{\xi})$.

Let $\mathbf{p}'$ be a point in $\omega(\boldsymbol{\xi})$. Then, by hypothesis, $\mathbf{p}'$ is not fixed. Since

$$\overline{\mathcal{O}}^+(\mathbf{p}') \subset \omega(\boldsymbol{\xi}) \subset \overline{\mathcal{O}}^+(\boldsymbol{\xi}),$$

$\overline{\mathcal{O}}^+(\mathbf{p}')$ is also compact and $\omega(\mathbf{p}') \neq \phi$.

Let $\boldsymbol{\zeta}$ be a point in $\omega(\mathbf{p}')$. Then $\boldsymbol{\zeta}$ is not fixed because $\omega(\mathbf{p}') \subset \omega(\boldsymbol{\xi})$. If $\boldsymbol{\zeta} = \mathbf{x}(t, \mathbf{p}')$ for some t, then $\mathbf{p}' = \mathbf{x}(-t, 0, \boldsymbol{\zeta}) \in \omega(\mathbf{p}')$ and $\mathbf{p}'$ would be periodic by *Theorem 6.9*. Thus $\boldsymbol{\zeta} \notin \mathcal{O}(\mathbf{p}')$.

Since $\boldsymbol{\zeta}$ is not a fixed point, there exists a local section L at $\boldsymbol{\zeta}$. Because $\boldsymbol{\zeta}$ is in $\omega(\mathbf{p}')$ and $\mathbf{p}'$ is not periodic, there exist distinct consecutive crossings $\mathbf{p} = \mathbf{x}(t_1, \mathbf{p}')$ and $\mathbf{q} = \mathbf{x}(t_2, \mathbf{p}')$ of L with $0 < t_1 < t_2$. Set

$$J = \{\mathbf{x}(t, \mathbf{p}') : t_1 \leq t \leq t_2\} \cup [\mathbf{p}, \mathbf{q}]_L$$

as shown in *Figure 6.3*. In this figure, S_P is the interior of J, but it is also possible for S_P to be the exterior of J as was the case in *Figure 6.2*. In either case, the argument leading to a contradiction is the same. The point $\boldsymbol{\zeta}$ lies somewhere on L, but plays no further role in the argument.

Since $\mathcal{O}(\mathbf{p}') \subset \omega(\boldsymbol{\xi})$, both $\mathbf{u} = \mathbf{x}(-\eta, \mathbf{p}) = \mathbf{x}(t_1 - \eta, \mathbf{p}')$ and $\mathbf{v} = \mathbf{x}(\eta, \mathbf{q}) = \mathbf{x}(t_2 + \eta, \mathbf{p}')$ are in $\omega(\boldsymbol{\xi})$. Because $\mathbf{v} = \mathbf{x}(\eta, \mathbf{q})$ is in S_P, there exists τ such that $\mathbf{x}(\tau, \boldsymbol{\xi})$ is in S_P, and therefore *Proposition 6.7* guarantees that $\mathbf{x}(t, \boldsymbol{\xi}) \in S_P$ for all $t \geq \tau$. In particular, it must be true that $\omega(\boldsymbol{\xi}) \subset \overline{S}_P = J \cup S_P$, contradicting the fact that $\mathbf{u} = \mathbf{x}(-\eta, \mathbf{q})$ is in both S_N and $\omega(\boldsymbol{\xi})$. □

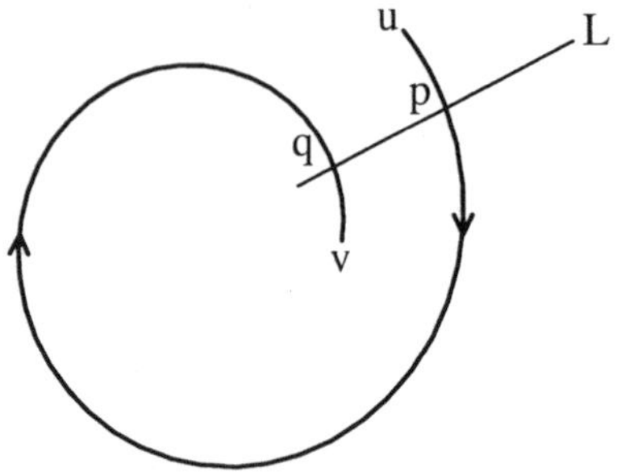

Figure 6.3: The Jordan curve used in the proof of *Theorem 6.10*.

When $\mathbf{x}(t, \boldsymbol{\xi})$ is defined for all $t \geq 0$, there are now two cases: either every point in $\omega(\boldsymbol{\xi})$ is periodic or there is at least one fixed point in $\omega(\boldsymbol{\xi})$. Both cases will be studied, but the results are more definitive in the first case.

The Poincaré return map will be used to study the case that $\omega(\boldsymbol{\xi})$ contains a periodic point by first investigating the behavior of the Poincaré return map near a periodic point.

Let $\mathbf{p}$ be a periodic point for $\dot{\mathbf{x}} = \mathbf{f}(\mathbf{x})$ and let $L = \boldsymbol{\psi}([-1, 1])$ be a local section at $\mathbf{p} = \boldsymbol{\psi}(0)$ of length 2η. Because $R(0) = 0$, where as usual R denotes the Poincaré return map, *Theorem 6.4* applies and there exists an open interval $a < s < b$, such that $a < 0 < b$ and R is continuous on (a, b) with $R[(a, b)] \subset (-1, 1)$. Furthermore, it follows from *Proposition 6.8* that $\boldsymbol{\psi}(s)$ is a periodic point for $\dot{\mathbf{x}} = \mathbf{f}(\mathbf{x})$ if and only if $R(s) = s$.

If T is the period of $\mathbf{p}$, then $\{\mathbf{x}(t, \mathbf{p}) : 0 \leq t \leq T\}$ is a Jordan curve and solutions cannot cross from the interior to the exterior or vice versa. It follows from *Proposition 6.8* that $\boldsymbol{\psi}([-1, 0))$ and $\boldsymbol{\psi}((0, 1])$ are on opposite sides of $\{\mathbf{x}(t, \mathbf{p}) : 0 \leq t \leq T\}$. Thus, $R(s) > 0$ when $s > 0$ and $R(s) < 0$ when $s < 0$. So it suffices to study R on the interval $0 \leq s < b$.

Consider s such that $0 < s < b$. If $R(s) \neq s$, then either $R(s) - s < 0$ or $R(s) - s > 0$. Because $R(s)$ is continuous, $R(s) - s$ can only change sign when $R(s) = s$ or $\boldsymbol{\psi}(s)$ is a periodic point.

Proposition 6.11 *If $R(s) - s < 0$ for $0 < s < b$, then for each s satisfying $0 < s < b$ the solution $\mathbf{x}(t, \boldsymbol{\psi}(s))$ is defined for all $t \geq 0$ and $\omega(\mathbf{x}(t, \boldsymbol{\psi}(s))) = \mathcal{O}(\mathbf{p})$.*

Proof. Consider a specific s_1 satisfying $0 < s_1 < b$. Clearly $0 < R(s_1) < s_1 < b$ and $R^2(s_1)$ is defined. By using induction, it is easy to show that $R^m(s_1)$ is defined for all positive integers m and that $R^m(s_1)$ is a decreasing sequence of positive numbers. It follows that $R^m(s_1)$ converges to some $s_2 \geq 0$. Using the continuity of R,

$$R(s_2) = R\left(\lim_{m\to\infty} R^m(s_1)\right) = \lim_{m\to\infty} R^{m+1}(s_1) = s_2.$$

Therefore, $s_2 = 0$.

It is also easy to check that there exists a sequence of positive numbers t_m such that $\boldsymbol{\psi}(R^m(s_1)) = \mathbf{x}(t_m, \boldsymbol{\psi}(s_1)) \in L^o$, and the points $\mathbf{x}(t_m, \boldsymbol{\psi}(s_1))$ and $\mathbf{x}(t_{m+1}, \boldsymbol{\psi}(s_1))$ are consecutive crossings of L. In fact,

$$t_{m+1} = t_m + \rho\big(\boldsymbol{\psi}^{-1}(\mathbf{x}(t_m, \boldsymbol{\psi}(s_1)))\big).$$

Proposition 6.2 implies that $t_{m+1} > t_m + 2\eta$, and consequently that $t_m \to \infty$ as $m \to \infty$. Therefore, the solution $\mathbf{x}(t, \boldsymbol{\psi}(s_1))$ must be defined for all $t \geq 0$ and $\mathbf{p} \in \omega(\boldsymbol{\psi}(s))$ because

$$\lim_{m\to\infty} \mathbf{x}(t_m, \boldsymbol{\psi}(s_1)) = \lim_{m\to\infty} \boldsymbol{\psi}(R^m(s_1)) = \boldsymbol{\psi}(0) = \mathbf{p}.$$

Hence, $\mathcal{O}(\mathbf{p}) \subset \omega(\boldsymbol{\psi}(s))$ because omega limit sets are invariant (*Proposition 5.5*).

It remains to prove that $\omega(\boldsymbol{\psi}(s)) \subset \mathcal{O}(\mathbf{p})$. If given $\varepsilon > 0$ there exits a positive integer M such that the distance between $\mathbf{x}(t, \boldsymbol{\psi}(s_1))$ and $\mathcal{O}(\mathbf{p}))$ is less than ε when $t \geq t_M$, then $\omega(\boldsymbol{\psi}(s_1)) \subset \mathcal{O}(\mathbf{p})$ and $\omega(\boldsymbol{\psi}(s_1)) = \mathcal{O}(\mathbf{p})$. Thus it suffices to show that given $\varepsilon > 0$ there exits M such that the distance between $\mathbf{x}(t, \boldsymbol{\psi}(s_1))$ and $\mathcal{O}(\mathbf{p}))$ is less than ε when $t \geq t_M$. The argument requires continuity in initial conditions again and is a little delicate.

Set $\mathbf{q}_1 = \boldsymbol{\psi}(s_1)$ for convenience and notice that the sequence $\|\mathbf{x}(t_m, \mathbf{q}_1) - \mathbf{p}\|$ is decreasing to 0. Choose ε' such that the Euclidean ball $\{\mathbf{x} : \|\mathbf{x} - \mathbf{p}\| < \varepsilon'\} \subset \Gamma^o$. Given $0 < \varepsilon < \varepsilon'$ there exist $\delta > 0$ such that $\|\mathbf{x}(t, \boldsymbol{\xi}) - \mathbf{x}(t, \mathbf{p})\| < \varepsilon$ for $0 \leq t \leq T + \eta$ when $\|\boldsymbol{\xi} - \mathbf{p}\| < \delta$. (As above, T is the period and 2η is the length of the flow box.)

There exists M such that $\|\mathbf{x}(t_M, \mathbf{q}_1) - \mathbf{p}\| < \delta$. Consider any $m \geq M$. Clearly, $\|\mathbf{x}(t_m, \mathbf{q}_1) - \mathbf{p}\| < \delta$. Then,

$$\|\mathbf{x}(t + t_m, \mathbf{q}_1) - \mathbf{x}(t, \mathbf{p})\| = \|\mathbf{x}(t, \mathbf{x}(t_m, \mathbf{q}_1)) - \mathbf{x}(t, \mathbf{p})\| < \varepsilon$$

for $0 \leq t \leq T + 2\eta$. In particular, $\|\mathbf{x}(T + t_m, \mathbf{q}_1) - \mathbf{p}\| < \varepsilon'$, and hence

$$t_{m+1} - t_m = \rho\big(\boldsymbol{\psi}^{-1}(\mathbf{x}(t_m, \psi(s_1)))\big) \leq T + \eta.$$

Therefore, if $m \geq M$, then

$$\|\mathbf{x}(t + t_m, \mathbf{q}_1) - \mathbf{x}(t, \mathbf{p})\| < \varepsilon$$

f or $0 \leq t \leq t_{m+1} - t_m$.

Every $t' \geq t_M$ can be written as $t' = t + t_m$ for some $m \geq M$ and t satisfying $0 \leq t \leq t_{m+1} - t_m$. Therefore, if $t' \geq t_M$, then for some t

$$\|\mathbf{x}(t', \boldsymbol{\psi}(s_1)) - \mathbf{x}(t, \mathbf{p})\| = \|\mathbf{x}(t + t_m, \mathbf{q}_1) - \mathbf{x}(t, \mathbf{p})\| < \varepsilon$$

and the proof is finished. □

Theorem 6.12 (Poincaré -Bendixson) *If $\overline{\mathcal{O}}^+(\boldsymbol{\xi})$ is a compact subset of Ω and $\overline{\mathcal{O}}^+(\boldsymbol{\xi}) \cap \mathcal{F} = \phi$, then $\omega(\boldsymbol{\xi})$ consists of a single periodic orbit.*

Proof. By *Theorem 6.10*, $\omega(\boldsymbol{\xi})$ contains only periodic points. The strategy will be to show that *Proposition 6.11* applies, and therefore the omega limit set consists of exactly one periodic orbit.

Let $\mathbf{p}$ be a periodic point in $\omega(\boldsymbol{\xi})$ and let $L = \boldsymbol{\psi}([-1, 1])$ be a local section at $\mathbf{p} = \boldsymbol{\psi}(0)$ of length 2η. As before $\boldsymbol{\psi}([-1, 0))$ and $\boldsymbol{\psi}((0, 1]$ are on opposite sides of $\{\mathbf{x}(t, \mathbf{p}) : 0 \leq t \leq T\}$. The Poincaré return map R maps some open interval (a, b) into $(-1, 1)$ and satisfies $R(s) > 0$ when $s > 0$ and $R(s) < 0$ when $s < 0$. Since $\mathbf{x}(t, \boldsymbol{\xi})$ must stay in either the interior of the Jordan curve $\mathcal{O}(\mathbf{p})$ or the exterior, it can be assumed without loss of generality that $\mathbf{x}(t, \boldsymbol{\xi})$ can only intersect $\boldsymbol{\psi}((0, 1])$.

Because $\mathbf{p} \in \omega(\boldsymbol{\xi})$, there exists an infinite sequence $\mathbf{x}(t_m, \boldsymbol{\xi}) = \boldsymbol{\psi}(s_m)$ of consecutive crossings of $\boldsymbol{\psi}((0, 1])$ at positive times t_m satisfying $2\eta \leq t_{m+1} - t_m$. If there does not exist M such that $0 < s_M < b$ and $R(s_M) - s_M < 0$, then $0 < \|\mathbf{x}(t_1, \boldsymbol{\xi}) - \mathbf{p}\| < \|\mathbf{x}(t_m, \boldsymbol{\xi}) - \mathbf{p}\|$ for $m > 1$ and $\mathbf{x}(t, \boldsymbol{\xi})$ does not enter the open set

$$\{\mathbf{x}(t, \boldsymbol{\psi}(s)) : -\eta < t < \eta \text{ and } -1 < s < s_1\}$$

containing $\mathbf{p}$. This is impossible because $\mathbf{p}$ is in $\omega(\boldsymbol{\xi})$. Therefore, there exists M such that $0 < s_M < b$ and $R(s_M) - s_M < 0$.

What can happen between 0 and s_M? Either $R(s) - s < 0$ when $0 < s \leq s_M$ or there exists a σ such that $R(\sigma) - \sigma = 0$. In the latter case, $\boldsymbol{\psi}(\sigma)$ is periodic and $\boldsymbol{\xi}$ and $\mathbf{p}$ are on opposites sides of the Jordan curve $\mathcal{O}(\boldsymbol{\psi}(\sigma))$, which is impossible because $\mathbf{p}$ is in $\omega(\boldsymbol{\xi})$. Therefore, $R(s) - s < 0$ when $0 < s \leq s_M$ and *Proposition 6.11* can be applied on $0 < s < b'$ for some b' slightly larger than s_M. Consequently, $\omega(\boldsymbol{\psi}(s_M)) = \mathcal{O}(\mathbf{p})$ and finally $\omega(\boldsymbol{\xi}) = \omega(\mathbf{x}(t_M, \boldsymbol{\xi})) = \omega(\boldsymbol{\psi}(s_M))$ because the omega limit set is the same for all points $\mathbf{x}(t, \boldsymbol{\xi})$. □

As a result of the above theorem a compact omega limit set in the plane is either a periodic point or contains fixed points. A periodic orbit that is the omega limit set for a point not on the periodic orbit is called a *limit cycle*. The next theorem provides further evidence of why limit cycles are of interest.

Theorem 6.13 *If $\mathbf{p}$ is a periodic point, then the set*

$$\{\boldsymbol{\xi} : \boldsymbol{\xi} \notin \mathcal{O}(\mathbf{p}) \text{ and } \omega(\boldsymbol{\xi}) = \mathcal{O}(\mathbf{p})\}$$

is an open set.

Proof. Let $\boldsymbol{\xi}$ be a point in Ω such that $\omega(\boldsymbol{\xi}) = \mathcal{O}(\mathbf{p})$. (If no such point exits, the result is trivial because the empty set is an open set.) The proof is a continuation of the proof of *Theorem 6.12* using exactly the same set up and notation

Specifically, $\mathbf{x}(t_M, \boldsymbol{\xi}) = \boldsymbol{\psi}(s_M)$ is in the set

$$U = \{\mathbf{x}(t, \boldsymbol{\psi}(s)) : -\eta < t < \eta \text{ and } 0 < s < b'\}$$

which is an open set by *Theorem 6.1.*

Because $R(s) - s < 0$ on $0 < s < b'$, it follows from *Proposition 6.11* that $\omega(\mathbf{q}) = \mathcal{O}(\mathbf{p})$ for every $\mathbf{q} \in U$. By continuity in initial conditions there exists $\delta > 0$ such that $\mathbf{x}(t_M, \mathbf{q}) \in U$, when $\|\mathbf{q} - \boldsymbol{\xi}\| < \delta$. Thus, $\omega(\mathbf{q}) = \mathcal{O}(\mathbf{p})$ when $\|\mathbf{q} - \boldsymbol{\xi}\| < \delta$. $\Box$

Theorem 6.14 *Suppose* $\overline{\mathcal{O}}^+(\boldsymbol{\xi})$ *is a compact subset of* Ω *and* $\omega(\boldsymbol{\xi}) \cap \mathcal{F} \neq \phi$. *If* $\mathbf{p} \in \omega(\boldsymbol{\xi})$ *and* $\mathbf{p} \notin \mathcal{F}$, *then both* $\omega(\mathbf{p})$ *and* $\alpha(\mathbf{p})$ *are subsets of* $\mathcal{F}$. *Moreover, if* $\omega(\boldsymbol{\xi}) \cap \mathcal{F}$ *is finite, then* $\omega(\mathbf{p})$ *and* $\alpha(\mathbf{p})$ *each consist of exactly one fixed point.*

Proof. Let $\mathbf{q}$ be a point in $\omega(\mathbf{p})$ with $\mathbf{p} \in \omega(\boldsymbol{\xi})$ and $\mathbf{p} \notin \mathcal{F}$. If $\mathbf{p}$ is periodic, then *Theorem 6.12* implies that $\omega(\boldsymbol{\xi}) = \mathcal{O}(\mathbf{p})$ and $\omega(\boldsymbol{\xi}) \cap \mathcal{F} = \phi$. Therefore, $\mathbf{p}$ is not periodic.

If $\mathbf{q} \in \mathcal{O}(\mathbf{p})$, then $\mathbf{p} \in \omega(\mathbf{p})$ and $\mathbf{p}$ is periodic by *Theorem 6.9*, but $\mathbf{p}$ cannot be periodic. Therefore, $\mathbf{q} \notin \mathcal{O}(\mathbf{p})$.

Suppose $\mathbf{q} \notin \mathcal{F}$. Then there exists a local section L at $\mathbf{q}$ of length 2η. Because $\mathbf{q} \in \omega(\mathbf{p})$, there exist consecutive crossings $\mathbf{p}_1 = \mathbf{x}(t_1, \mathbf{p})$ and $\mathbf{p}_2 = \mathbf{x}(t_2, \mathbf{p})$ with $t_1 < t_2$, and

$$J = \{\mathbf{x}(t, \mathbf{p}) : t_1 \leq t \leq t_2\} \cup [\mathbf{p}_1, \mathbf{p}_2]_L$$

is a Jordan curve.

Observe that $\mathbf{x}(t_2 + \eta, \mathbf{p})$ is in $S_P \cap \omega(\boldsymbol{\xi})$ and that $\mathbf{x}(t_1 - \eta, \mathbf{p})$ is in $S_N \cap \omega(\boldsymbol{\xi})$. Hence, there exists τ such that $\mathbf{x}(\tau, \boldsymbol{\xi}) \in S_P$. Consequently, $\mathbf{x}(t + \tau, \boldsymbol{\xi}) \in S_P$ for all $t \geq 0$ and $\omega(\boldsymbol{\xi}) \subset \overline{S}_P$. But this contradicts $\mathbf{x}(t_1 - \eta, \mathbf{p}) \in S_N \cap \omega(\boldsymbol{\xi})$. Therefore, $\mathbf{q}$ is a fixed point and $\omega(\mathbf{p}) \subset \mathcal{F}$. A similar argument shows that $\alpha(\mathbf{p}) \subset \mathcal{F}$.

The second part follows immediately from the fact that $\omega(\boldsymbol{\xi})$ is connected. (See *Proposition 5.6* and *Exercise 4* on page 156.) $\Box$

When $\omega(\mathbf{p}) \cap \mathcal{F}$ is not finite, $\omega(\mathbf{q})$ and $\alpha(\mathbf{q})$ can be complicated sets that are not very amenable to analysis because there are no local sections to put to good use.

There are, of course, results for the alpha limit sets parallel to those in this section for the omega limit sets. For example, if $\boldsymbol{\xi} \in \alpha(\boldsymbol{\xi})$, then $\boldsymbol{\xi}$ is either a fixed point or a periodic point.

EXERCISES

1. Let $\mathbf{f} : \Omega \to \mathbb{R}^2$ be a locally Lipschitz vector field on Ω an open subset of $\mathbb{R}^2$, the usual set up in this section and these exercises. Suppose the solution $\mathbf{x}(t, \boldsymbol{\xi})$ of $\dot{\mathbf{x}} = \mathbf{f}(\mathbf{x})$ is defined on $\mathbb{R}$. Show that if $\omega(\boldsymbol{\xi}) \cap \alpha(\boldsymbol{\xi}) \neq \phi$, then $\omega(\boldsymbol{\xi}) \cap \alpha(\boldsymbol{\xi}) \subset \mathcal{F}$.

2. Under the hypothesis of *Proposition 6.8*, show that $\mathbf{x}(t_2 + \eta, \boldsymbol{\xi})$ lies in S_P and that $\mathbf{x}(t_1 - \eta, \boldsymbol{\xi})$ lies in S_N.

3. Prove the corollary on page 196.

4. Show that there are at most a countable number of limit cycles for a given locally Lipschitz vector field on $\Omega \subset \mathbb{R}^2$.

5. Consider the planar autonomous system of differential equations

$$\begin{aligned} \dot{x} &= y + x + \psi(r^2) \\ \dot{y} &= -x + y\psi(r^2) \end{aligned}$$

where $x^2 + y^2 = r^2$ and ψ is a real-valued continuous function defined on $[0, \infty)$ with a continuous derivative for $r > 0$. Use type system to construct examples of the following:

   (a) A system with exactly two limit cycles and no other periodic orbits.

   (b) A system with a sequence of limit cycles such that every point on the plane is inside a limit cycle.

   (c) A system with a periodic orbit that is not a limit cycle but is the limit of limit cycles.

6. Show that the system

$$\begin{aligned} \dot{x} &= x + yx^3 \\ \dot{y} &= -x + y - y^3 \end{aligned}$$

has at least one periodic orbit. Hint: Let $V(x, y) = x^2 + y^2$ and use $\dot{V}$ to help apply the results in the section.

7. The planar system of differential equations

$$\begin{aligned} \dot{x} &= \sin x \cos y - (\sin x \cos x)/2 \\ \dot{y} &= -\sin y \cos x + (\sin y \cos y)/2, \end{aligned}$$

appeared in a paper by Anosov and defines a flow on $\mathbb{R}^2$.

   (a) Show that the fixed points are the union of the two sets $\{(m\pi, n\pi) : m, n \in \mathbb{Z}\}$ and $\{(m\pi + \pi/2, n\pi + \pi/2) : m, n \in \mathbb{Z}\}$.

   (b) Show that the lines $x = k\pi$ and $y = k\pi$ are invariant sets for every integer k and construct the phase portrait on them.

(c) Use the function $V(x, y) = \sin x \sin y$ and the ideas from Section 5.1 to complete the phase portrait of this system.

8. Consider the system in the previous exercise restricted to the invariant set $\Omega = \{(x, y) : 0 < y < 1\}$. Use the change of variables $(x, y) = (u, \pi/2 + \arctan v)$ to construct a system of differential equations on $\mathbb{R}^2$ and construct its phase portrait. Determine the omega and alpha limit sets of the orbits and show that they are not all connected. Why doesn't this example violate the conclusions of *Theorem 5.6*?

9. A subset W of $\mathbb{R}^d$ is arcwise connected if for any two points $\mathbf{p}$ and $\mathbf{q}$ in W there exist a curve $\gamma : [0, 1] \to W$ such that $\gamma(0) = \mathbf{p}$ and $\gamma(1) = \mathbf{q}$. (For $\mathbb{R}^d$ this definition is equivalent to the more usual definition requiring that γ also be one-to-one, and easier to work with in the present context. For a full discussion see these ideas see Chapter 3 of [19].) Let U be an open set in $\mathbb{R}^d$ and let $\mathbf{p}$ be a point in W. Show that the set

$$\{\mathbf{q} \in U : \text{ there exists } \gamma : [0, 1] \to U \text{ with } \gamma(0) = \mathbf{p} \text{ and } \gamma(1) = \mathbf{q}\}$$

is also an open set. Prove that every open set in $\mathbb{R}^d$ is the disjoint union of open arcwise connected sets.

10. With $\mathbf{f}$ as above, suppose $\boldsymbol{\xi}$ is periodic and the interior of $\mathcal{O}(\boldsymbol{\xi})$ is contained in Ω. Show that if there are no periodic points in the interior of $\mathcal{O}(\boldsymbol{\xi})$, then there is a fixed point in the interior of $\mathcal{O}(\boldsymbol{\xi})$. (Proving that there is always a fixed point inside a periodic orbit in the plane is best done with the Brouwer fixed-point theorem and *Exercise 5* on page 240.)

6.3 Recurrence

The phenomenon of solutions eventually returning close to their initial positions goes under the name *recurrence.* There are many variations of recurrence that have been investigated in different aspects of the study of dynamical systems. This section introduces a few of the simpler notions of recurrence through concrete examples. The goal is to show that the behavior of solutions in general is much richer and more interesting than what is possible on the plane.

The most basic forms of recurrent behavior are fixed points, where a solution never leaves its starting point, and periodic points, where a solution regularly comes back precisely to its starting point. The first potentially interesting form of recurrence is when $\boldsymbol{\xi} \in \omega(\boldsymbol{\xi})$ and $\boldsymbol{\xi}$ is not periodic. The results in the previous section show that this possibility does not occur in the plane because $\boldsymbol{\xi} \in \omega(\boldsymbol{\xi})$ implies that $\boldsymbol{\xi}$ is fixed or periodic. Things are quite different in higher dimensions and this concept deserves a name.

The point $\boldsymbol{\xi}$ is said to be *positively recurrent* if $\boldsymbol{\xi} \in \omega(\boldsymbol{\xi})$, *negatively recurrent* if $\boldsymbol{\xi} \in \alpha(\boldsymbol{\xi})$, and *recurrent* if $\boldsymbol{\xi} \in \alpha(\boldsymbol{\xi}) \cap \omega(\boldsymbol{\xi})$. This definition includes fixed points and periodic points as recurrent points and like many concepts of recurrence is hierarchial. The main results of this section are that various kinds of recurrent points that are not periodic or fixed do occur in the study of autonomous differential equations and specific examples will be exhibited. Before constructing any examples, it is important to note that points on a solution have the same recurrence properties as the initial point.

Remark *If $\boldsymbol{\xi}$ is positively recurrent, negatively recurrent, or recurrent, then $\mathbf{x}(t, \boldsymbol{\xi})$ is positively recurrent, negatively recurrent, or recurrent accordingly.*

Proof. Suppose $\boldsymbol{\xi}$ is positively recurrent. Then using *Proposition 5.5*, $\boldsymbol{\xi} \in \omega(\boldsymbol{\xi})$ implies that $\mathbf{x}(t, \boldsymbol{\xi}) \in \omega(\boldsymbol{\xi}) = \omega(\mathbf{x}(t, \boldsymbol{\xi}))$ and $\mathbf{x}(t, \boldsymbol{\xi})$ is positively recurrent. The rest of the proof is similar. $\square$

Turning next to the construction of examples, consider the linear differential equation $\dot{\mathbf{x}} = A\mathbf{x}$ with constant coefficients on $\mathbb{R}^4$ and

$$A = \begin{bmatrix} 0 & 2\pi\alpha & 0 & 0 \\ -2\pi\alpha & 0 & 0 & 0 \\ 0 & 0 & 0 & 2\pi \\ 0 & 0 & -2\pi & 0 \end{bmatrix}. \tag{6.7}$$

The matrix A is skew symmetric, so e^{At} is orthogonal and the results about orthogonal matrices in Section 4.5 will be helpful. For example, it follows from the lemma on page 143 that

$$e^{At} = \begin{bmatrix} \cos(2\pi\alpha t) & \sin(2\pi\alpha t) & 0 & 0 \\ -\sin(2\pi\alpha t) & \cos(2\pi\alpha t) & 0 & 0 \\ 0 & 0 & \cos(2\pi t) & \sin(2\pi t) \\ 0 & 0 & -\sin(2\pi t) & \cos(2\pi t) \end{bmatrix} \tag{6.8}$$

because it was shown at the end of Section 4.5 that

$$\exp \begin{bmatrix} 0 & \theta \\ -\theta & 0 \end{bmatrix} = \begin{bmatrix} \cos\theta & \sin\theta \\ -\sin\theta & \cos\theta \end{bmatrix}. \tag{6.9}$$

The solutions to $\dot{\mathbf{x}} = A\mathbf{x}$ are, of course, defined on $\mathbb{R}$ and $\mathbf{x}(t, \boldsymbol{\xi}) = e^{At}\boldsymbol{\xi}$ defines a flow on $\mathbb{R}^4$. It will be convenient to use the coordinates (x, y, u, v) on $\mathbb{R}^4$.

Let $\boldsymbol{\xi}' = (0, 0, 1, 0)$. The subspace V of $\mathbb{R}^4$ orthogonal to $A\boldsymbol{\xi}'$ is the x, y, u subspace or $V = \{(x, y, u, v) : v = 0\}$. Note that $\boldsymbol{\xi}' \in V$. *Theorem 6.1* implies that for some $r > 0$ the set $\{(x, y, u, 0) : x^2 + y^2 + (u - 1)^2 \leq r^2\}$ is a local cross section at $\boldsymbol{\xi}'$. In this case, however, there is a much larger section and an easily calculated Poincaré return map that can be used. Moreover, $\boldsymbol{\xi}'$ is a periodic point and not really of any interest.

Set $L = \{(x, y, u, v) : u > 0 \text{ and } v = 0\}$ and let $\mathbf{w} = (x, y, u, 0)$ be a generic point in L. Then, $e^{At}\mathbf{w}$ is precisely

$$(x\cos(2\pi\alpha t) + y\sin(2\pi\alpha t), -x\sin(2\pi\alpha t) + y\cos(2\pi\alpha t), u\cos(2\pi t), -u\sin(2\pi t)).$$

Consequently, $e^{At}\mathbf{w} \in L$ if and only if $t \in \mathbb{Z}$, the integers. Although L is not a local section in the exact sense of Section 6.1, solutions do cross L from the $v < 0$ side to the $v > 0$ side. In fact, $e^{At}\mathbf{w}$ is on the negative side for $-\pi < t < 0$ and on the positive side for $0 < t < \pi$.

The next step is to calculate the Poincaré return map for L using the notations and definitions from Section 6.1. We can let $\boldsymbol{\psi}(x, y, u) = T(x, y, u) = (x, y, u, 0)$ because $\boldsymbol{\xi}'$ is in the subspace V. Instead of defining the Poincaré return map just on a disk, here it can be defined on $D = \{(x, y, u) : u > 0\} \subset \mathbb{R}^3$. It follows from $e^{At}\mathbf{w} \in L$ if and only if $t \in \mathbb{Z}$ that the first return time $\rho(x, y, u) \equiv 1$ on D. Therefore,

$$\begin{aligned} R(x,y,u) &= \boldsymbol{\psi}^{-1}\mathbf{x}(1, \boldsymbol{\psi}(x,y,u)) \\ &= \boldsymbol{\psi}^{-1}\left(e^{A}\boldsymbol{\psi}(x,y,u)\right) \\ &= \boldsymbol{\psi}^{-1}\big(x\cos(2\pi\alpha) + y\sin(2\pi\alpha), -x\sin(2\pi\alpha) + y\cos(2\pi\alpha), u, 0)\big) \\ &= \big(x\cos(2\pi\alpha) + y\sin(2\pi\alpha), -x\sin(2\pi\alpha) + y\cos(2\pi\alpha), u\big) \\ &= \begin{bmatrix} \cos(2\pi\alpha) & \sin(2\pi\alpha) & 0 \\ -\sin(2\pi\alpha) & \cos(2\pi\alpha) & 0 \\ 0 & 0 & 1 \end{bmatrix} \begin{pmatrix} x \\ y \\ u \end{pmatrix}. \end{aligned}$$

Therefore, the Poincaré return map on D is a rotation of $2\pi\alpha$ radians about the u-axis.

Since the Poincaré return map fixes the points on the u-axis, its behavior is really governed by the planar rotation

$$R_\alpha(x,y) = \begin{bmatrix} \cos(2\pi\alpha) & \sin(2\pi\alpha) \\ -\sin(2\pi\alpha) & \cos(2\pi\alpha) \end{bmatrix} \begin{pmatrix} x \\ y \end{pmatrix}. \tag{6.10}$$

The next lemma will be used to analyze the Poincaré return map.

Lemma *If $R_\alpha^m(a,b) = (a,b)$ for some $(a,b) \neq (0,0)$ and $m \neq 0$, then α is rational and $R_\alpha^m = \mathrm{I}$.*

Proof. The equation *(6.9)* implies two important facts about R_α^m. First,

$$R_\alpha^m(x,y) = \begin{bmatrix} \cos(2\pi\alpha m) & \sin(2\pi\alpha m) \\ -\sin(2\pi\alpha m) & \cos(2\pi\alpha m) \end{bmatrix} \begin{pmatrix} x \\ y \end{pmatrix}$$

because $(e^A)^m = e^{mA}$. Second, the eigenvalues of the linear transformation R_α^m are $\cos(2\pi\alpha m) \pm i\sin(2\pi\alpha m)$ because $e^{\pm i2\pi\alpha m} = \cos(2\pi\alpha m) \pm i\sin(2\pi\alpha m)$.

If $R_\alpha^m(a,b) = (a,b)$ and $(a,b) \neq (0,0)$, then 1 is an eigenvalue of R_α^m implying that $\cos(2\pi\alpha m) = 1$ and $\sin(2\pi\alpha m) = 0$. Hence, $2\pi\alpha m = 2k\pi$ for some $k \in \mathbb{Z}$ and $\alpha = k/m$. $\square$

Returning to the differential equation $\dot{\mathbf{x}} = A\mathbf{x}$ where A is given by *(6.7)*, if the point $(a,b,c,0) \in L$ is periodic and $a^2 + b^2 \neq 0$, then $R^m(a,b,c) = (a,b,c)$ and $R_\alpha^m(a,b) = (a,b)$ for some m. The lemma applies because $a^2 + b^2 \neq 0$.

Consequently, a single periodic point of the form $(a, b, c, 0) \in L$ with $a^2 + b^2 \neq 0$ implies that α is rational and every point in $\mathbb{R}^4$ is periodic.

Therefore, when α is irrational, the solution starting at a point $\boldsymbol{\xi} = (a, b, c, 0)$ in L with $a^2 + b^2 \neq 0$ is not periodic and cannot leave the compact set

$$\{(x, y, u, v) : x^2 + y^2 + u^2 + v^2 = a^2 + b^2 + c^2\}$$

because the solution is of the form $e^{At}\boldsymbol{\xi}$ with e^{At} orthogonal. Determining the alpha and omega limit sets of such points will be the focus of our attention.

Proposition 6.15 *If α is irrational and $(a, b) \neq (0, 0)$, then*

$$\overline{\{R_\alpha^m(a, b) : m \in \mathbb{Z}\}} = \{(x, y) : x^2 + y^2 = a^2 + b^2\} = C.$$

Proof. If the conclusion is false, there exists an open arc I of the circle C such that $R_\alpha^m(a, b) \notin I$ for all $m \in \mathbb{Z}$. Furthermore, there is clearly a largest such arc in the circle C. Assume I is such a maximal interval. The proof depends heavily on the facts that $R_\alpha(C) = C$ and that rotations preserve arc length; in particular, $R_\alpha^m(I)$ and I are arcs of C of equal length.

Consider a nonzero integer n. It is easy to see that $R_\alpha^m(a, b) \notin I$ for all $m \in \mathbb{Z}$ if and only if $R_\alpha^m(a, b) \notin R_\alpha^n(I)$ for all $m \in \mathbb{Z}$. If $R_\alpha^n(I) \cap I \neq \phi$, then $R_\alpha^n(I) \cup I$ is an interval and it follows that $R_\alpha^m(a, b) \notin R_\alpha^n(I) \cup I$ because I is assumed to have this property. Since I is a maximal interval with this property, $R_\alpha^n(I)$ would coincide with I, if $R_\alpha^n(I) \cap I \neq \phi$. Therefore, either $R_\alpha^n(I) = I$ or $R_\alpha^n(I) \cap I = \phi$.

Suppose that $R_\alpha^n(I) = I$ for some $n \neq 0$. Clearly, R_α^n must map end points of I to end points of I. Let (a_1, b_1) and (a_2, b_2) be the end points of $I = R_\alpha^n(I)$. There are two possibilities. Either $R_\alpha^n(a_1, b_1) = (a_1, b_1)$ and $R_\alpha^n(a_2, b_2) = (a_2, b_2)$ or $R_\alpha^n(a_1, b_1) = (a_2, b_2)$ and $R_\alpha^n(a_2, b_2) = (a_1, b_1)$. The latter implies that $R_\alpha^{2n}(a_1, b_1) = (a_1, b_1)$ and $R_\alpha^{2n}(a_2, b_2) = (a_2, b_2)$. Thus in both cases the lemma implies that α is rational and consequently $R_\alpha^n(I) \cap I = \phi$ for all nonzero integers n.

Next suppose that $R_\alpha^m(I) \cap R_\alpha^n(I) \neq \phi$ for some $m \neq n$. It follows that $R_\alpha^{m-n}(I) \cap I \neq \phi$, which is impossible. Therefore, $R_\alpha^m(I)$, $m \in \mathbb{Z}$, is an infinite set of disjoint arcs of C each having the same positive arc length, contradicting the fact that the circumference of C is finite. □

When the 4×4 matrix A is given by equation *(6.7)*, then the functions

$$E_1(x, y, u, v) = x^2 + y^2$$

and

$$E_2(x, y, u, v) = u^2 + v^2$$

are both integrals of $\dot{\mathbf{x}} = A\mathbf{x}$. It is easy to check that if either of these integrals is zero along a solution, then the solution is periodic.

Let r_1 and r_2 be two positive real numbers and define the circles:

$$C_1 = \{(x, y) : x^2 + y^2 = r_1^2\}$$

in the xy-plane and

$$C_2 = \{(u,v) : u^2 + v^2 = r_2^2\}$$

in the uv-plane. Then

$$\begin{aligned} C_1 \times C_2 &= \{(x,y,u,v) : x^2 + y^2 = r_1^2 \text{ and } u^2 + v^2 = r_2^2\} \\ &= \{(x,y,u,v) : E_1(x,y,u,v) = r_1^2 \text{ and } E_2(x,y,u,v) = r_2^2\}. \end{aligned}$$

If $\boldsymbol{\xi} \in C_1 \times C_2$, then $\mathbf{x}(t,\boldsymbol{\xi}) \in C_1 \times C_2$ for all t. Clearly $C_1 \times C_2$ is closed and bounded, and thus compact. It follows that

$$\overline{\mathcal{O}}(\boldsymbol{\xi}) \subset C_1 \times C_2$$

and $\overline{\mathcal{O}}(\boldsymbol{\xi})$ is a compact set.

The Cartesian product of two circles like $C_1 \times C_2$ is called a *torus*. Because circles are one dimensional, the torus is two dimensional but is not planar. When imbedded in $\mathbb{R}^3$ it looks like a doughnut.

The intersection of the torus $C_1 \times C_2$ and the section L is the circle

$$C = \{(x,y,u,0) : x^2 + y^2 = r_1^2 \text{ and } u = r_2\}$$

in $\mathbb{R}^4$. If $\boldsymbol{\xi} = (a,b,c,d)$ is a point in $C_1 \times C_2$, then (c,d) can be written as $r_2(\cos\theta, \sin\theta)$ with $0 < \theta \le 2\pi$. Setting $\tau = (\theta - 2\pi)/2\pi$, it is easy to check that $\mathbf{x}(\tau,\boldsymbol{\xi}) \in C$ and $-1 < \tau \le 0$. Since $\mathbf{x}(-\tau, \mathbf{x}(\tau,\boldsymbol{\xi})) = \boldsymbol{\xi}$,

$$C_1 \times C_2 = \{\mathbf{x}(t,\mathbf{w}) : \mathbf{w} \in C \text{ and } 0 \le t < 1\}.$$

Proposition 6.16 *Let A be a 4×4 matrix given by equation (6.7) with α irrational, let $\mathbf{x}(t,\boldsymbol{\xi}) = e^{At}\boldsymbol{\xi}$ be the flow defined on $\mathbb{R}^4$ by $\dot{\mathbf{x}} = A\mathbf{x}$, and let $\boldsymbol{\xi} = (a,b,c,d)$ be a point in $\mathbb{R}^4$. If $E_1(a,b,c,d) = r_1 > 0$ and $E_2(a,b,c,d) = r_2 > 0$, then*

$$\overline{\mathcal{O}}(\boldsymbol{\xi}) = C_1 \times C_2.$$

Moreover, $\overline{\mathcal{O}}(\boldsymbol{\xi})$ is a fixed point, periodic orbit, or a torus according as both $E_1(a,b,c,d) = 0$ and $E_2(a,b,c,d) = 0$, either $E_1(a,b,c,d) = 0$ or $E_2(a,b,c,d) = 0$ but not both, or both $E_1(a,b,c,d) \neq 0$ and $E_2(a,b,c,d) \neq 0$.

Proof. The preceding characterization of $C_1 \times C_2$ as $\{\mathbf{x}(t,\mathbf{w}) : \mathbf{w} \in C \text{ and } 0 \le t < 1\}$ simplifies the proof in two ways. First, it is sufficient to prove the result when $c = r_2$ and $d = 0$ because $\mathcal{O}(\boldsymbol{\xi}) = \mathcal{O}(\mathbf{x}(t,\boldsymbol{\xi}))$. Second, it then suffices to show that

$$C \subset \overline{\mathcal{O}}(\boldsymbol{\xi})$$

because $\mathbf{x}(t,\mathbf{p})$ is in $\overline{\mathcal{O}}(\boldsymbol{\xi})$ when $\mathbf{p}$ is in $\overline{\mathcal{O}}(\boldsymbol{\xi})$.

It follows by induction from the defining equation for the Poincaré return map,

$$R(x,y,u) = \boldsymbol{\psi}^{-1}\mathbf{x}(1, \boldsymbol{\psi}(x,y,u)),$$

that

$$R^m(x,y,u) = \boldsymbol{\psi}^{-1}(\mathbf{x}(m, \boldsymbol{\psi}(x,y,u))$$

or

$$\psi(R^m(x,y,u)) = \mathbf{x}(m,(x,y,u,0))$$

because $\mathbf{x}(m,\boldsymbol{\xi}) = \mathbf{x}(1,\mathbf{x}(m-1,\boldsymbol{\xi}))$.

Given $(x,y,r_2,0) \in C$, by *Proposition 6.15* there exists a sequence of integers m_k such that

$$\lim_{k\to\infty} R^{m_k}(a,b,r_2) = (x,y,r_2)$$

hence

$$\lim_{k\to\infty} \mathbf{x}(m_k,(a,b,r_2,0)) = \lim_{k\to\infty} \psi\big(R^{m_k}(a,b,r_2)\big) = (x,y,r_2,0).$$

Thus every point in C is in the orbit closure of $(a,b,r_2,0)$ and

$$C \subset \overline{\mathcal{O}}(a,b,r_2,0).$$

The second part of the statement of the proposition just gathers together the information about all the orbit closures in a single statement. □

The set $C_1 \times C_2$ as above has the property that

$$C_1 \times C_2 = \overline{\mathcal{O}}(\boldsymbol{\xi})$$

for every $\boldsymbol{\xi} \in C_1 \times C_2$. In other words, there are no smaller orbit closures within $C_1 \times C_2$, which is, in fact, a strong form of recurrence, that will be discussed in the context of flows.

Let $\Phi : \mathbb{R} \times \Omega \to \Omega$ be a flow on Ω. (See page 65 for the definition of a flow.) The *orbit* of a point $\mathbf{p}$ for a flow Φ is the obvious generalization, namely, $\mathcal{O}(\mathbf{p}) = \{\Phi(t,\mathbf{p}) : t \in \mathbb{R}\}$. The definitions of the omega and alpha limit sets on page 152 also extend to flows in the obvious way. A subset W of Ω is an *invariant set* for the flow provided $\Phi(t,\mathbf{x}) \in W$ for all t whenever $\mathbf{x} \in W$.

A nonempty set M is a *minimal set* for a flow if

$$\overline{\mathcal{O}}(\mathbf{p}) = M$$

for every $\mathbf{p} \in M$. Obviously, fixed points and periodic orbits are minimal sets. Moreover, the orbit of $\mathbf{p}$ is minimal if both $\alpha(\mathbf{p})$ and $\omega(\mathbf{p})$ are empty. Consequently, compact minimal sets are of the most interest. As a consequence of *Proposition 6.16* there are compact minimal sets other than periodic orbits and fixed points. Minimal sets have been studied extensively, often in the more general setting of a topological group acting on a metric space or a compact Hausdorff space. See [3] for example.

Remark *If M is a compact minimal set of a flow, then the following hold:*

(a) If W is a closed nonempty invariant subset of M, then $W = M$.

(b) $\omega(\mathbf{p}) = M = \alpha(\mathbf{p})$, for every $\mathbf{p} \in M$.

(c) Every point in M is recurrent.

Proof. Exercise. □

A point $\mathbf{p}$ of Ω is an *almost periodic point* if given $\varepsilon > 0$ there exists $T > 0$ such that every interval of length T in $\mathbb{R}$ contains a τ such that $\|\Phi(\tau, \mathbf{p})-\mathbf{p}\| < \varepsilon$. Almost periodicity is intimately connected with minimality.

Proposition 6.17 *Let Φ be a flow on Ω and let $\mathbf{p}$ be a point in Ω whose orbit closure is a compact subset of Ω. Then $\overline{\mathcal{O}}(\mathbf{p})$ is a minimal set if and only if $\mathbf{p}$ is an almost periodic point.*

Proof. Suppose that $M = \overline{\mathcal{O}}(\mathbf{p})$ is a minimal set and let $\varepsilon > 0$. If $\mathbf{x}$ is any other point in $\overline{\mathcal{O}}(\mathbf{p})$, then there exists t such that $\|\Phi(t, \mathbf{x}) - \mathbf{p}\| < \varepsilon$ because $M = \overline{\mathcal{O}}(\mathbf{x})$ for all $\mathbf{x} \in M$. Consequently,

$$M \subset \bigcup_{t\in\mathbb{R}} \{\mathbf{x} : \|\Phi(t, \mathbf{x}) - \mathbf{p}\| < \varepsilon\}.$$

For each t, the set $\{\mathbf{x} :: \|\Phi(t, \mathbf{x}) - \mathbf{p}\| < \varepsilon\}$ is an open set because the map $\mathbf{x} \to \Phi(t, \mathbf{x})$ is continuous. Since M is compact, there exist a finite set of real numbers, $t_1, \ldots, t_k$, such that

$$M \subset \bigcup_{j=1}^{k} \{\mathbf{x} : \|\Phi(t_j, \mathbf{x}) - \mathbf{p}\| < \varepsilon\}.$$

Set $T = 2\max\{|t_1|, \ldots, |t_k|\} + 1$.

Let $[\tau - T/2, \tau + T/2]$ be an arbitrary closed interval of length T. Then for some j,

$$\Phi(\tau, \mathbf{p}) \in \{\mathbf{x} : \|\Phi(t_j, \mathbf{x}) - \mathbf{p}\| < \varepsilon\}.$$

Hence,

$$\Phi(\tau + t_j, \mathbf{p}) \in \{\mathbf{x} : \|\mathbf{x} - \mathbf{p}\| < \varepsilon\}.$$

or $\|\Phi(\tau + t_j, \mathbf{p}) - \mathbf{p}\| < \varepsilon$. Clearly, $|t_j| < T/2$ and $\tau + t_j$ is in $[\tau - T/2, \tau + T/2]$ proving that minimality implies almost periodicity.

Suppose $\mathbf{p}$ is almost periodic and let $\mathbf{q}$ be an arbitrary point in $\overline{\mathcal{O}}(\mathbf{p})$. Let $\varepsilon > 0$. It suffices to show that $\|\Phi(\tau, \mathbf{q}) - \mathbf{p}\| \leq \varepsilon$ for some $\tau \in \mathbb{R}$ because then $\mathbf{p}$ is in $\overline{\mathcal{O}}(\mathbf{q})$ and $\overline{\mathcal{O}}(\mathbf{p}) = \overline{\mathcal{O}}(\mathbf{q})$. Let T be given by the almost periodicity of $\mathbf{p}$ for ε.

Because $\mathbf{q} \in \overline{\mathcal{O}}(\mathbf{p})$, there exists a sequence of real numbers t_k such that $\lim_{k\to\infty} \Phi(t_k, \mathbf{p}) = \mathbf{q}$. For each t_k there exists τ_k such that $0 \leq \tau_k \leq T$ and $\|\Phi(t_k + \tau_k, \mathbf{p}) - \mathbf{p}\| < \varepsilon$. Without loss of generality it can be assumed that τ_k converges to τ. It follows from the definition of a flow that $\Phi(t_k + \tau_k, \mathbf{p}) = \Phi(\tau_k, \Phi(t_k, \mathbf{p}))$ converges to $\Phi(\tau, \mathbf{q})$. Finally,

$$\|\Phi(\tau, \mathbf{q}) - \mathbf{p}\| = \lim_{k\to\infty} \|\Phi(t_k + \tau_k, \mathbf{p}) - \mathbf{p}\| \leq \varepsilon$$

to complete the proof. $\square$

Because orthogonal linear transformations preserve distance, the example has an even stronger form of almost periodic behavior.

Proposition 6.18 *Let A be a 4×4 matrix given by equation (6.7) with α irrational, let $\mathbf{x}(t, \boldsymbol{\xi}) = e^{At}\boldsymbol{\xi}$ be the flow defined on $\mathbb{R}^4$ by $\dot{\mathbf{x}} = A\mathbf{x}$, and let $\mathbf{p}$ be a point in $\mathbb{R}^4$ such that $E_1(\boldsymbol{\xi}) = r_1 > 0$ and $E_2(\boldsymbol{\xi}) = r_2 > 0$. Then, given $\varepsilon > 0$, there exists $T > 0$ such that every interval of length T in $\mathbb{R}$ contains a τ such that $\|\mathbf{x}(\tau, \boldsymbol{\xi}) - \boldsymbol{\xi}\| < \varepsilon$ for all $\boldsymbol{\xi} \in \overline{\mathcal{O}}(\mathbf{p})$.*

Proof. Let $\varepsilon > 0$. By *Proposition 6.17*, $\mathbf{p}$ is an almost periodic point and there exists $T > 0$ with the required property for the point $\mathbf{p}$. Let $\boldsymbol{\xi}$ be an arbitrary point in $\overline{\mathcal{O}}(\mathbf{p})$. Then there exist a sequence of real numbers t_k such $\lim_{k\to\infty} \mathbf{x}(t_k, \mathbf{p}) = \boldsymbol{\xi}$.

Let τ be any real number such that $\|\mathbf{x}(\tau, \mathbf{p}) - \mathbf{p}\| < \varepsilon$. It suffices to show that $\|\mathbf{x}(\tau, \boldsymbol{\xi}) - \boldsymbol{\xi}\| \leq \varepsilon$. Since $\mathbf{x}(\tau, \boldsymbol{\xi}) = e^{At}\boldsymbol{\xi}$ is an orthogonal linear transformation, on the one hand,

$$\begin{aligned} \|\mathbf{x}(t_k + \tau, \mathbf{p}) - \mathbf{x}(t_k, \mathbf{p})\| &= \|\mathbf{x}(t_k, \mathbf{x}(\tau, \mathbf{p})) - \mathbf{x}(t_k, \mathbf{p})\| \\ &= \|\mathbf{x}(\tau, \mathbf{p}) - \mathbf{p}\| < \varepsilon. \end{aligned}$$

On the other hand,

$$\begin{aligned} \lim_{k\to\infty} \|\mathbf{x}(t_k + \tau, \mathbf{p}) - \mathbf{x}(t_k, \mathbf{p})\| &= \lim_{k\to\infty} \|\mathbf{x}(\tau, \mathbf{x}(t_k, \mathbf{p})) - \mathbf{x}(t_k, \mathbf{p})\| \\ &= \|\mathbf{x}(\tau, \boldsymbol{\xi}) - \boldsymbol{\xi}\|. \end{aligned}$$

Therefore, $\|\mathbf{x}(\tau, \boldsymbol{\xi}) - \boldsymbol{\xi}\| \leq \varepsilon$. $\square$

It is not true that an orbit closure is a minimal set if every point in it is recurrent. Consequently, recurrent points need not be almost periodic points. A counterexample will be constructed by modifying the $\dot{\mathbf{x}} = A\mathbf{x}$ example. Let $\boldsymbol{\xi}_1 = (1, 0, 1, 0)$ so $r_1 = 1 = r_2$ and $\overline{\mathcal{O}}(\boldsymbol{\xi}) = C_1 \times C_2$ where both C_1 and C_2 are unit circles. Set $\boldsymbol{\xi}_2 = \mathbf{x}(1, \boldsymbol{\xi}) = e^{A}\boldsymbol{\xi}$ and $\boldsymbol{\xi}_3 = \mathbf{x}(-1, \boldsymbol{\xi}) = e^{-A}\boldsymbol{\xi}$. By the remark on page 208

$$\omega(\boldsymbol{\xi}_2) = C_1 \times C_2$$

and

$$\alpha(\boldsymbol{\xi}_3) = C_1 \times C_2.$$

Also set $\boldsymbol{\xi}_4 = (-1, 0, 1, 0)$ and note that $\boldsymbol{\xi}_4 \notin \mathcal{O}(\boldsymbol{\xi}_1)$.

The strategy of the example is to change the speed of the solutions so that $\boldsymbol{\xi}_1$ becomes a fixed point, but the other orbits remain unchanged. (See *Proposition 2.5* and the surrounding discussion in Section 2.2.) Define a real-valued function $h : \mathbb{R}^4 \to \mathbb{R}$ by

$$h(x, y, u, v) = \frac{(x-1)^2 + y^2 + (u-1)^2 + v^2}{1 + (x-1)^2 + y^2 + (u-1)^2 + v^2}.$$

Clearly $h(x, y, u, v) = 0$ if and only if $(x, y, u, v) = (1, 0, 1, 0) = \boldsymbol{\xi}_1$. In addition, it has continuous first partial derivatives and $0 \leq h(x, y, u, v) < 1$.

Consider the nonlinear system

$$\begin{pmatrix} \dot{x} \\ \dot{y} \\ \dot{u} \\ \dot{v} \end{pmatrix} = h(x, y, u, v) \begin{bmatrix} 0 & 2\pi\alpha & 0 & 0 \\ -2\pi\alpha & 0 & 0 & 0 \\ 0 & 0 & 0 & 2\pi\alpha \\ 0 & 0 & -2\pi\alpha & 0 \end{bmatrix} \begin{pmatrix} x \\ y \\ u \\ v \end{pmatrix}.$$

Obviously the right-hand side has continuous first partial derivatives and is locally Lipschitz, which guarantees continuity initial conditions. If $\mathbf{w}$ is used to denote a generic (x, y, u, v) in $\mathbb{R}^4$, then this equation has the form $\dot{\mathbf{w}} = h(\mathbf{w})A\mathbf{w}$ and has sublinear growth in $\mathbf{w}$ because

$$\|h(\mathbf{w})A\mathbf{w}\| \leq |h(\mathbf{w})| \, \|A\| \, \|\mathbf{w}\| \leq \|A\| \, \|\mathbf{w}\|.$$

By *Theorem 2.10*, the solutions of $\dot{\mathbf{w}} = h(\mathbf{w})A\mathbf{w}$ are defined on $\mathbb{R}$ and $\mathbf{w}(t, \boldsymbol{\xi})$ defines a flow on $\mathbb{R}^4$.

Although the speed of every orbit has been altered, the orbits remain unchanged, except for the orbit of $\boldsymbol{\xi}_1$. The orbit of $\boldsymbol{\xi}_1$ is broken into three orbits. The point $\boldsymbol{\xi}_1$ is now fixed and by itself is an orbit. The other two orbits are those of $\boldsymbol{\xi}_2$ and $\boldsymbol{\xi}_3$ and they can be described as sets in terms of the original linear flow as follows:

$$\mathcal{O}(\boldsymbol{\xi}_2) = \{e^{At}\boldsymbol{\xi}_2 : -1 < t < \infty\}$$

and

$$\mathcal{O}(\boldsymbol{\xi}_3) = \{e^{At}\boldsymbol{\xi}_3 : -\infty < t < 1\}.$$

In other words, as t ranges from $-\infty$ to ∞ the new solution through $\boldsymbol{\xi}_2$ will have the same range as the old solution had with t ranging from -1 to ∞. Notice that because $h(x, y, u, v) \geq 0$, the new solutions and old solutions move in the same direction.

It follows from the above characterization of the orbit of $\boldsymbol{\xi}_2$ that

$$\omega(\boldsymbol{\xi}_2) = C_1 \times C_2$$

and

$$\alpha(\boldsymbol{\xi}_2) = \{\boldsymbol{\xi}_1\}.$$

Therefore $\boldsymbol{\xi}_2$ is positively recurrent but not negatively recurrent. Similarly, $\boldsymbol{\xi}_3$ is negatively recurrent and not positively recurrent. Because $\boldsymbol{\xi}_4 \notin \mathcal{O}(\boldsymbol{\xi}_1)$ for the original linear flow, the orbit of $\boldsymbol{\xi}_4$ and its alpha and omega limit sets are the same for both flows, that is,

$$\overline{\mathcal{O}}(\boldsymbol{\xi}_4) = \alpha(\boldsymbol{\xi}_4) = \omega(\boldsymbol{\xi}_4) = C_1 \times C_2$$

for both flows. Therefore, $\boldsymbol{\xi}_4$ is still a recurrent point, but $\overline{\mathcal{O}}(\boldsymbol{\xi}_4)$ is not a minimal set in the modified flow because $\boldsymbol{\xi}_1 \in \overline{\mathcal{O}}(\boldsymbol{\xi}_4)$ and

$$\overline{\mathcal{O}}(\boldsymbol{\xi}_1) = \{\boldsymbol{\xi}_1\} \neq \overline{\mathcal{O}}(\boldsymbol{\xi}_4).$$

To summarize, a fixed point is the strongest form of recurrence followed by a periodic point. Next comes the almost periodic points that are not periodic. Within this class there are also gradations of almost periodicity. For example, the linear example in this section had a uniform almost periodic behavior, which is the strongest form of recurrence short of being periodic. Recurrent points are a still weaker form of recurrence because the orbit closure of a recurrent point need not be minimal. The weakest forms of recurrence in this section were the positively and negatively recurrent points in the final example. In the plane, however, this whole hierarchy of recurrence collapses to fixed point and periodic points.

EXERCISES

1. Let A be given by *(6.7)*. Show that if α is irrational, then there are exactly 2 periodic orbits of $\dot{\mathbf{x}} = A\mathbf{x}$ lying on each sphere of the form $\{\mathbf{x} : \|\mathbf{x}\| = r\}$ with $r > 0$.

2. Let A be a 4×4 real skew symmetric matrix. Find necessary and sufficient for every orbit of $\dot{\mathbf{x}} = A\mathbf{x}$ to be a periodic orbit or a fixed point.

3. Prove the remark on page 208.

4. Construct a flow on the plane with a noncompact minimal set.

5. Prove the following: If $\mathbf{p}$ is an almost periodic point for a flow on $\mathbb{R}^d$, then the orbit closure of $\mathbf{p}$ is a compact set.

6. Construct a flow with a recurrent point whose orbit closure contains a point which is neither positively nor negatively recurrent.

7. Let $\Phi : \mathbb{R} \times \mathbb{R}^d \to \mathbb{R}^d$ be a flow on $\mathbb{R}^d$ with at least one fixed point and such that $\|\Phi(t, \mathbf{x}) - \Phi(t, \mathbf{y})\| = \|\mathbf{x} - y\|$ for all $t \in \mathbb{R}$ and $\mathbf{x}$, $\mathbf{y} \in \mathbb{R}^d$. Prove that $\overline{\mathcal{O}}(\boldsymbol{\xi})$ is a compact minimal set for every $\boldsymbol{\xi} \in \mathbb{R}^d$.

8. Let A be a real skew symmetric matrix and let $\mathbf{x}(t, \boldsymbol{\xi}) = e^{At}\boldsymbol{\xi}$ be the flow determined $\dot{\mathbf{x}} = A\mathbf{x}$. Prove that every point in $\mathbb{R}^d$ is an almost periodic point.

9. A continuous function $g : \mathbb{R} \to \mathbb{R}$ is an *almost periodic function* if given $\varepsilon > 0$ there exists $L > 0$ such that every interval of length L contains at least one number τ for which $|g(t + \tau) - g(t)| < \varepsilon$ for every $t \in \mathbb{R}$. (The classical treatment of almost periodic functions can be found in [5].) Let A be a 4×4 matrix given by equation *(6.7)* with α irrational, let $\mathbf{x}(t, \boldsymbol{\xi}) = e^{At}\boldsymbol{\xi}$ be the flow defined on $\mathbb{R}^4$ by $\dot{\mathbf{x}} = A\mathbf{x}$, and let $\boldsymbol{\xi}$ be a point in $\mathbb{R}^4$ such that $E_1(\boldsymbol{\xi}) = r_1 > 0$ and $E_2(\boldsymbol{\xi}) = r_2 > 0$. Prove that if $G : \mathbb{R}^4 \to \mathbb{R}$ is a continuous function, then $g(t) = G(\mathbf{x}(t, \boldsymbol{\xi}))$ is an almost periodic function.

Chapter 7

Smooth Vector Fields

A major shift in emphasis begins with this chapter. In the earlier chapters, differentiability of the function $f(t, \mathbf{x})$ played only a minor role. Chapters 1 and 2, which provided the basic historical context, analytic results, and examples, rested almost entirely on continuity and Lipschitz hypotheses. The use of linear algebra was the prominent feature of Chapters 3 and 4. Then in the context of locally Lipschitz functions, stability and the Poincaré return map were studied in Chapters 5 and 6 to introduce the basic dynamical problem of determining the eventual behavior of solutions as time evolves.

Beginning in this chapter and continuing throughout the rest of the book, differentiability will play a central role. The assumption that $\mathbf{f}$ is differentiable will be used to linearize the differential equation and then apply ideas from the previous chapters. In particular, in subsequent chapters, significant dynamical results will be obtained for the autonomous differential equation $\dot{\mathbf{x}} = \mathbf{f}(\mathbf{x})$ when $\mathbf{f}$ is differentiable.

Functions with continuous first partial derivatives on an open domain are especially important. They are not only locally Lipschitz, but they are differentiable in the sense of having very good linear approximations at each point. The term smooth is often associated with functions having continuous partial derivatives, but there is some variation in the terminology. In the remaining chapters, a *smooth function* will always mean a function $\mathbf{f} : E \longrightarrow \mathbb{R}^n$ with continuous first partial derivatives on an open set $E \subset \mathbb{R}^m$.

In the context of autonomous systems, the function $\mathbf{f}$ is often thought of as assigning a vector $\mathbf{f}(\mathbf{x})$ to the point $\mathbf{x}$ and is called a *vector field*. The chapter title refers to vector fields that are also smooth functions, and hence called *smooth vector fields*. Smooth dynamics is the study of flows determined by smooth vector fields and the iteration of smooth maps from a space into itself.

Because of the importance of differentiable and smooth functions in the remaining chapters, this chapter begins with a discussion of the derivative as a linear transformation, the simplest smooth mapping. The main result, differentiability in initial conditions, is the focus of the second section. This result can be stated very simply for the autonomous differential equation $\dot{\mathbf{x}} = \mathbf{f}(\mathbf{x})$.

Specifically, if $\mathbf{f}$ is a smooth function, then $\mathbf{x}(t, \boldsymbol{\xi})$ is a smooth function. For the general equation $\dot{\mathbf{x}} = \mathbf{f}(t, \mathbf{x})$, it is sufficient to assume that $\mathbf{f}(t, \mathbf{x})$ is smooth in just the space variable $\mathbf{x}$ to prove that the solution $\mathbf{x}(t, \tau, \boldsymbol{\xi})$ is a smooth function on $D_{\mathbf{f}}$.

The last two sections of the chapter are devoted to a few basic linearization results for smooth vector fields to illustrate this concept. Linearization refers to the process of determining properties of the solutions of an autonomous differential equation $\dot{\mathbf{x}} = \mathbf{f}(\mathbf{x})$ from a linear equation $\dot{\mathbf{x}} = A\mathbf{x}$ where A is the Jacobian matrix of $\mathbf{f}$ at some point.

7.1 Differentiable Functions

The presentation of the material in this section has its roots in *Calculus on Manifolds* by Spivak [35].

Let E be an open subset of $\mathbb{R}^m$ and let $\mathbf{g} : \mathbb{R}^m \to \mathbb{R}^n$ be a vector valued function defined on E. The function $\mathbf{g}$ is *differentiable* at $\boldsymbol{\xi}$ in E if there exists a linear transformation $T : \mathbb{R}^m \to \mathbb{R}^n$ such that

$$\lim_{\mathbf{h} \to \mathbf{0}} \frac{|\mathbf{g}(\boldsymbol{\xi} + \mathbf{h}) - \mathbf{g}(\boldsymbol{\xi}) - T(\mathbf{h})|}{|\mathbf{h}|} = 0.$$

Note that this definition is independent of the norm used and can be written

$$\lim_{\mathbf{x} \to \boldsymbol{\xi}} \frac{|\mathbf{g}(\mathbf{x}) - \mathbf{g}(\boldsymbol{\xi}) - T(\mathbf{x} - \boldsymbol{\xi})|}{|\mathbf{x} - \boldsymbol{\xi}|} = 0.$$

It is also easy to check that when $m = 1 = n$ and E is just an open interval in $\mathbb{R}$, the above definition agrees with usual definition by setting $T(h) = g'(\xi)h$.

If this is the right way to think about differentiability for functions of several variables, then there should be only one such T, it should be calculable from the partial derivatives of g, and differentiability should imply continuity. The first three propositions neatly address these issues.

Proposition 7.1 *If the function $\mathbf{g}$ is differentiable at $\boldsymbol{\xi}$ in E, then there exists unique linear transformation T such that*

$$\lim_{\mathbf{h} \to \mathbf{0}} \frac{|\mathbf{g}(\boldsymbol{\xi} + \mathbf{h}) - \mathbf{g}(\boldsymbol{\xi}) - T(\mathbf{h})|}{|\mathbf{h}|} = 0.$$

Proof. Suppose T_1 and T_2 both satisfy the above equation. For $\mathbf{h} \neq \mathbf{0}$ and $\boldsymbol{\xi} + \mathbf{h}$ in E, we have

$$\begin{aligned} 0 &\leq \frac{|T_1(\mathbf{h}) - T_2(\mathbf{h})|}{|\mathbf{h}|} \\ &= \frac{|T_1(\mathbf{h}) + \mathbf{g}(\boldsymbol{\xi}) - \mathbf{g}(\boldsymbol{\xi} + \mathbf{h}) + \mathbf{g}(\boldsymbol{\xi} + \mathbf{h}) - \mathbf{g}(\boldsymbol{\xi}) - T_2(\mathbf{h})|}{|\mathbf{h}|} \\ &\leq \frac{|T_1(\mathbf{h}) + \mathbf{g}(\boldsymbol{\xi}) - \mathbf{g}(\boldsymbol{\xi} + \mathbf{h})|}{|\mathbf{h}|} + \frac{|\mathbf{g}(\boldsymbol{\xi} + \mathbf{h}) - \mathbf{g}(\boldsymbol{\xi}) - T_2(\mathbf{h})|}{|\mathbf{h}|} \end{aligned}$$

from which it follows that

$$\lim_{\mathbf{h}\to\mathbf{0}} \frac{|T_1(\mathbf{h}) - T_2(\mathbf{h})|}{|\mathbf{h}|} = 0.$$

In particular, for any nonzero $\mathbf{v}$ in $\mathbb{R}^m$,

$$\lim_{s\to 0} \frac{|T_1(s\mathbf{v}) - T_2(s\mathbf{v})|}{|s\mathbf{v}|} = 0.$$

Since

$$\frac{|T_1(\mathbf{v}) - T_2(\mathbf{v})|}{|\mathbf{v}|} = \frac{|T_1(s\mathbf{v}) - T_2(s\mathbf{v})|}{|s\mathbf{v}|},$$

we have

$$\lim_{s\to 0} \frac{|T_1(\mathbf{v}) - T_2(\mathbf{v})|}{|\mathbf{v}|} = 0,$$

which implies $T_1(\mathbf{v}) = T_2(\mathbf{v})$. □

When $\mathbf{g}$ is differentiable at $\boldsymbol{\xi}$ in E, the unique linear transformation T satisfying

$$\lim_{\mathbf{h}\to\mathbf{0}} \frac{|\mathbf{g}(\boldsymbol{\xi}+\mathbf{h}) - \mathbf{g}(\boldsymbol{\xi}) - T(\mathbf{h})|}{|\mathbf{h}|} = 0$$

is called *the derivative* of $\mathbf{g}$ at $\boldsymbol{\xi}$ and will be denoted by $\mathrm{D}\mathbf{g}(\boldsymbol{\xi})$, Since $\mathrm{D}\mathbf{g}(\boldsymbol{\xi})$ is linear transformation depending on $\boldsymbol{\xi}$, its value at a vector $\mathbf{v}$ is denoted by $\mathrm{D}\mathbf{g}(\boldsymbol{\xi})(\mathbf{v})$.

Remark *If $T : \mathbb{R}^m \to \mathbb{R}^n$ is a linear transformation, then T is differentiable and $\mathrm{DT}(\boldsymbol{\xi}) = \mathrm{T}$ for all $\boldsymbol{\xi}$.*

Proof. The limit

$$\lim_{\mathbf{h}\to\mathbf{0}} \frac{|T(\boldsymbol{\xi}+\mathbf{h}) - T(\boldsymbol{\xi}) - T(\mathbf{h})|}{|\mathbf{h}|} = \lim_{\mathbf{h}\to\mathbf{0}} \frac{0}{|\mathbf{h}|} = 0$$

and $\mathrm{DT}(\boldsymbol{\xi}) = \mathrm{T}$ by the definition of the derivative. □

The most common way of calculating the value of a linear transformation T at $\mathbf{x}$ is to use the matrix of T with respect to the standard basis $\mathbf{e}_1, \ldots, \mathbf{e}_m$ so that $T(\mathbf{x}) = A\mathbf{x}$. Determining the matrix of $\mathrm{D}\mathbf{g}(\boldsymbol{\xi})$ is an important calculation and one that has a unifying answer. Following the one-dimensional notation, it is often convenient to denote the matrix of $\mathrm{D}\mathbf{g}(\boldsymbol{\xi})$ with respect to the standard basis by simply $\mathbf{g}'(\boldsymbol{\xi})$. It is the simplest notation for the matrix of $\mathrm{D}\mathbf{g}(\boldsymbol{\xi})$; the other common and useful notations for this matrix are included in the statement of the next proposition.

Proposition 7.2 *If the function* $\mathbf{g}$ *is differentiable at* $\boldsymbol{\xi}$ *in* E*, then all the first partial derivatives of* $\mathbf{g}$ *exist at* $\boldsymbol{\xi}$ *and the matrix of* $\mathrm{D}\mathbf{g}(\boldsymbol{\xi})$ *with respect to the standard basis is the Jacobian matrix of partial derivatives*

$$\mathbf{g}'(\boldsymbol{\xi}) = \frac{\partial \mathbf{g}}{\partial \mathbf{x}}(\boldsymbol{\xi}) = \left[\frac{\partial g_i(\boldsymbol{\xi})}{\partial x_j}\right] = \begin{bmatrix} \frac{\partial g_1(\boldsymbol{\xi})}{\partial x_1} & \cdots & \frac{\partial g_1(\boldsymbol{\xi})}{\partial x_m} \\ \vdots & \ddots & \vdots \\ \frac{\partial g_n(\boldsymbol{\xi})}{\partial x_1} & \cdots & \frac{\partial g_n(\boldsymbol{\xi})}{\partial x_m} \end{bmatrix},$$

where $\mathbf{g} = (g_1, \ldots, g_n)$ *and* $\mathbf{x} = (x_1, \ldots, x_m)$ *as usual.*

Proof. Let $A = (a_{ij})$ denote the matrix of $\mathrm{D}\mathbf{g}(\boldsymbol{\xi})$ with respect to the standard basis $\mathbf{e}_1, \ldots, \mathbf{e}_m$ of $\mathbb{R}^m$ and $\mathbf{e}_1, \ldots, \mathbf{e}_n$ of $\mathbb{R}^n$. It must be shown that

$$a_{ij} = \frac{\partial g_i(\boldsymbol{\xi})}{\partial x_j} = \lim_{h \to 0} \frac{g_i(\boldsymbol{\xi} + h\mathbf{e}_j) - g_i(\boldsymbol{\xi})}{h}.$$

Observe that

$$\begin{aligned} 0 &\le \left| \frac{g_i(\boldsymbol{\xi} + h\mathbf{e}_j) - g_i(\boldsymbol{\xi})}{h} - a_{ij} \right| \\ &\le \left| \frac{\mathbf{g}(\boldsymbol{\xi} + h\mathbf{e}_j) - \mathbf{g}(\boldsymbol{\xi})}{h} - A\mathbf{e}_j \right| \\ &= \frac{|\mathbf{g}(\boldsymbol{\xi} + h\mathbf{e}_j) - \mathbf{g}(\boldsymbol{\xi}) - \mathrm{D}\mathbf{g}(\boldsymbol{\xi})(\mathrm{h}\mathbf{e}_\mathrm{j})|}{|h\mathbf{e}_j|}. \end{aligned}$$

It follows that

$$\lim_{h \to 0} \left| \frac{g_i(\boldsymbol{\xi} + h\mathbf{e}_j) - g_i(\boldsymbol{\xi})}{h} - a_{ij} \right| = 0.$$

Therefore, $\frac{\partial g_i}{\partial x_j}$ exists at $\boldsymbol{\xi}$ and equals a_{ij}. □

For a real-valued function g that is differentiable at $\boldsymbol{\xi}$, it follows that the matrix of Dg is just $\nabla g(\boldsymbol{\xi})$ and $\mathrm{D}g(\boldsymbol{\xi})(\mathbf{v}) = \nabla \mathrm{g}(\boldsymbol{\xi}) \cdot \mathbf{v}$, a frequently used elementary fact.

The converse of the preceding proposition is not true. The existence of partial derivatives does not imply that a function is differentiable for several variables because partial derivatives only indicate differentiable behavior in the finite coordinate directions, and there are infinitely many other directions. For example, define $g : \mathbb{R}^2 \to \mathbb{R}$ by

$$g(x, y) = \frac{xy}{x^2 + y^2}$$

when $(x, y) \neq (0, 0)$ and set $g(0, 0) = 0$. Clearly,

$$\frac{\partial g(0,0)}{\partial x} = 0 = \frac{\partial g(0,0)}{\partial y}.$$

The function g is not even continuous at $(0, 0)$ because $\lim_{s \to 0} g(s, s) = 1/2$ and $\lim_{s \to 0} g(s, 2s) = 2/5$. It will follow from the next proposition that g is not differentiable at $(0, 0)$.

Proposition 7.3 *If the function* $\mathbf{g}$ *is differentiable at* $\boldsymbol{\xi}$ *in* E, *then* $\mathbf{g}$ *is continuous at* $\boldsymbol{\xi}$.

Proof. By definition of differentiable there exists $\eta > 0$ such that

$$|\mathbf{g}(\mathbf{x}-\boldsymbol{\xi}) - \mathbf{g}(\boldsymbol{\xi}) - \mathrm{D}\mathbf{g}(\boldsymbol{\xi})(\mathbf{x}-\boldsymbol{\xi})| < |\mathbf{x}-\boldsymbol{\xi}|$$

when $|\mathbf{x}-\boldsymbol{\xi}| < \eta$. Then

$$\begin{aligned} |\mathbf{g}(\mathbf{x}) - \mathbf{g}(\boldsymbol{\xi})| &\leq |\mathbf{g}(\mathbf{x}-\boldsymbol{\xi}) - \mathbf{g}(\boldsymbol{\xi}) - \mathrm{D}\mathbf{g}(\boldsymbol{\xi})(\mathbf{x}-\boldsymbol{\xi})| + |\mathrm{D}\mathbf{g}(\boldsymbol{\xi})(\mathbf{x}-\boldsymbol{\xi})| \\ &\leq |\mathbf{x}-\boldsymbol{\xi}| + |\mathrm{D}\mathbf{g}(\boldsymbol{\xi})|\,|\mathbf{x}-\boldsymbol{\xi}| \\ &= |\mathbf{x}-\boldsymbol{\xi}|(1 + |\mathrm{D}\mathbf{g}(\boldsymbol{\xi})|) \end{aligned}$$

when $|\mathbf{x}-\boldsymbol{\xi}| < \eta$. Given $\varepsilon > 0$, set δ equal to the minimum of the two numbers η and $\varepsilon/(1+|\mathrm{D}\mathbf{g}(\boldsymbol{\xi})|)$. It is obvious that $|\mathbf{g}(\mathbf{x}) - \mathbf{g}(\boldsymbol{\xi})| < \varepsilon$ when $|\mathbf{x}-\boldsymbol{\xi}| < \delta$. □

Not surprisingly, the differentiability of $\mathbf{g} : \mathbb{R}^m \to \mathbb{R}^n$ is equivalent to the differentiability of its coordinate functions.

Proposition 7.4 *The function* $\mathbf{g}(\mathbf{x}) = (g_1(\mathbf{x}), g_2(\mathbf{x}), \ldots, g_n(\mathbf{x}))$ *defined on the open set* $E \subset \mathbb{R}^m$ *is differentiable at* $\boldsymbol{\xi} \in E$ *if and only if* g_i *is differentiable at* $\boldsymbol{\xi}$ *for* $i = 1, \ldots, n$.

Proof. Exercise.

The chain rule is fundamentally one of the most important results about the differentiation. In the present context, it can be stated elegantly in terms of the composition of linear maps or the multiplication of matrices. For convenience and clarity, the notation $\mathbf{h} \circ \mathbf{g}$ and $T \circ S$ will be used to denote the composition of two functions or two linear transformations.

Theorem 7.5 (Chain Rule) *Let* $\mathbf{g} : E \to \mathbb{R}^n$ *and* $\mathbf{h} : F \to \mathbb{R}^p$ *be vector valued function defined on open sets* E *and* F *in* $\mathbb{R}^m$ *and* $\mathbb{R}^n$, *respectively. If* $\mathbf{g}$ *is differentiable at* $\mathbf{p}$ *in* E *and* $\mathbf{h}$ *is differentiable at* $\mathbf{q} = \mathbf{g}(\mathbf{p})$, *an element of* F, *then the composition* $\mathbf{h} \circ \mathbf{g}$ *is differentiable at* $\mathbf{p}$,

$$\mathrm{D}(\mathbf{h} \circ \mathbf{g})(\mathbf{p}) = \mathrm{D}\mathbf{h}(\mathbf{g}(\mathbf{p})) \circ \mathrm{D}\mathbf{g}(\mathbf{p}),$$

and

$$(\mathbf{h} \circ \mathbf{g})'(\mathbf{p}) = \mathbf{h}'(\mathbf{g}(\mathbf{p}))\mathbf{g}'(\mathbf{p}).$$

Proof. Following Spivak [35], let $S = \mathrm{D}\mathbf{g}(\mathbf{p})$, and let $T = \mathrm{D}\mathbf{h}(\mathbf{q})$. It must be shown that

$$\lim_{\mathbf{x}\to\mathbf{p}} \frac{|\mathbf{h} \circ \mathbf{g}(\mathbf{x}) - \mathbf{h} \circ \mathbf{g}(\mathbf{p}) - T \circ S(\mathbf{x}-\mathbf{p})|}{|\mathbf{x}-\mathbf{p}|} = 0.$$

This will require adding and subtracting terms to the numerator and breaking the problem into two separate limits. To this end it is convenient to introduce some notation for the numerators of all three derivative limits. Set

$$\begin{aligned}\boldsymbol{\theta}(\mathbf{x}) &= \mathbf{g}(\mathbf{x}) - \mathbf{g}(\mathbf{p}) - S(\mathbf{x}-\mathbf{p})\\ \boldsymbol{\psi}(\mathbf{y}) &= \mathbf{h}(\mathbf{y}) - \mathbf{h}(\mathbf{q}) - T(\mathbf{y}-\mathbf{q})\\ \boldsymbol{\rho}(\mathbf{x}) &= \mathbf{h}\circ\mathbf{g}(\mathbf{x}) - \mathbf{h}\circ\mathbf{g}(\mathbf{p}) - T\circ S(\mathbf{x}-\mathbf{p}).\end{aligned}$$

The next step is to rewrite $\boldsymbol{\rho}(\mathbf{x})$ by adding and subtracting $T(\mathbf{g}(\mathbf{x})-\mathbf{q})$. Then,

$$\begin{aligned}\boldsymbol{\rho}(\mathbf{x}) &= \mathbf{h}\circ\mathbf{g}(\mathbf{x}) - \mathbf{h}\circ\mathbf{g}(\mathbf{p}) - T\circ S(\mathbf{x}-\mathbf{p}) \pm T(\mathbf{g}(\mathbf{x})-\mathbf{q})\\ &= \mathbf{h}\big(\mathbf{g}(\mathbf{x})\big) - \mathbf{h}\big(\mathbf{g}(\mathbf{p})\big) - T\big(\pm(\mathbf{g}(\mathbf{x})-\mathbf{q}) + S(\mathbf{x}-\mathbf{p})\big)\\ &= \mathbf{h}\big(\mathbf{g}(\mathbf{x})\big) - \mathbf{h}(\mathbf{q}) - T\big(\mathbf{g}(\mathbf{x}) - \mathbf{q} - \boldsymbol{\theta}(\mathbf{x})\big)\\ &= \boldsymbol{\psi}\big(\mathbf{g}(\mathbf{x})\big) + T\big(\boldsymbol{\theta}(\mathbf{x})\big).\end{aligned}$$

Therefore, it suffices to show that

$$\lim_{\mathbf{x}\to\mathbf{p}} \frac{|\boldsymbol{\psi}(\mathbf{g}(\mathbf{x}))|}{|\mathbf{x}-\mathbf{p}|} = 0$$

and

$$\lim_{\mathbf{x}\to\mathbf{p}} \frac{|T(\boldsymbol{\theta}(\mathbf{x}))|}{|\mathbf{x}-\mathbf{p}|} = 0.$$

The second limit is easy because

$$|T(\boldsymbol{\theta}(\mathbf{x}))| \le |T|\,|\boldsymbol{\theta}(\mathbf{x})|$$

and

$$\lim_{\mathbf{x}\to\mathbf{p}} \frac{|\boldsymbol{\theta}(\mathbf{x})|}{|\mathbf{x}-\mathbf{p}|} = \lim_{\mathbf{x}\to\mathbf{p}} \frac{|\mathbf{g}(\mathbf{x}) - \mathbf{g}(\mathbf{p}) - S(\mathbf{x}-\mathbf{p})|}{|\mathbf{x}-\mathbf{p}|} = 0$$

by the hypothesis. The first limit will require a more delicate analysis.

Let $\varepsilon > 0$. Since

$$\lim_{\mathbf{y}\to\mathbf{q}} \frac{|\boldsymbol{\psi}(\mathbf{y})|}{|\mathbf{y}-\mathbf{q}|} = \lim_{\mathbf{y}\to\mathbf{q}} \frac{|\mathbf{h}(\mathbf{y}) - \mathbf{h}(\mathbf{q}) - T(\mathbf{y}-\mathbf{q})|}{|\mathbf{y}-\mathbf{q}|} = 0$$

by hypothesis, there exists $\delta_1 > 0$ such that

$$|\boldsymbol{\psi}(\mathbf{y})| < \varepsilon|\mathbf{y}-\mathbf{q}|$$

when $|\mathbf{y}-\mathbf{q}| < \delta_1$. The differentiability of $\mathbf{g}$ at $\mathbf{p}$ implies that $\mathbf{g}$ is continuous at $\mathbf{p}$ and there exists $\delta_2 > 0$ such that $|\mathbf{g}(\mathbf{x})-\mathbf{q}| < \delta_1$ when $|\mathbf{x}-\mathbf{p}| < \delta_2$. Thus,

$$|\boldsymbol{\psi}(\mathbf{g}(\mathbf{x}))| < \varepsilon|\mathbf{g}(\mathbf{x})-\mathbf{q}|$$

when $|\mathbf{x}-\mathbf{p}| < \delta_2$.

Write $\mathbf{g}(\mathbf{x}) - \mathbf{q} = \boldsymbol{\theta}(\mathbf{x}) - S(\mathbf{x} - \mathbf{p})$ by adding and subtracting $S(\mathbf{x} - \mathbf{p})$. Hence, $|\mathbf{g}(\mathbf{x}) - \mathbf{q}| \leq |\boldsymbol{\theta}(\mathbf{x})| + |S|\,|\mathbf{x} - \mathbf{p}|$ and

$$|\boldsymbol{\psi}(\mathbf{g}(\mathbf{x}))| < \varepsilon\left(|\boldsymbol{\theta}(\mathbf{x})| + |S|\,|\mathbf{x} - \mathbf{p}|\right)$$

when $|\mathbf{x} - \mathbf{p}| < \delta_2$. Since

$$\lim_{\mathbf{x}\to\mathbf{p}} \frac{|\boldsymbol{\theta}(\mathbf{x})|}{|\mathbf{x} - \mathbf{p}|} = 0,$$

there exists $\delta_3 > 0$ such that $\delta_3 < \delta_2$ and

$$\frac{|\boldsymbol{\theta}(\mathbf{x})|}{|\mathbf{x} - \mathbf{p}|} < 1$$

when $|\mathbf{x} - \mathbf{p}| < \delta_3$. Therefore,

$$\begin{aligned}\frac{|\boldsymbol{\psi}(\mathbf{g}(\mathbf{x}))|}{|\mathbf{x} - \mathbf{p}|} &< \varepsilon\left(\frac{|\boldsymbol{\theta}(\mathbf{x})|}{|\mathbf{x} - \mathbf{p}|} + |S|\right)\\ &< \varepsilon(1 + |S|)\end{aligned}$$

when $|\mathbf{x} - \mathbf{p}| < \delta_3$ and

$$\lim_{\mathbf{x}\to\mathbf{p}} \frac{|\boldsymbol{\psi}(\mathbf{g}(\mathbf{x}))|}{|\mathbf{x} - \mathbf{p}|} = 0,$$

which completes the proof. □

The calculation of the derivative of a bilinear function provides a concrete example and a formula that will be useful in the last two sections of this chapter. A real-valued function $\beta : \mathbb{R}^m \times \mathbb{R}^m \to \mathbb{R}$ is said to be *bilinear* if for all $\mathbf{u}$, $\mathbf{v}$, and $\mathbf{x}$ in $\mathbb{R}^m$ and $s \in \mathbb{R}$

$$\beta(\mathbf{u} + \mathbf{v}, \mathbf{x}) = \beta(\mathbf{u}, \mathbf{x}) + \beta(\mathbf{v}, \mathbf{x}),$$

$$\beta(\mathbf{x}, \mathbf{u} + \mathbf{v}) = \beta(\mathbf{x}, \mathbf{u}) + \beta(\mathbf{x}, \mathbf{v}),$$

and

$$\beta(s\mathbf{u}, \mathbf{v}) = s\beta(\mathbf{u}, \mathbf{v}) = \beta(\mathbf{u}, s\mathbf{v}).$$

The familiar dot product $\mathbf{x} \cdot \mathbf{y}$ is a bilinear function. More generally, any matrix $B = (b_{jk}) \in \mathcal{M}_m(\mathbb{R})$ defines by a bilinear function by setting

$$\beta(\mathbf{x}, \mathbf{y}) = \mathbf{x} \cdot (B\mathbf{y}) = \sum_{j=1}^{m} \sum_{k=1}^{m} x_j b_{jk} y_k.$$

Conversely, every bilinear function β has the form $\beta(\mathbf{x}, \mathbf{y}) = \mathbf{x} \cdot (B\mathbf{y})$ with $b_{jk} = \beta(\mathbf{e}_j, \mathbf{e}_k)$.

Proposition 7.6 *If $\beta : \mathbb{R}^m \times \mathbb{R}^m \to \mathbb{R}$ is a bilinear function, then its derivative at $(\boldsymbol{\xi}, \boldsymbol{\rho}) \in \mathbb{R}^m \times \mathbb{R}^m$ is*

$$\mathrm{D}\beta(\boldsymbol{\xi}, \boldsymbol{\rho})(\mathbf{u}, \mathbf{v}) = \beta(\boldsymbol{\xi}, \mathbf{v}) + \beta(\mathbf{u}, \boldsymbol{\rho})$$

Proof. The derivative of β at $(\boldsymbol{\xi}, \boldsymbol{\rho})$ is a linear map $\mathrm{D}\beta(\boldsymbol{\xi}, \boldsymbol{\rho}) : \mathbb{R}^{\mathrm{m}} \times \mathbb{R}^{\mathrm{m}} \to \mathbb{R}$ not the value of the linear map. The formula defines this linear map by giving its value at an arbitrary point $(\mathbf{u}, \mathbf{v}) \in \mathbb{R}^m \times \mathbb{R}^m$. Using the Euclidean norm will make the calculations a little easier.

It must be shown that

$$\lim_{(\mathbf{h}_1,\mathbf{h}_2)\to(\mathbf{0},\mathbf{0})} \frac{|\beta(\boldsymbol{\xi}+\mathbf{h}_1, \boldsymbol{\rho}+\mathbf{h}_2) - \beta(\boldsymbol{\xi}, \boldsymbol{\rho}) - \beta(\boldsymbol{\xi}, \mathbf{h}_2) - \beta(\mathbf{h}_1, \boldsymbol{\rho})|}{\|(\mathbf{h}_1, \mathbf{h}_2)\|} = 0.$$

The numerator simplifies to $\|\beta(\mathbf{h}_1, \mathbf{h}_2)\|$ because β is bilinear. This reduces the problem to showing that

$$\lim_{(\mathbf{h}_1,\mathbf{h}_2)\to(\mathbf{0},\mathbf{0})} \frac{|\beta(\mathbf{h}_1, \mathbf{h}_2)|}{\|(\mathbf{h}_1, \mathbf{h}_2)\|} = 0.$$

Because $\beta(\mathbf{x}, \mathbf{y}) = \mathbf{x} \cdot (B\mathbf{y})$ for some $B \in \mathcal{M}_m(\mathbb{R})$, the Cauchy-Schwarz inequality (*Exercise 1* on page 11) can be used to estimate $|\beta(\mathbf{h}_1, \mathbf{h}_2)|$. Specifically,

$$|\beta(\mathbf{h}_1, \mathbf{h}_2)| = |\mathbf{h}_1 \cdot (B\mathbf{h}_2)| \le \|\mathbf{h}_2\| \, \|B\mathbf{h}_2\| \le \|\mathbf{h}_1\| \, \|B\| \, \|\mathbf{h}_2\|.$$

The denominator can be written as

$$\|(\mathbf{h}_1, \mathbf{h}_2)\| = \sqrt{\|\mathbf{h}_1\|^2 + \|\mathbf{h}_2\|^2}$$

and

$$\frac{|\beta(\mathbf{h}_1, \mathbf{h}_2)|}{\|(\mathbf{h}_1, \mathbf{h}_2)\|} \le \frac{\|\mathbf{h}_1\| \, \|B\| \, \|\mathbf{h}_2\|}{\sqrt{\|\mathbf{h}_1\|^2 + \|\mathbf{h}_2\|^2}}.$$

Therefore, it suffices to show that

$$\lim_{(\mathbf{h}_1,\mathbf{h}_2)\to(\mathbf{0},\mathbf{0})} \frac{\|\mathbf{h}_1\| \, \|\mathbf{h}_2\|}{\sqrt{\|\mathbf{h}_1\|^2 + \|\mathbf{h}_2\|^2}} = 0.$$

The simple inequality for positive real numbers that $2ab \le a^2 + b^2$ implies that

$$\frac{\|\mathbf{h}_1\| \, \|\mathbf{h}_2\|}{\sqrt{\|\mathbf{h}_1\|^2 + \|\mathbf{h}_2\|^2}} \le \frac{1}{2}\sqrt{\|\mathbf{h}_1\|^2 + \|\mathbf{h}_2\|^2}$$

Therefore, the limit equals zero as required. □

Corollary *If $\beta(\mathbf{x}, \mathbf{y}) = \mathbf{x} \cdot (B\mathbf{x})$ is a bilinear function and $V(\mathbf{x}) = \beta(\mathbf{x}, \mathbf{x})$, then $V(\mathbf{x})$ is differentiable at all $\mathbf{x} \in \mathbb{R}^m$,*

$$\mathrm{D}V(\boldsymbol{\xi})(\mathbf{v}) = \beta(\boldsymbol{\xi}, \mathbf{v}) + \beta(\mathbf{v}, \boldsymbol{\xi}),$$

and

$$\nabla V(\boldsymbol{\xi}) = (B + B^{\mathrm{t}})\boldsymbol{\xi}.$$

Proof. The function $V(\mathbf{x})$ is the composition of the linear function $T : \mathbb{R}^m \to \mathbb{R}^m \times \mathbb{R}^m$ defined by $T(\mathbf{x}) = (\mathbf{x}, \mathbf{x})$ and $\beta(\mathbf{x}.\mathbf{y})$. By the proposition and the chain rule it is differentiable for all $\mathbf{x}$ and

$$\mathrm{DV}(\boldsymbol{\xi}) = \mathrm{D}(\beta \circ \mathrm{T})(\boldsymbol{\xi}) = \mathrm{D}\beta(\boldsymbol{\xi}, \boldsymbol{\xi}) \circ \mathrm{DT}(\boldsymbol{\xi}) = \mathrm{D}\beta(\boldsymbol{\xi}, \boldsymbol{\xi}) \circ \mathrm{T}$$

and then

$$\mathrm{DV}(\boldsymbol{\xi})(\mathbf{v}) = \mathrm{D}\beta(\boldsymbol{\xi}, \boldsymbol{\xi})(\mathbf{v}, \mathbf{v}) = \beta(\boldsymbol{\xi}, \mathbf{v}) + \beta(\mathbf{v}, \boldsymbol{\xi}).$$

The above simplifies as follows

$$\begin{aligned} \mathrm{DV}(\boldsymbol{\xi})(\mathbf{v}) &= \beta(\boldsymbol{\xi}, \mathbf{v}) + \beta(\mathbf{v}, \boldsymbol{\xi}) \\ &= \boldsymbol{\xi} \cdot (B\mathbf{v}) + \mathbf{v} \cdot (B\boldsymbol{\xi}) \\ &= (B^{\mathrm{t}}\boldsymbol{\xi}) \cdot \mathbf{v} + (B\boldsymbol{\xi}) \cdot \mathbf{v} \\ &= \big((B + B^{\mathrm{t}})\boldsymbol{\xi}\big) \cdot \mathbf{v}. \end{aligned}$$

Since $\mathrm{DV}(\boldsymbol{\xi})(\mathbf{v}) = \nabla \mathrm{V}(\boldsymbol{\xi}) \cdot \mathbf{v}$, it follows that $\nabla V(\boldsymbol{\xi}) = (B + B^{\mathrm{t}})\boldsymbol{\xi}$. □

Since the existence of first partial derivatives does not imply that a function is differentiable, is there a simple test for differentiability? The answer is yes and it is the same test we use for locally Lipschitz, namely, continuous first partial derivatives or a smooth function.

Theorem 7.7 *If the function* $\mathbf{g} : E \to \mathbb{R}^n$ *has continuous first partial derivatives in a neighborhood of* $\boldsymbol{\xi}$ *in* E, *then* $\mathbf{g}$ *is differentiable at* $\boldsymbol{\xi}$.

Proof. It suffices to consider a real-valued function $g : E \to \mathbb{R}$ by *Proposition 7.4*. By hypothesis there, exists $r > 0$ such that the first partial derivatives of g are continuous on the set $\{\mathbf{x} : |\mathbf{x} - \boldsymbol{\xi}| < r\} \subset E$. By the mean-value theorem (See the discussion preceding the proof of *Theorem 1.12* on page 24 in Chapter 1.)

$$g(\mathbf{x}) - \mathbf{g}(\boldsymbol{\xi}) = \nabla g(\theta\mathbf{x} + (1 - \theta)\boldsymbol{\xi}) \cdot (\mathbf{x} - \boldsymbol{\xi})$$

for some θ, $0 < \theta < 1$ when $|\mathbf{x} - \boldsymbol{\xi}| < r$. Subtract $\nabla g(\boldsymbol{\xi}) \cdot (\mathbf{x} - \boldsymbol{\xi})$ from both sides to produce

$$g(\mathbf{x}) - g(\boldsymbol{\xi}) - \nabla g(\boldsymbol{\xi}) \cdot (\mathbf{x} - \boldsymbol{\xi}) = [\nabla g(\theta\mathbf{x} + (1 - \theta)\boldsymbol{\xi}) - \nabla g(\boldsymbol{\xi})] \cdot (\mathbf{x} - \boldsymbol{\xi}).$$

Then apply the Cauchy-Schwartz, inequality and divide by $\|\mathbf{x} - \boldsymbol{\xi}\|$ to obtain

$$\frac{|g(\mathbf{x}) - g(\boldsymbol{\xi}) - \nabla g(\boldsymbol{\xi}) \cdot (\mathbf{x} - \boldsymbol{\xi})|}{\|\mathbf{x} - \boldsymbol{\xi}\|} \le \|\nabla g(\theta\mathbf{x} + (1 - \theta)\boldsymbol{\xi}) - \nabla g(\boldsymbol{\xi})\|.$$

Clearly, the right-hand side goes to zero as $\mathbf{x}$ approaches $\boldsymbol{\xi}$ because the first partial derivatives are continuous at $\boldsymbol{\xi}$. Therefore, g is differentiable at $\boldsymbol{\xi}$ with $\mathrm{Dg}(\boldsymbol{\xi})(\mathbf{v}) = \nabla \mathrm{g}(\boldsymbol{\xi}) \cdot \mathbf{v}$. □

Corollary *A smooth function* $\mathbf{g} : E \to \mathbb{R}^n$ *on an open set* E *of* $\mathbb{R}^m$ *is differentiable at every point of* E.

Since the definition of a smooth function is equivalent to $\mathbf{g}' : E \to \mathcal{M}_m(\mathbb{R})$ being a continuous function, smooth is synonymous with continuously differentiable in the sense that the function has a derivative at every point of its domain and the derivative is a continuous function.

EXERCISES

1. Show that a real-valued function $f : \mathbb{R}^m \to \mathbb{R}$ is differentiable at $\mathbf{0}$ if for some $\alpha > 1$ the function satisfies the inequality
$$|f(\mathbf{x})| \leq |\mathbf{x}|^\alpha$$
for all $\mathbf{x} \in \mathbb{R}^m$.

2. Let $\boldsymbol{\psi}(t) = (\psi_1(t), \ldots, \psi_m(t))$ be a curve defined on an open interval I with values in $\mathbb{R}^m$. Suppose that the coordinate functions ψ_j are all differentiable in the usual sense for a real-valued function. Show that $\boldsymbol{\psi}$ is differentiable as defined on page 214 and that $\mathrm{D}\boldsymbol{\psi}(\mathrm{t})(1) = \dot{\psi}(\mathrm{t})$.

3. Prove *Proposition 7.4*.

4. Determine the set of points in $\mathbb{R}^2$ at which the continuous real-valued function $f(x, y) = \sqrt{|xy|}$ is differentiable.

5. Let B be an invertible $d \times d$ matrix and let $\dot{\mathbf{x}} = \mathbf{f}(\mathbf{x})$ be an autonomous differential equation on $\mathbb{R}^d$. Determine the differential equation $\dot{\mathbf{y}} = g(\mathbf{y})$ for the change of variables $\mathbf{x} = B\mathbf{y}$.

6. Show that the real-valued function on $\mathbb{R}^2$ defined by
$$g(x, y) = (x^2 + y^2) \sin\left(\frac{1}{\sqrt{x^2 + y^2}}\right)$$
when $(x, y) \neq (0, 0)$ and 0 when $(x, y) = (0, 0)$ is differentiable at $(0, 0)$, but does not have continuous first partial derivatives on $\{(x, y) : \|(x, y)\| < r\}$ for any $r > 0$.

7. Let $F(\mathbf{x})$ and $G(\mathbf{y})$ be real-valued functions defined on open sets E_1 and E_2 of $\mathbb{R}^m$and $\mathbb{R}^n$ and define a real function $H : E_1 \times E_2 \to \mathbb{R}$ by $H(\mathbf{x}, \mathbf{y}) = F(\mathbf{x})G(\mathbf{y})$. Show that if F is differentiable at $\mathbf{x}$ and G is differentiable at $\mathbf{y}$, then H is differentiable at $(\mathbf{x}, \mathbf{y})$ and
$$\mathrm{D(H)}(\mathbf{x}, \mathbf{y})(\mathbf{u}, \mathbf{v}) = \mathrm{G}(\mathbf{y})\nabla\mathrm{F}(\mathbf{x}) \cdot \mathbf{u} + \mathrm{F}(\mathbf{x})\nabla\mathrm{G}(\mathbf{y}) \cdot \mathbf{v}.$$

8. Let $\beta(\mathbf{x}, \mathbf{y})$ be a bilinear function on $\mathbb{R}^{2d}$ and let $\mathbf{f}$ and $\mathbf{g}$ be smooth vector fields on $\mathbb{R}^d$. Derive a matrix formula for the gradient of $\beta(\mathbf{f}(\mathbf{x}), \mathbf{g}(\mathbf{x}))$.

9. Show that Det is a smooth function on $\mathcal{M}_d(\mathbb{R})$ and calculate its derivative. Then find a formula for the derivative of Det $[F(t)]$, when $F : \mathbb{R} \to \mathcal{M}_d(\mathbb{R})$ is a smooth function.

7.2 Differentiation in Initial Conditions

When $\mathbf{f}(t, \mathbf{x})$ is locally Lipschitz on D, the solutions of $\mathbf{x} = \mathbf{f}(t, \mathbf{x})$ define a global solution $\mathbf{x}(t, \tau, \boldsymbol{\xi})$ on an open set $D_{\mathbf{f}} \subset \mathbb{R}^{n+2}$. Recall that $\mathbf{x}(t, \tau, \boldsymbol{\xi})$ is the maximal solution satisfying $\mathbf{x}(\tau, \tau, \boldsymbol{\xi}) = \boldsymbol{\xi}$ and

$$D_{\mathbf{f}} = \{(t, \tau, \boldsymbol{\xi}) : (\tau, \boldsymbol{\xi}) \in D \text{ and } t \text{ is in the domain of } \mathbf{x}(\cdot, \tau, \boldsymbol{\xi})\}.$$

In Chapter I (*Theorem 1.23*), it was shown that $\mathbf{x}(t, \tau, \boldsymbol{\xi})$ is continuous on $D_{\mathbf{f}}$, and that $D_{\mathbf{f}}$ is an open set. We are now ready to study the differentiability of the global solution. Specifically, focus will be on proving that when $\mathbf{f}(t, \mathbf{x})$ is smooth, then $\mathbf{x}(t, \tau, \boldsymbol{\xi})$ is smooth.

Actually it is only necessary to assume that $\mathbf{f}(t, \mathbf{x})$ has continuous partial derivatives only with respect to the space variables $x_1, \ldots, x_n$ to prove that $\mathbf{x}(t, \tau, \boldsymbol{\xi})$ is smooth. The fact that t does not enter the hypothesis is not surprising, since the partial of $\mathbf{x}(t, \tau, \boldsymbol{\xi})$ with respect to t is just the time derivative of the solution. It always exists and is continuous when $x(t, \tau, \boldsymbol{\xi})$ is continuous because

$$\frac{\partial x}{\partial t} = \dot{\mathbf{x}}(t, \tau, \boldsymbol{\xi}) = \mathbf{f}(t, \mathbf{x}(t, \tau, \boldsymbol{\xi})).$$

Getting a handle on the other partial derivatives of $\mathbf{x}(t, \tau, \boldsymbol{\xi})$ is technically rather difficult. The strategy for obtaining the partial derivatives of $\mathbf{x}(t, \tau, \boldsymbol{\xi})$ with respect to τ and $\xi_1, \ldots, \xi_n$ is to take the time derivative of the appropriate difference quotient and show that it looks like a solution of a linear system with an error term. By proving that the error term goes to zero uniformly with the difference, we see not only that the partial derivatives exist, but that they are also solutions to the linear equation. Finally, the continuity follows from the uniform convergence.

If $\mathbf{g}$ is differentiable at $\mathbf{y}$, then $\mathbf{g}(\mathbf{x})$ can be approximated by $\mathbf{g}(\mathbf{y}) + g'(\mathbf{y})(\mathbf{y} - \mathbf{x})$ for $\mathbf{x}$ near $\mathbf{y}$. How large is the error? As one would expect the error is small when the partial derivatives are nearly constant. This can be made precise using the mean-value theorem.

Proposition 7.8 *Let* $\mathbf{g} : E \to \mathbb{R}^n$ *be a smooth function on the open set* $E \subset \mathbb{R}^m$ *and let* $\mathbf{y} \in E$. *If the line segment joining* $\mathbf{y}$ *and* $\mathbf{x}$ *lies in* E, *then*

$$\mathbf{g}(\mathbf{x}) = \mathbf{g}(\mathbf{y}) + [\mathbf{g}'(\mathbf{y}) + \Gamma(\mathbf{x})]\,(\mathbf{x} - \mathbf{y}),$$

where $\Gamma(\mathbf{x})$ *is a* $n \times m$ *matrix such that*

$$|\Gamma| \leq \sum_{i=1}^{m} \sum_{j=1}^{n} \sup \left\{ \left| \frac{\partial g_i(s\mathbf{x} + (1-s)\mathbf{y})}{\partial x_j} - \frac{\partial g_i(\mathbf{y})}{\partial x_j} \right| : 0 \leq s \leq 1 \right\}.$$

Proof. For each j, the following holds

$$g_j(x) = g_j(y) + \nabla g_j(\theta_j x \,+\, (1 - \theta_j) y) \cdot (x - y)$$

for some θ_j such that $0 < \theta_j < 1$. (See the discussion of the mean-value theorem beginning on page 24.) Adding and subtracting $\nabla g(y) \cdot (y - x)$,

$$g_j(x) = g_j(y) + (\nabla g(y) + \nabla g(\theta_j x + (1 - \theta_j)y) - \nabla g(y)) \cdot (x - y)$$

which is just the one variable version of the desired result with $\Gamma(x)$ equal

$$\nabla g_j(\theta_j x + (1 - \theta_j)y) - \nabla g_j(y).$$

To get the general, result let $\Gamma(x)$ be the matrix whose jth row is given by the above expression. The upper bound for $|\Gamma(y)|$ now follows from *Proposition 3.2.*
□

The proof of the next and main theorem of this section requires several careful estimates. The first one makes use of the previous proposition. Two additional estimates use the result (*Theorem 2.13* on page 82) that even approximate solutions of a Lipschitz differential equation have a separation rate that is at most exponential.

Theorem 7.9 *If* $\mathbf{f}(t, x)$ *has continuous first partial derivatives with respect to the variables* $x_1, \cdots, x_d$ *on* D*, then* $\mathbf{x}(t, \tau, \boldsymbol{\xi})$ *has continuous first partial derivatives with respect to the variables* $t, \tau, \boldsymbol{\xi}_1, \cdots, \boldsymbol{\xi}_d$ *on* $D_{\mathbf{f}}$*.*

Proof. This proof is fairly arduous and requires several subarguments to establish key facts needed in the main line of reasoning. To make it easier to see how the pieces fit together, these essential facts requiring a slight digression are clearly stated as claims and proved as separate subitems.

It has already been pointed out that $\mathbf{x}(t, \tau, \boldsymbol{\xi})$ has a continuous partial derivative with respect to t. The main argument will show that $\mathbf{x}(t, \tau, \boldsymbol{\xi})$ has continuous first partial derivatives with respect to $\xi_1, \ldots, \xi_d$. The final step will be to show how to modify the argument to prove that $\mathbf{x}(t, \tau, \boldsymbol{\xi})$ also has a continuous partial derivative with respect to τ.

Because $\mathbf{f}$ has continuous first partial derivatives with respect to $x_1, \cdots, x_d$ on D, the function $\mathbf{f}(t, \mathbf{x})$ is locally Lipschitz on D and thus theorems requiring locally Lipschitz as a hypothesis can and will be used in the proof.

The first step is to construct the set structure needed for the argument. Let $(t_0, \tau_0, \boldsymbol{\xi}_0)$ be a point in D_f. For convenience, let $\boldsymbol{\psi}(t) = \mathbf{x}(t, \tau_0, \boldsymbol{\xi}_0)$ and let $[a, b]$ be a closed interval in the domain of $\boldsymbol{\psi}(t)$ such that $a < t_0 < b$ and $a < \tau_0 <$ b. The compact set

$$C_0 = \{(t, \tau, \boldsymbol{\psi}(\tau)) : a \le t \le b \text{ and } a \le \tau \le b\}$$

is contained in $D_{\mathbf{f}}$ because $\mathbf{x}(t, \tau, \boldsymbol{\psi}(\tau)) = \boldsymbol{\psi}(t)$ by uniqueness of solutions. Because $D_{\mathbf{f}}$ is open there exists and open set V that contains C_0 and that is contained in a compact subset of $D_{\mathbf{f}}$. Specifically, there exists $\alpha > 0$ such that the compact set

$$C = \{(t, \tau, \boldsymbol{\xi}) : a \le t \le b,\ a \le \tau \le b, \text{ and } |\boldsymbol{\xi} - \boldsymbol{\psi}(\tau)| \le 2\alpha\}$$

is also contained in D_f. Set

$$V = \{(t, \tau, \boldsymbol{\xi}) \,:\, a < t < b,\ a < \tau < b, \text{ and } |\boldsymbol{\xi} - \boldsymbol{\psi}(\tau)| < \alpha\},$$

which is an open set. We now have $C_0 \subset V \subset C \subset D_{\mathbf{f}}$. Since no restrictions were placed on the point $(t_0, \tau_0, \boldsymbol{\xi}_0)$ in $D_{\mathbf{f}}$, it can be any point in $D_{\mathbf{f}}$ and hence it suffices to show that $\mathbf{x}(t, \tau, \boldsymbol{\xi})$ has continuous first partial derivatives on V.

Computing partial derivatives requires working with $\mathbf{x}(t, \tau, \boldsymbol{\xi}+h\mathbf{e}_j)$ where, as usual, $\mathbf{e}_1, \ldots, \mathbf{e}_d$ will denote the standard basis of $\mathbb{R}^d$. For this purpose, observe that $(t, \tau, \boldsymbol{\xi} + h\mathbf{e}_j) \in C$ when $(t, \tau, \boldsymbol{\xi}) \in V$ and $|h| < \alpha$ by the construction of C.

To make full use of the local Lipschitz property, an appropriate subset of D is required. For any point $(t, \tau, \boldsymbol{\xi}) \in D_{\mathbf{f}}$, the solution $\mathbf{x}(s, \tau, \boldsymbol{\xi})$ must be defined on an open interval containing the closed interval between τ and t, which can be either the closed interval $[\tau, t]$ or $[t, \tau]$ according as $\tau < t$ or $t < \tau$. For convenience, let $I_{t,\tau}$ denote this closed interval. For $(t, \tau, \boldsymbol{\xi}) \in D_{\mathbf{f}}$, the set $\{(s, \tau, \boldsymbol{\xi}) : s \in I_{t,\tau}\} \subset D_{\mathbf{f}}$. It follows from this observation and the compactness of both C and $I_{t,\tau}$ that

$$C_1 = \{(s, \tau, \boldsymbol{\xi}) : (t, \tau, \boldsymbol{\xi}) \in C \text{ and } s \in I_{t,\tau}\}$$

is a compact subset of $D_{\mathbf{f}}$ and $V \subset C \subset C_1$.

The map $(t, \tau, \boldsymbol{\xi}) \to (t, \mathbf{x}(t, \tau, \boldsymbol{\xi}))$ is a continuous map of $D_{\mathbf{f}}$ onto D (*Theorem 1.23*). Thus the image of C_1 under this map is a compact subset C_2 of D and

$$C_2 = \{(s, \mathbf{x}(s, \tau, \boldsymbol{\xi})) : (t, \tau, \boldsymbol{\xi}) \in C \text{ and } s \in I_{t,\tau}\}\,.$$

The compact subset C_2 of D must also be enlarged. There exists $\beta > 0$ such that

$$C' = \{(s, \mathbf{x}) : |s - t| \leq \beta \text{ and } |\mathbf{x} - \mathbf{y}| \leq \beta \text{ for some } (t, \mathbf{y}) \in C_2\} \subset D.$$

Because C' is compact, $\mathbf{f}(t, \mathbf{x})$ satisfies a Lipschitz condition on C' by *Theorem 1.14*. Set

$$V' = \{(s, \mathbf{x}) : |s - t| < \beta \text{ and } |\mathbf{x} - \mathbf{y}| < \beta \text{ for some } (t, \mathbf{y}) \in C_2\}.$$

Thus V' is an open set such that $C_2 \subset V' \subset C' \subset D$ and on V' the differential equation satisfies a Lipschitz condition. The rest of the proof will take place on V and V' and make critical use of the following fact about the trajectories: If $(t, \tau, \boldsymbol{\xi})$ is in V and $|h| < \alpha$, then

$$\{(s, \mathbf{x}(s, \tau, \boldsymbol{\xi})) : s \in I_{t,\tau}\} \subset V'$$

and

$$\{(s, \mathbf{x}(s, \tau, \boldsymbol{\xi} + h\mathbf{e}_j)) : s \in I_{t,\tau}\} \subset V'.$$

The set structure is now in place.

The next step is to set up the difference quotient formula needed in the argument. The difference quotient $\boldsymbol{\theta} : V \times \{h : 0 < |h| < \alpha\} \to \mathbb{R}^d$ by

$$\boldsymbol{\theta}(t, \tau, \boldsymbol{\xi}, h) = \frac{\mathbf{x}(t, \tau, \boldsymbol{\xi} + h\mathbf{e}_j) - \mathbf{x}(t, \tau, \boldsymbol{\xi})}{h}$$

is well defined for each j by the choice of α.

Clearly, $\boldsymbol{\theta}$ is continuous in all variables and can be differentiated with respect to t. If $\boldsymbol{\theta}(t,\tau,\boldsymbol{\xi},h)$ converges to a continuous function on V as h tends to zero, then the partial derivative of each component of $\mathbf{x}(t,\tau,\boldsymbol{\xi})$ with respect to ξ_j exists and is continuous on V. In other words, the first partial derivatives of all d components of $\mathbf{x}(t,\tau,\boldsymbol{\xi})$ with respect to ξ_j can be obtained simultaneously.

The derivative of $\boldsymbol{\theta}$ with respect to t is

$$\dot{\boldsymbol{\theta}}(t,\tau,\boldsymbol{\xi},h) = \frac{\mathbf{f}(t,\mathbf{x}(t,\tau,\boldsymbol{\xi}+h\mathbf{e}_j)) - \mathbf{f}(t,\mathbf{x}(t,\tau,\boldsymbol{\xi}))}{h}.$$

The preceding proposition can be applied to the numerator with $\mathbf{y} = \mathbf{x}(t,\tau,\boldsymbol{\xi})$ and $\mathbf{x} = \mathbf{x}(t,\tau,\boldsymbol{\xi}+h\mathbf{e}_j)$ to obtain

$$\dot{\boldsymbol{\theta}}(t,\tau,\boldsymbol{\xi},h) = \frac{1}{h}\left[\mathbf{f}'_{\mathbf{x}}(t,\mathbf{y}) + \Gamma(\mathbf{x})\right](\mathbf{x}-\mathbf{y}),$$

where $\mathbf{f}'_{\mathbf{x}}$ denotes the $d \times d$ Jacobian matrix of $\mathbf{f}(t,\mathbf{x})$ with respect to the space variable $\mathbf{x} = (x_1,\ldots,x_d)$. Since $\boldsymbol{\theta} = (1/h)(\mathbf{x}-\mathbf{y})$, the above can be rewritten as

$$\dot{\boldsymbol{\theta}}(t,\tau,\boldsymbol{\xi},h) = \left[\mathbf{f}'_{\mathbf{x}}(t,\mathbf{x}(t,\tau,\boldsymbol{\xi})) + \Gamma(t,\tau,\boldsymbol{\xi},h)\right]\boldsymbol{\theta}(t,\tau,\boldsymbol{\xi},h)$$

or more suggestively as

$$\dot{\boldsymbol{\theta}} = \mathbf{f}'_{\mathbf{x}}\boldsymbol{\theta} + \Gamma\boldsymbol{\theta}.$$

In other words, if Γ goes to zero as h goes to zero, then $\boldsymbol{\theta}$ should be converging to the solution of a system of linear differential equations, but using this idea requires considerable further analysis.

Set

$$A(t,\tau,\boldsymbol{\xi}) = \mathbf{f}'_{\mathbf{x}}(t,\mathbf{x}(t,\tau,\boldsymbol{\xi})).$$

By hypothesis, $A_{\mathbf{f}}(t,\tau,\boldsymbol{\xi})$ is a continuous function from $D_{\mathbf{f}}$ to $\mathcal{M}_d(\mathbb{R})$ and

$$\dot{\mathbf{w}} = A(t,\tau,\boldsymbol{\xi})\mathbf{w}$$

can be thought of as a system of linear differential equations parameterized by τ and $\boldsymbol{\xi}$. Let $\mathbf{w}(t,\tau,\boldsymbol{\xi})$ be the solution of $\dot{\mathbf{w}} = A(t,\tau,\boldsymbol{\xi})\mathbf{w}$ such that $\mathbf{w}(\tau,\tau,\boldsymbol{\xi}) = \mathbf{e}_j$. The function $\mathbf{w}(t,\tau,\boldsymbol{\xi})$ is continuous in all variables by *Theorem 1.25.* Thus it suffices to show that $\boldsymbol{\theta}(t,\tau,\boldsymbol{\xi},h)$ converges to $\mathbf{w}(t,\tau,\boldsymbol{\xi})$ on V as h goes to zero. Moreover, it will then follow that

$$\frac{\partial \mathbf{x}}{\partial \boldsymbol{\xi}}(t,\tau,\boldsymbol{\xi}) = W(t,\tau,\boldsymbol{\xi}),$$

where $W(t,\tau,\boldsymbol{\xi})$ is the principal matrix solution of $\dot{\mathbb{W}} = A_{\mathbf{f}}(t,\tau,\boldsymbol{\xi})\mathbb{W}$ at τ. This program will be carried out in a series of three estimates.

Claim 1 *Given $\varepsilon > 0$ there exists $\eta > 0$ such that $|\Gamma(t,\tau,\boldsymbol{\xi},h)| < \varepsilon$ when $(t,\tau,\boldsymbol{\xi}) \in V$ and $|h| < \eta$.*

Proof of Claim 1. Let $(t, \tau, \boldsymbol{\xi})$ be a point in V and let $|h| < \alpha$. The estimate of $|\Gamma|$ provided by *Proposition 7.8* will be used to prove *Claim 1*. Because all the first partial derivatives of $\mathbf{f}(t, \mathbf{x})$ with respect to $x_1, \ldots, x_d$ are uniformly continuous on C', given $\varepsilon > 0$ there exists $\eta' > 0$, such that for $(t, \mathbf{x})$ and $(t, \mathbf{y})$ in C' and $0 \leq s \leq 1$ we have

$$\left| \frac{\partial f_i(t, s\mathbf{x} + (1-s)\mathbf{y})}{\partial x_j} - \frac{\partial f_i(t, \mathbf{y})}{\partial x_j} \right| < \frac{\varepsilon}{d^2},$$

when $|(s\mathbf{x}+(1-s)\mathbf{y})-\mathbf{y}| = s|\mathbf{x}-\mathbf{y}| \leq |\mathbf{x}-\mathbf{y}| < \eta'$. It follows that $|\Gamma(t, \tau, \boldsymbol{\xi}, h)| < \varepsilon$ for $(t, \tau, \boldsymbol{\xi}) \in V$ when $|\mathbf{x}(t, \tau, \boldsymbol{\xi} + h\mathbf{e}_j) - \mathbf{x}(t, \tau, \boldsymbol{\xi})| < \eta'$.

Because $\mathbf{x}(t, \tau, \boldsymbol{\xi})$ is continuous on $D_{\mathbf{f}}$ by *Theorem 1.23*, it is uniformly continuous on C and there exists $\eta > 0$ such that $|\mathbf{x}(t, \tau, \boldsymbol{\xi} + h\mathbf{e}_j) - \mathbf{x}(t, \tau, \boldsymbol{\xi})| = |\mathbf{x} - \mathbf{y}| < \eta'$ when $|h| < \eta$. Therefore, $|\Gamma(t, \tau, \boldsymbol{\xi}, h)| < \varepsilon$ when $|h| < \eta$. □

Claim 2 *Let L be the Lipschitz constant for $\mathbf{f}(t, \mathbf{x})$ on V'. If $(t, \tau, \boldsymbol{\xi}) \in V$ and $|h| < \alpha$, then*

$$|\boldsymbol{\theta}(s, \tau, \boldsymbol{\xi}, h)| < e^{L(b-a)}$$

for $s \in I_{t,\tau}$.

Proof of Claim 2. Let $(t, \tau, \boldsymbol{\xi})$ be a point in V and let $|h| < \alpha$. *Theorem 2.13* on page 82 can be applied on V' to the solutions $\mathbf{x}(s, \tau, \boldsymbol{\xi})$ and $\mathbf{x}(s, \tau, \boldsymbol{\xi} + h\mathbf{e}_j)$ on the interval $I_{t,\tau}$ starting at τ. Since $\mathbf{x}(s, \tau, \boldsymbol{\xi})$ and $\mathbf{x}(s, \tau, \boldsymbol{\xi} + h\mathbf{e}_j)$ are actual solutions, $\varepsilon_1 = 0 = \varepsilon_2$ and $\delta = |\mathbf{x}(\tau, \tau, \boldsymbol{\xi} + h\mathbf{e}_j) - \mathbf{x}(\tau, \tau, \boldsymbol{\xi})| = |\boldsymbol{\xi} + h\mathbf{e}_j - \boldsymbol{\xi}| = |h|$. Consequently,

$$|\mathbf{x}(s, \tau, \boldsymbol{\xi} + h\mathbf{e}_j) - \mathbf{x}(s, \tau, \boldsymbol{\xi})| \leq |h| e^{L|s-\tau|} \leq |h| e^{L(b-a)}$$

for $s \in I_{t,\tau} \subset [a, b]$ and dividing by $|h|$ gives the desired estimate. □

Claim 3 *If $(t, \tau, \boldsymbol{\xi})$ is in V and $|h| < \alpha$, then there exists a positive constant L' such that*

$$|\boldsymbol{\theta}(t, \tau, \boldsymbol{\xi}, h) - \mathbf{w}(t, \tau, \boldsymbol{\xi})| \leq \sup_{s \in I_{t,\tau}} |\Gamma(s, \tau, \boldsymbol{\xi}, h)| e^{L(b-a)} \left[\frac{e^{L'(b-a)} - 1}{L'} \right].$$

Proof of Claim 3. Let $(t, \tau, \boldsymbol{\xi})$ be a point in V and let $|h| < \alpha$. *Theorem 2.13* will be applied again, but this time to $\dot{\mathbf{w}} = A(t, \tau, \boldsymbol{\xi})\mathbf{w}$ for fixed τ and $\boldsymbol{\xi}$. [Remember that $\boldsymbol{\xi}$ is a parameter here not an initial position for $\mathbf{w}(t, \tau, \boldsymbol{\xi})$ and τ is both the initial time and a parameter because $\mathbf{w}(\tau, \tau, \boldsymbol{\xi}) = \mathbf{e}_j$.] Since $(t, \tau, \boldsymbol{\xi}) \in C$ implies that $\{(s, \tau, \boldsymbol{\xi}) : s \in I_{t,\tau}\} \subset C_1$, the linear differential equation $\dot{\mathbf{w}} = A(t, \tau, \boldsymbol{\xi})\mathbf{w}$ and its solution $\mathbf{w}(t, \tau, \boldsymbol{\xi})$ are defined on an open interval containing $I_{t,\tau}$. Furthermore, the appropriate Lipschitz constant for $\dot{\mathbf{w}} = A(t, \tau, \boldsymbol{\xi})\mathbf{w}$ with $(t, \tau, \boldsymbol{\xi}) \in C$ is

$$L' = \sup\{|A(t, \tau, \boldsymbol{\xi})| : (t, \tau, \boldsymbol{\xi}) \in C_1\}.$$

The initial time for the estimate is τ, and thus $\delta = 0$ because

$$\boldsymbol{\theta}(\tau, \tau, \boldsymbol{\xi}, h) = \mathbf{e}_j = \mathbf{w}(\tau, \tau, \boldsymbol{\xi}).$$

For $s \in I_{t,\tau}$, the earlier equation

$$\dot{\boldsymbol{\theta}}(t, \tau, \boldsymbol{\xi}, h) = [\mathbf{f}'_{\mathbf{x}}(t, \mathbf{x}(t, \tau, \boldsymbol{\xi})) + \Gamma(t, \tau, \boldsymbol{\xi}, h)]\, \boldsymbol{\theta}(t, \tau, \boldsymbol{\xi}, h)$$

can be rewritten as

$$\dot{\boldsymbol{\theta}}(s, t, \tau, \boldsymbol{\xi}) - A(s, \tau, \boldsymbol{\xi})\boldsymbol{\theta}(s, \tau, \boldsymbol{\xi}) = \Gamma(s, \tau, \boldsymbol{\xi})\boldsymbol{\theta}(s, \tau, \boldsymbol{\xi}, h).$$

It was shown in *Claim 2* that $|\boldsymbol{\theta}(s, \tau, \boldsymbol{\xi})|$ is uniformly bounded by $e^{L(b-a)}$ for $(t, \tau, \boldsymbol{\xi}) \in V$, $|h| < \alpha$, and $s \in I_{t,\tau}$. Consequently,

$$\begin{aligned} |\dot{\boldsymbol{\theta}}(s, t, \tau, \boldsymbol{\xi}) - A(s, \tau, \boldsymbol{\xi})\boldsymbol{\theta}(s, \tau, \boldsymbol{\xi})| &= |\Gamma(s, \tau, \boldsymbol{\xi})\boldsymbol{\theta}(s, \tau, \boldsymbol{\xi}, h)| \\ &\le |\Gamma(s, \tau, \boldsymbol{\xi})|\, |\boldsymbol{\theta}(s, \tau, \boldsymbol{\xi}, h)| \\ &\le \sup_{s \in I_{t,\tau}} \{|\Gamma(s, \tau, \boldsymbol{\xi}, h)|\} e^{L(b-a)} \\ &= \varepsilon_1. \end{aligned}$$

Because $\mathbf{w}(t, \tau, \boldsymbol{\xi})$ is a solution of $\dot{\mathbf{w}} = A(t, \tau, \boldsymbol{\xi})\mathbf{w}$, set $\varepsilon_2 = 0$. The conclusion is now an immediate consequence of *Theorem 2.13.* □

The pieces can now be put together. Given $\varepsilon > 0$, let $\eta > 0$ be determined by *Claim 1*. It follows that

$$\sup_{s \in I_{t,\tau}} |\Gamma(s, \tau, \boldsymbol{\xi}, h)| < \varepsilon$$

when $|h| < \eta$ and $(t, \tau, \boldsymbol{\xi}) \in V$. Then by *Claim 3*,

$$|\boldsymbol{\theta}(t, \tau, \boldsymbol{\xi}, h) - \mathbf{w}(t, \tau, \boldsymbol{\xi})| \le \varepsilon e^{L(b-a)} \left[\frac{e^{L'(b-a)} - 1}{L'} \right]$$

when $|h| < \eta$ and $(t, \tau, \boldsymbol{\xi}) \in V$. Therefore, $\boldsymbol{\theta}(t, \tau, \boldsymbol{\xi}, h)$ converges uniformly to $\mathbf{w}(t, \tau, \boldsymbol{\xi})$ on V as h goes to zero because the constants a, b, L, and L' depend only on the sets V and V'. This proves that $\mathbf{x}(t, \tau, \boldsymbol{\xi})$ has continuous first partial derivatives with respect to $\xi_1, \ldots, \xi_d$ on V.

For the partial with respect to τ, set

$$\boldsymbol{\theta}(t, \tau, \boldsymbol{\xi}, h) = \frac{\mathbf{x}(t, \tau + h, \boldsymbol{\xi}) - \mathbf{x}(t, \tau, \boldsymbol{\xi})}{h}$$

and proceed as above with two minor modifications based on the formula

$$\mathbf{x}(t, \tau + h, \boldsymbol{\xi}) - \mathbf{x}(t, \tau, \boldsymbol{\xi}) = \int_{\tau}^{\tau+h} \mathbf{f}(s, \mathbf{x}(s, \tau + h, \boldsymbol{\xi}))ds.$$

First, $|\boldsymbol{\theta}|$ is estimated differently using

$$|\mathbf{x}(t, \tau + h, \boldsymbol{\xi}) - \mathbf{x}(t, \tau, \boldsymbol{\xi})| \le \int_{\tau}^{\tau+h} |\mathbf{f}(s, \mathbf{x}(s, \tau + h, \boldsymbol{\xi}))|ds \le M|h|,$$

where M is a bound for $|\mathbf{f}(t, \mathbf{x})|$ on the compact set C. Hence, $|\boldsymbol{\theta}| \leq M$ on C'. Second, choose $\mathbf{w}(t, \tau, \boldsymbol{\xi})$ to be the solution of $\dot{\mathbf{w}} = A_{\mathbf{f}}(t, \tau, \boldsymbol{\xi})\mathbf{w}$ such that $\mathbf{w}(\tau, \tau, \boldsymbol{\xi}) = -\mathbf{f}(\tau, \boldsymbol{\xi})$ because

$$\lim_{h \to 0} \boldsymbol{\theta}(t, \tau, \boldsymbol{\xi}) = \lim_{h \to 0} \frac{1}{h} \int_{\tau}^{\tau+h} \mathbf{f}(s, \mathbf{x}(s, \tau + h, \boldsymbol{\xi}))ds = -\mathbf{f}(\tau, \boldsymbol{\xi}).$$

With these two modifications in place, the proof that $\mathbf{x}(t, \tau, \boldsymbol{\xi})$ has continuous first partial derivatives with respect $\boldsymbol{\xi}_1, \ldots, \boldsymbol{\xi}_d$ also proves that $\mathbf{x}(t, \tau, \boldsymbol{\xi})$ has a continuous partial derivative with respect to τ, the initial time. □

Corollary *If $\mathbf{f}(t, \mathbf{x})$ has continuous first partial derivatives on D with respect to $x_1, \ldots, x_d$, then*

$$\lim_{\mathbf{h} \to \mathbf{0}} \frac{|\mathbf{x}(t, \tau, \boldsymbol{\xi} + \mathbf{h}) - \mathbf{x}(t, \tau, \boldsymbol{\xi}) - W(t, \tau, \boldsymbol{\xi})\mathbf{h}|}{|\mathbf{h}|} = 0$$

where $W(t, \tau, \boldsymbol{\xi})$ is the principal matrix solution at τ of

$$\dot{W} = \left[\frac{\partial \mathbf{f}}{\partial \mathbf{x}}(t, \mathbf{x}(t, \tau, \boldsymbol{\xi}))\right] W.$$

Proof. This is an immediate consequence of *Theorems 7.7 and 7.9* and the definition of a differentiable function. □

The proof of *Theorem 7.9* established substantially more than stated in the theorem. The next theorem records the additional facts that emerged from the proof of *Theorem 7.9.*

Theorem 7.10 *If $\mathbf{f}(t, \mathbf{x})$ has continuous first partial derivatives with respect $x_1, \ldots, x_d$ on D, then the following hold:*

(a) The global solution $\mathbf{x}(t, \tau, \boldsymbol{\xi})$ is smooth on its domain $D_{\mathbf{f}}$.

(b) For fixed τ and $\boldsymbol{\xi}$ the Jacobian matrix

$$\left[\frac{\partial \mathbf{x}}{\partial \boldsymbol{\xi}}(t, \tau, \boldsymbol{\xi})\right]$$

as a function of t is the principal matrix solution at τ of the linear matrix differential equation

$$\dot{W} = \left[\frac{\partial \mathbf{f}}{\partial \mathbf{x}}(t, \mathbf{x}(t, \tau, \boldsymbol{\xi}))\right] W.$$

(c) For fixed τ and $\boldsymbol{\xi}$ the first partial derivative of $\mathbf{x}(t,\tau,\boldsymbol{\xi})$ with respect to τ is

$$\frac{\partial\, \mathbf{x}(t,\tau,\boldsymbol{\xi})}{\partial\, \tau} = \left[\frac{\partial\, \mathbf{x}}{\partial\, \boldsymbol{\xi}}(t,\tau,\boldsymbol{\xi})\right](-\mathbf{f}(\tau,\boldsymbol{\xi}))$$

and is a solution of the linear differential equation

$$\dot{\mathbf{w}} = \left[\frac{\partial\, \mathbf{x}}{\partial\, \boldsymbol{\xi}}(t,\tau,\boldsymbol{\xi})\right]\mathbf{w}.$$

(d)

$$\text{Det}\left[\frac{\partial\, \mathbf{x}}{\partial\, \boldsymbol{\xi}}(t,\tau,\boldsymbol{\xi})\right] = \exp\left(\int_{\tau}^{t} \text{Tr}\ [\mathbf{f}'_{\mathbf{x}}(t,\mathbf{x}(t,\tau,\boldsymbol{\xi}))]\, ds\right).$$

Proof. The first three parts were established in the proof of *Theorem 7.9* and the last one follows from part (b) and *Theorem 3.8* (Abel's formula). □

It is also easy to extend *Theorem 7.9* to smooth systems with parameters.

Theorem 7.11 *If $\mathbf{f}(t,\mathbf{x},\boldsymbol{\mu})$ has continuous first partial derivatives on D with respect to the space variables $x_1,\ldots,x_d$ and the parameters $\mu_1,\ldots,\mu_k$, then the general solution $\mathbf{x}(t,\tau,\boldsymbol{\xi},\boldsymbol{\mu})$ of $\dot{\mathbf{x}} = \mathbf{f}(t,\mathbf{x},\boldsymbol{\mu})$ is smooth on $D_{\mathbf{f}}$.*

Proof. The main theorem (*Theorem 7.9*) applies to the smooth system

$$\begin{aligned}
\dot{x}_1 &= f_1(t,\mathbf{x},\boldsymbol{\mu}) \\
&\vdots \\
\dot{x}_d &= f_d(t,\mathbf{x},\boldsymbol{\mu}) \\
\dot{\mu}_1 &= 0 \\
&\vdots \\
\dot{\mu}_k &= 0
\end{aligned}$$

and implies that $\mathbf{x}(t,\tau,\boldsymbol{\xi},\boldsymbol{\mu})$ is a smooth function on $D_{\mathbf{f}}$. □

A function with continuous first partial derivatives or a *smooth function* is also called a C^1 function. More generally, a *C^r function* is a function that has continuous partial derivatives of all orders up to and including the rth order partial derivatives. Naturally, a C^∞ function is a function that has continuous partial derivatives of all orders. Some authors define smooth functions to be C^∞ functions.

Theorem 7.12 *If $\mathbf{f}(t,\mathbf{x})$ has a continuous first partial derivative with respect to t and continuous first and second partial derivatives with respect to $x_1,\ldots,\mathbf{x}_d$, then $\mathbf{x}(t,\tau,\boldsymbol{\xi})$ is a C^2 function on its domain $D_{\mathbf{f}}$.*

Proof. Clearly, *Theorem 7.11* applies to

$$\dot{\mathbf{w}} = \left[\frac{\partial \mathbf{f}}{\partial \mathbf{x}}(t, \mathbf{x}(t, \tau, \boldsymbol{\xi}))\right] \mathbf{w}.$$

Therefore, its solutions

$$\frac{\partial \mathbf{x}(t, \tau, \boldsymbol{\xi})}{\partial \xi_j} \quad \text{and} \quad \frac{\partial \mathbf{x}(t, \tau, \boldsymbol{\xi})}{\partial \tau}$$

have continuous first partial derivatives with respect to t and the parameters τ and $\xi_1, \ldots, \xi_d$. □

What was not said in the last theorem was that the second partial derivatives of $\mathbf{x}(t, \tau, \boldsymbol{\xi})$ would be solutions of a large linear system of differential equations whose coefficients where second partial derivatives of $\mathbf{f}(t, \mathbf{x})$. If $\mathbf{f}$ had continuous third partial derivatives, the argument could be repeated. It should now be intuitively clear that, if $\mathbf{f}$ is a C^r function, then $\mathbf{x}(t, \tau, \boldsymbol{\xi})$ is also a C^r function, but we shall not attempt to state and prove such a result here.

EXERCISES

1. Prove that the set C_1 in the proof of *Theorem 7.9* is a compact set.

2. Complete the proof of part (c) of *Theorem 7.10* as outlined at the end of the proof of *Theorem 7.9.*

3. Consider the scalar differential equation $\dot{x} = f(t, x) = tx^2$.

   (a) Directly calculate the general solution $x(t, \tau, \xi)$ and its first partial derivative with respect to ξ.

   (b) Calculate the solution $w(t, \tau, \xi)$ of the linear differential equation

   $$\dot{w} = \frac{\partial f}{\partial x}(t, \mathbf{x}(t, \tau, \xi))\, w$$

   for $f(t, x) = tx^2$.

   (c) Verify that

   $$\frac{\partial x(t, \tau, \xi)}{\partial \xi} \equiv w(1, \tau, \xi).$$

4. Let $\mathbf{f}(\mathbf{x})$ be a smooth vector field on Ω with a fixed point at $\mathbf{p}$. Use *Theorem 7.10* to calculate

   $$\left[\frac{\partial \mathbf{x}}{\partial \boldsymbol{\xi}}(t, \mathbf{p})\right].$$

5. Let $\dot{x} = f(t,x)$ be a scalar differential equation defined on the infinite strip $D = \{(t,x) : a < t < b\}$ and suppose there exist a constant C such that
$$\left|\frac{\partial f(t,x)}{\partial x}\right| \le C.$$
Use *Theorem 7.10* to prove that
$$|x(t,\tau,\xi)| \le |\xi| e^{C(|t-\tau|)}$$
and that $x(t,\tau,\xi)$ is defined on $a < t < b$.

6. Consider the scalar differential equation $\dot{x} = f(t,x)$ defined on $D = \{(t,x) : a < t < \infty\}$ as above and suppose there exists a constants K and L such that
$$K \le \frac{\partial f(t,x)}{\partial x} \le L.$$
Use the mean-value theorem and *Theorem 7.10* to prove that
$$|\xi - \xi'| e^{K(t-\tau)} \le |x(t,\tau,\xi) - x(t,\tau,\xi')| \le |\xi - \xi'| e^{L(t-\tau)}$$
for $t \ge \tau$. Then prove the following:
   (a) If $L = 0$, then every solution is uniformly stable.
   (b) If $L < 0$, then every solution is uniformly asymptotically stable.
   (c) If $K > 0$, then no solution is Lyapunov stable.

7.3 Linearization

Consider an autonomous differential equation $\dot{\mathbf{x}} = \mathbf{f}(\mathbf{x})$ on an open set Ω in $\mathbb{R}^d$ and suppose that $\mathbf{f}$ is differentiable at $\boldsymbol{\xi} \in \Omega$. For $\mathbf{x}$ near $\boldsymbol{\xi}$ we can approximate $\mathbf{f}(\mathbf{x})$ by $\mathbf{f}(\boldsymbol{\xi}) + \mathrm{D}\mathbf{f}(\boldsymbol{\xi})(\mathbf{x} - \boldsymbol{\xi})$. The linear transformation $\mathrm{D}\mathbf{f}(\boldsymbol{\xi})$ is called the *linear part* of $\mathbf{f}$ at $\boldsymbol{\xi}$ and contains information about $\mathbf{f}$. Naturally, $\mathbf{f} - \mathrm{D}\mathbf{f}(\boldsymbol{\xi})$ is called the *nonlinear part* of $\mathbf{f}$ at $\boldsymbol{\xi}$. Broadly speaking linearization is the process of studying $\dot{\mathbf{x}} = \mathbf{f}(\mathbf{x})$ through its linear parts—at particular points or collectively. This section contains an example of each.

All the vector fields $\mathbf{f}(\mathbf{x})$ in this section will be smooth on Ω, and hence differentiable at every point $\mathbf{x} \in \Omega$. Since smooth vector fields are locally Lipschitz (*Theorem 1.12*), their solutions of $\dot{\mathbf{x}} = \mathbf{f}(\mathbf{x})$ have many desirable properties. Because only the autonomous differential equation $\dot{\mathbf{x}} = \mathbf{f}(\mathbf{x})$ will be considered in this section, the simpler notation $\mathbf{x}(t,\boldsymbol{\xi})$ will be used instead of $\mathbf{x}(t,0,\boldsymbol{\xi})$.

The primary linearization theorem in this section uses only $\mathrm{D}\mathbf{f}$ at a fixed point $\mathbf{p}$ of $\mathbf{f}$ to obtain a sufficient condition for the asymptotic stability of the fixed point. It requires some preliminary results about positive definite matrices so that Lyapunov's method (*Theorem 5.1* on page 148) can be applied .

A bilinear function $\beta : \mathbb{R}^d \times \mathbb{R}^d \to \mathbb{R}$ is *positive definite* if $\beta(\mathbf{x}, \mathbf{x}) \geq 0$ for all $\mathbf{x} \in \mathbb{R}^d$ and $\beta(\mathbf{x}, \mathbf{x}) = 0$ if and only if $\mathbf{x} = \mathbf{0}$. (See page 219 for the definition of bilinear function.) A real $d \times d$ matrix B is *a positive definite matrix* if the bilinear function $\beta(\mathbf{x}, \mathbf{y}) = \mathbf{x} \cdot (B\mathbf{y})$ is a positive definite bilinear function. If a bilinear function is positive definite, then $V(\mathbf{x}) = \beta(\mathbf{x}, \mathbf{x})$ is a positive definite function at $\mathbf{0}$ as defined for Lyapunov functions on page 148 and is a natural candidate for a Lyapunov function.

Recall that for an autonomous differential equation $\dot{\mathbf{x}} = \mathbf{f}(\mathbf{x})$, the real-valued function $\dot{V}(\mathbf{x}) = \nabla V(\mathbf{x}) \cdot \mathbf{f}(\mathbf{x})$ was the essential ingredient in Lyapunov's method. Assuming as usual that V has continuous first partial derivatives or equivalently is a smooth function, V is differentiable at every $\mathbf{x}$ in its domain and $\mathrm{DV}(\mathbf{x})(\mathbf{v}) = \nabla \mathrm{V}(\mathbf{x}) \cdot \mathbf{v}$. Hence, $\dot{V}(\mathbf{x})$ can be written as

$$\dot{V}(\mathbf{x}) = \mathrm{DV}(\mathbf{x})(\mathbf{f}(\mathbf{x}))$$

If $\beta(\mathbf{x}, \mathbf{y})$ is a bilinear function and $V(\mathbf{x}) = \beta(\mathbf{x}, \mathbf{x})$, then the corollary to *Proposition 7.6* implies that

$$\dot{V}(\mathbf{x}) = \beta(\mathbf{x}, \mathbf{f}(\mathbf{x})) + \beta(\mathbf{f}(\mathbf{x}), \mathbf{x}).$$

Suppose, in addition, that $\mathbf{f}(\mathbf{x}) = A\mathbf{x}$ for some $A \in \mathcal{M}_d(\mathbb{R})$ and that $\beta(\mathbf{x}, \mathbf{y}) = \mathbf{x} \cdot (B\mathbf{y})$ is an arbitrary bilinear form given by $B \in \mathcal{M}_d(\mathbb{R})$. If $V(\mathbf{x}) = \beta(\mathbf{x}, \mathbf{x}) = \mathbf{x} \cdot (B\mathbf{x})$, then

$$\begin{aligned} \dot{V}(\mathbf{x}) &= \mathrm{DV}(\mathbf{x})(\mathrm{A}\mathbf{x})) \\ &= \mathbf{x} \cdot (BA\mathbf{x}) + (A\mathbf{x}) \cdot (B\mathbf{x}) \\ &= \mathbf{x} \cdot (BA\mathbf{x}) + \mathbf{x} \cdot A^{\mathrm{t}}(B\mathbf{x}) \\ &= \mathbf{x} \cdot (BA\mathbf{x}) + \mathbf{x} \cdot (A^{\mathrm{t}}B\mathbf{x}) \\ &= \mathbf{x} \cdot ((BA + A^{\mathrm{t}}B)\mathbf{x}). \end{aligned}$$

Therefore, to use $V(\mathbf{x}) = \mathbf{x} \cdot (B\mathbf{x})$ to show that $\mathbf{0}$ is a positively asymptotically stable fixed point of $\dot{\mathbf{x}} = A\mathbf{x}$, both bilinear functions $\beta(\mathbf{x}, \mathbf{y}) = \mathbf{x} \cdot (B\mathbf{y})$ and

$$\gamma(\mathbf{x}, \mathbf{y}) = -\mathbf{x} \cdot \big((BA + A^{\mathrm{t}}B)\mathbf{x}\big)$$

must be positive definite.When does this happen? The answer is surprising.

Theorem 7.13 *Let A be a real $d \times d$ matrix. The following are equivalent:*

(a) There exist positive definite matrices B and C such that

$$BA + A^{\mathrm{t}}B = -C.$$

(b) The real parts of all the eigenvalues of A are less than zero.

(c) The matrix equation

$$BA + A^{\mathrm{t}}B = -C$$

has a positive definite (symmetric) solution B for every positive definite (symmetric) matrix C.

Proof. The first step is to show that (b) follows from (a). Assuming B and C are positive definite matrices such that

$$BA + A^{t}B = -C,$$

set $V(\mathbf{x}) = \mathbf{x}\cdot(B\mathbf{x})$. From the discussion preceding the statement of the theorem,

$$\dot{V}(\mathbf{x}) = \mathbf{x}\cdot((BA + A^{t}B)\mathbf{x}) = -\mathbf{x}\cdot(C\mathbf{x}).$$

Consequently, both $V(\mathbf{x})$ and $-\dot{V}(\mathbf{x})$ are positive definite functions at the fixed point $\mathbf{0}$ for $\dot{\mathbf{x}} = A\mathbf{x}$. By *Theorem 5.1*, the fixed point $\mathbf{0}$ is an asymptotically stable fixed point and by *Theorem 5.9* the real parts of the eigenvalues of A are all less than zero. Thus (a) implies (b).

Obviously (c) implies (a). So it remains to prove that (b) implies (c), which is the more substantive step in the proof. Suppose the eigenvalues of A have negative real parts. By *Theorem 5.8* there exist positive constants K and α such that for $t \geq 0$,

$$|e^{At}| \leq Ke^{-\alpha t}$$

and

$$|e^{A^{t}t}| \leq Ke^{-\alpha t}.$$

Given a positive definite matrix C, set

$$B = \int_0^\infty e^{A^{t}s}Ce^{As}ds.$$

It is easy to check that this improper integral converges to a positive definite matrix B. Moreover, B is symmetric, if C is symmetric. Notice that

$$\lim_{b\to\infty}\int_0^b \frac{d[e^{A^{t}s}Ce^{As}]}{ds}ds = \lim_{b\to\infty} e^{A^{t}b}Ce^{Ab} - C = -C,$$

and hence

$$\int_0^\infty \frac{d[e^{A^{t}s}Ce^{As}]}{ds}ds = -C.$$

The above integral is also equal to

$$\int_0^\infty \left(A^{t}e^{A^{t}s}Ce^{As} + e^{A^{t}s}Ce^{As}A\right) ds$$

$$= A^{t}\int_0^\infty e^{A^{t}s}Ce^{As}ds + \int_0^\infty e^{A^{t}s}Ce^{As}dsA = A^{t}B + BA.$$

Thus $A^{t}B + BA = -C$ and the proof is finished. □

For an asymptotically stable linear systems, this theorem guarantees the existence of a Lypunov function that meets the hypothesis of *Theorem 5.1* for asymptotic stability. The obvious strategy is to try to show that this same Lyapunov function will work for $\dot{\mathbf{x}} = \mathbf{f}(\mathbf{x})$ when the linear part of $\mathbf{f}$ at a fixed point $\mathbf{p}$ is $A\mathbf{x}$ and $\dot{\mathbf{x}} = A\mathbf{x}$ is asymptotically stable.

Theorem 7.14 *Let* $\mathbf{f}(\mathbf{x})$ *be a smooth vector field on* Ω *with a fixed point at* $\mathbf{p}$. *If all the eigenvalues of* $\mathbf{f}'(\mathbf{p})$ *have negative real parts, then* $\mathbf{p}$ *is a positively asymptotically stable fixed point.*

Proof. Without loss of generality we can assume that $\mathbf{p} = \mathbf{0}$ (The change of variables $\mathbf{x} = \mathbf{y} + \mathbf{p}$ accomplishes this without changing the derivative of the vector field at the fixed point.) Set $A = \mathbf{f}'(\mathbf{0})$. By *Theorem 7.13* there exists a positive definite symmetric matrix B such that $A^t B + BA = -I$. Then $V(\mathbf{x}) = \mathbf{x} \cdot (B\mathbf{x})$ is positive definite at $\mathbf{0}$ and it suffices by Lyapunov's method (*Theorem 5.1*) to show that $-\dot{V}(\mathbf{x})$ is also positive definite at $\mathbf{0}$.

Because $\mathbf{f}$ is smooth and $\mathbf{f}(\mathbf{0}) = \mathbf{0}$,

$$\mathbf{f}(\mathbf{x}) = \mathrm{D}\mathbf{f}(\mathbf{0})(\mathbf{x}) + \mathbf{n}(\mathbf{x}) = \mathrm{A}\mathbf{x} + \mathbf{n}(\mathbf{x}),$$

where the error term $\mathbf{n}(\mathbf{x})$ satisfies

$$\lim_{\mathbf{x}\to\mathbf{0}} \frac{\|\mathbf{n}(\mathbf{x})\|}{\|\mathbf{x}\|} = 0.$$

Using these formulae, $C = I$, and $B^t = B$, it follows from the discussion preceding *Theorem 7.13* that

$$\begin{aligned}
\dot{V}(\mathbf{x}) &= \mathrm{DV}(\mathbf{x})(\mathrm{A}\mathbf{x} + \mathbf{n}(\mathbf{x})) \\
&= \mathrm{DV}(\mathbf{x})(\mathrm{A}\mathbf{x}) + \mathrm{DV}(\mathbf{x})(\mathbf{n}(\mathbf{x})) \\
&= \mathbf{x} \cdot ((BA + A^t B)\mathbf{x}) + \mathbf{x} \cdot (B\mathbf{n}(\mathbf{x})) + \mathbf{n}(\mathbf{x}) \cdot (B\mathbf{x}) \\
&= -\mathbf{x} \cdot \mathbf{x} + 2\mathbf{x} \cdot (B\mathbf{n}(\mathbf{x})).
\end{aligned}$$

Therefore, $-\dot{V}(\mathbf{x}) = \|\mathbf{x}\|^2 - 2\mathbf{x} \cdot (B\mathbf{n}(\mathbf{x}))$.

The next step is to show that

$$\lim_{\mathbf{x}\to\mathbf{0}} \frac{|\mathbf{x} \cdot (B\mathbf{n}(\mathbf{x}))|}{\|\mathbf{x}\|^2} = 0.$$

By the Cauchy-Schwartz inequality,

$$|\mathbf{x} \cdot (B\mathbf{n}(\mathbf{x}))| \le \|\mathbf{x}\| \, \|B\mathbf{n}(\mathbf{x})\| \le \|\mathbf{x}\| \, \|B\| \, \|\mathbf{n}(\mathbf{x})\|,$$

and hence

$$0 \le \frac{|\mathbf{x}B\mathbf{n}(\mathbf{x})|}{\|\mathbf{x}\|^2} \le \frac{\|B\| \, \|\mathbf{n}(\mathbf{x})\|}{\|\mathbf{x}\|},$$

which in turn implies that

$$\lim_{\mathbf{x}\to\mathbf{0}} \frac{|\mathbf{x}B\mathbf{n}(\mathbf{x})|}{\|\mathbf{x}\|^2} = 0$$

because

$$\lim_{\mathbf{x}\to\mathbf{0}} \frac{\|\mathbf{n}(\mathbf{x})\|}{\|\mathbf{x}\|} = 0.$$

The following lemma applies to prove that $\dot{V}(\mathbf{x}) = \|\mathbf{x}\|^2 - 2\mathbf{x} \cdot (B(\mathbf{n}(\mathbf{x}))$ is positive definite at $\mathbf{0}$ and complete the proof. $\square$

Lemma *Suppose C is a positive definite $d \times d$ matrix and $g(\mathbf{x})$ is a continuous real-valued function in a neighborhood of $\mathbf{0}$ in $\mathbb{R}^d$. If*

$$\lim_{\mathbf{x}\to\mathbf{0}} \frac{|g(\mathbf{x})|}{\|\mathbf{x}\|^2} = 0,$$

then

$$W(\mathbf{x}) = \mathbf{x} \cdot (C\mathbf{x}) + g(\mathbf{x})$$

is a positive definite function at $\mathbf{0}$.

Proof. Set $\alpha = \min\{\mathbf{x} \cdot (C\mathbf{x}) : \|\mathbf{x}\| = 1\}$, which is positive because C is positive definite. The hypothesis that

$$\lim_{\mathbf{x}\to\mathbf{0}} \frac{|g(\mathbf{x})|}{\|\mathbf{x}\|^2} = 0,$$

implies that there exist $\delta > 0$ such that $|g(\mathbf{x})|/\|\mathbf{x}\|^2 < \alpha/2$ when $\|\mathbf{x}\| < \delta$. Then, for $0 < \|\mathbf{x}\| < \delta$,

$$\begin{aligned}
W(\mathbf{x}) &\geq \mathbf{x} \cdot (C\mathbf{x}) - |g(\mathbf{x})| \\
&\geq \mathbf{x} \cdot (C\mathbf{x}) - \frac{\alpha\|\mathbf{x}\|^2}{2} \\
&= \|\mathbf{x}\|^2 \left(\frac{\mathbf{x}}{\|\mathbf{x}\|} \cdot \left(C\frac{\mathbf{x}}{\|\mathbf{x}\|} \right) \right) - \frac{\alpha\|\mathbf{x}\|^2}{2} \\
&\geq \|\mathbf{x}\|^2 \alpha - \frac{\alpha\|\mathbf{x}\|^2}{2} \\
&= \frac{\alpha\|\mathbf{x}\|^2}{2} \\
&> 0.
\end{aligned}$$

Since $g(\mathbf{x})$ goes to 0 faster than $\|\mathbf{x}\|^2$, it follows that $g(\mathbf{0}) = 0$ and $W(\mathbf{0}) = 0$. Thus $W(\mathbf{x})$ is a positive definite function at $\mathbf{0}$. $\square$

It is not possible, however, to infer anything about the stability of a fixed point when the linear part is only positively stable. Positive stability is very fragile as the following example shows: Consider the system

$$\begin{aligned}
\dot{x} &= y + \mu x^3 \\
\dot{y} &= -x + \mu y^3
\end{aligned}$$

for real μ. Then $(0,0)$ is the only fixed point and the linear part of the vector field at this fixed point for all values of μ is

$$\begin{bmatrix} 0 & 1 \\ -1 & 0 \end{bmatrix}.$$

Since the eigenvalues are $\pm i$, the linear system is positively stable but not asymptotically stable at $(0,0)$. In fact, every point except $(0,0)$ is periodic. Note that the linear system is the same as the original system with $\mu = 0$.

Let $V(x,y) = x^2+y^2$, which is positive definite. Then $\dot{V}(x,y) = 2\mu(x^4+y^4)$, which is positive or negative definite according as μ is positive or negative. Thus the stability of $(0,0)$ in the original system changes from positively asymptotically stable to negatively asymptotically stable as μ moves from negative to positive values while the linear part remains the same. In other words, there is a dramatic change in the qualitative behavior of the solutions as the parameter μ cross the origin that is not predicted by the linear part. Furthermore, the converse of *Theorem 7.14* is false.

The second linearization result uses the linear part of the vector field at every point in Ω to determine when $\mathbf{x}(t,\boldsymbol{\xi})$ is volume preserving, but first it is necessary to provide some background.

In the present context, volume will mean $\int_C 1\, d\mathbf{x}$ provided the Riemann integral exists. Sets for which the Riemann integral $\int_C 1\, d\mathbf{x}$ exists are usually called *Jordan measurable sets* and include many familiar bounded sets. We will restrict our attention to Jordan measurable sets.

The volume of the unit cube $C = [0,1] \times \ldots \times [0,1] = \{\mathbf{x} : 0 \le x_i \le 1\}$ is obviously 1. If $T : \mathbb{R}^d \to \mathbb{R}^d$ is an invertible linear transformation, then $T(C)$ is a parallel pipette of volume $|\text{Det}\,[T]|$. For more general sets C, the volume of $T(C)$ is the volume of C times the $|\text{Det}\,[T]|$. Thus the volume of $T(C)$ can be written as

$$\int_{T(C)} 1\, d\mathbf{x} = |Det[T]| \int_C 1\, d\mathbf{x}.$$

In particular, a linear transformation $T : \mathbb{R}^d \to \mathbb{R}^d$ preserves the volume of sets if and only if $\text{Det}\,[T] = \pm 1$. The above formula for the volume of $T(C)$ can be extended to more general maps using the Jacobian.

A *diffeomorphism* is a smooth map $\mathbf{g} : E \to E'$ with a smooth inverse $g^{-1} : E' \to E$ where E and E' are open subsets of $\mathbb{R}^d$. If g is a diffeomorphism, then by the chain rule

$$\begin{aligned}
\text{Det}\left[(\mathbf{g}^{-1})'(\mathbf{g}(\mathbf{x}))\right]\text{Det}\left[\mathbf{g}'(\mathbf{x})\right] &= \text{Det}\left[(\mathbf{g}^{-1})'(\mathbf{g}(\mathbf{x}))\mathbf{g}'(\mathbf{x})\right] \\
&= \text{Det}\left[(\mathbf{g}^{-1} \circ \mathbf{g})'(\mathbf{x})\right] \\
&= \text{Det}\,[\mathrm{I}] = 1
\end{aligned}$$

and thus $\text{Det}\,[\mathbf{g}(\mathbf{x})] \neq 0$ for all $\mathbf{x} \in E$.

In the setting of a diffeomorphism, $\mathbf{g} : E \to E'$, the change of variable theorem for integration of functions of several variables states that

$$\int_{\mathbf{g}(C)} h(\mathbf{x}) d\mathbf{x} = \int_C h(g(\mathbf{x}))\big|\text{Det}\,[\mathbf{g}'(\mathbf{x})]\big|\, d\mathbf{x}$$

when $h : \mathbf{g}(C) \to \mathbb{R}$ is Riemann integrable. In particular,

$$\int_{\mathbf{g}(C)} 1\, d\mathbf{x} = \int_C \big|\text{Det}\,[\mathbf{g}'(\mathbf{x})]\big| d\mathbf{x},$$

and $\mathbf{g}$ is volume preserving if $|\text{Det}\,[\mathbf{g}'(\mathbf{x})]| = 1$ for all $\mathbf{x} \in E$. Moreover, the change variable theorem for integration can be viewed as a linearization result in its own right. (See [21] for a full discussion of Jordan measurable sets and the change of variable theorem for integration of functions of several variables.)

There is a family of diffeomorphisms naturally associated with a smooth vector field $\mathbf{f}(\mathbf{x})$ on an open set Ω. For $t \in \mathbb{R}$, set $\Omega_t = \{\boldsymbol{\xi} : (t, 0, \boldsymbol{\xi}) \in D_{\mathbf{f}}\}$ and note that Ω_t is an open set because $D_{\mathbf{f}}$ is an open set. The map $\mathbf{g}_t : \Omega_t \to \Omega$ defined by $\mathbf{g}_t(\boldsymbol{\xi}) = \mathbf{x}(t, \boldsymbol{\xi})$ is called *a transition map*. By uniqueness, $\mathbf{x}(s - t, \mathbf{x}(t, \boldsymbol{\xi})) = \mathbf{x}(s, \boldsymbol{\xi})$ because the two solutions agree when $s = t$. Hence, $\mathbf{x}(-t, \mathbf{x}(t, \boldsymbol{\xi})) = \mathbf{x}(0, \boldsymbol{\xi}) = \boldsymbol{\xi}$ by setting $s = 0$. It follows that $\Omega_{-t} = \mathbf{g}_t(\Omega_t)$ and $\mathbf{g}_t^{-1} = \mathbf{g}_{-t}$. This much is true for any locally Lipschitz vector field on Ω. If the vector field is smooth, then by *Theorem 7.10* both $\mathbf{g}_t$ and $\mathbf{g}_t^{-1} = \mathbf{g}_{-t}$ are smooth on their domains and are examples of diffeomorphisms.

Before applying change of variable theorem to determine when transition maps in general are volume preserving, consider the special case of a linear system with constant real coefficients, $\dot{\mathbf{x}} = A\mathbf{x}$. Since $\mathbf{x}(t, \tau, \boldsymbol{\xi}) = e^{A(t-\tau)}\boldsymbol{\xi}$, the map $\mathbf{g}_t$ is the linear transformation defined by $\mathbf{g}_t(\boldsymbol{\xi}) = e^{At}\boldsymbol{\xi}$. It is volume preserving if and only if $\text{Det}\,[e^{At}] = e^{\text{Tr}\,[A]} = 1$ if and only if $\text{Tr}\,[A] = 0$. This leads naturally to the general result.

Theorem 7.15 *Suppose* $\mathbf{f}$ *is smooth vector field on an open set* Ω. *If* $\text{div}\,\mathbf{f} \equiv 0$ *on* Ω, *then the transition maps for* $\dot{\mathbf{x}} = \mathbf{f}(\mathbf{x})$ *are all volume preserving.*

Proof. Recall that

$$\text{div}\,\mathbf{f} = \sum_{i=1}^{d} \frac{\partial f_i}{\partial x_i} = \text{Tr}\,[\,\mathbf{f}'].$$

Since

$$\frac{\partial \mathbf{g}_t}{\partial \boldsymbol{\xi}}(\boldsymbol{\xi}) = \frac{\partial \mathbf{x}}{\partial \boldsymbol{\xi}}(t, \boldsymbol{\xi}),$$

it follows from part (d) of *Theorem 7.10* that

$$\text{Det}\,[\mathbf{g}_t'(\boldsymbol{\xi})] = \exp\left(\int_0^t \text{div}[\mathbf{f}(\mathbf{x}(s, \boldsymbol{\xi})]ds\right) = e^0 = 1.$$

Consequently,

$$\int_{\mathbf{g}_t(C)} 1 d\mathbf{x} = \int_C 1 d\mathbf{x}$$

when C and $g_t(C)$ are both Jordan measurable. □

If the solutions of $\dot{\mathbf{x}} = \mathbf{f}(\mathbf{x})$ are defined for all t, then $D_{\mathbf{f}} = \mathbb{R}^2 \times D$ and $\Omega_t = \Omega$ for all t. In particular, the transition maps $\mathbf{g}_t$ are diffeomorphisms from Ω onto Ω. In this case, $\mathbf{x}(t, \boldsymbol{\xi})$ defines a smooth flow on Ω because $\mathbf{x}(t, \boldsymbol{\xi})$ is a smooth function satisfying the flow conditions on page 65. The simplest examples of smooth flows come from the linear autonomous systems $\dot{\mathbf{x}} = A\mathbf{x}$ on $\mathbb{R}^d$ and are frequently models for studying the local behavior of smooth flows in general.

If $\mathbf{f}(\mathbf{x})$ is a smooth vector field defining a flow on Ω and $\mathrm{div}\mathbf{f} \equiv 0$ on Ω, then the transition maps are all volume preserving diffeomorphisms of Ω onto itself and $\mathbf{x}(t, \boldsymbol{\xi})$ defines a volume preserving flow on Ω. In this setting, the long-term behavior of orbits can be studied from a statistical perspective. For example, what fraction of time does $\mathbf{x}(t, \boldsymbol{\xi})$ spend in a set U?

To be a little more specific, assume that the volume of Ω is finite and let U be a Jordan measurable open set in Ω. Define a real-valued function χ by

$$\chi(\mathbf{x}) = \begin{cases} 1 & \text{if } \mathbf{x} \in U \\ 0 & \text{if } \mathbf{x} \notin U. \end{cases}$$

Then

$$\frac{1}{T}\int_0^T \chi(\mathbf{x}(t, \boldsymbol{\xi}))dt$$

is the fraction of time that $\mathbf{x}(t, \boldsymbol{\xi})$ is in U for $0 \le t \le T$. What happens as T goes to infinity? It can be shown that for most points

$$\lim_{T\to\infty} \frac{1}{T}\int_0^T \chi(\mathbf{x}(t, \boldsymbol{\xi}))dt$$

actually exists and represents the fraction of time that the positive orbit spends in the set U. Volume preserving flows for which time average equals the space average or

$$\lim_{t\to\infty} \frac{1}{T}\int_0^T \chi(\mathbf{x}(t, \boldsymbol{\xi}))dt = \frac{\int_U 1 d\mathbf{x}}{\int_\Omega 1 d\mathbf{x}}$$

are especially important. In other words, the fraction of time that a typical orbit spends in a set U is the same as the fraction of the space Ω that U occupies. Ergodic theory is the study of this phenomena and related ones. It is a very different and fertile approach to understanding the limiting behavior of orbits in a volume preserving flow. For an introduction to ergodic theory see [36].

EXERCISES

1. Show that the improper integral

$$\int_0^\infty e^{A^t s} C e^{As} ds$$

converges to a matrix $B \in \mathcal{M}_d(\mathbb{R})$ when A and C are in $\mathcal{M}_d(\mathbb{R})$ and the eigenvalues of A have negative real parts.

2. Show that $\dot{\mathbf{x}} = A\mathbf{x}$ is not volume preserving if $\mathbf{0}$ is positively asymptotically stable or negatively asymptotically stable.

3. Prove the converse of *Theorem 7.15.*

4. Define the transition maps $\mathbf{g}_t(\boldsymbol{\xi})$ for a flow $\Phi : \mathbb{R} \times \Omega \to \Omega$ on Ω and show that $\mathbf{g}_t^n(\boldsymbol{\xi}) = \mathbf{g}_{nt}(\boldsymbol{\xi})$ for all integers n.

5. Let W be a compact invariant set for a flow $\Phi : \mathbb{R} \times \Omega \to \Omega$ on Ω. Prove that if for each positive real number τ

$$\{\mathbf{x} \in \Omega : \mathbf{g}_\tau(\mathbf{x}) = \mathbf{x}\} \neq \phi,$$

then there exists a fixed point $\mathbf{p} \in W$ such that $\Phi(t, \mathbf{p}) = \mathbf{p}$ for all $t \in \mathbb{R}$. Hint: The set

$$\bigcup_{n=1}^{\infty} \{k/2^n : k \in \mathbb{Z}\}$$

is dense in $\mathbb{R}$.

7.4 Hamiltonian Systems

When writing $F = ma$ or $\mathbf{F}(\mathbf{x}) = m\ddot{\mathbf{x}}$ as a system of first-order differential equations, the choice of additional variables is not prescribed. There are other choices besides the coordinates $\dot{\mathbf{x}}$ or velocity. One particularly interesting choice is momentum or the derivative of $m\mathbf{x}$ because the resulting first-order system lies in a class of differential equations with a rich mathematical structure that can be exploited. These differential equations are called Hamiltonian systems and is extensively studied independent of its roots in Newtonian mechanics. This section provides a brief introduction to Hamiltonian systems and their linear parts. For a full treatment of Hamiltonian differential equations, see [25]

Let Ω be an open set in $\mathbb{R}^{2d}$ and let $H : \Omega \to \mathbb{R}$ be a C^2 function (continuous first and second partial derivatives). Write as $\mathbf{x} = (\mathbf{q}, \mathbf{p})$, where $\mathbf{q}$ and $\mathbf{p}$ are in $\mathbb{R}^d$. Then the autonomous system of differential equations:

$$\begin{aligned}
\dot{q}_1 &= \frac{\partial H}{\partial p_1} \\
\vdots &\quad \vdots \\
\dot{q}_d &= \frac{\partial H}{\partial p_d} \\
\dot{p}_1 &= -\frac{\partial H}{\partial q_1} \\
\vdots &\quad \vdots \\
\dot{p}_d &= -\frac{\partial H}{\partial q_d}
\end{aligned}$$

is called a *Hamiltonian system* and the function $H(\mathbf{q}, \mathbf{p})$ is called a *Hamiltonian.* The vectors $\mathbf{q}$ and $\mathbf{p}$ are usually thought of a position and momentum, their classical origins. The requirement that the function H have continuous first and second partial derivatives guarantees that the vector field of a Hamiltonian system is smooth.

Remark *A Hamiltonian system is volume preserving.*

Proof. The divergence of a Hamiltonian vector filed is

$$\sum_{j=1}^{d} \frac{\partial^2 H}{\partial q_j \partial p_j} - \sum_{j=1}^{d} \frac{\partial^2 H}{\partial p_j \partial q_j} = 0$$

because the order of differentiation in second partial derivatives can be interchanged for a C^2 function. □

If we define a $2d \times 2d$ matrix J by

$$J = \begin{bmatrix} \mathrm{O} & \mathrm{I} \\ -\mathrm{I} & \mathrm{O} \end{bmatrix},$$

where as usual I is the $d \times d$ identity matrix and O represents the $d \times d$ matrix of all zeros, then the Hamiltonian system can also be written in the more compact form

$$\dot{\mathbf{x}} = J \nabla H(\mathbf{x}). \tag{7.1}$$

(Here it is understood that as with other vectors, the vector $\nabla H(\mathbf{x})$ is interpreted as being vertical because it is being multiplied by the matrix J on its left.) In particular, if H is a Hamiltonian, then the Hamiltonian vector field is simply $J \nabla H(\mathbf{x})$.

The bilinear function

$$\sigma(\mathbf{x}, \mathbf{y}) = \mathbf{x} \cdot (J\mathbf{y}) \tag{7.2}$$

is called the *symplectic bilinear function.* It plays a large role in the study of Hamiltonian systems. Note that $\sigma(\mathbf{x}, \mathbf{y})$ is skew symmetric, that is,$\sigma(\mathbf{x}, \mathbf{y}) = -\sigma(\mathbf{y}, \mathbf{x})$, and consequently $\sigma(\mathbf{x}, \mathbf{x}) = 0$ for all $\mathbf{x}$.

Remark *The Hamiltonian H is an integral for the Hamiltonian differential equation $\dot{\mathbf{x}} = J \nabla H(\mathbf{x})$.*

Proof. Just note that $\dot{H}(\mathbf{x}) = \nabla H(\mathbf{x}) \cdot (J \nabla H(\mathbf{x})) = \sigma\big(\nabla H(\mathbf{x}), \nabla H(\mathbf{x})\big) \equiv 0$.
□

Proposition 7.16 *If the Hamiltonian H has an isolated local minimum (maximum) at $\mathbf{y}$, then $\mathbf{y}$ is a positively and negatively stable fixed point of $\dot{\mathbf{x}} = J \nabla H(\mathbf{x})$.*

Proof. Clearly, $\nabla H(\mathbf{y}) = \mathbf{0}$ and $J\nabla H(\mathbf{y}) = \mathbf{0}$, so $\mathbf{y}$ is a fixed point. There exists $r > 0$ such that $H(\mathbf{y}) < H(\mathbf{x})$ when $0 < \|\mathbf{x} - \mathbf{y}\| \leq r$. Given $0 < \varepsilon < r$, set

$$\alpha = \min\{H(\mathbf{x}) : \|\mathbf{x} - \mathbf{y}\| = \varepsilon\}.$$

There exists $\delta > 0$ such that $H(\mathbf{x}) < \alpha/2$ when $\|\mathbf{x} - \mathbf{y}\| < \delta$. Obviously $\delta < \varepsilon$.

Since H is an integral for the system, $H(\mathbf{x}(t, \boldsymbol{\xi})) = H(\boldsymbol{\xi}) < \alpha/2$ for all t in the domain of $\mathbf{x}(t, \boldsymbol{\xi})$, when $\|\boldsymbol{\xi} - \mathbf{y}\| < \delta$. It follows that $\|\mathbf{x}(t, \boldsymbol{\xi}) - \mathbf{y}\| < \varepsilon$ for all t in its domain, when $\|\boldsymbol{\xi} - \mathbf{y}\| < \delta$. Consequently, $\mathbf{x}(t, \boldsymbol{\xi})$ is defined for all $t \in \mathbb{R}$ and $\|\mathbf{x}(t, \boldsymbol{\xi}) - \mathbf{y})\| < \varepsilon$ for all $t \in \mathbb{R}$ when $\|\boldsymbol{\xi} - \mathbf{y}\| < \delta$. □

As an example, consider the second-order scalar equation $\ddot{w} + g(w) = 0$. Let $G(w)$ be any real-valued function such that $G' = g$ and set

$$H(q, p) = \frac{1}{2}p^2 + G(q).$$

Then the Hamiltonian system is

$$\begin{aligned} \dot{q} &= \frac{\partial H}{\partial p} = p \\ \dot{p} &= -\frac{\partial H}{\partial q} = -g(q). \end{aligned}$$

It follows that $\ddot{q} = -g(q)$ and we have recaptured the original system. In other words, there is a first-order version of $\ddot{w} + g(w) = 0$ that is is a Hamiltonian system.

The above example can be generalized to the second-order system $M\ddot{\mathbf{x}} + \nabla G(\mathbf{x}) = 0$, where M ia an invertible symmetric $d \times d$ matrix and G is a C^2 real-valued function defined on $\mathbb{R}^d$. Let $\mathbf{q} = \mathbf{x}$ and $\mathbf{p} = M\dot{\mathbf{x}}$, and set

$$H(\mathbf{q}, \mathbf{p}) = \frac{\mathbf{p} \cdot (M^{-1}\mathbf{p})}{2} + G(\mathbf{q})$$

Using the corollary to *Theorem 7.6* on page 219 to help calculate $\nabla H(\mathbf{q}, \mathbf{p})$ yields

$$\nabla H(\mathbf{q}, \mathbf{p}) = (\nabla G(\mathbf{q}), M^{-1}\mathbf{p})$$

because M^{-1} is also symmetric. Hence,

$$J\nabla H(\mathbf{q}, \mathbf{p}) = \begin{pmatrix} M^{-1}\mathbf{p} \\ -\nabla G(\mathbf{q}) \end{pmatrix},$$

and

$$\begin{aligned} \dot{\mathbf{q}} &= M^{-1}\mathbf{p} \\ \dot{\mathbf{p}} &= -\nabla G(\mathbf{q}). \end{aligned}$$

As in the scalar version of this example, it readily follows that

$$M\ddot{\mathbf{q}} = \dot{\mathbf{p}} = -\nabla G(\mathbf{q}).$$

This observation will be applied at the end of this section to the *n-body problem* with M a diagonal matrix of masses.

Next, we want to calculate the linear part of a Hamiltonian system and determine the linear Hamiltonian systems. By thinking of $J\nabla H$ as the composition of the linear function $\mathbf{y} \to J\mathbf{y}$ and the smooth function $\mathbf{x} \to \nabla H(\mathbf{x})$, the chain rule can be applied and the linear part of $\dot{\mathbf{x}} = J\nabla H(\mathbf{x})$ is given by

$$J\left[\frac{\partial^2 H}{\partial x_i \partial x_j}\right].$$

Because H is a C^2 function, the order of partial differentiation can be interchanged in the second-order partial derivatives and the matrix

$$\left[\frac{\partial^2 H}{\partial x_i \partial x_j}\right]$$

is symmetric. Thus the linear part of a Hamiltonian system has the form JB where B is symmetric. The converse is also true.

Remark *If B is a symmetric $2d \times 2d$ matrix, then the linear system $\dot{\mathbf{x}} = JB\mathbf{x}$ is a Hamiltonian system with Hamiltonian $H(\mathbf{x}) = \mathbf{x} \cdot (B\mathbf{x})/2$.*

Proof. The corollary to *Theorem 7.6* can be applied, noting that $B + B^{t} = 2B$. It follows that $\nabla H(\mathbf{x}) = B\mathbf{x}$. □

A $2d \times 2d$ real matrix of the form JB with B symmetric is called *Hamiltonian matrix*. Hamiltonian matrices can be characterized in several ways.

Proposition 7.17 *The following are equivalent:*

(a) $A = JB$ *with* B *symmetric.*

(b) $JA + A^{t}J = 0$.

(c) JA *is symmetric.*

Proof. The proof hinges on the simple observation that $J^{-1} = -J = J^{t}$, so J is both an orthogonal matrix and a skew symmetric matrix.

For (a) implies (b), note that $A = JB$ with B symmetric implies that $JA = -B$ and then $B = B^{t} = \big(-JA\big)^{t} = \big(J^{t}A\big)^{t} = A^{t}J$. Hence, $JA + A^{t}J = 0$.

Next, assume that $JA + A^{t}J = 0$ or $JA = -A^{t}J$. It follows that $(JA)^{t} = (-A^{t}J)^{t} = - - JA = JA$ to show that (b) implies (c).

Finally, (c) implies (a) by setting $B = -JA$ and multiplying by J. □

So a matrix is a Hamiltonian matrix if and only if it satisfies any one of these equivalent conditions. A real $2d \times 2d$ matrix B is called a *symplectic matrix* if $B^{t}JB = J$. The connection between Hamiltonian and symplectic matrices is the content of the next proposition.

Proposition 7.18 *The linear system $\dot{\mathbf{x}} = A\mathbf{x}$ is a Hamiltonian system if and only if e^{At} is a sympletic matrix for all t*

Proof. If $\dot{\mathbf{x}} = A\mathbf{x}$ is a Hamiltonian system, then $JA + A^{\mathrm{t}}J = 0$ or $J^{-1}A^{\mathrm{t}}J = -A$. It suffices to show $\left(e^{At}\right)^{\mathrm{t}}J = Je^{-At}$ or $J^{-1}\left(e^{At}\right)^{\mathrm{t}}J = e^{-At}$. Using the properties of e^{At},

$$J^{-1}\left(e^{At}\right)^{\mathrm{t}}J = J^{-1}e^{A^{\mathrm{t}}t}J = e^{J^{-1}A^{\mathrm{t}}Jt} = e^{-At}.$$

If $\left(e^{At}\right)^{\mathrm{t}}Je^{At} = J$ for all t, then $\left(e^{A^{\mathrm{t}}t}\right)Je^{At} = J$ for all t. Differentiating this expression with respect to t and setting $t = 0$, yields $A^{\mathrm{t}}J + JA = 0$. □

For an example of a Hamiltonian matrix, let C be a real skew symmetric matrix, that is, $C^{\mathrm{t}} = -C$, and set

$$A = \begin{bmatrix} C & \mathrm{O} \\ \mathrm{O} & C \end{bmatrix}.$$

Then A is a Hamiltonian matrix because

$$\begin{aligned} JA + A^{\mathrm{t}}J &= \begin{bmatrix} \mathrm{O} & C \\ -C & \mathrm{O} \end{bmatrix} + \begin{bmatrix} -C & \mathrm{O} \\ \mathrm{O} & -C \end{bmatrix} J \\ &= \begin{bmatrix} \mathrm{O} & C \\ -C & \mathrm{O} \end{bmatrix} + \begin{bmatrix} \mathrm{O} & -C \\ C & \mathrm{O} \end{bmatrix} \\ &= \mathrm{O}. \end{aligned}$$

Proposition 7.19 *If A is a real Hamiltonian matrix and $g(\lambda)$ is its characteristic polynomial, then $g(-\lambda) = g(\lambda)$. In particular, if λ is an eigenvalue of a real Hamiltonian matrix A, then $-\lambda$, $\overline{\lambda}$, and $-\overline{\lambda}$ are also eigenvalues of A.*

Proof. The following calculation proves the first part using $A = JA^{\mathrm{t}}J$:

$$\begin{aligned} g(\lambda) &= \mathrm{Det}\,[A - \lambda \mathrm{I}] \\ &= \mathrm{Det}\,[JA^{\mathrm{t}}J + \lambda JJ] \\ &= \mathrm{Det}\,[J]\,\mathrm{Det}\,[A^{\mathrm{t}} + \lambda \mathrm{I}]\,\mathrm{Det}\,[J] \\ &= \mathrm{Det}\,[A^{\mathrm{t}} + \lambda \mathrm{I}] \\ &= \mathrm{Det}\,[A - -\lambda \mathrm{I}] \\ &= g(-\lambda). \end{aligned}$$

Because $g(\lambda)$ is a polynomial with real coefficients, $g(\overline{\lambda}) = \overline{g(\lambda)}$. Hence, $g(\lambda) = 0$ if and only if $g(-\lambda) = 0$ if and only if $g(\overline{\lambda}) = 0$ if and only if $g(-\overline{\lambda}) = 0$. □

Theorem 7.20 *Let A be a real Hamiltonian matrix and let $\lambda_1, \ldots, \lambda_\rho$ be the distinct eigenvalues (real and complex) of A. The following hold:*

(a) The fixed point $\mathbf{0}$ of $\dot{\mathbf{x}} = A\mathbf{x}$ is neither positively nor negatively asymptotically stable.

(b) *The fixed point* $\mathbf{0}$ *of* $\dot{\mathbf{x}} = A\mathbf{x}$ *is positively stable if and only if the eigenvalues are all simple and pure imaginary, that is,* $Re\lambda_j = 0$ *for* $j = 1, \ldots, \rho$.

(c) *The fixed point* $\mathbf{0}$ *of* $\dot{\mathbf{x}} = A\mathbf{x}$ *is positively stable if and only if it is negatively stable.*

Proof. It follows from the previous proposition that a real Hamiltonian matrix has an eigenvalue with negative real part if and only if it has an eigenvalue with positive real part. Thus the origin can be neither a positively asymptotically stable fixed point nor a negatively asymptotically stable fixed point and part (a) holds. Furthermore, by *Theorem 5.9* the only way it can be positively stable is for every eigenvalue to be a simple pure imaginary eigenvalue. Hence, parts (b) and (c) hold. □.

Hamiltonian systems have a variety of special properties that arise from $\sigma(\nabla H_1, \nabla H_2)$. The following result is a simple example of such a result.

Theorem 7.21 (Noether) *Let* H_1 *and* H_2 *be* C^2 *functions on an open set* Ω *in* $\mathbb{R}^{2d}$. *Then* H_2 *is an integral for* $\dot{\mathbf{x}} = J\nabla H_1(\mathbf{x})$ *if and only if* H_1 *is an integral for* $\dot{\mathbf{x}} = J\nabla H_2(\mathbf{x})$.

Proof. The function H_j is an integral of $\dot{\mathbf{x}} = J\nabla H_j(\mathbf{x})$ if and only if

$$\nabla H_j(\mathbf{x}) \cdot (J\nabla H_i(\mathbf{x})) \equiv 0$$

on Ω, or

$$\sigma(\nabla H_j(\mathbf{x}), \nabla H_i(\mathbf{x})) \equiv 0$$

on Ω. Since

$$\sigma(\nabla H_j(\mathbf{x}), \nabla H_i(\mathbf{x})) = -\sigma(\nabla H_i(\mathbf{x}), \nabla H_j(\mathbf{x})),$$

we have $\sigma(\nabla H_j(\mathbf{x}), \nabla H_i(\mathbf{x})) \equiv 0$ on Ω if and only if $\sigma(\nabla H_i(\mathbf{x}), \nabla H_j(\mathbf{x})) \equiv 0$ on Ω, which completes the proof. □

The n-body problem provides a good illustration of how Noether's theorem can be used to obtain integrals. In particular, it will be used to show that total momentum and angular momentum of the n-body system is conserved. The idea is to observe that for the n-body problem the Hamiltonian H_1 is invariant under translation and rotation in $\mathbb{R}^3$ where the bodies reside. And then to show that these translations and rotations can be viewed as transition maps for another Hamiltonian system given by H_2. Thus the n-body Hamiltonian will be an integral for the $\dot{\mathbf{x}} = J\nabla H_2(\mathbf{x})$, and then by Noether's theorem H_2 is an integral for the n-body problem.

The first step is to write the n-body problem as a Hamiltonian system. In Section 2.1 on page 56, the n-body problem was set up with $m_1, \ldots, m_n$ denoting the masses of the bodies, $\mathbf{x}_1, \ldots, \mathbf{x}_n$ denoting their positions, and $\mathbf{v}_1, \ldots, \mathbf{v}_n$ denoting their velocities. The same notation will be used here.

Since each $\mathbf{x}_i$ is an arbitrary point of $\mathbb{R}^3$, the natural space in which to set up the second-order equation is on a subset U of $\left(\mathbb{R}^3\right)^n = \mathbb{R}^{3n}$ that avoids two or more of the masses occupying the point in the space. Then the n-body problem can be written as n systems of differential equations each of the form $m_j\ddot{\mathbf{x}} = \mathbf{F}_j$, where $\mathbf{F}_j(x)$ is the sum of the forces the masses m_j at $\mathbf{x}_j$, $j \neq i$, exert on m_i at $\mathbf{x}_i$.

It was also demonstrated in Section 2.1 that the total force $\mathbf{F} = (\mathbf{F}_1, \ldots, \mathbf{F}_n)$ was conservative by setting

$$G(\mathbf{x}) = -\gamma \sum_{i<j} \frac{m_i m_j}{\|\mathbf{x}_i - \mathbf{x}_j\|}$$

and showing that $-\nabla G = \mathbf{F}$. The entire problem can now be stated as single second-order equation. If

$$M = \text{Diag}\,\{m_1, m_1, m_1, m_2, m_2, m_2, \ldots, m_n, m_n, m_n\},$$

then the n-body problem has the form

$$M\ddot{\mathbf{x}} + \nabla G(\mathbf{x}) = 0.$$

It was pointed earlier in this section that such an equation can always be written as a Hamiltonian system. Let $\mathbf{q}_j = \mathbf{x}_j$ and $\mathbf{p}_j = m_j \mathbf{v}_j$, the momentum of the jth mass. Then

$$H_1(\mathbf{q}, \mathbf{p}) = \frac{\mathbf{p} \cdot (M^{-1}\mathbf{p})}{2} - \gamma \sum_{i<j} \frac{m_i m_j}{\|\mathbf{q}_i - \mathbf{q}_j\|}$$

is the Hamiltonian for the n-body problem on $\mathbb{R}^{6n}$.

If we use the convention that

$$\frac{\partial H_1}{\partial \mathbf{p}_i} = \left(\frac{\partial H_1}{\partial p_{i1}}, \frac{\partial H_1}{\partial p_{i2}}, \frac{\partial H_1}{\partial p_{i3}} \right)$$

and likewise for the partial derivatives of H with respect to $\mathbf{q}_i$, then

$$\begin{aligned} \dot{\mathbf{q}}_i &= \frac{\partial H_1}{\partial \mathbf{p}_i} \\ \dot{\mathbf{p}}_i &= -\frac{\partial H_1}{\partial \mathbf{q}_i} \end{aligned}$$

is a Hamiltonian written as a system of systems. By calculating the partial derivatives, the Hamiltonian version of the classical n-body problem written as a system of systems is

$$\begin{aligned} \dot{\mathbf{q}}_i &= \frac{\mathbf{p}_i}{m_i} \\ \dot{\mathbf{p}}_i &= -\sum_{j \neq i} \frac{\gamma m_i m_j (\mathbf{q}_j - \mathbf{q}_i)}{\|\mathbf{q}_j - \mathbf{q}_i\|^3}. \end{aligned}$$

For the first application of Noether's theorem let $\mathbf{v}$ be any non zero vector in $\mathbb{R}^3$. Clearly,

$$H_1(\mathbf{q}_1 + t\mathbf{v}, \ldots, \mathbf{q}_n + t\mathbf{v}, \mathbf{p}_1, \ldots, \mathbf{p}_n) = H_1(\mathbf{q}_1, \ldots, \mathbf{q}_n, \mathbf{p}_1, \ldots, \mathbf{p}_n)$$

for real t. Set

$$\varphi(t) = (\mathbf{q}_1 + t\mathbf{v}, \ldots, \mathbf{q}_n + t\mathbf{v}, \mathbf{p}_1, \ldots, \mathbf{p}_n).$$

Obviously, $H_1(\varphi(t)) = H_1(\mathbf{q}, \mathbf{p})$ for all t. Since $\varphi(t)$ is a typical solution of the system

$$\begin{aligned} \dot{\mathbf{q}}_i &= \mathbf{v} \text{ for } 1 \le i \le n \\ \dot{\mathbf{p}}_i &= 0 \text{ for } 1 \le i \le n \end{aligned}$$

and H_1 is an integral for this system. Moreover, this system is also a Hamiltonian system with $H_2(\mathbf{q}, \mathbf{p}) = \mathbf{v} \cdot (\mathbf{p}_1 + \cdots + \mathbf{p}_n)$. Therefore, by Noether's theorem $H_2(\mathbf{q}, \mathbf{p})$ is an integral for the n-body problem. In particular, taking $\mathbf{v} = \mathbf{e}_1 = (1, 0, 0)$ shows that the total momentum of the n-body system in the $\mathbf{e}_1$ direction is conserved because

$$\sum_{j=1}^{n} \mathbf{e}_1 \cdot \mathbf{p}_j = \sum_{j=1}^{n} p_{j1}$$

is an integral. Similarly the total momentum in the $\mathbf{e}_2$ and $\mathbf{e}_3$ directions are conserved and we have obtained three of the classical integrals for the n-body problem using Noether's theorem.

There are also three classical integrals of angular momentum for the n-body problem. Deriving them requires the use of orthogonal matrices. Let $\hat{B}$ be an orthogonal 3×3 matrix with determinant 1. Because the complex eigenvalues come in conjugate pairs for a real matrix, $\hat{B}$ must have at least one real eigenvalue. Moreover, it is easy to see by checking cases that, in fact, 1 must be an eigenvalue of multiplicity 1 if $\hat{B}$ is not the identity.

By *Theorem 4.17*, there exists a skew symmetric matrix $\hat{C} \in \mathcal{M}_3(\mathbb{R})$ such that $\hat{B} = e^{\hat{C}}$. In fact, $\hat{C}t$ is skew symmetric and $e^{\hat{C}t}$ is orthogonal for all $t \in \mathbb{R}$. Set $C = \text{Diag}\,\{\hat{C}, \ldots, \hat{C}\} \in \mathcal{M}_{3n}(\mathbb{R})$. Clearly, C is also skew symmetric and

$$e^{Ct} = \text{Diag}\,\{e^{\hat{C}t}, \ldots, e^{\hat{C}t}\}$$

is orthogonal for all $t \in \mathbb{R}$.

The next step is to show that

$$H_1(e^{Ct}\mathbf{q}, e^{Ct}\mathbf{p}) = H_1(\mathbf{q}, \mathbf{p}).$$

Since e^{Ct} is orthogonal, $e^{Ct}\mathbf{x} \cdot e^{Ct}\mathbf{y} = \mathbf{x} \cdot \mathbf{y}$ and $\|e^{Ct}\mathbf{x} - e^{Ct}\mathbf{y}\| = \|\mathbf{x} - \mathbf{y}\|$ for all $\mathbf{x}$ and $\mathbf{y}$ in $\mathbb{R}^{3n}$. It follows that

$$-\gamma \sum_{i<j} \frac{m_i m_j}{\|\mathbf{e}^{Ct}\mathbf{q}_i - e^{Ct}\mathbf{q}_j\|} = -\gamma \sum_{i<j} \frac{m_i m_j}{\|\mathbf{q}_i - \mathbf{q}_j\|}.$$

To show that $e^{Ct}\mathbf{p} \cdot (M^{-1}e^{Ct}\mathbf{p}) = \mathbf{p} \cdot (M^{-1}\mathbf{p})$ holds, first observe that $\text{Diag}\{m_j, m_j, m_j\}e^{\hat{C}t} = e^{\hat{C}t}\text{Diag}\{m_j, m_j, m_j\}$ and hence $M^{-1}e^{Ct} = e^{Ct}M^{-1}$. It follows that

$$e^{Ct}\mathbf{p} \cdot (M^{-1}e^{Ct}\mathbf{p}) = e^{Ct}\mathbf{p} \cdot (e^{Ct}M^{-1}\mathbf{p}) = \mathbf{p} \cdot (M^{-1}\mathbf{p})$$

and $H_1(e^{Ct}\mathbf{q}, e^{Ct}\mathbf{p}) = H_1(\mathbf{q}, \mathbf{p})$. Therefore, H_1 is an integral for the system

$$\begin{aligned} \dot{\mathbf{q}} &= C\mathbf{q} \\ \dot{\mathbf{p}} &= C\mathbf{p}. \end{aligned}$$

It is also a Hamiltonian linear system like that in the example on page 244. The Hamiltonian is $H_3(\mathbf{x})) = \mathbf{x} \cdot (-JA)\mathbf{x}/2$ where

$$A = \begin{bmatrix} C & O \\ O & C \end{bmatrix}.$$

Because

$$-JA = -J\begin{bmatrix} C & O \\ O & C \end{bmatrix} = \begin{bmatrix} O & -C \\ C & O \end{bmatrix},$$

the Hamiltonian is

$$H_3(\mathbf{q}, \mathbf{p}) = \left(\frac{1}{2}\right)(\mathbf{q}, \mathbf{p})\begin{bmatrix} O & -C \\ C & O \end{bmatrix}\begin{pmatrix} \mathbf{q} \\ \mathbf{p} \end{pmatrix}$$

which simplifies to

$$H_3(\mathbf{q}, \mathbf{p}) = \mathbf{q} \cdot (C\mathbf{p}).$$

Again by Noether's theorem H_3 is an integral for the n-body problem. Using the three matrices

$$\begin{bmatrix} 0 & 0 & 0 \\ O & 0 & 1 \\ 0 & -1 & 0 \end{bmatrix}, \begin{bmatrix} 0 & 0 & 1 \\ 0 & 0 & 0 \\ -1 & 0 & 0 \end{bmatrix} \text{ and } \begin{bmatrix} 0 & 1 & 0 \\ -1 & 0 & 0 \\ 0 & 0 & 0 \end{bmatrix}$$

produces the classical conservation of angular momentum around the coordinate axis.

EXERCISES

1. Prove that a fixed point of $\dot{\mathbf{x}} = J\nabla H(\mathbf{x})$ with $H(\mathbf{x})$ a C^2 function is never positively asymptotically stable.

2. Show that if B is a symplectic matrix, then $\sigma(B\mathbf{x}, B\mathbf{y}) = \sigma(\mathbf{x}, \mathbf{y})$.

3. Let A, B, C, and D be in $\mathcal{M}_d(\mathbb{R})$ and consider the $2d \times 2d$ matrix with block form

$$Q = \begin{bmatrix} A & B \\ C & D \end{bmatrix}.$$

(a) Show that Q is a Hamiltonian matrix if and only if $A^t + D = O$ and both B and C are symmetric.

(b) Show that Q is a symplectic matrix if and only if $A^tD - C^tB = I$ and both A^tC and B^tD are symmetric.

4. Find necessary and sufficient conditions for a 2×2 real matrix to be Hamiltonain.

5. Find necessary and sufficient conditions for a 2×2 real matrix to be symplectic.

6. Show that the set of symplectic matrices is a subgroup of $GL_{2d}(\mathbb{R})$. Is it a compact set?

7. Let F, G, and H be C^2 real-valued functions defined on $\mathbb{R}^{2d}$. The *Poisson bracket* is the C^∞ function $\{F, G\} = \sigma(\nabla F, \nabla G)$. Show that

$$\{F, \{G, H\}\} + \{G, \{H, F\}\} + \{H, \{F, G\}\} \equiv 0.$$

8. Let F, G, and H be C^2 real-valued functions defined on $\mathbb{R}^{2d}$. Show that if F and G are integrals for the Hamiltonian system $\dot{\mathbf{x}} = J\nabla H(\mathbf{x})$, then the Poisson bracket $\{F, G\}$ is also an integral for $\dot{\mathbf{x}} = J\nabla H(\mathbf{x})$.

Chapter 8

Hyperbolic Phenomenon

The final result in this chapter is a deep linearization theorem for the behavior of solutions near a fixed point. The path that leads to this theorem is interesting in its own right. It includes the continuity of eigenvalues, the classification of linear hyperbolic flows, metric spaces, and the contraction mapping principle.

The primary goal is a stable manifold theorem. The article "a" is used deliberately because there are various stable manifold theorems. Their common features are a lower dimensional family of solutions, the stable manifold, that are decaying exponentially in some way as time increases and a complementary family of solutions, the unstable manifold, exhibiting the same behavior as time decreases. The dimensions of the stable and unstable manifolds are obtained from a derivative and sum to the dimension of the phase space. Either one can be trivial.

The main theorem is in fact a generalization of *Theorem 7.14* that a fixed point is positively asymptotically stable, if all the eigenvalues of the matrix of first partial derivatives have negative real parts. In this case, the unstable manifold is just the fixed point and the stable manifold includes a neighborhood of the fixed point. The challenge is to analyze the dynamical behavior near a fixed point that is the omega limit set for some solutions and the alpha limit set for other solutions. Or for that matter showing that the derivative at a fixed point can even imply such dynamic behavior.

The odyssey begins with the study of hyperbolic linear vector fields. They are the linear models used in the stable manifold theorem. They have two very nice properties—small changes in the vector field do not alter the dynamical behavior of the resulting flow and there are linear hyperbolic vector fields arbitrarily close to every linear vector field. In addition, the hyperbolic linear flows have an elegant classification based on their eigenvalues.

The shift from linear to nonlinear systems is accomplished by a careful analysis of perturbed linear hyperbolic systems and the introduction of a new tool - the contraction mapping principle. It is an abstract fixed-point theorem, and by showing that it can be used to prove the existence of a certain family of solutions a rather complicated integral equation becomes manageable. Finally, all these ideas come together in the final section to prove a stable manifold theorem.

8.1 Hyperbolic Linear Vector Fields

The first step is to identify a suitable class of linear systems to model the qualitative behavior of solutions near a fixed points. Some restrictions are necessary because pure imaginary eigenvalues of $\mathbf{f}'(\mathbf{p})$, the matrix of first partial derivatives, contain very little information about the properties of nearby solutions. At the same time, the restrictions on the linear systems should be mild enough to ensure that these systems are very prevalent. Specifically, they should be an open and dense set of $\mathcal{M}_d(\mathbb{R})$, the set of all real $d \times d$ matrices. In this sense, the excluded matrices should be rare.

It is essential that the open and dense set be constructed so that slight perturbations of the matrices in it do not change the qualitative behavior of the solutions of $\dot{\mathbf{x}} = A\mathbf{x}$. In other words, the qualitative picture should be insensitive to slight measurement errors. Simply excluding pure imaginary eigenvalues is sufficient to accomplish these goals. This section is devoted to the study of linear systems $\dot{\mathbf{x}} = A\mathbf{x}$ without pure imaginary eigenvalues and the qualitative behavior of their solutions.

A *hyperbolic matrix* is a real matrix whose eigenvalues all have nonzero real parts. It follows that 0 cannot be an eigenvalue of a hyperbolic matrix, and thus every hyperbolic matrix is invertible. The linear vector field $\mathbf{f}(\mathbf{x}) = A\mathbf{x}$ determined by a real $d \times d$ hyperbolic matrix A is called a *hyperbolic linear vector field.* Let $\mathcal{H}$ denote the subset of $\mathcal{M}_d(\mathbb{R})$ consisting of the $d \times d$ real hyperbolic matrices. So the linear vector field $\mathbf{f}(\mathbf{x}) = A\mathbf{x}$ is hyperbolic if and only if $A \in \mathcal{H}$.

The solutions of $\dot{\mathbf{x}} = A\mathbf{x}$ are given by $\mathbf{x}(t, \boldsymbol{\xi}) = e^{At}\boldsymbol{\xi}$ and define a flow on $\mathbb{R}^d$. The transition maps, $g_t(\boldsymbol{\xi}) = e^{At}\boldsymbol{\xi}$ are linear isomorphisms of $\mathbb{R}^d$ onto itself. Recall that if the distinct eigenvalues of A are $\lambda_1, \ldots, \lambda_\rho$, then $e^{\lambda_1}, \ldots, e^{\lambda_\rho}$ are the eigenvalues of e^A with some duplication possible in the list. In particular, if A is hyperbolic, then e^A does not have any eigenvalues on the unit circle, $S^1 = \{z \in \mathbb{C} : |z| = 1\}$, and conversely.

The adjective hyperbolic is also used to describe a linear isomorphism of $\mathbb{R}^d$ onto itself none of whose eigenvalues lie on the unit circle. Specifically, a *hyperbolic automorphism* is a linear isomorphism of $\mathbb{R}^d$ onto itself with no eigenvalues lying on S^1. Thus the linear vector field $\mathbf{f}(\mathbf{x}) = A\mathbf{x}$ is hyperbolic if and only if the transition maps $g_t(\boldsymbol{\xi}) = e^{At}\boldsymbol{\xi}$ are hyperbolic linear automorphisms for $t \neq 0$.

For a hyperbolic linear vector field, the qualitative behavior of the solutions determines a decomposition of $\mathbb{R}^d$ into two subspaces usually called the stable and unstable manifolds. Consequently, linear algebra can be used to help analyze the phase portrait. These two subspaces will in fact determine the basic properties of the phase portrait of $\dot{\mathbf{x}} = A\mathbf{x}$. The first theorem establishes this fundamental decomposition, which in various forms is present in all so-called hyperbolic phenomenon.

The choice of terminology is, however, slightly unfortunate because the solutions not in these subspaces exhibit unstable behavior in both directions. We will refer to these subspaces as the positively stable and the negatively stable manifolds. The first theorem establishes their existence and explains our choice of terminology.

Theorem 8.1 *If the linear vector field* $\mathbf{f}(\mathbf{x}) = A\mathbf{x}$ *is hyperbolic, then there exist subspaces* E^+ *and* E^- *such that*

$$\begin{aligned} \mathbb{R}^d &= E^+ \oplus E^- \\ E^+ &= \{\boldsymbol{\xi} : \lim_{t\to\infty} \mathbf{x}(t,\boldsymbol{\xi}) = \mathbf{0}\} \\ E^- &= \{\boldsymbol{\xi} : \lim_{t\to-\infty} \mathbf{x}(t,\boldsymbol{\xi}) = \mathbf{0}\}. \end{aligned}$$

The subspaces E^+ *and* E^- *are invariant under the linear transformation* $T(\mathbf{x}) = A\mathbf{x}$. *In addition, there exist positive constants* K *and* α *such that when* $\mathbf{v} \in E^+$,

$$|e^{At}\mathbf{v}| \le e^{-K\alpha t}|\mathbf{v}|$$

for $t \ge 0$, *and when* $\mathbf{v} \in E^-$,

$$|e^{At}\mathbf{v}| \le e^{K\alpha t}|\mathbf{v}|$$

for $t \le 0$.

Proof. Let $\lambda_1, \ldots, \lambda_\rho$ be the distinct eigenvalues (real and complex) of A and start in $\mathbb{C}^d$ by setting

$$\begin{aligned} V^N &= \sum_{\lambda_j+\overline{\lambda}_j<0} M(\lambda_j) \\ V^P &= \sum_{\lambda_j+\overline{\lambda}_j>0} M(\lambda_j). \end{aligned}$$

Clearly, $\mathbb{C}^d = V^N \oplus V^P$ by *Theorem 4.6* because $A \in \mathcal{H}$ if and only if $\lambda_j + \overline{\lambda}_j \ne 0$ for $j = 1, \ldots, \rho$. Since each generalized eigenspace $M(\lambda_j)$ is invariant under the linear transformation $T(\mathbf{z}) = A\mathbf{z}$, so are V^P and V^N. Moreover, if $\boldsymbol{\zeta}$ is a generalized eigenvector for λ_j, then $\overline{\boldsymbol{\zeta}}$ is a generalized eigenvector for $\overline{\lambda}_j$. Thus $\mathbf{w} \in V^N$ if and only if $\overline{\mathbf{w}} \in V^N$ and likewise for V^P.

Set

$$\begin{aligned} E^+ &= \{\boldsymbol{\xi} \in \mathbb{R}^d : \boldsymbol{\xi} \in V^N\} \\ E^- &= \{\boldsymbol{\xi} \in \mathbb{R}^d : \boldsymbol{\xi} \in V^P\} \end{aligned}$$

because eigenvalues with negative real parts imply positive stability and positive real parts imply negative stability. In other words, E^+ and E^- are composed of the vectors in V^N and V^P with real coordinates. Clearly, E^+ and E^- are subspaces of $\mathbb{R}^d$ and $E^+ \cap E^- = \mathbf{0}$. Since A is a real matrix, it follows that the subspaces E^+ and E^- are invariant under the linear transformation $T(\mathbf{x}) = A\mathbf{x}$.

Let $\boldsymbol{\xi}$ be an arbitrary vector in $\mathbb{R}^d$. Because $\mathbb{R}^d \subset \mathbb{C}^d$, there exist unique vectors $\boldsymbol{\zeta}_1 \in V^N$ and $\boldsymbol{\zeta}_2 \in V^P$ such that $\boldsymbol{\xi} = \boldsymbol{\zeta}_1 + \boldsymbol{\zeta}_2$. Since

$$\boldsymbol{\xi} = \overline{\boldsymbol{\xi}} = \overline{\boldsymbol{\zeta}}_1 + \overline{\boldsymbol{\zeta}}_2,$$

it follows that $\boldsymbol{\zeta}_1 = \overline{\boldsymbol{\zeta}}_1$ and $\boldsymbol{\zeta}_2 = \overline{\boldsymbol{\zeta}}_2$. Hence $\boldsymbol{\zeta}_1 \in E^+$ and $\boldsymbol{\zeta}_2 \in E^-$, and

$$\mathbb{R}^d = E^+ \oplus E^-.$$

To prove that

$$E^+ = \{\boldsymbol{\xi} \in \mathbb{R}^d : \lim_{t\to\infty} \mathbf{x}(t,\boldsymbol{\xi}) = \mathbf{0}\}$$

it suffices to show that

$$V^N = \{\boldsymbol{\zeta} \in \mathbb{C}^d : \lim_{t\to\infty} \mathbf{z}(t,\boldsymbol{\zeta}) = \mathbf{0}\}$$

because A is a real matrix.

Let $\boldsymbol{\zeta}$ be in V^N and index the eigenvalues so that $\lambda_j + \overline{\lambda}_j < 0$ if and only if $1 \le j \le \sigma$. Recall that by *Theorem 4.7*

$$\mathbf{z}(t,\boldsymbol{\zeta}) = \sum_{j=1}^{\sigma} e^{\lambda_j t}\left[\sum_{k=0}^{r(\lambda_j)-1} (A-\lambda_j \mathrm{I})^k \frac{t^k}{k!}\right]\boldsymbol{\zeta}_j$$

where $\boldsymbol{\zeta}_j \in M(\lambda_j)$. Then, as in the proof of *Theorem 5.8*, there exist positive constants K and α such that for all $t > 0$

$$|z(t,\boldsymbol{\zeta})| \le K e^{-\alpha t}|\boldsymbol{\zeta}|$$

because $Re\lambda_j = (\lambda_j + \overline{\lambda_j})/2 < 0$ for $j = 1,\ldots,\sigma$. Therefore, $\lim_{t\to\infty} \mathbf{z}(t,\boldsymbol{\zeta}) = \mathbf{0}$ if $\boldsymbol{\zeta} \in V^N$ and

$$V^N \subset \{\boldsymbol{\zeta} \in \mathbb{C}^d : \lim_{t\to\infty} \mathbf{z}(t,\boldsymbol{\zeta}) = \mathbf{0}\}.$$

Suppose that $\lim_{t\to\infty} \mathbf{z}(t,\boldsymbol{\zeta}) = \mathbf{0}$ and using *Theorem 4.7* again, write

$$\mathbf{z}(t,\boldsymbol{\zeta}) = \sum_{j=1}^{\rho} e^{\lambda_j t}\left[\sum_{k=0}^{r(\lambda_j)-1} (A-\lambda_j \mathrm{I})^k \frac{t^k}{k!}\right]\boldsymbol{\zeta}_j$$

where $\boldsymbol{\zeta}_j \in M(\lambda_j)$. It follows using *Theorem 4.6* that

$$\lim_{t\to\infty} e^{\lambda_j t}\left[\sum_{k=0}^{r(\lambda_j)-1} (A-\lambda_j \mathrm{I})^k \frac{t^k}{k!}\right]\boldsymbol{\zeta}_j = \mathbf{0}.$$

If $\boldsymbol{\zeta}_j \neq \mathbf{0}$, then it is a generalized eigenvector of degree $p > 0$ and the vectors $(A-\lambda_j \mathrm{I})^k \boldsymbol{\zeta}_j$, $k = 0,\ldots,p-1$ are linearly independent (*Proposition 4.4*). Consequently, the fact that the previous limit was $\mathbf{0}$ implies that

$$\lim_{t\to\infty} e^{\lambda_j t}\frac{t^k}{k!} = 0$$

for $k = 0, \ldots, p-1$. With $k = 0$ it is apparent that $Re\lambda_j < 0$ when $\boldsymbol{\zeta} \neq \mathbf{0}$. Therefore, $\boldsymbol{\zeta} \in V^N$ and

$$\{\boldsymbol{\zeta} \in \mathbb{C}^d : \lim_{t\to\infty} \mathbf{z}(t, \boldsymbol{\zeta}) = \mathbf{0}\} \subset V^N.$$

A similar argument proves that when $\boldsymbol{\zeta}$ is in V^N,

$$|z(t, \boldsymbol{\zeta})| \leq K e^{\alpha t} |\boldsymbol{\zeta}|$$

for $t \leq 0$ and that

$$V^P = \{\boldsymbol{\zeta} \in \mathbb{C}^d : \lim_{t\to-\infty} \mathbf{z}(t, \boldsymbol{\zeta}) = \mathbf{0}\}.$$

Finally, the required inequalities hold for $\mathbf{v}$ in E^+ or E^- because they hold for V^P and V^N, which contain E^+ and E^- respectively. $\square$

When A is in $\mathcal{H}$, the subspaces E^+ and E^- will be called the *positively stable manifold* and the *negatively stable manifold* respectively. These subspaces are also invariant under the flow $\mathbf{x}(t, \boldsymbol{\xi}) = e^{At}\boldsymbol{\xi}$. Specifically, $\mathbf{x}(t, \boldsymbol{\xi})$ is in E^+ or E^- for all $t \in \mathbb{R}$ according as $\boldsymbol{\xi}$ is in E^+ or E^-. This is an immediate consequence of the preceding theorem and the elementary formula $\mathbf{x}(t, \mathbf{x}(\tau, \boldsymbol{\xi})) = \mathbf{x}(t+\tau, \boldsymbol{\xi})$.

Proposition 8.2 *If A is in $\mathcal{H}$ and has distinct eigenvalues $\lambda_1, \ldots, \lambda_\rho$ with multiplicities $m_1, \ldots, m_\rho$, then the dimension of the stable manifold E^+ is given by*

$$\text{Dim}\, E^+ = \sum \{m_j : Re\lambda_j < 0\}.$$

Proof. Recall that the dimension of $M(\lambda_j)$ is m_j (*Theorem 4.6*). Since V^N is the direct sum of the subspaces $M(\lambda_j)$ such that $Re\lambda_j < 0$, it follows that the dimension of V^N is $\sum\{m_j : Re\lambda_j < 0\}$. Thus, it suffices to show that the dimension of V^N over $\mathbb{C}$ is the same as the dimension of E^+ over $\mathbb{R}$.

If $\mathbf{w}_1, \ldots, \mathbf{w}_{m_j}$ is a basis for $M(\lambda_j)$, then because A is real $\overline{\mathbf{w}}_1, \ldots, \overline{\mathbf{w}}_{m_j}$ is a basis for $M(\overline{\lambda}_j)$. As before the vectors $\mathbf{w}_k + \overline{\mathbf{w}}_k$ and $i(\mathbf{w}_k - \overline{\mathbf{w}}_k)$ for $k = 1, \ldots, m_j$ are linearly independent vectors in $\mathbb{R}^d$ and form a basis for

$$\{\mathbf{v} \in \mathbb{R}^d : \mathbf{v} \in M(\lambda_j) \oplus M(\overline{\lambda}_j)\}.$$

In this way, a basis for V^N can be constructed that will produce a basis of the same cardinality for E^+. $\square$

Remark *Let $\mathbf{f}(\mathbf{x}) = A\mathbf{x}$ be a hyperbolic linear vector field.*

(*a*) *If $\boldsymbol{\xi}$ lies in E^+, then* $\lim_{t\to-\infty} |\mathbf{x}(t, \boldsymbol{\xi})| = \infty$.

(*b*) *If $\boldsymbol{\xi}$ lies in E^-, then* $\lim_{t\to\infty} |\mathbf{x}(t, \boldsymbol{\xi})| = \infty$.

(*c*) *If $\boldsymbol{\xi}$ is in neither E^+ nor E^-, then* $\lim_{t\to\pm\infty} |\mathbf{x}(t, \boldsymbol{\xi})| = \infty$.

Proof. Exercise.

Thus the phase portrait for a hyperbolic linear system when $E^+ \neq \mathbf{0}$ and $E^- \neq \mathbf{0}$ is a higher dimensional version of the familiar phase portrait for the 2×2 hyperbolic matrix

$$\begin{bmatrix} 1 & 0 \\ 0 & -1 \end{bmatrix}.$$

In fact, the latter is often used as a schematic representation of the former as in *Figure 8.1*, but this ignores the dimensions of the stable and unstable manifolds. In this figure, the dimension of the subspace E^+ pictured as a line running from the upper left to lower right ranges from 1 to $d-1$. In the extreme case when $E^- = \mathbf{0}$, the origin is positively asymptotically stable and the phase portrait is very different and schematically like *Figure 5.4* on page 163. This raises the natural questions of what does it mean for the phase portraits of two linear hyperbolic systems to be the same? When are they the same and when are they different?

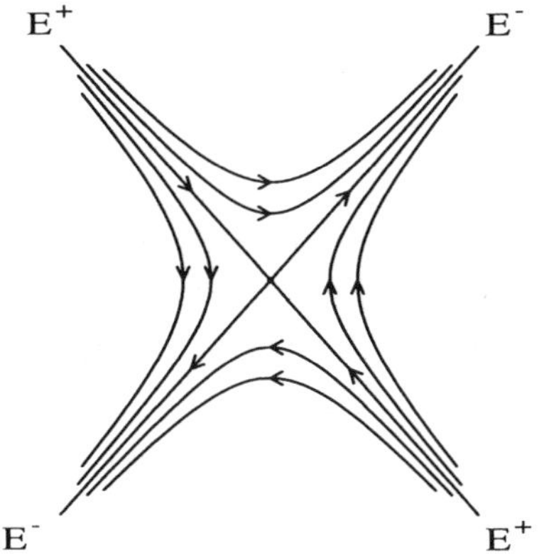

Figure 8.1: Schematic phase portrait of a typical hyperbolic linear system.

The categorization and classification of mathematical objects can be a fruitful endeavor but the classification is often out of reach. The categorization is usually done by defining when two objects are essentially the same in some useful way. For example, two circles in the plane are congruent if a rigid motion maps one onto the other. (A rigid motion is a composition of translations, rotations, and reflections.) Two $d \times d$ matrices A and B are *similar* if there exist an invertible matrix P such that $A = PBP^{-1}$. And two norms on a vector space are equivalent if there exists positive constants α and β satisfying $\alpha\|\mathbf{x}\|_1 \leq \|\mathbf{x}\|_2 \leq \beta\|\mathbf{x}\|_1$ for all $\mathbf{x}$. There are many other examples of this type.

Classification is much harder. The idea is to establish necessary and sufficient conditions for two objects to be in the same category and use them to determine a unique representative of each category. For example, two circles in the plane are congruent if and only if they have the same radius. All circles with a common center provides a set of representatives for the congruence classes of circles because every circle is congruent to exactly one of them. The classification of equivalent norms on finite dimensional vector spaces is trivial, they are all equivalent to the Euclidean norm, and impossible on infinite-dimensional spaces. The Jordan canonical form is a classification scheme for similar matrices, albeit a complicated one. Two matrices are similar if and only if their Jordan canonical forms have the same set of diagonal blocks, but not necessarily in the same order.

The classification of all flows is probably impossible, but there are interesting categorizations of flows to study. There are also special classes of flows for which classification is possible. One such class is the flows on $\mathbb{R}^d$ obtained from hyperbolic linear vector fields. This classification is based on the notion of when two flows are isomorphic. For hyperbolic linear flows, it depends only on the dimension of the stable manifold.

Two flows $\Phi_1 : \mathbb{R} \times \Omega_1 \to \Omega_1$ and $\Phi_2 : \mathbb{R} \times \Omega_2 \to \Omega_2$ are *isomorphic* if there exists a continuous, map $\mathbf{h}$ of Ω_1 onto Ω_2 with a continuous inverse such that for all $\mathbf{x} \in \Omega_1$ and $t \in \mathbb{R}$ the following holds:

$$\mathbf{h}(\Phi_1(t, \mathbf{x})) = \Phi_2(\mathbf{h}(\mathbf{x}), t). \tag{8.1}$$

In other words, moving along the orbit of $\mathbf{x}$ for t units of time and then applying $\mathbf{h}$ is the same as applying $\mathbf{h}$ to $\mathbf{x}$ and then moving along the orbit of $\mathbf{h}(\mathbf{x})$ for t units of time. It is easy to see that the inverse of an isomorphism is an isomorphism and that the composition of two isomorphisms is another isomorphism. The next proposition shows how an isomorphism preserves the dynamical structure of a flow.

Proposition 8.3 *If $\mathbf{h}$ is an isomorphism between two flows, then the following hold:*

(a) $\mathbf{h}(\mathcal{O}(\mathbf{x})) = \mathcal{O}(\mathbf{h}(\mathbf{x}))$,

(b) $\mathbf{x}$ *is a fixed point if and only if* $\mathbf{h}(\mathbf{x})$ *is a fixed point,*

(c) $\mathbf{x}$ *is a periodic point of period T if and only if* $\mathbf{h}(\mathbf{x})$ *is a periodic point of period T,*

(d) $\mathbf{x}$ *is an almost periodic point if and only if* $\mathbf{h}(\mathbf{x})$ *is an almost periodic point,*

(e) $\mathbf{h}(\omega(\mathbf{x})) = \omega(\mathbf{h}(\mathbf{x}))$,

(f) $\mathbf{h}(\alpha(\mathbf{x})) = \alpha(\mathbf{h}(\mathbf{x}))$,

(g) $\mathbf{x}$ *is a recurrent point if and only if* $\mathbf{h}(\mathbf{x})$ *is a recurrent point,*

(h) $\mathbf{h}(\overline{\mathcal{O}}(\mathbf{x})) = \overline{\mathcal{O}}(\mathbf{h}(\mathbf{x}))$,

(i) M *is a minimal set if and only if* $\mathbf{h}(M)$ *is a minimal set.*

Proof. Exercise

If the autonomous differential equations $\dot{\mathbf{x}} = \mathbf{f}(\mathbf{x})$ and $\dot{\mathbf{x}} = \mathbf{g}(\mathbf{x})$ determine isomorphic flows on Ω_1 and Ω_2, respectively, then the isomorphism maps orbits-to-orbits and preserves their dynamical properties. Thus their phase portraits are indistinguishable by means of their dynamical properties. So isomorphism of flows is one way to equate phase portraits. It is, however, a coarser classification of phase portraits than we made in the two-dimensional case at the end of Section 5.2.

For hyperbolic linear systems, it is possible to determine precisely when the resulting flows on $\mathbb{R}^d$ are isomorphic. The first step is to show that all positively asymptotically stable linear systems on $\mathbb{R}^d$ produce isomorphic flows. This will require a bit of hard work and will make use of the elementary fact that similar matrices produce isomorphic flows.

Remark *If A and B are similar $d \times d$ real matrices, then the flows $\mathbf{x}(t, \boldsymbol{\xi}) = e^{At}\boldsymbol{\xi}$ and $\mathbf{y}(t, \boldsymbol{\xi}) = e^{Bt}\boldsymbol{\xi}$ are isomorphic.*

Proof. There exists an invertible matrix P such that $B = PAP^{-1}$. Let $\mathbf{h}(\boldsymbol{\xi}) = P\boldsymbol{\xi}$. Clearly, $\mathbf{h}^{-1}(\boldsymbol{\xi}) = P^{-1}\boldsymbol{\xi}$ and both $\mathbf{h}$ and $\mathbf{h}^{-1}$ are continuous. Then,

$$\mathbf{h}(e^{At}\boldsymbol{\xi}) = Pe^{At}\boldsymbol{\xi} = Pe^{At}P^{-1}P\boldsymbol{\xi} = e^{PAP^{-1}t}\mathbf{h}(\boldsymbol{\xi}) = e^{Bt}\mathbf{h}(\boldsymbol{\xi})$$

and $\mathbf{h}$ is an isomorphism of flows. □

Theorem 8.4 *Let A and A' be $d \times d$ real matrices. If all the eigenvalues of both A and A' have negative real parts, then the flows determined by $\dot{\mathbf{x}} = A\mathbf{x}$ and $\dot{\mathbf{y}} = A'\mathbf{y}$ are isomorphic.*

Proof. Without loss of generality, it can be assumed that $A' = -\mathrm{I}$. The flows determined by $\dot{\mathbf{x}} = A\mathbf{x}$ and $\dot{\mathbf{y}} = -\mathbf{y}$ are then given by

$$\begin{aligned} \mathbf{x}(t, \boldsymbol{\xi}) &= e^{At}\boldsymbol{\xi} \\ \mathbf{y}(t, \boldsymbol{\xi}) &= e^{-t}\boldsymbol{\xi}. \end{aligned}$$

It suffices to construct a continuous map $\mathbf{h} : \mathbb{R}^d \to \mathbb{R}^d$ with a continuous inverse such that

$$\mathbf{h}(e^{-t}\boldsymbol{\xi}) = e^{At}\mathbf{h}(\boldsymbol{\xi}).$$

Notice that if $\boldsymbol{\xi} \neq \mathbf{0}$, then the solution $\mathbf{y}(t, \boldsymbol{\xi}) = e^{-t}\boldsymbol{\xi}$ crosses the unit sphere $\{\mathbf{y} : \|\mathbf{y}\| = 1\}$ exactly once when $t = \log(\|\boldsymbol{\xi}\|)$. (The unit sphere is a very nice global section for the flow on $\mathbb{R}^d \setminus \{\mathbf{0}\}$.) The strategy of the proof is to find an analogue of the unit sphere for the flow determined by $\dot{\mathbf{x}} = A\mathbf{x}$, map the unit sphere onto it, and then let the flow map orbits to orbits.

By *Theorem 7.13*, part (c) there exists a positive definite matrix B such that $BA + A^{\mathrm{t}}B = -\mathrm{I}$. Because B is positive definite, $\mathbf{x} \star \mathbf{y} = \mathbf{x} \cdot (B\mathbf{y})$ has the same properties as the usual inner product $\mathbf{x} \cdot \mathbf{y}$ and $\|\mathbf{x}\|_B = \sqrt{\mathbf{x} \star \mathbf{x}}$ defines a norm on $\mathbb{R}^d$. The Gram-Schmidt process on page 140 can also be applied to $\mathbf{x} \star \mathbf{x}$ to obtain a basis $\mathbf{w}_1, \ldots, \mathbf{w}_d$ of $\mathbb{R}^d$ such that $\mathbf{w}_j \star \mathbf{w}_k = 0$ for $j \neq k$ and $\mathbf{w}_j \star \mathbf{w}_j = 1$. The set $\{\mathbf{x} : \|\mathbf{x}\|_B = 1\}$ is the appropriate unit sphere for $\dot{\mathbf{x}} = A\mathbf{x}$.

Define a map $\theta : \{\mathbf{y} : \|\mathbf{y}\| = 1\} \to \{\mathbf{x} : \|\mathbf{x}\|_B = 1\}$ by

$$\theta(\mathbf{y}) = \sum_{j=1}^{d} y_j \mathbf{w}_j.$$

It is easy to check that θ is continuous and has a continuous inverse. And then define $\mathbf{h}$ by

$$\mathbf{h}(\mathbf{y}) = e^{-A \log(\|y\|)} \theta(\mathbf{y}/\|\mathbf{y}\|)$$

for $\mathbf{y} \neq \mathbf{0}$ and $\mathbf{h}(\mathbf{0}) = \mathbf{0}$. Obviously, $\mathbf{h}$ is continuous on $\mathbb{R}^d \setminus \{\mathbf{0}\}$. The continuity at $\mathbf{0}$ follows from the existence of positive constants K and α such that $|e^{At}| \leq Ke^{-\alpha t}$ for $t \geq 0$. The following calculation shows that $\mathbf{h}$ satisfies the isomorphism equation $\mathbf{h}(e^{-t}\boldsymbol{\xi}) = e^{At}\mathbf{h}(\boldsymbol{\xi})$:

$$\begin{aligned}
\mathbf{h}(e^{-t}\boldsymbol{\xi}) &= \exp[-A \log(\|e^{-t}\boldsymbol{\xi}\|)]\, \theta\left(\frac{e^{-t}\boldsymbol{\xi}}{\|e^{-t}\boldsymbol{\xi}\|}\right) \\
&= \exp[-A(\log e^{-t} + \log \|\boldsymbol{\xi}\|)]\, \theta\left(\frac{\boldsymbol{\xi}}{\|\boldsymbol{\xi}\|}\right) \\
&= e^{At}\mathbf{h}(\boldsymbol{\xi}).
\end{aligned}$$

The stage is now set for the central part of the proof, showing first that $\mathbf{h}$ is one-to-one and onto, and thus has an inverse. And the final step will be showing that $\mathbf{h}^{-1}$ is continuous.

Let $V(\mathbf{x}) = \mathbf{x} \cdot (B\mathbf{x}) = \|\mathbf{x}\|_B^2$. It follows from the calculation of $\dot{V}(\mathbf{x})$ on page 233 that $\dot{V}(\mathbf{x}) = -\|\mathbf{x}\|^2$ because $BA + A^{t}B = -\mathrm{I}$. For any solution $\varphi(t)$ of $\dot{\mathbf{x}} = A\mathbf{x}$, the function $V(\varphi(t))$ is strictly decreasing on $\mathbb{R}$. The positive asymptotic stability implies that

$$\lim_{t \to \infty} V(\varphi(t)) = \mathbf{0}.$$

Since all norms on $\mathbb{R}^d$ are equivalent, there exists M such that $\|\mathbf{x}\|^2 \geq M\|\mathbf{x}\|_B^2 = MV(\mathbf{x})$, and hence $-\dot{V}(\varphi(t)) = \|\varphi(t)\|^2 \geq MV(\varphi(t))$. If $t \leq 0$, then the minimum value of $V(\varphi(t))$ on $[t, 0]$ is $V(\varphi(0))$ and

$$\begin{aligned}
V(\varphi(t)) &= V(\varphi(0)) + \int_0^t \dot{V}(\varphi(s))ds \\
&= V(\varphi(0)) + \int_t^0 -\dot{V}(\varphi(s))ds \\
&\geq V(\varphi(0)) + M \int_t^0 V(\varphi(s))ds \\
&\geq V(\varphi(0)) + M|t|V(\varphi(0)).
\end{aligned}$$

For $t \leq 0$

$$V(\varphi(t)) \geq V(\varphi(0))(1 + M|t|) \tag{8.2}$$

and

$$\lim_{t \to -\infty} V(\varphi(t)) = \infty.$$

Therefore for every nonzero solution $\varphi(t) = e^{At}\boldsymbol{\xi}$ there exists a unique τ such that

$$V(\varphi(\tau)) = \|\varphi(\tau))\|_B^2 = 1$$

and a unique point $\varphi(\tau)$ where the solution intersects $\{\mathbf{x} : \|\mathbf{x}\|_B = 1\}$. This observation will play a critical role in the rest of the proof.

To prove that $\mathbf{h}$ is onto consider $\mathbf{x} \neq \mathbf{0}$ and τ such that $\|e^{A\tau}\mathbf{x}\| = 1$. Because the map θ is onto, there exists $\mathbf{y}$ with $\|\mathbf{y}\| = 1$ such that $\theta(\mathbf{y}) = e^{A\tau}\mathbf{x}$. The isomorphism equation implies that

$$\mathbf{h}\big(e^{-(-\tau)}\mathbf{y}\big) = e^{-A\tau}\theta(\mathbf{y}) = e^{-A\tau}e^{A\tau}\mathbf{x} = \mathbf{x}$$

and thus $\mathbf{h}$ is onto.

If $\mathbf{h}(\mathbf{y}_1) = \mathbf{h}(y_2)$, then

$$\begin{aligned} \exp[-A\log(\|y_1\|)]\,\theta(\mathbf{y}_1/\|\mathbf{y}_1\|) &= \exp[-A\log(\|y_2\|)]\,\theta(\mathbf{y}_2/\|\mathbf{y}_2\|) \\ \theta(\mathbf{y}_1/\|\mathbf{y}_1\|) &= \exp[-A\log(\|y_2\|/\|y_1\|)]\,\theta(\mathbf{y}_2/\|\mathbf{y}_2\|). \end{aligned}$$

Since $\|\theta(\mathbf{y}_j/\|\mathbf{y}_j\|)\|_B = 1$, the previous equation can only hold if

$$\log(\|y_2\|/\|y_1\|) = 0$$

and

$$\theta(\mathbf{y}_1/\|\mathbf{y}_1\|) = \theta(\mathbf{y}_2/\|\mathbf{y}_2\|.$$

These two equations force $\mathbf{y}_1 = \mathbf{y}_2$ and thus $\mathbf{h}$ is one-to-one.

For $\mathbf{x} \neq \mathbf{0}$, set $\psi(\mathbf{x}) = \tau$ such that $\|e^{A\tau}\|_B = 1$. It was shown above that there is a unique such τ and so ψ is a well-defined function. It is easy to verify that

$$\mathbf{h}^{-1}(\mathbf{x}) = e^{\psi(\mathbf{x})}\theta^{-1}(e^{A\psi(\mathbf{x})}\mathbf{x}).$$

If ψ is continuous, then $\mathbf{h}^{-1}$ is continuous on $\mathbb{R}^d \setminus \{\mathbf{0}\}$. The continuity of $\mathbf{h}^{-1}$ at $\mathbf{0}$ is proved in the same way as that the continuity of $\mathbf{h}$ at $\mathbf{0}$ was proved. The following lemma completes the proof of the theorem. □

Lemma *The real-valued function ψ is continuous on $\mathbb{R}^d \setminus \{\mathbf{0}\}$.*

Proof. Let $\mathbf{x}_n$ be a sequence of points converging to $\mathbf{x}$ in $\mathbb{R}^d \setminus \{\mathbf{0}\}$. So there exist positive constants β_1 and β_2 such that $\beta_1 \leq \|\mathbf{x}_n\|_B$ and $|\mathbf{x}_n| \leq \beta_2$ for all n. It suffices to show that $\psi(\mathbf{x}_n)$ converges $\psi(\mathbf{x})$.

When $\psi(\mathbf{x}_n) \geq 0$, the following estimate shows that the sequence $\psi(\mathbf{x}_n)$ must be bounded above:

$$1 = \|e^{A\psi(\mathbf{x}_n)}\mathbf{x}_n\|_B \leq MKe^{-\alpha\psi(\mathbf{x}_n)}|\mathbf{x}_n| \leq MKe^{-\alpha\psi(\mathbf{x}_n)}\beta_2.$$

When $\psi(\mathbf{x}_n) \leq 0$,

$$1 = \|e^{A\psi(\mathbf{x}_n)}\mathbf{x}_n\|_B^2 = V(e^{A\psi(\mathbf{x}_n)}\mathbf{x}_n) \geq V(\mathbf{x}_n)(1 + M\psi(\mathbf{x}_n))$$

because *(8.2)* can be applied to $\mathbf{x}(\psi(\mathbf{x}_n), \mathbf{x}_n) = e^{A\psi(\mathbf{x}_n)}\mathbf{x}_n$. So the sequence $\psi(\mathbf{x}_N)$ must be bounded below. Therefore, $\psi(\mathbf{x}_n)$ is bounded and has convergent subsequences.

If $\psi(\mathbf{x}_n)$ does not converge to $\psi(\mathbf{x})$, then there exists a subsequence $\psi(\mathbf{x}_{n_k})$ converging to $\sigma \neq \psi(\mathbf{x})$. By continuity in initial conditions,

$$\lim_{n\to\infty} e^{A\psi(\mathbf{x}_{n_k})}\mathbf{x}_{n_k} = e^{A\sigma}\mathbf{x}.$$

Furthermore, $\|e^{A\sigma}\mathbf{x}\|_B = 1$ because $\|e^{A\psi(\mathbf{x}_{n_k})}\mathbf{x}_{n_k}\|_B = 1$. But this is impossible because $\psi(\mathbf{x})$ is the unique time for which $\|e^{A\psi(\mathbf{x})}\mathbf{x}\|_B = 1$ and $\sigma \neq \psi(\mathbf{x})$. This contradiction completes the proof of the lemma. □

Corollary *Let A and A' be $d \times d$ real matrices. If all the eigenvalues of both A and A' have positive real parts, then the flows determined by $\dot{\mathbf{x}} = A\mathbf{x}$ and $\dot{\mathbf{y}} = A'\mathbf{y}$ are isomorphic.*

Proof. Apply the theorem to $-A$ and $-A'$ and observe that

$$\mathbf{h}(e^{At}\boldsymbol{\xi}) = \mathbf{h}(e^{-A(-t)}\boldsymbol{\xi}) = e^{-A'(-t)}\mathbf{h}(\boldsymbol{\xi}) = e^{A't}\mathbf{h}(\boldsymbol{\xi})$$

to complete the proof. □

Theorem 8.4 shows that the dynamics of the phase portraits of positively asymptotically stable linear systems are all the same. With the corollary the classification of the flows by isomorphism for positively and negatively asymptotically stable linear systems is complete. There are only two classes and they are represented by −I and I. The next step is to extend this classification to hyperbolic linear systems.

Given A in $\mathcal{H}$ with distinct eigenvalues $\lambda_1, \ldots, \lambda_\rho$ and with multiplicities $m_1, \ldots, m_\rho$, define the *index* of A by

$$\iota(A) = \sum\{m_j : Re\lambda_j < 0\}.$$

Clearly, the values of $\iota(A)$ are the integers $0, 1, \ldots, d$ and $\iota(A)$ equals the dimension of the stable manifold E^+ for $\dot{\mathbf{x}} = A\mathbf{x}$ by *Proposition 8.2*. Setting

$$\mathcal{H}_j = \{A : \iota(A) = j\}$$

partitions $\mathcal{H}$ into $d+1$ disjoint sets.

Theorem 8.5 *Let A and B be in $\mathcal{H}$. The flows $\mathbf{x}(t, \boldsymbol{\xi}) = e^{At}\boldsymbol{\xi}$ and $\mathbf{y}(t, \boldsymbol{\xi}) = e^{Bt}\boldsymbol{\xi}$ determined by the differential equations $\dot{\mathbf{x}} = A\mathbf{x}$ and $\dot{\mathbf{y}} = B\mathbf{y}$, respectively, are isomorphic if and only if $\iota(A) = \iota(B)$.*

Proof. Suppose $\iota(A) = \iota(B) = k$. By the Remark on page 257, A can be replaced by any matrix similar to A. Because $R^d = E^+ \oplus E^-$ and both these subspaces are invariant under the linear transformation $T(\mathbf{v}) = A\mathbf{v}$, the matrix A is similar to a real matrix of the form

$$\begin{bmatrix} A^+ & O \\ O & A^- \end{bmatrix}$$

such that λ_j is an eigenvalue for A^+ if and only if $\lambda_j + \overline{\lambda}_j < 0$ and A^+ is a $k \times k$ matrix. Likewise for B. Replace A and B by matrices in this block form. The first k elements of the standard basis $\mathbf{e}_1, \ldots, \mathbf{e}_d$ of $\mathbb{R}^d$ are a basis for E^+ and the last $d - k$ are a basis for E^- for both A and B because $\iota(A) = \iota(B) = k$.

Theorem 8.4 can now be applied to $\dot{\mathbf{x}} = A^+\mathbf{x}$ and $\dot{\mathbf{y}} = B^+\mathbf{y}$ to obtain an isomorphism $\mathbf{h}^+ : \mathbb{R}^k \to \mathbb{R}^k$ satisfying $\mathbf{h}^+(e^{A^+t}\boldsymbol{\xi}) = e^{B^+t}\mathbf{h}^+(\boldsymbol{\xi})$. By the corollary to *Theorem 8.4* there exists an isomorphism $\mathbf{h}^- : \mathbb{R}^{d-k} \to \mathbb{R}^{d-k}$ satisfying $\mathbf{h}^-(e^{A^-t}\boldsymbol{\xi}) = e^{B^-t}\mathbf{h}^-(\boldsymbol{\xi})$.

The required isomorphism $\mathbf{h} : \mathbb{R}^d \to \mathbb{R}^d$ will be constructed using $\mathbb{R}^d = E^+ \oplus E^d$ to combine $\mathbf{h}^+$ and and $\mathbf{h}^-$. Specifically, let $\mathbf{x}_+ = (x_1, \ldots, x_k)$ and $\mathbf{x}_- = (x_{k+1}, \ldots, x_d)$ so that $\mathbf{x}_+$ is a generic point in E^+, $\mathbf{x}_- = (x_{k+1}, \ldots, x_d)$ is a generic point in E^-, and $\mathbf{x} = (\mathbf{x}_+, \mathbf{x}_-)$ is a generic point of $\mathbb{R}^d$. Set

$$\mathbf{h}(\mathbf{x}) = (\mathbf{h}^+(\mathbf{x}_P), \mathbf{h}^-(x_U))$$

and note that

$$e^{At}\mathbf{x} = (e^{A^+t}\mathbf{x}_P, e^{A^-t}\mathbf{x}_N).$$

It is obvious that $\mathbf{h}$ is the required isomorphism.

Now suppose $\mathbf{h}$ is an isomorphism between the flows determined by $\dot{\mathbf{x}} = A\mathbf{x}$ and $\dot{\mathbf{y}} = B\mathbf{y}$. Since $\mathbf{0}$ is the only fixed point of either flow, by *Proposition 8.3* $\mathbf{h}(\mathbf{0}) = \mathbf{0}$. *Theorem 8.1* implies that $E^+ = \{\boldsymbol{\xi} : \omega(\boldsymbol{\xi}) = \mathbf{0}\}$, and hence using *Proposition 8.3* again $\mathbf{h}$ maps the positively stable manifold E_A^+ for A onto the positively stable manifold E_B^+ for B. And $\mathbf{h}^{-1}$ does the reverse. Therefore, restricting $\mathbf{h}$ to E_A^+ yields a continuous map of E_A^+ onto E_B^+ with a continuous inverse. Because the stable manifold is just a copy of $\mathbb{R}^{\iota(A)}$ sitting in $\mathbb{R}^d$, there exists a continuous map $\mathbf{g}$ of $\mathbb{R}^{\iota(A)}$ onto $\mathbb{R}^{\iota(B)}$ with a continuous inverse. To conclude that $\iota(A) = \iota(B)$, we must invoke a substantive theorem from topology. Specifically, the invariance of domain theorem (See [34] for a proof.) states that if $\mathbf{g}$ is a continuous map from an open set of R^m onto an open set of R^n with a continuous inverse, then $m = n$. Therefore, the existence of the map $\mathbf{g}$ obtained from the isomorphism $\mathbf{h}$ implies that $\iota(A) = \iota(B)$ □

As a consequence of the above theorem for hyperbolic linear vector fields on $\mathbb{R}^d$ there are exactly $d + 1$ isomorphism classes of flows and the matrices in each $\mathcal{H}_j$ determine a class of isomorphic flows. Thus, when A and B are both in $\mathcal{H}_j$ the solutions of $\dot{\mathbf{x}} = A\mathbf{x}$ and $\dot{\mathbf{y}} = B\mathbf{y}$ have the same dynamical behavior. Some features, however, are lost in this classification. For example, it does not distinguish between solutions spiraling toward the origin and those approaching the origin along a ray.

For each of the $d + 1$ isomorphism classes for linear hyperbolic flows, it is easy to pick a representative element of the isomorphism class. For example,

$$\begin{aligned} A_0 &= \text{Diag}\{1, 1, 1, \ldots, 1\} \\ A_1 &= \text{Diag}\{-1, 1, 1, \ldots, 1\} \\ A_2 &= \text{Diag}\{-1, -1, 1, \ldots, 1\} \\ &\vdots \\ A_d &= \text{Diag}\{-1, -1, -1, \ldots, -1\} \end{aligned}$$

is a set of matrices such that A_j is in $\mathcal{H}_j$. Therefore, every hyperbolic linear flow on $\mathbb{R}^d$ is isomorphic to $\mathbf{x}(t, \boldsymbol{\xi}) = e^{A_j t}\boldsymbol{\xi}$ for precisely one j. These representative flows are easy to calculate explicitly.

Although there are only $d+1$ isomorphism classes of linear hyperbolic flows on $\mathbb{R}^d$, there are infinitely many isomorphism classes of linear hyperbolic linear flows because flows on $\mathbb{R}^m$ and $\mathbb{R}^n$ cannot be isomorphic by the invariance of domain theorem.

Two important questions remain. How sensitive is the classification of hyperbolic linear vector fields to small perturbations? Not every linear vector field $\mathbf{f}(\mathbf{x}) = A\mathbf{x}$ is in $\mathcal{H}$, but in some reasonable sense are most of them in $\mathcal{H}$? A better understanding of how eigenvalues depend continuously on matrices is necessary to answer these questions. We begin with a few simple facts about eigenvalues.

Remark *Let A be a $d \times d$ complex matrix. If λ is an eigenvalue of A, then*

$$|\lambda| \leq |A| = \sup\{|Az| : |z| = 1\}.$$

Proof. Let $\boldsymbol{\zeta}$ be an eigenvector for λ such that $|\boldsymbol{\zeta}| = 1$. Then

$$|\lambda| = |\lambda\boldsymbol{\zeta}| = |A\boldsymbol{\zeta}| \leq |A|\,|\boldsymbol{\zeta}| = |A|$$

as required. □

Proposition 8.6 *Let A_k be a sequence in $\mathcal{M}_d(\mathbb{C})$ converging to A. If λ_k is an eigenvalue of A_k for each k and and λ_k converges to λ, then λ is an eigenvalue of A*

Proof. Let $\boldsymbol{\zeta}_k$ be an eigenvector of A_k for the eigenvalue λ_k such that $|\boldsymbol{\zeta}_k| = 1$. Because $\{\mathbf{z} \in \mathbb{C}^d : |\mathbf{z}| = 1\}$ is compact, it can be assumed without loss of generality that $\boldsymbol{\zeta}_k$ converges to $\boldsymbol{\zeta}$. Clearly, $|\boldsymbol{\zeta}| = 1$ and $\boldsymbol{\zeta} \neq \mathbf{0}$. Then

$$\begin{aligned} |\lambda - A\boldsymbol{\zeta}| &= |\lambda\,\boldsymbol{\zeta} - \lambda_k\,\boldsymbol{\zeta}_k + A_k\boldsymbol{\zeta} - A\boldsymbol{\zeta}| \\ &\leq |\lambda\,\boldsymbol{\zeta} - \lambda_k\,\boldsymbol{\zeta}_k| + |A_k\boldsymbol{\zeta}_k - A\boldsymbol{\zeta}| \end{aligned}$$

and it suffices to show that both terms on the right go to zero as k goes to infinity.

The two terms on the right can be estimated as follows:

$$\begin{aligned} |\lambda\,\zeta - \lambda_k\,\zeta_k| &\le |\lambda\,\zeta - \lambda\zeta_k| + |\lambda\,\zeta_k - \lambda_k\zeta_k| \\ &\le |\lambda|\,|\zeta - \zeta_k| + |\lambda - \lambda_k|\,|\zeta_k| \\ &= |\lambda|\,|\zeta - \zeta_k| + |\lambda - \lambda_k| \end{aligned}$$

and

$$\begin{aligned} |A_k\zeta_k - A\zeta| &\le |A_k\zeta_k - A\zeta_k| + |A\zeta_k - A\zeta| \\ &\le |A_k - A|\,|\zeta| + |A|\,|\zeta_k - \zeta| \\ &= |A_k - A| + |A|\,|\zeta_k - \zeta|. \end{aligned}$$

Therefore,

$$|\lambda - A\zeta| \le |\lambda|\,|\zeta - \zeta_k| + |\lambda - \lambda_k| + |A_k - A| + |A|\,|\zeta_k - \zeta|$$

and all four terms on the right go to zero as k goes to infinity. □

An eigenvalue of multiplicity greater than 1 can decompose into several eigenvalues as the coefficients of the matrix vary, or two or more eigenvalues can coalesce into a single eigenvalue of multiplicity greater than 1. Consequently, trying to express an eigenvalue as a continuous function of a general matrix is impossible. There is, however, a way of explaining how the eigenvalues collectively depend continuously on the matrices in $\mathcal{M}_d(\mathbb{C})$ and it can be used to show that each $\mathcal{H}_j$ is an open set in $\mathcal{M}_d(\mathbb{R})$. Before proving such a result, some notation is needed.

Let A_k be a sequence of matrices in $\mathcal{M}_d(\mathbb{C})$. Let $\sigma(k)$ denote the number of distinct eigenvalues of A_k, so $1 \le \sigma(k) \le d$. Let $\mu(1,k), \ldots, \mu(\sigma(k),k)$ be an indexed list of the eigenvalues for A_k and denote a generic eigenvalue of A_k by $\mu(j,k)$. Let $m(j,k)$ denote the multiplicity of $\mu(j,k)$. Note that $m(1,k) + \ldots + m(\sigma(k),k) = d$. The desired theorem can now be stated as follows:

Theorem 8.7 *Let A_k be a sequence of matrices in $\mathcal{M}_d(\mathbb{C})$ converging to A and let $\lambda_1, \ldots, \lambda_\rho$ be the distinct eigenvalues of A with multiplicities $m_1, \ldots, m_\rho$. Let $\varepsilon > 0$. If for $1 \le q \le \rho$ the disks*

$$B_\varepsilon(\lambda_q) = \{z \in \mathbb{C} : |z - \lambda_q| < \varepsilon\}$$

in the complex plane are disjoint and

$$\Lambda(q,k) = \{j : \mu(j,k) \in B_\varepsilon(\lambda_q)\},$$

then there exists K such that

$$\sum_{j\in\Lambda(q,k)} m(j,k) = m_q$$

for $k \ge K$.

Proof. Because A_k is convergent it is bounded, and there exists $\alpha > 0$ such that $|A_k| < \alpha$ for all k. By the previous remark, $|\mu(j,k)| < \alpha$ for all k. *Proposition 8.6* guarantees that for large k every $\mu(j,k)$ is in some $B_\varepsilon(\lambda_q)$. [If not, a subsequence of $\mu(j,k)$ would converge to an eigenvalue of A not in the list $\lambda_1, \ldots, \lambda_q$.] Since the disks $B_\varepsilon(\lambda_q)$ are disjoint, $\Lambda(1,k), \ldots, \Lambda(\rho,k)$ is a partition of $1, \ldots, \sigma(k)$ into disjoint sets for large k.

Set

$$\gamma(q,k) = \sum_{j \in \Lambda(q,k)} m(j,k)$$

and notice that because the sets $\Lambda(q,k)$ are a partition of $1, \ldots, \sigma(k)$

$$\sum_{q=1}^{\rho} \gamma(q,k) = \sum_{j=1}^{\sigma(k)} m(j,k) = d = \sum_{q=1}^{\rho} m_q.$$

Therefore it suffices to prove that $\gamma(q,k) \leq m_q$ for $q = 1, \ldots, \rho$.

We will assume that $\gamma(j,k) > m_p$ infinitely often for a specific p and derive a contradiction. From now on p is this specific integer. By passing to a subsequence it can be assumed that $\gamma(p,k) = \gamma > m_p$ is constant over k.

In the remainder of the argument it will be necessary simply "to pass to a subsequence" several more times without modifying the notation. This avoids notation like k_{m_n} and worse. But first it is necessary to show that the whole sequence can be replaced by one with distinct eigenvalues and $\gamma(p,k) = \gamma > m_p$.

From the Jordan canonical form, it is clear that we can find A'_k such that for each k the following hold:

(a) $|A_k - A'_k| < 1/k$;

(b) A'_k has d distinct eigenvalues;

(c) precisely $\gamma(q,k)$ of the eigenvalues of A'_k lie in $B_\varepsilon(\lambda_q)$;

(d) $\gamma(p,k) = \gamma > m_p$.

Since A'_k also converges to A, it can now be assumed that each A_k has d distinct eigenvalues, $\sigma(k) = d$, and $m(j,k) = 1$ without changing the numbers $\gamma(q,k)$ and $\gamma(p,k) = \gamma > n_p$. Furthermore, the eigenvalues of A_k can be indexed so that $\Lambda(p,k) = \{1, 2, \ldots, \gamma\}$. *Proposition 8.6* implies that $\mu(j,k)$ converges to λ_p for $j = 1, \ldots, \gamma$.

Set

$$B_k = \prod_{j=1}^{\gamma} \left[A_k - \mu(j,k)\mathrm{I}\right].$$

It follows that dimension of the null space of each B_k is γ because the matrices $A_k - \mu(j,k)\mathrm{I}$ commute with each other. The sequence of matrices B_k converges to

$$(A - \lambda_p \mathrm{I})^\gamma.$$

Now, construct an orthonormal basis $\mathbf{e}(1,k),\ldots,\mathbf{e}(\gamma,p)$ for the null space of B_k. By passing to a subsequence again, we can assume that $\mathbf{e}(j,k)$ converges to $\mathbf{e}(j)$ as k goes to infinity for $1 \leq j \leq \gamma$. It is easily verified that $\{\mathbf{e}(1),\ldots,\mathbf{e}(\gamma)\}$ is an orthonormal, and thus linearly independent, set of vectors in the null space of $(A-\lambda_p \mathrm{I})^\gamma$. Therefore the dimension of the null space of $(A-\lambda_p \mathrm{I})^\gamma$ is at least γ.

The dimension of $M(\lambda_p)$ is m_p by *Theorem 4.6*. Hence, $r(\lambda_p) \leq m_p < \gamma$ by the lemma on page 116 and by the same lemma the null space of $(A-\lambda_p \mathrm{I})^\gamma$ is the generalized eigenspace $M(\lambda_p)$ for A. Therefore, the dimension of the null space of $(A-\lambda_p \mathrm{I})^\gamma$ equals $m_p < \gamma$, a contradiction. □

Corollary *Let A be a given matrix in $\mathcal{M}_d(\mathbb{R})$ having distinct eigenvalues $\lambda_1,\ldots,\lambda_\rho$ with multiplicities $m_1,\ldots,m_\rho$, and let $B \in \mathcal{M}_d(\mathbb{R})$ be an arbitrary matrix with $\sigma(B)$ distinct eigenvalues $\mu(1,B),\ldots,\mu(\sigma(B),B)$ with multiplicities $m(1,B),\ldots,m(\sigma(B),B)$. If for $1 \leq q \leq \rho$ and some $\varepsilon > 0$ the disks*

$$B_\varepsilon(\lambda_q) = \{z \in \mathbb{C} : |z-\lambda_q| < \varepsilon\}$$

in the complex plane $\mathbb{C}$ are disjoint and

$$\Lambda(q,B) = \{j : \mu(j,B) \in B_\varepsilon(\lambda_q)\},$$

then there exists δ such that

$$\sum_{j\in\Lambda(q,B)} m(j,B) = m_q$$

when $|B-A| < \delta$.

Proof. If no such $\delta > 0$ exists, then there exists a sequence that violates the conclusion of the theorem. □

Theorem 8.8 *The sets $\mathcal{H}_0,\ldots,\mathcal{H}_d$ are open subsets of $\mathcal{M}_d(\mathbb{R})$.*

Proof. Let A be in $\mathcal{H}_j$ with distinct eigenvalues $\lambda_1,\ldots,\lambda_\rho$. Choose $\varepsilon > 0$ such that the sets $B_\varepsilon(\lambda_q)$ are disjoint and do not intersect the imaginary axis. Let $\delta > 0$ be given by the previous corollary. It follows that B is in $\mathcal{H}$ and $\iota(B) = \iota(A)$ when $|B-A| < \delta$. Thus

$$\{B : |B-A| < \delta\} \subset \mathcal{H}_j$$

and $\mathcal{H}_j$ is open □

It follows from this theorem that if A is in $\mathcal{H}_j$, then there exists $\delta > 0$ such that when $|B-A| < \delta$, the flows $\mathbf{x}(t,\boldsymbol{\xi}) = e^{At}\boldsymbol{\xi}$ and $\mathbf{x}(t,\boldsymbol{\xi}) = e^{Bt}\boldsymbol{\xi}$ are isomorphic. This property is an elementary example of what is called *structural stability* and has been studied in a variety of settings. For example, there is a substantive discussion of structural stability in [29]

Theorem 8.9 *The set $\mathcal{H}$ is an open and dense subset of $\mathcal{M}_d(\mathbb{R})$.*

Proof. The set $\mathcal{H} = \mathcal{H}_0 \cup \ldots \cup \mathcal{H}_d$ is the union of open sets by the previous theorem and thus open. For the density part, let A be an arbitrary element of $\mathcal{M}_d(\mathbb{R})$ with distinct eigenvalues $\lambda_1, \ldots, \lambda_p$. Suppose A is not in $\mathcal{H}$.

Consider $A + \mu\mathrm{I}$ for small positive μ. Observe that λ is an eigenvalue for A if and only if $\lambda + \mu$ is an eigenvalue for $A + \mu\mathrm{I}$. Set

$$\beta = \inf\{|\lambda_j + \overline{\lambda}_j|/2 \, : \, 0 < |\lambda_j + \overline{\lambda}_j|\}.$$

Consequently, $A + \mu\mathrm{I}$ is in $\mathcal{H}$ when $0 < \mu < \beta$. Since $|A - (A + \mu\mathrm{I})| = \mu$, the set $\mathcal{H}$ is dense. □

EXERCISES

1. Prove the remark on page 254.

2. Prove *Proposition 8.3*

3. Let $\mathbf{u}$ and $\mathbf{v}$ be nonzero vectors in $\mathbb{R}^d$. Prove that $\dot{\mathbf{x}} = \mathbf{u}$ and $\dot{\mathbf{y}} = \mathbf{v}$ determine isomorphic flows on $\mathbb{R}^d$.

4. Show that $\|\mathbf{x}\|_B = \sqrt{\mathbf{x} \star \mathbf{x}}$ defines a norm on $\mathbb{R}^d$ when B is positive definite.

5. Let $p(x)$ be a polynomial with real coefficients. Show that

$$\dot{x} = \frac{p(x)}{1 + p(x)^2}$$

defines a flow on $\mathbb{R}$ with a finite number of fixed points. Construct a polynomial so that the flow determined as above has positively asymptotically stable fixed points at $x = 1$ and $x = 5$, a fixed point that is negatively asymptotically stable at $x = 3$ and fixed points at $x = 2$ and $x = 4$ that are neither positively nor negatively asymptotically stable.

6. Classify the isomorphic flows on $\mathbb{R}$ with a finite number of fixed points.

7. Classify the isomorphic linear flows determined by $\dot{\mathbf{x}} = A\mathbf{x}$ when

   (a) A is a 2×2 real matrix;

   (b) A is a 3×3 invertible real matrix.

8. Let A be a real matrix such that an eigenvalue of A is simple when its real part is zero and consider the flow determined by $\dot{\mathbf{x}} = A\mathbf{x}$. Prove that

$$E^c = \{\mathbf{x} : \overline{\mathcal{O}}(\mathbf{x}) \text{ is compact }\}$$

is an invariant subspace and $\mathbb{R}^d = E^+ \oplus E^c \oplus E^-$ where E^+ and E^- are defined as in *Theorem 8.1*.

8.2 Perturbed Hyperbolic Systems

Having thoroughly analyzed the dynamics of hyperbolic linear vector fields in Section 8.1, it is possible to start analyzing nonlinear hyperbolic phenomenon. The goal is to use the dynamics of hyperbolic linear vector fields as models for the behavior of solutions near critical points. We begin by studying perturbations of linear hyperbolic vector fields in this section. Although the results about hyperbolic linear vector fields in the first section were of interest in their own right, the purpose of this section is primarily to set the stage for the last two sections of this chapter.

This section will focus on vector fields of the form $\mathbf{f}(\mathbf{x}) = A\mathbf{x} + \mathbf{g}(\mathbf{x})$ with $A \in \mathcal{H}$. The generalized variation of constants formula (*Proposition 5.18* on page 174) will be applied to find an integral equation for bounded solutions of these perturbed hyperbolic systems. This integral equation will then be used to introduce a different perspective on solutions; they can be viewed as fixed points. In addition, several necessary technical estimates for this integral equation will be obtained for later use. These estimates will ensure that the fixed point theorem in the Section 8.3 can be applied to prove the existence of a local stable manifolds in the final section.

Let A be in $\mathcal{H}$. Let E^+ and E^- be its positively stable and negatively stable manifolds given by *Theorem 8.1.* Given $\mathbf{x} \in \mathbb{R}^d$ there exist unique vectors $\mathbf{x}_+ \in E^+$ and $\mathbf{x}_- \in E^-$ such that $\mathbf{x} = \mathbf{x}_+ + \mathbf{x}_-$ because $\mathbb{R}^d = E^+ \oplus E^-$. (We will consistently use $+$ and $-$ subscripts and superscripts to denote mathematical objects that pertain to the behavior of solutions for positive and negative time, respectively.) The maps $P_+ : \mathbb{R}^d \to \mathbb{R}^d$ and $P_- : \mathbb{R}^d \to \mathbb{R}^d$ defined by $P_+(\mathbf{x}) = \mathbf{x}_+$ and $P_-(\mathbf{x}) = \mathbf{x}_-$, respectively, are linear maps called *projections* and have the following properties:

(a) $\mathrm{I} = P_+ + P_-$;

(b) $P_+P_- = \mathrm{O} = P_-P_+$;

(c) $P_+(\mathbf{x}) = \mathbf{x}$ if and only if $\mathbf{x} \in E^+$;

(d) $P_-(\mathbf{x}) = \mathbf{x}$ if and only if $\mathbf{x} \in E^-$.

They are all immediate consequences of the definitions. (In the second item, the notation O for the matrix of all zeros is also being used for the linear transformation that maps every vector to $\mathbf{0}$ because its matrix with respect to any basis is always the matrix of all zeros.)

The invariance of E^+ and E^- under the linear transformation $T(\mathbf{x}) = A\mathbf{x}$ guarantees the additional properties:

$$\begin{aligned} P_+(A\mathbf{x}) &= AP_+(\mathbf{x}) \\ P_-(A\mathbf{x}) &= AP_-(\mathbf{x}). \end{aligned}$$

It then follows that

$$\begin{aligned} P_+\left(e^{At}\mathbf{x}\right) &= e^{At}P_+\left(\mathbf{x}\right) \\ P_-\left(e^{At}\mathbf{x}\right) &= e^{At}P_-\left(\mathbf{x}\right). \end{aligned}$$

The eigenvalues of A restricted to E^+ all have negative real parts. Of course, the real parts are positive when A is restricted to E^-. By *Theorem 8.1* there exist positive constants K and α such that

$$|e^{At}\mathbf{v}| \le Ke^{-\alpha t}|\mathbf{v}|$$

for $\mathbf{v} \in E^+$ and $t \ge 0$ and

$$|e^{At}\mathbf{v}| \le Ke^{\alpha t}|\mathbf{v}|$$

for $\mathbf{v} \in E^-$ and $t \le 0$. All the above notation will be kept fixed whenever A is in $\mathcal{H}$.

Theorem 8.10 *Let $\mathbf{g} : \mathbb{R}^d \to \mathbb{R}^d$ be a continuous function, let A be in $\mathcal{H}$, and let $\varphi : [0,\infty) \to \mathbb{R}^d$ be a continuous and bounded function. Then the improper integral*

$$\int_0^\infty e^{-As}P_-\left[\mathbf{g}(\varphi(t+s))\right]ds$$

converges to a value in E^-. The function φ is a solution of $\dot{\mathbf{x}} = A\mathbf{x} + \mathbf{g}(\mathbf{x})$ if and only if there exists $\boldsymbol{\xi}_+ \in E^+$ such that for $t \ge 0$ the function φ satisfies

$$\begin{aligned}\varphi(t) &= e^{At}\boldsymbol{\xi}_+ + \int_0^t e^{A(t-s)}P_+\left[\mathbf{g}(\varphi(s))\right]ds - \int_0^\infty e^{-As}P_-\left[\mathbf{g}(\varphi(t+s))\right]ds \\ &= e^{At}\boldsymbol{\xi}_+ + \int_0^t e^{A(t-s)}P_+\left[\mathbf{g}(\varphi(s))\right]ds - \int_t^\infty e^{A(t-s)}P_-\left[\mathbf{g}(\varphi(s))\right]ds.\end{aligned}$$

Proof. By the hypothesis, there exist positive constants β and γ such that $|\varphi(t)| \le \beta$ and $|\mathbf{g}(\varphi(t))| \le \gamma$ for $t \ge 0$. For $0 \le \sigma' < \sigma$ the following estimate holds:

$$\left|\int_0^\sigma e^{-As}P_-\left[\mathbf{g}(\varphi(t+s))\right]ds - \int_0^{\sigma'} e^{-As}P_-\left[\mathbf{g}(\varphi(t+s))\right]ds\right| \le$$

$$\int_{\sigma'}^\sigma \left|e^{A(-s)}P_-\left[\mathbf{g}(\varphi(t+s))\right]\right|ds \le$$

$$\int_{\sigma'}^\sigma Ke^{\alpha(-s)}\left|P_-\left[\mathbf{g}(\varphi(t+s))\right]\right|ds \le$$

$$K|P_-|\gamma\int_{\sigma'}^\sigma e^{-\alpha s}ds \le$$

$$\frac{K|P_-|\gamma}{\alpha}\left(e^{-\alpha\sigma'} - e^{-\alpha\sigma}\right).$$

For σ' and σ large the last term is small and

$$\int_0^\infty e^{-As}P_-\left[\mathbf{g}(\varphi(t+s))\right]ds = \lim_{\sigma\to\infty}\int_0^\sigma e^{-As}P_-\left[\mathbf{g}(\varphi(t+s))\right]ds$$

exists. Furthermore, by setting $\sigma' = 0$ and letting σ go to infinity, it follows from the above estimate that

$$\left| \int_0^\infty e^{-As} P_- \left[\mathbf{g}(\varphi(t+s)) \right] ds \right| \leq \frac{K|P_-|\gamma}{\alpha}.$$

The second formula for φ can be obtained from the first by making the change of variable $v = t + s$, so the formulas are interchangeable. Moreover, the second formula can also be written as

$$\varphi(t) = e^{At}\boldsymbol{\xi}_+ + e^{At} \int_0^t e^{-As} P_+ \left[\mathbf{g}(\varphi(s)) \right] ds + e^{At} \int_\infty^t e^{-As} P_- \left[\mathbf{g}(\varphi(s)) \right] ds.$$

If φ satisfies this integral equation, then a straightforward differentiation of this formula shows that φ is a solution of $\dot{\mathbf{x}} = A\mathbf{x} + \mathbf{g}(\mathbf{x})$.

For the final step of the proof, assume that φ is a solution of the differential equation $\dot{\mathbf{x}} = A\mathbf{x} + \mathbf{g}(\mathbf{x})$. By an earlier generalization of the variation of constants formula found in *Proposition 5.18* on page 174

$$\begin{aligned} \varphi(t) &= e^{A(t-\tau)}\varphi(\tau) + \int_\tau^t e^{A(t-s)} \mathbf{g}(\varphi(s)) ds \\ &= e^{A(t-\tau)}\varphi(\tau) + \int_{\tau-t}^0 e^{-As} \mathbf{g}(\varphi(s+t)) ds. \end{aligned}$$

Again, these equations are interchangeable by a simple change of variables.

Applying the projection P_+ to the first version of the above formula with $\tau = 0$, yields

$$P_+ \left[\varphi(t) \right] = e^{At} P_+ \left[\varphi(0) \right] + \int_0^t e^{A(t-s)} P_+ \left[\mathbf{g}(\varphi(s)) \right] ds.$$

The right-hand side of this equation provides the first two terms of the right-hand side of the equation in the theorem by setting $\boldsymbol{\xi}_+ = P_+[\varphi(0)]$.

Since $\mathrm{I} = P_+ + P_-$, to complete the proof it suffices to show that

$$P_- \left[\varphi(t) \right] = -\int_0^\infty e^{-As} P_- \left[\mathbf{g}(\varphi(t+s)) \right] ds = \int_\infty^0 e^{-As} P_- \left[\mathbf{g}(\varphi(t+s)) \right] ds.$$

To do this, apply P_- to the second version of the above generalized variation of constants formula and obtain

$$P_- \left[\varphi(t) \right] = e^{A(t-\tau)} P_- \left[\varphi(\tau) \right] + \int_{\tau-t}^0 e^{-As} P_- \left[\mathbf{g}(\varphi(t+s)) \right] ds.$$

If t is fixed and τ goes to infinity, then $\int_{\tau-t}^0 e^{-As} P_- \left[\mathbf{g}(\varphi(t+s)) \right] ds$ converges to the required term. It remains to show that

$$\lim_{\tau \to \infty} e^{A(t-\tau)} P_- \left[(\varphi(\tau) \right] = 0.$$

The norm of $e^{A(t-\tau)}P_-[\varphi(\tau)]$ can be estimated as follows:

$$\begin{aligned}\left|e^{A(t-\tau)}P_-[\varphi(\tau)]\right| &\leq \left|e^{At}\right|\left|e^{A(-\tau)}\right|\left|P_-[\varphi(\tau)]\right| \\ &\leq \left|e^{At}\right|Ke^{\alpha(-\tau)}\left|P_-[\varphi(\tau)]\right| \\ &\leq \left|e^{At}\right|Ke^{-\alpha\tau}\left|P_-\right|\beta\end{aligned}$$

Clearly, the last term goes to zero as τ goes to infinity to complete the proof. □

Corollary *If $\varphi : [0,\infty) \to \mathbb{R}^d$ is a bounded solution of $\dot{\mathbf{x}} = A\mathbf{x} + \mathbf{g}(\mathbf{x})$ with $A \in \mathcal{H}$ and $\mathbf{g} : R^d \to \mathbb{R}^d$ a continuous function, then*

$$\begin{aligned}P_+[\varphi(t)] &= e^{At}P_+[\varphi(0)] + \int_0^t e^{A(t-s)}P_+[\mathbf{g}(\varphi(s))]\,ds \\ P_-[\varphi(t)] &= -\int_0^\infty e^{-As}P_-[\mathbf{g}(\varphi(t+s))]\,ds \\ &= -\int_t^\infty e^{A(t-s)}P_-[\mathbf{g}(\varphi(s))]\,ds.\end{aligned}$$

Proof. Apply the projections to the formula for $\varphi(t)$ in the theorem and use their properties to simplify the expressions. □

Given $\boldsymbol{\xi}_+ \in E^+$, the function $e^{At}\boldsymbol{\xi}_+$ is bounded on the interval $[0,\infty)$. If as above $\varphi : [0,\infty) \to \mathbb{R}^d$ is a bounded function, then so is $\varphi(t)+e^{At}\boldsymbol{\xi}_+$, which will be the desired form of the solutions being sought. It follows from the theorem that

$$\int_0^\infty e^{-As}P_-\left[\mathbf{g}\left(\varphi(t+s)+e^{A(t+s)}\right)\right]ds = \int_t^\infty e^{A(t-s)}P_-\left[\mathbf{g}\left(\varphi(s)+e^{As}\right)\right]ds$$

exists.

In addition to φ being bounded for $t \geq 0$, suppose that $P_+[\varphi(0)] = \mathbf{0}$. *Theorem 8.10* implies that $\varphi(t) + e^{At}\boldsymbol{\xi}_+$ is a solution of $\dot{\mathbf{x}} = A\mathbf{x} + \mathbf{g}(\mathbf{x})$ if and only if

$$\begin{aligned}\varphi(t) + e^{At}\boldsymbol{\xi}_+ &= e^{At}\boldsymbol{\xi}_+ + \int_0^t e^{A(t-s)}P_+\left[\mathbf{g}\left(\varphi(s)+e^{As}\right)\right]ds \\ &\quad - \int_t^\infty e^{A(t-s)}P_-\left[\mathbf{g}\left(\varphi(s)+e^{As}\right)\right]ds,\end{aligned}$$

which is the same as

$$\begin{aligned}\varphi(t) &= \int_0^t e^{A(t-s)}P_+\left[\mathbf{g}\left(\varphi(s)+e^{As}\right)\right]ds \\ &\quad - \int_t^\infty e^{A(t-s)}P_-\left[\mathbf{g}\left(\varphi(s)+e^{As}\right)\right]ds.\end{aligned} \tag{8.3}$$

Note that if φ satisfies this equation, then $P_+[\varphi(0)] = \mathbf{0}$ and it will not be necessary to carry this condition along as a hypothesis in what follows.

The right side of the above equation can be used to define a map that assigns to a function and a vector a new function. Specifically, let $\mathbf{F}(\varphi, \boldsymbol{\xi}_+)$ be the function defined by

$$\begin{aligned}\mathbf{F}(\varphi, \boldsymbol{\xi}_+)(t) &= \int_0^t e^{A(t-s)} P_+ \left[\mathbf{g}\left(\varphi(s) + e^{As}\boldsymbol{\xi}_+\right)\right] ds \\ &- \int_t^\infty e^{A(t-s)} P_- \left[\mathbf{g}\left(\varphi(s) + e^{As}\boldsymbol{\xi}_+\right)\right] ds. \qquad (8.4)\end{aligned}$$

Now *(8.3)* above can be written simply as

$$\mathbf{F}(\varphi, \boldsymbol{\xi}_+) = \varphi, \qquad (8.5)$$

which is a fixed-point equation. The question is given $\boldsymbol{\xi}_+$, does $\mathbf{F}$ have a fixed point? A general theorem about the existence of fixed points that can be applied to $\mathbf{F}$ will be proved in the next section. The strategy will then be to use it to obtain solutions to $\mathbf{F}(\varphi, \boldsymbol{\xi}_+) = \varphi$ and add $e^{At}\boldsymbol{\xi}_+$ to them to obtain a family of solutions to $\dot{\mathbf{x}} = A\mathbf{x} + \mathbf{g}(\mathbf{x})$ that depend on $\boldsymbol{\xi}_+$.

In the meantime, additional properties of $\mathbf{F}$ must be established in this section to show that $\mathbf{F}$ satisfies the hypothesis of the fixed-point theorem.

Estimating and controlling terms like $|\mathbf{F}(\varphi_1, \boldsymbol{\xi}_+) - \mathbf{F}(\varphi_2, \boldsymbol{\xi}_+)|$ will require a variation of the Lipschitz condition. This variation will seem more natural, if placed in the right context before the formal definition is presented. To this end suppose $\mathbf{f}$ is a smooth vector field on $\mathbb{R}^d$ with $\mathbf{f}(\mathbf{0}) = \mathbf{0}$. Let $A = \mathbf{f}'(\mathbf{0})$, the $d \times d$ matrix of first partial derivatives of $\mathbf{f}$, and set $\mathbf{g}(\mathbf{x}) = \mathbf{f}(\mathbf{x}) - A\mathbf{x}$. Then $\mathbf{g}$ is a smooth function such that $\mathbf{g}(\mathbf{0}) = \mathbf{0}$ and $\mathbf{g}'(\mathbf{0}) = \mathrm{O}$, the $d \times d$ zero matrix.

Theorem 1.12 implies that $\mathbf{g}$ is locally Lipschitz and tells us how to calculate the Lipschitz constant for $B_r(\mathbf{0}) = \{\mathbf{x} : \|\mathbf{x}\| < r\}$. The proof of this theorem shows that $|\mathbf{g}(\mathbf{x}) - \mathbf{g}(\mathbf{y})| \leq dM|\mathbf{x} - \mathbf{y}|$, where

$$M = \sum_{j=1}^{d} M_j = \sum_{j=1}^{d} \sup\{\|\nabla g_j(\mathbf{x})\| : \|\mathbf{x}\| \leq r\},$$

but ignores the dependency of M on r. It is an exercise to show that

$$\sup\{\|\nabla g_j(\mathbf{x})\| : \|\mathbf{x}\| \leq r\}$$

is a continuous nondecreasing function of r. Consequently,

$$\psi(r) = d\sum_{j=1}^{d} \sup\{\|\nabla g_j(\mathbf{x})\| : \|\mathbf{x}\| \leq r\},$$

is also a continuous nondecreasing function of r. Because $\mathbf{g}$ is a smooth function with $\mathbf{g}'(\mathbf{0}) = 0$, it follows that $\lim_{r\to 0}\psi(r) = 0$ and ψ can be extended to 0 by setting $\psi(0) = 0$. Therefore,

$$|\mathbf{g}(\mathbf{x}) - \mathbf{g}(\mathbf{y})| \leq \psi(r)\|\mathbf{x} - \mathbf{y}|$$

when $\|\mathbf{x}\| \leq r$ and $\|\mathbf{y}\| \leq r$ for ψ a continuous nondecreasing function on $[0, \infty)$ such that $\psi(0) = 0$. This is essentially the definition of $\mathcal{L}(\psi)$.

Given a continuous nondecreasing function $\psi : [0, \infty) \to \mathbb{R}$ with $\psi(0) = 0$, let $\mathcal{L}(\psi)$ denote the set of all continuous functions $\mathbf{g} : \mathbb{R}^d \to \mathbb{R}^d$ such that

$$\begin{aligned} \mathbf{g}(\mathbf{0}) &= \mathbf{0} \\ |\mathbf{g}(\mathbf{x}) - \mathbf{g}(\mathbf{y})| &\leq \psi(r)|\mathbf{x} - \mathbf{y}|, \end{aligned}$$

when $|\mathbf{x}| \leq r$ and $|\mathbf{y}| \leq r$. There are two simple observations about any $\mathbf{g} \in \mathcal{L}(\psi)$ that are worth making. First, $\mathbf{g}$ is locally Lipschitz on $\mathbb{R}^d$ because every $\mathbf{x}$ is in some $B_r(\mathbf{0})$. Second, if $\|\mathbf{x}\| < r$, then $|\mathbf{g}(\mathbf{x})| \leq \psi(r)|\mathbf{x}|$ because $\mathbf{g}(\mathbf{0}) = \mathbf{0}$.

The discussion preceding the definition of $\mathcal{L}(\psi)$ can now be summarized in a proposition for future use.

Proposition 8.11 *Let* $\mathbf{g} : \mathbb{R}^d \to \mathbb{R}^d$ *be a smooth function. If* $\mathbf{g}(\mathbf{0}) = \mathbf{0}$ *and* $\mathbf{g}'(\mathbf{0}) = \mathrm{O}$, *then*

$$\psi(r) = \begin{cases} d\sum_{j=1}^{d} \sup\{\|\nabla g_j(\mathbf{x})\| : \|\mathbf{x}\| \leq r\} & \text{for } r > 0 \\ 0 & \text{for } r = 0 \end{cases}$$

is a continuous nondecreasing function with $\psi(0) = 0$ *and* $\mathbf{g}$ *is in* $\mathcal{L}(\psi)$.

Returning to the general context of this section, A is a hyperbolic $d \times d$ matrix with stable and unstable manifolds E^+ and E^-. The projections onto E^+ and E^- are denoted by P_+ and P_-, respectively. The positive constants K and α are determined by *Theorem 8.1*. Set $L = \max\{|P_+|, |P_-|\}$. This notation will be used in the next three propositions.

As before, $\boldsymbol{\varphi}$ will continue to denote a bounded real-valued function on $[0, \infty)$. The following notation will help streamline the remaining results in this section and will be critical in the next section. Set

$$|\boldsymbol{\varphi}| = \sup\{|\boldsymbol{\varphi}(t)| : t \geq 0\}.$$

This norm will be fully discussed in the next section. For now it is just a convenient notation for the least upper bound of norms of the values of a bounded function $\boldsymbol{\varphi}$.

Since both $\boldsymbol{\varphi}$ and $e^{At}\boldsymbol{\xi}_+$ are bounded on $[0, \infty)$, there exists $\beta > 0$ such that $|\boldsymbol{\varphi}(t)| \leq \beta$ and $|e^{At}\boldsymbol{\xi}_+| \leq \beta$ for $t \geq 0$. It is important to note that β is governed by the choice of $\boldsymbol{\varphi}$ and $\boldsymbol{\xi}_+ \in E^+$. These symbols will also be used consistently as described in the next three propositions and most importantly $\mathbf{F}(\boldsymbol{\varphi}, \boldsymbol{\xi}_+)$ will always refer to formula *(8.4)*.

Proposition 8.12 *If* $\mathbf{g}$ *is in* $\mathcal{L}(\psi)$, *then*

$$|\mathbf{F}(\boldsymbol{\varphi}, \boldsymbol{\xi}_+)| \leq \frac{4\beta K L}{\alpha}\, \psi(2\beta).$$

Proof. The required estimate for $|\mathbf{F}(\varphi, \boldsymbol{\xi}_+)|$ will be obtained by estimating the two integrals that make up $\mathbf{F}(\varphi, \boldsymbol{\xi}_+)(t)$ and then bringing the results together.

First,

$$\begin{aligned}
\left| \int_0^t e^{A(t-s)} P_+ \left[\mathbf{g} \left(\varphi(s) + e^{As} \boldsymbol{\xi}_+ \right) \right] ds \right| &\leq \\
\int_0^t \left| e^{A(t-s)} P_+ \left[\mathbf{g} \left(\varphi(s) + e^{As} \boldsymbol{\xi}_+ \right) \right] \right| ds &\leq \\
\int_0^t \left| e^{A(t-s)} \right| |P_+| \left| \mathbf{g} \left(\varphi(s) + e^{As} \boldsymbol{\xi}_+ \right) \right| ds &\leq \\
\int_0^t K e^{-\alpha(t-s)} L \, \psi(2\beta) \, 2\beta \, ds &= \\
2\beta K L \, \psi(2\beta) \frac{1 - e^{-\alpha t}}{\alpha} &< \\
\frac{2\beta K L}{\alpha} \psi(2\beta). &
\end{aligned}$$

Second,

$$\begin{aligned}
\left| \int_t^\infty e^{A(t-s)} P_- \left[\mathbf{g} \left(\varphi(s) + e^{As} \boldsymbol{\xi}_+ \right) \right] ds \right| &\leq \\
\int_t^\infty \left| e^{A(t-s)} P_- \left[\mathbf{g} \left(\varphi(s) + e^{As} \boldsymbol{\xi}_+ \right) \right] \right| ds &\leq \\
\int_t^\infty \left| e^{A(t-s)} \right| |P_-| \left| \mathbf{g} \left(\varphi(s) + e^{As} \boldsymbol{\xi}_+ \right) \right| ds &\leq \\
\int_t^\infty K e^{\alpha(t-s)} L \, \psi(2\beta) \, 2\beta \, ds &= \\
\frac{2\beta K L}{\alpha} \psi(2\beta). &
\end{aligned}$$

Both calculations lead to the same bound and their sum is desired bound as stated. □

Note that the bound on $|\mathbf{F}(\varphi, \boldsymbol{\xi})|$ goes to zero as β goes to zero and is, therefore, under our control. The proofs of the last two propositions make use of similar estimates. They are technical, tedious, and essential.

Proposition 8.13 *If* $\mathbf{g}$ *is in* $\mathcal{L}(\psi)$, *then*

$$\left| \mathbf{F}(\varphi_1, \boldsymbol{\xi}_+) - \mathbf{F}(\varphi_2, \boldsymbol{\xi}_+) \right| \leq \frac{2KL}{\alpha} \psi(2\beta) \left| \varphi_1 - \varphi_2 \right|.$$

Proof. The argument uses the same estimating techniques used in the previous proof, but leaves the term $|\varphi_1 - \varphi_2|$ intact.

To begin, observe that

$$\begin{aligned}
\mathbf{F}(\varphi_1, \boldsymbol{\xi})(t) - \mathbf{F}(\varphi_2, \boldsymbol{\xi})(t) &= \\
\int_0^t e^{A(t-s)} P_+ \left[\mathbf{g}\left(\varphi_1(s) + e^{As}\boldsymbol{\xi}_+\right) - \mathbf{g}\left(\varphi_2(s) + e^{As}\boldsymbol{\xi}_+\right)\right] ds &- \\
\int_t^\infty e^{A(t-s)} P_- \left[\mathbf{g}\left(\varphi_1(s) + e^{As}\boldsymbol{\xi}_+\right) - \mathbf{g}\left(\varphi_2(s) + e^{As}\boldsymbol{\xi}_+\right)\right] ds.
\end{aligned}$$

Again it makes sense to estimate each integral separately. The inequalities are very similar to those in the previous proposition.

First,

$$\begin{aligned}
\left| \int_0^t e^{A(t-s)} P_+ \left[\mathbf{g}\left(\varphi_1(s) + e^{As}\boldsymbol{\xi}_+\right) - \mathbf{g}\left(\varphi_2(s) + e^{As}\boldsymbol{\xi}_+\right)\right] ds \right| &\le \\
\int_0^t K e^{-\alpha(t-s)} L \left| \mathbf{g}\left(\varphi_1(s) + e^{As}\boldsymbol{\xi}_+\right) - \mathbf{g}\left(\varphi_2(s) + e^{As}\boldsymbol{\xi}_+\right)\right| ds &\le \\
\int_0^t K e^{-\alpha(t-s)} L\, \psi(2\beta) \left| \varphi_1(s) - \varphi_2(s) \right| ds &\le \\
K L\, \psi(2\beta) \left| \varphi_1 - \varphi_2 \right| \int_0^t e^{-\alpha(t-s)} ds &\le \\
K L\, \psi(2\beta) \left| \varphi_1 - \varphi_2 \right| \frac{1 - e^{-\alpha t}}{\alpha} &< \\
\frac{KL}{\alpha}\, \psi(2\beta) \left| \varphi_1 - \varphi_2 \right|.
\end{aligned}$$

Second,

$$\begin{aligned}
\left| \int_t^\infty e^{A(t-s)} P_- \left[\mathbf{g}\left(\varphi_1(s) + e^{As}\boldsymbol{\xi}_+\right) - \mathbf{g}\left(\varphi_2(s) + e^{As}\boldsymbol{\xi}_+\right)\right] ds \right| &\le \\
\int_t^\infty K e^{\alpha(t-s)} L \left| \mathbf{g}\left(\varphi_1(s) + e^{As}\boldsymbol{\xi}_+\right) - \mathbf{g}\left(\varphi_2(s) + e^{As}\boldsymbol{\xi}_+\right)\right| ds &\le \\
\int_t^\infty K e^{\alpha(t-s)} L\, \psi(2\beta) \left| \varphi_1(s) - \varphi_2(s) \right| ds &\le \\
K L\, \psi(2\beta) \left| \varphi_1 - \varphi_2 \right| \int_t^\infty e^{\alpha(t-s)} ds &\le \\
\frac{KL}{\alpha}\, \psi(2\beta) \left| \varphi_1 - \varphi_2 \right|.
\end{aligned}$$

Once again the sum of the two estimates is the desired result. □

Proposition 8.14 *If* $\mathbf{g}$ *is in* $\mathcal{L}(\psi)$, *then*

$$\left| \mathbf{F}(\varphi, \boldsymbol{\xi}_+) - \mathbf{F}(\varphi, \boldsymbol{\xi}'_+) \right| \le \frac{2K^2 L}{\alpha}\, \psi(2\beta) \left| \boldsymbol{\xi}_+ - \boldsymbol{\xi}'_+ \right|$$

Proof. Exercise.

EXERCISES

1. State and prove the companion theorem to *Theorem 8.10* for a continuous and bounded function $\varphi : [0, -\infty) \to \mathbb{R}^d$.

2. Let $g : \mathbb{R}^d \to \mathbb{R}$ be a smooth real-valued function such that $g(\mathbf{0}) = 0$ and $\nabla g(\mathbf{0}) = \mathbf{0}$. Show that

$$\psi(r) = \begin{cases} \sup\{\|\nabla g(\mathbf{x})\| : \|\mathbf{x}\| \le r\} & \text{for } r > 0 \\ 0 & \text{for } r = 0 \end{cases}$$

is a continuous non decreasing function.

3. Prove *Proposition 8.14*

4. Prove that if $\mathbf{g}$ is in $\mathcal{L}(\psi)$, then $\mathbf{g}$ is differentiable at $\mathbf{0}$ and $\mathbf{g}'(\mathbf{0}) = O$.

5. Show that the set of continuous bounded functions φ from $[0, \infty)$ into $\mathbb{R}^m$ denoted by $\mathcal{C}([0, \infty), \mathbb{R}^m)$ is a vector space and $|\varphi| = \sup\{|\varphi(t)| : t \ge 0\}$ satisfies the definition of a norm on page 7.

8.3 The Contraction Mapping Principle

The natural context for the contraction mapping principle is a complete metric space. The space $\mathbb{R}^m$ that has been the setting for everything here-to-fore is the standard example of a complete metric space, but it is no longer general enough for our purposes. The transition from $\mathbb{R}^m$ to complete metric spaces is fortunately a relatively easy one. The first step is define a metric space and show how all the standard concepts for $\mathbb{R}^m$ carry over to metric spaces.

A *metric space* is simply a set X and a function $d : X \times X \to \mathbb{R}$, called a *metric*, satisfying the following conditions:

(a) $d(x, y) \ge 0$ for all x and y in W,

(b) $d(x, y) = 0$ if and only if $x = y$,

(c) $d(x, y) = d(y, x)$ for all x and y in X,

(d) $d(x, y) \le d(x, z) + d(z, y)$ for all x, y, and z in X.

Not surprisingly the fourth condition is called the triangle inequality.

If $\|\cdot\|_a$ is any norm on any $\mathbb{R}^m$, then $d(\mathbf{x}, \mathbf{y}) = \|\mathbf{x} - \mathbf{y}\|_a$ is a metric on $\mathbb{R}^m$. In particular, the usual Euclidean distance $\|\mathbf{x} - \mathbf{y}\|$ is a metric. Norms, of course, are only defined on vector spaces, but a metric does not require an underlying vector space structure. Consequently, if X is any subset of $\mathbb{R}^m$, then

$d(\mathbf{x}, \mathbf{y}) = \|\mathbf{x} - \mathbf{y}\|_a$ for $\mathbf{x}$ and $\mathbf{y}$ in X defines a metric on X. Thus every subset X of $\mathbb{R}^m$ is naturally a metric space.

It is easy to mimic the definitions of open, closed, and compact subsets for metric spaces. A subset U of a metric space is an *open set* if for every $y \in U$ there exists $r > 0$ such that

$$B_r(y) = \{x : d(x, y) < r\} \subset U.$$

A subset C of X is a *closed set* if and only if $X - C = \{x \in X : x \notin C\}$ is open. It is an exercise to show that $B_r(y)$ is always an open set and that

$$\overline{B}_r(y) = \{x \in X : d(x, y) \leq r\}$$

is always a closed set. A set C is *compact* provided that whenever $\{U_\lambda\}$, $\lambda \in \Lambda$, is a family of open sets indexed by Λ such that

$$C \subset \bigcup_{\lambda \in \Lambda} U_\lambda$$

then there exists a finite set of indices $\lambda_1, \ldots, \lambda_k$ such that

$$C \subset \bigcup_{i=1}^{k} U_{\lambda_i}.$$

It is another exercise to show that a compact subset of a metric space is a closed set.

If X is a subset of $\mathbb{R}^m$ and is a metric space with metric $d(\mathbf{x}, \mathbf{y}) = \|\mathbf{x} - \mathbf{y}\|_a$, then

$$\{\mathbf{x} \in X : d(\mathbf{x}, \mathbf{y}) < r\} = \{\mathbf{x} \in \mathbb{R}^m : \|\mathbf{x} - \mathbf{y}\|_a < r\} \cap X.$$

It follows that U is an open subset of X if and only if there exists an open set V of $\mathbb{R}^m$ such that $U = V \cap X$. Thus the concept of a metric space gives meaning to the concept of an open subset of X in a way that is closely linked to the open sets of $\mathbb{R}^m$.

Heretofore, open set meant an open set of $\mathbb{R}^m$ for some m, but now it is possible to consider an open set of X for any subset X of $\mathbb{R}^m$. Unless X is itself an open set of $\mathbb{R}^m$, an open subset of X is quite different from an open subset of $\mathbb{R}^m$. When a set is open in a metric space other than $\mathbb{R}^m$, we will be careful to specify the space in which it is open.

For example, if

$$X = \{(x, y, z) : x^2 + y^2 + z^2 = 1\},$$

the unit sphere in $\mathbb{R}^3$, then the upper hemisphere,

$$\begin{aligned} U &= \{(x, y, z) : x^2 + y^2 + z^2 = 1 \text{ and } z > 0\} \\ &= \{(x, y, z) : z > 0\} \cap X, \end{aligned}$$

is an open subset of the unit sphere but not of $\mathbb{R}^3$. Moreover, U is the graph of the function

$$g(x, y) = \sqrt{1 - x^2 - y^2}$$

on the open unit disk, $\{(x, y) : x^2 + y^2 < 1\}$ of the xy-plane, and hence U is a curved two-dimensional disk as would be hoped because the unit sphere in $\mathbb{R}^3$ is a surface, and hence two dimensional.

The notion of a metric space also provides a host of new spaces that are not subsets of $\mathbb{R}^m$. One that will be important in the next section is the space of continuous bounded functions φ from $[0, \infty)$ into $\mathbb{R}^m$ denoted by $\mathcal{C}([0, \infty), \mathbb{R}^m)$. It is a vector space and setting

$$|\varphi| = \sup\{|\varphi(t)| : t \geq 0\}$$

defines a norm on it (*Exercise 5* on page 275). Hence, $\mathcal{C}([0, \infty), \mathbb{R}^m)$ is a metric space but more importantly every subspace of it is also a metric space. Applying the contraction mapping principle to prove that $F(\varphi, \boldsymbol{\xi}_S) = \varphi$ has a solution, will require carefully selecting the right subset of $\mathcal{C}([0, \infty), \mathbb{R}^m)$.

Let X and Y be metric spaces with metrics d_X and d_Y. Then

$$X \times Y = \{(x, y) : x \in X \text{ and } y \in Y\}$$

is also a metric space with metric

$$d((x, y), (x', y')) = d_X(x, x') + d_Y(y, y').$$

It is also easy to define continuity in this context. A function $g : X \to Y$ is *continuous* at x, if given $\varepsilon > 0$, there exists $\delta > 0$ such that $d_Y(f(z), f(x)) < \varepsilon$, whenever $d_X(z, x) < \delta$. Not surprisingly, f is continuous at every point of X if and only if $f^{-1}(U)$ is an open set of X whenever U is an open set of Y.

As with flows, it is natural to ask when should two metric spaces be regarded as the same. The usual notion of sameness or equivalence for metric spaces is homeomorphic spaces. Two metric spaces X and Y are *homeomorphic* if there exists a continuous map $h : X \to Y$ called a *homeomorphism* with a continuous inverse $h^{-1} : Y \to X$. In particular, a homeomorphism must be one-to-one and onto. If h is a homeomorphism between two metric spaces X and Y, then U is open in X if and only if $h(U)$ is open in Y.

Homeomorphisms were already implicit in the first section of this chapter. It follows from the definition of isomorphic flows (page 256) that the underlying metric spaces must be homeomorphic. And the proof of *Theorem 8.5* made use invariance of domain, which states that open sets of $\mathbb{R}^m$ and $\mathbb{R}^n$ cannot be homeomorphic if $m \neq n$. In other words, invariance of domain implies that open sets in Euclidean spaces of different dimensions are fundamentally different as metric spaces. Homeomorphisms will be used again in the next section to describe the local stable manifolds.

The metric space $\mathbb{R}^m$ has the nice property that every Cauchy sequence $\mathbf{x}_k$ in $\mathbb{R}^m$ converges to some $\mathbf{x} \in \mathbb{R}^m$. (See *Exercise 9* on page 12.) Although the definition of a Cauchy sequence carries over to metric spaces, not every metric space has the property that every Cauchy sequence converges. For example, consider, $X = \{x \in \mathbb{R} : x > 0\}$ with the metric $d(x, y) = |x - y|$. Clearly, $x_k = 1/k$ is a Cauchy sequence, but it does not converge to a point in X. This property is quite sensitive to the choice of metrics. If the metric $d(x, y) = |x - y|$ on X is replaced with the metric $d'(x, y) = |\log x - \log y|$, then every Cauchy sequence for d' does converge to a point in X. The open sets of X arising from d and d', however, are identical, but $1/k$ is not a Cauchy sequence for the metric d'.

A sequence x_k in a metric space is a *Cauchy sequence* if given $\varepsilon > 0$ there exists $N > 0$ such that $d(x_j, x_k) < \varepsilon$ whenever $j > N$ and $k > N$. A metric space X is *complete* if every Cauchy sequence in X converges to a point in X. The spaces $\mathbb{R}^m$ and their closed subsets are complete metric spaces for any metric $d(\mathbf{x}, \mathbf{y}) = \|\mathbf{x} - \mathbf{y}\|_a$ obtained from a norm $\|\cdot\|_a$.

The space $\mathcal{C}([0, \infty), \mathbb{R}^m)$ is a vector space under the addition and scalar multiplication of functions, but it is not finite dimensional. The function $|\varphi| = \sup\{|\varphi(t)| : t \geq 0\}$ that assigns a nonnegative real number to $\varphi \in \mathcal{C}([0, \infty), \mathbb{R}^m)$ is a norm (as defined on page 7) on $\mathcal{C}([0, \infty), \mathbb{R}^m)$ making it a *normed linear space.* A norm always defines a metric by setting $d(x, y) = \|x - y\|_a$, and when the normed linear space is complete with respect to this metric, it is called a *Banach space.*

Proposition 8.15 *The space $\mathcal{C}([0, \infty), \mathbb{R}^m)$ is a Banach space and each of its closed subsets with the metric*

$$d(\varphi_1, \varphi_2) = |\varphi_1 - \varphi_2| = \sup\{|\varphi_1(t) - \varphi_2(t)| : t \geq 0\}$$

is a complete metric space.

Proof. Suppose φ_k is a sequence in a closed subset C of $\mathcal{C}([0, \infty), \mathbb{R}^m)$. If φ_k converges in $\mathcal{C}([0, \infty), \mathbb{R}^m)$, then its limit must be in C because C is closed. Hence, it suffices to show that $\mathcal{C}([0, \infty), \mathbb{R}^m)$ is complete.

Let φ_k be a Cauchy sequence in $\mathcal{C}([0, \infty), \mathbb{R}^m)$. The inequality

$$|\varphi_j(t) - \varphi_k(t)| \leq |\varphi_j - \varphi_k| = d(\varphi_j, \varphi_k)$$

implies that $\varphi_k(t)$ is a Cauchy sequence in $\mathbb{R}^m$ for every t, and therefore $\varphi_k(t)$ converges to some function $\varphi(t)$.

Given $\varepsilon > 0$, there exists N such that $|\varphi_j - \varphi_k| < \varepsilon$ when $j > N$ and $k > N$. Fixing $k > N$ and letting j go to infinity in $|\varphi_j(t) - \varphi_k(t)| \leq |\varphi_j - \varphi_k|$ shows that

$$|\varphi(t) - \varphi_k(t)| < \varepsilon$$

for $t \geq 0$ when $k > N$. Therefore, $\varphi_k(t)$ converges to $\varphi(t)$ uniformly on $[0, \infty)$ and φ is continuous.

It also follows from the previous inequality that

$$|\varphi(t)| \leq |\varphi(t) - \varphi_k(t)| + |\varphi_k(t)| < \varepsilon + |\varphi_k|$$

and φ is a bounded function. Therefore φ is in $\mathcal{C}([0,\infty),\mathbb{R}^m)$ and $|\varphi - \varphi_k| < \varepsilon$ when $k > N$. $\square$.

A word of caution about normed linear spaces, two norms on the same infinite-dimensional vector space need not be equivalent, not all normed linear spaces are Banach spaces, and closed and bounded sets need not be compact even in a Banach space. In spite of these limitations, normed linear spaces and Banach spaces are quite common and useful. Accessible and more extensive introductions to metric spaces and Banach spaces can be found in both [22] and [31].

Another class of Banach spaces that is useful in differential equations is the space of vector valued functions on a compact subset of $\mathbb{R}^m$. Specifically, let C be a compact subset of $\mathbb{R}^m$ and let $\mathcal{C}(C,\mathbb{R}^n)$ denote the vector space of all continuous functions on C with values in $\mathbb{R}^n$. Because C is compact, every function in $\mathcal{C}(C,\mathbb{R}^n)$ is automatically bounded and $|g| = \sup\{|g(\mathbf{x})| : \mathbf{x} \in C\}$ defines a norm on $\mathcal{C}(C,\mathbb{R}^n)$. Now the proof of the previous proposition can be repeated almost verbatim to prove that $\mathcal{C}(C,\mathbb{R}^n)$ is also a Banach space and its closed sets are complete metric spaces.

A function $F : X \to X$ from a metric space to itself is a *contraction* if there exists a real number λ with $0 < \lambda < 1$ such that

$$d(F(x),F(y)) < \lambda d(x,y)$$

for all $x, y \in X$. A contraction is not just continuous but also uniformly continuous with $\delta = \varepsilon/\lambda$. A contraction is squeezing things closer and closer together. In a complete metric space, this is only possible if every point is being squeezed closer and closer to a unique fixed point as the next theorem shows.

Theorem 8.16 (The Contraction Mapping Principle) *Let X be a complete metric space. If $F : X \to X$ is a contraction, then F has a unique fixed point x_o and for any $x \in X$ the sequence $F^n(x)$ converges to x_o.*

Proof. Since F maps X to itself, it can be iterated for $x \in X$, and $F^2(x) = F(F(x))$, $F^3(x) = F(F^2(x)), \ldots, F^n(x) = F(F^{n-1}(x)), \ldots$ is a sequence in X. The first task is to show that it is a Cauchy sequence. Set $x_n = F^n(x)$ for $n \geq 1$ and note that

$$\begin{aligned} d(x_n, x_{n+1}) &= d\big(F(F^{n-1}(x)), F(F^n(x))\big) \\ &\leq \lambda d\big(F^{n-1}(x), F^n(x)\big) \end{aligned}$$

and so by induction

$$d(x_n, x_{n+1}) \leq \lambda^n d(x, x_1).$$

For $n > m$, the following estimate holds:

$$d(x_m, x_n) \leq \sum_{k=m}^{n-1} d(x_k, x_{k+1})$$

$$\begin{aligned} &\leq d(x,x_1)\sum_{k=m}^{n-1}\lambda^k \\ &= d(x,x_1)\frac{\lambda^m-\lambda^n}{1-\lambda} \\ &\leq \frac{\lambda^m d(x,x_1)}{1-\lambda}. \end{aligned}$$

Because λ^m converges to zero as n goes to infinity, x_n is a Cauchy sequence and x_n converges to some point y in X because X is complete.

Then $F(x_n)$ converges to $F(y)$ by continuity. Since $F(x_n) = x_{n+1}$, the sequence $F(x_n)$ converges to y. Thus $F(y) = y$ and F has at least one fixed point in X. Moreover, $F^n(x)$ converges to a fixed point for every $x \in X$.

Suppose y and y' are both fixed points of F. Then

$$d(y,y') = d(F(y),F(y')) \leq \lambda d(y,y') < d(y,y')$$

which is impossible if $y \neq y'$. □

Iterating a map $F : X \to X$ is another type of dynamical system that is frequently studied with an emphasis on analyzing the behavior of the sequences $F^n(x)$ as n goes to infinity. Even when X is a closed interval of the real line, the dynamics can be very interesting. For example, see [12] for an introduction to iterative dynamics in familiar spaces. The dynamics of a contraction, however, are not very interesting. For every x, the sequence $F^n(x)$ converges to a unique fixed point and there is little else to say. In fact, the dynamics are very similar to those of the flow $\mathbf{x}(t,\boldsymbol{\xi}) = e^{-t}\boldsymbol{\xi}$ on $\mathbb{R}$.

Solving the equation $e^{-x} = x$ provides a simple example in which one can see how $F^n(x)$ converges to the fixed point. Let X be the closed interval $[0.1, 2]$. (When a metric is not explicitly given, it should be understood that it is the obvious one.) It is easy to check that $F(x) = e^{-x}$ maps X into itself. The absolute value of the derivative of F is also e^{-x} and its maximal value on X is $e^{-0.1}$, which is less than 1. By the mean-value theorem

$$|F(x) - F(y)| \leq e^{-0.1}|x - y|$$

on X. Hence, F is a contraction on X and by *Theorem 8.16*, $F^n(x)$ converges to the solution of $e^{-x} = x$ for all x satisfying $0.1 \leq x \leq 2$. The convergence is quite rapid and easily observed using a scientific calculator.

If this problem is broadened to solving $e^{-\gamma x} = x$ when $|\gamma - 1|$ is small, $F(x,\gamma) = e^{-\gamma x}$ will still be a contraction for each γ and we would expect the solution to depend continuously on γ. The next theorem establishes this as a general principle.

Theorem 8.17 *Let X and Y be metric spaces and let $F : X \times Y \to X$ be a function such that $F_x(y) = F(x,y)$ is a continuous function from Y into X for each $x \in X$. If X is complete and there exists λ with $0 < \lambda < 1$ such that*

$$d(F(x,y), F(x',y)) \leq \lambda d(x,x')$$

for all x and x' in X and y in Y, then there exists a continuous function $h : Y \to X$ such that

$$F(h(y), y) = h(y).$$

Proof. For each $y \in Y$, the function $F_y(x) = F(x,y)$ is a contraction by hypothesis and by the contraction mapping principle there exists a unique fixed point that depends on y, and hence defines a function from Y to X. Denote the fixed point of $F_y(x)$ by $h(y)$ so $F_y(h(y)) = h(y)$ or $F(h(y), y) = h(y)$. It only remains to prove that $h : Y \to X$ is continuous.

Given $w \in Y$, the distance between $h(y)$ and $h(w)$ can be estimated as follows:

$$\begin{aligned} d\big(h(y), h(w)\big) &= d\big(F(h(y), y), F(h(w), w)\big) \\ &\le d\big(F(h(y), y), F(h(w), y)\big) + d(F(h(w), y), F(h(w), w)\big) \\ &\le \lambda d\big(h(y), h(w)\big) + d\big(F_{h(w)}(y), F_{h(w)}(w)\big). \end{aligned}$$

Solving the above inequality for $d\big(h(y), h(w)\big)$ produces

$$d\big(h(y), h(w)\big) \le \frac{1}{1-\lambda}\, d\big(F_{h(w)}(y), F_{h(w)}(w)\big). \tag{8.6}$$

By the hypothesis of the theorem, the function $F_{h(w)}(\cdot)$ is continuous at w. Thus given $\varepsilon > 0$ there exists $\delta > 0$ such that

$$d\big(F_{h(w)}(y), F_{h(w)}(w)\big) < (1-\lambda)\varepsilon$$

when $d(y, w) < \delta$. Hence, $d(y, w) < \delta$ implies that

$$d\big(h(y), h(w)\big) < \varepsilon$$

and the proof is complete. □

For a preview of how the contraction mapping principle can be applied to differential equations, consider the autonomous differential equation $\dot{\mathbf{x}} = \mathbf{f}(\mathbf{x})$ and assume that $\mathbf{f}(\mathbf{x})$ is locally Lipschitz on Ω an open subset of $\mathbb{R}^d$. Let Y be a compact set contained in Ω. There exists $b > 0$ such that

$$R = \{\mathbf{x} : |\mathbf{x} - \boldsymbol{\xi}| \le b \text{ for some } \boldsymbol{\xi} \in Y\} \subset \Omega.$$

Let K be the Lipschitz constant for $\mathbf{f}(\mathbf{x})$ on R, and let $M = \sup\{|\mathbf{f}(\mathbf{x})| : \mathbf{x} \in R\}$. Set

$$\alpha = \min\left\{b, \frac{b}{M}, \frac{1}{2K}\right\}.$$

The complete metric space will be a closed subset of the Banach space $\mathcal{C}([-\alpha, \alpha], \mathbb{R}^d)$ with the norm $|\varphi| = \sup\{|\varphi(t)| : |t| \le \alpha\}$. Explicitly, set

$$X = \{\varphi \in \mathcal{C}([-\alpha, \alpha], \mathbb{R}^d) : |\varphi| \le b\}$$

and define $F : X \times Y \to X$ by

$$F(\boldsymbol{\varphi}, \boldsymbol{\xi})(t) = \int_0^t \mathbf{f}(\boldsymbol{\varphi}(s) + \boldsymbol{\xi})ds.$$

If $F(\boldsymbol{\varphi}, \boldsymbol{\xi}) = \boldsymbol{\varphi}$, then

$$\boldsymbol{\varphi}(t) = \int_0^t \mathbf{f}(\boldsymbol{\varphi}(s) + \boldsymbol{\xi})ds,$$

which is the same as

$$\boldsymbol{\varphi}(t) + \boldsymbol{\xi} = \boldsymbol{\xi} + \int_0^t \mathbf{f}(\boldsymbol{\varphi}(s) + \boldsymbol{\xi})ds.$$

Therefore, $F(\boldsymbol{\varphi}, \boldsymbol{\xi}) = \boldsymbol{\varphi}$ implies that $\boldsymbol{\varphi}(t) + \boldsymbol{\xi}$ is a solution of the initial-value problem

$$\begin{aligned} \dot{\mathbf{x}} &= \mathbf{f}(\mathbf{x}) \\ \boldsymbol{\xi} &= \mathbf{x}(\tau). \end{aligned}$$

To apply *Theorem 8.17*, three properties of $F(\boldsymbol{\varphi}, \boldsymbol{\xi})$ must be verified. It must be shown first that $F(\boldsymbol{\varphi}, \boldsymbol{\xi})$ is in X when $(\boldsymbol{\varphi}, \boldsymbol{\xi})$ is in $X \times Y$, second that there exist λ between 0 and 1 such that when $\boldsymbol{\xi}$ is fixed $F_{\boldsymbol{\xi}}(\boldsymbol{\varphi})$ is a contraction on X with respect to λ, and third that for fixed $\boldsymbol{\varphi}$ the function $F_{\boldsymbol{\varphi}}(\boldsymbol{\xi})$ it is a continuous function from Y into X. These three steps are the same steps that will be used in the beginning of the next section (Section 8.4) to prove the existence of local stable manifolds with the estimates (*Propositions 8.12, 8.13, and 8.14*) from the Section 8.2.

To begin, note that $\boldsymbol{\varphi}(t) + \boldsymbol{\xi}$ is in R when $(\boldsymbol{\varphi}, \boldsymbol{\xi})$ is in $X \times Y$. Hence,

$$\begin{aligned} |F(\boldsymbol{\varphi}, \boldsymbol{\xi})(t)| &= \left| \int_0^t \mathbf{f}(\boldsymbol{\varphi}(s) + \boldsymbol{\xi})ds \right| \\ &\leq \int_0^t |\mathbf{f}(\boldsymbol{\varphi}(s) + \boldsymbol{\xi})|\, ds \\ &\leq \int_0^t M\, ds \\ &\leq \alpha M < b \end{aligned}$$

and $F(\boldsymbol{\varphi}, \boldsymbol{\xi})$ is in X as required.

Next,

$$\begin{aligned} |F(\boldsymbol{\varphi}, \boldsymbol{\xi})(t) - F(\boldsymbol{\psi}, \boldsymbol{\xi})(t)| &\leq \int_0^t |\mathbf{f}(\boldsymbol{\varphi}(s) + \boldsymbol{\xi}) - \mathbf{f}(\boldsymbol{\psi}(s) + \boldsymbol{\xi})|\, ds \\ &\leq \int_0^t K\, |\boldsymbol{\varphi}(s) - \boldsymbol{\psi}(s)|\, ds \\ &\leq \alpha K\, |\boldsymbol{\varphi} - \boldsymbol{\psi}| \\ &\leq \frac{1}{2}\, |\boldsymbol{\varphi} - \boldsymbol{\psi}| \end{aligned}$$

and the contraction condition is satisfied because the last term does not depend on t and is thus an upper bound for the norm of $F(\varphi, \boldsymbol{\xi}) - F(\psi, \boldsymbol{\xi})$.

Finally,

$$\begin{aligned} \left|F(\varphi, \boldsymbol{\xi})(t) - F(\varphi, \boldsymbol{\xi}')(t)\right| &\leq \int_0^t \left|\mathbf{f}(\varphi(s) + \boldsymbol{\xi}) - \mathbf{f}(\varphi(s) + \boldsymbol{\xi}')\right| ds \\ &\leq \alpha K |\boldsymbol{\xi} - \boldsymbol{\xi}'| < \frac{1}{2}|\boldsymbol{\xi} - \boldsymbol{\xi}'| \end{aligned}$$

and $|F_{\varphi}(\boldsymbol{\xi}) - F_{\varphi}(\boldsymbol{\xi}')| < \varepsilon$ when $|\boldsymbol{\xi} - \boldsymbol{\xi}'| < 2\varepsilon$.

Therefore, by *Theorem 8.17* there exists a continuous function h from Y into $X \subset \mathcal{C}([-\alpha, \alpha], \mathbb{R}^d)$ such that $h(\boldsymbol{\xi}) + \boldsymbol{\xi}$ is a solution of the initial problem

$$\begin{aligned} \dot{\mathbf{x}} &= \mathbf{f}(\mathbf{x}) \\ \boldsymbol{\xi} &= \mathbf{x}(\tau) \end{aligned}$$

and $|h(\boldsymbol{\xi})(t) - h(\boldsymbol{\xi}')(t)|$ is small for $|t| \leq \alpha$ when $|\boldsymbol{\xi} - \boldsymbol{\xi}'|$ is sufficiently small.

It has just been demonstrated how the contraction mapping principle can be used to prove an existence theorem for locally Lipschitz vector fields, but there is nothing new here. This is the same iterative process that was used in Section 6 of Chapter 1. With a little more care in the choice of Y, the preceding can be converted into a full proof of *Theorem 1.22* on page 43 for autonomous differential equations.

For fixed $\boldsymbol{\xi}$, *Theorem 8.16* applies to $F(\varphi) = F_{\boldsymbol{\xi}}(\varphi)$ and $F^n(\varphi)$ converges to a solution of the above initial-value problem by *Theorem 8.16*. Starting with $\varphi(t) \equiv \mathbf{0}$, the sequence of functions $\boldsymbol{\xi} + F^n(\varphi)$ is the same sequence of functions that appears in the Picard-Lindelöf theorem on page 46.

Thus fixed points and the contraction mapping principle provide a more conceptual way of thinking about the existence of solutions. Although theorems like the Picard-Lindelöf theorem are not that difficult to prove directly, in more complicated situations by using the contraction mapping principle we can avoid the morass of notation needed to prove that a complicated iterative process converges to a solution. The construction of local stable manifolds is an excellent example of how the contraction mapping principle can make a proof more understandable.

EXERCISES

1. Let X be a metric space with metric $d(x, y)$. Show that given $x \in X$ and $r > 0$ the set $\{y \in X : d(y, x) < r\}$ is an open set of X and the set $\{y \in X : d(y, x) \leq r\}$ is a closed set of X. Show that if x and y are distinct points in X, then there exist disjoint open sets U and V of X such that $\mathbf{x} \in U$ and $y \in V$.

2. Let C be a compact subset of a metric space X. Prove that C is a closed subset of X. Also prove that if $f : X \to \mathbb{R}$ is a continuous real-valued function on X, then $f(C)$ is a compact subset of $\mathbb{R}$.

3. Let X be a metric space. Show that the limit of a convergent sequence x_k in X is unique, that is, if y_1 and y_2 are points in X for which given $\varepsilon > 0$ there exists $N > 0$ such that $d(x_k, y_j) < \varepsilon$ when $k > N$, then $y_1 = y_2$.

4. Let h be a homeomorphism of the metric space X onto the metric space Y. Prove that U is an open set in X if and only if $h(U)$ is an open set in Y. Prove the same result for closed sets and for compact sets.

5. Let X be a complete metric space. Prove that if $F : X \to X$ is a function such that F^N is a contraction, then F has a unique fixed point.

6. Let X be a compact metric space. If $F : X \to X$ is a function such that $d(F(x), F(y)) < d(x, y)$ for all x and y in X, then F has a unique fixed point. Hint: First show that the function $x \to d(x, F(x))$ assumes its minimum value.

8.4 Local Stable Manifolds

The idea of the local stable manifold theorem is to show that the phase portrait near a fixed point looks like the phase portrait near the origin of a linear system with constant coefficients. If $\mathbf{p}$ is a fixed point for $\dot{\mathbf{x}} = \mathbf{f}(\mathbf{x})$, then the linear model for the qualitative behavior of the solutions of $\dot{\mathbf{x}} = \mathbf{f}(\mathbf{x})$ near $\mathbf{p}$ should be $\dot{\mathbf{x}} = \mathbf{f}'(\mathbf{p})\mathbf{x}$ near $\mathbf{0}$. Since the nonlinear part of $\mathbf{f}$ at $\mathbf{p}$ can dramatically alter the behavior of solutions near $\mathbf{p}$ when $\mathbf{f}'(\mathbf{p})$ has pure imaginary eigenvalues, restrictions on $\mathbf{f}'(\mathbf{p})$ are necessary. Requiring that $\mathbf{f}'(\mathbf{p})$ be a hyperbolic matrix for a fixed point $\mathbf{p}$ is sufficient to prove a linearization result that completely determines the qualitative behavior of solutions near such a fixed point. The analogues of E^+ and E^- near $\mathbf{p}$ are the local stable manifolds and the results of the previous two sections will be used to prove they exist.

The simple change of variable $\mathbf{x} = \mathbf{y} + \mathbf{p}$ allows us to assume that the fixed point of $\dot{\mathbf{x}} = \mathbf{f}(\mathbf{x})$ occurs at $\mathbf{0}$ without altering the linear part of $\mathbf{f}$ at the fixed point. Moreover, $\dot{\mathbf{x}} = \mathbf{f}(\mathbf{x})$ can be written as $\dot{\mathbf{x}} = A\mathbf{x} + \mathbf{g}(\mathbf{x})$, where $A = \mathbf{f}'(\mathbf{0})$ and the nonlinear part, $\mathbf{g}(\mathbf{x})$, is in $\mathcal{L}(\psi)$ for some ψ by *Proposition 8.11*. Consequently, studying the qualitative behavior of $\dot{\mathbf{x}} = \mathbf{f}(\mathbf{x})$ near a fixed point with a hyperbolic linear part is the same as studying the qualitative behavior near the origin of a hyperbolic linear system perturbed by a function in $\mathcal{L}(\psi)$. The latter point of view will be used to establish the main result.

As before, A will be a hyperbolic real matrix with positively stable and negatively stable manifolds E^+ and E^-. The linear projections of $\mathbb{R}^d$ onto E^+ and E^- will be denoted by P_+ and P_-, respectively, and $L = \max\{|P_+|, |P_-|\}$. The positive constants K and α will continue to be determined by *Theorem 8.1*.

The Banach space $\mathcal{C}([0, \infty), \mathbb{R}^d)$ of continuous bounded functions φ from $[0, \infty)$ into $\mathbb{R}^d$ with the norm $|\varphi| = \sup\{|\varphi(t)| : t \geq 0\}$ will be used to apply *Theorem 8.17* to the function $\mathbf{F}(\varphi, \boldsymbol{\xi})$ defined by equation *(8.4)*. For this purpose, set

$$X_\delta = \{\varphi \in \mathcal{C}([0,\infty),\mathbb{R}^d) : |\varphi| \le \delta\}.$$

Then X_δ is a closed subset of $\mathcal{C}([0,\infty),\mathbb{R}^d)$, and hence a complete metric space by *Proposition 8.15.*

Similarly, let $Y_\gamma = \{\boldsymbol{\xi}_+ \in E^+ : |\boldsymbol{\xi}_+| \le \gamma\}$, which is also a metric space. The preparatory results in Sections 2 and 3 can now be brought together. For simplicity, it will be assumed that all the vector fields are defined on $\mathbb{R}^d$. This is not essential; it just makes the statements of results and their proofs a little simpler without altering the main ideas and arguments.

Proposition 8.18 *Consider* $\dot{\mathbf{x}} = A\mathbf{x} + \mathbf{g}(\mathbf{x})$ *with* $A \in \mathcal{H}$ *and* $\mathbf{g}$ *in* $\mathcal{L}(\psi)$. *If* $\delta > 0$ *and* $\gamma > 0$ *satisfy the following conditions:*

$$\begin{aligned} \psi(2\delta) &< \frac{\alpha}{4KL} \\ \gamma &< \frac{\delta}{K}, \end{aligned}$$

then

$$\begin{aligned} \mathbf{F}(\varphi,\boldsymbol{\xi}_+)(t) &= \int_0^t e^{A(t-s)}P_+[\mathbf{g}(\varphi(s) + e^{As}\boldsymbol{\xi}_+)]ds \\ &- \int_t^\infty e^{A(t-s)}P_-[\mathbf{g}(\varphi(s) + e^{As}\boldsymbol{\xi}_+)]ds \end{aligned}$$

defines a function $\mathbf{F}(\varphi,\boldsymbol{\xi}_+)$ *of* $X_\delta \times Y_\gamma$ *into* X_δ *that satisfies the hypothesis of Theorem 8.17.*

Proof. The proof follows the same three steps used in the example at the end of Section 3 and uses each one of the final three propositions in Section 2 for one of the steps. Those propositions required a common bound β for the functions φ and $e^{At}\boldsymbol{\xi}_+$ in $\mathcal{C}([0,\infty),\mathbb{R}^d)$.

Note that if $\boldsymbol{\xi}_+$ is inY_γ, then for $t \ge 0$

$$|e^{At}\boldsymbol{\xi}_+| \le Ke^{-\alpha t}|\boldsymbol{\xi}_+| \le K\gamma < \delta.$$

Hence, in the present context those propositions can be applied with $\beta = \delta$.

The first step is to show that $\mathbf{F}(\varphi,\boldsymbol{\xi})$ maps $X_\delta \times Y_\gamma$ into X_δ. By *Proposition 8.12* and the hypothesis

$$\begin{aligned} |\mathbf{F}(\varphi,\boldsymbol{\xi}_+)| &\le \frac{4\delta KL}{\alpha}\psi(2\delta) \\ &\le \frac{4\delta KL}{\alpha}\frac{\alpha}{4KL} \\ &= \delta. \end{aligned}$$

Thus $\mathbf{F}(\varphi,\boldsymbol{\xi}_+)$ is in X_δ when $(\varphi,\boldsymbol{\xi}_+)$ is in $X_\delta \times Y_\gamma$.

The second step is to show that $\mathbf{F}(\varphi, \boldsymbol{\xi})$ is a contraction for fixed $\boldsymbol{\xi}_+$ and that the contraction coefficient is independent of $\boldsymbol{\xi}_+$. By *Proposition 8.13* and the hypothesis

$$\begin{aligned} \left|\mathbf{F}(\varphi_1, \boldsymbol{\xi}_+) - \mathbf{F}(\varphi_2, \boldsymbol{\xi}_+)\right| &\leq \frac{2KL}{\alpha}\psi(2\delta)\left|\varphi_1 - \varphi_2\right| \\ &\leq \frac{2KL}{\alpha}\frac{\alpha}{4KL}\left|\varphi_1 - \varphi_2\right| \\ &= \frac{1}{2}\left|\varphi_1 - \varphi_2\right| \end{aligned}$$

as required.

Finally, *Proposition 8.14* implies that

$$\left|\mathbf{F}(\varphi, \boldsymbol{\xi}_+) - \mathbf{F}(\varphi, \boldsymbol{\xi}'_+)\right| \leq \frac{K}{2}\left|\boldsymbol{\xi}_+ - \boldsymbol{\xi}'_+\right| \tag{8.7}$$

and $\boldsymbol{\xi}_+ \to \mathbf{F}(\varphi, \boldsymbol{\xi})$ is a continuous function from Y_γ into X_δ when φ is held fixed. □

Because ψ is continuous and $\psi(0) = 0$, there exist $\delta > 0$ and $\gamma > 0$ satisfying the hypothesis of the previous proposition, in fact, all sufficiently small $\delta > 0$ and $\gamma > 0$ do so. For the rest of this section, δ and γ will satisfy such conditions.

By *Theorem 8.17*, there exists a continuous function $\mathbf{h} : Y_\gamma \to X_\delta$ such that

$$\mathbf{F}(\mathbf{h}(\boldsymbol{\xi}_+), \boldsymbol{\xi}_+) = \mathbf{h}(\boldsymbol{\xi}_+). \tag{8.8}$$

Since $\mathbf{h}(\boldsymbol{\xi}_+)$ is itself a function of t and part of a solution of $\dot{\mathbf{x}} = A\mathbf{x} + \mathbf{g}(\mathbf{x})$, it will be convenient to use the notation $\mathbf{h}(t, \boldsymbol{\xi}_+)$ rather than $\mathbf{h}(\boldsymbol{\xi}_+)(t)$ to denote the value of $\mathbf{h}(\boldsymbol{\xi})$ at t.

Using the full definition of $\mathbf{F}(\varphi, \boldsymbol{\xi}_+)$, the above equation can also be written as:

$$\begin{aligned} \mathbf{h}(t, \boldsymbol{\xi}_+) &= \int_0^t e^{A(t-s)} P_+[\mathbf{g}(\mathbf{h}(s, \boldsymbol{\xi}_+) + e^{As}\boldsymbol{\xi}_+)]\, ds \\ &- \int_t^\infty e^{A(t-s)} P_-[\mathbf{g}(\mathbf{h}(s, \boldsymbol{\xi}_+) + e^{As}\boldsymbol{\xi}_+)]\, ds, \end{aligned} \tag{8.9}$$

and adding $e^{At}\boldsymbol{\xi}_+$ to both sides produces

$$\begin{aligned} \mathbf{h}(t, \boldsymbol{\xi}_+) + e^{At}\boldsymbol{\xi}_+ &= e^{At}\boldsymbol{\xi}_+ + \int_0^t e^{A(t-s)} P_+[\mathbf{g}(\mathbf{h}(s, \boldsymbol{\xi}_+) + e^{As}\boldsymbol{\xi}_+)]\, ds \\ &- \int_t^\infty e^{A(t-s)} P_-[\mathbf{g}(\mathbf{h}(ts, \boldsymbol{\xi}_+) + e^{As}\boldsymbol{\xi}_+)]\, ds. \end{aligned} \tag{8.10}$$

Therefore, by *Proposition 8.10*

$$\boldsymbol{\theta}(t, \boldsymbol{\xi}_+) = \mathbf{h}(t, \boldsymbol{\xi}_+) + e^{At}\boldsymbol{\xi}_+ \tag{8.11}$$

is a solution of $\dot{\mathbf{x}} = A\mathbf{x} + \mathbf{g}(\mathbf{x})$ for each $\boldsymbol{\xi}_+ \in Y_\delta$.

There is a subtle point here that needs to be addressed; namely, what affect does the choice of δ and γ have on the function $\mathbf{h}(t, \boldsymbol{\xi}_+)$, and consequently the solution $\boldsymbol{\theta}(t, \boldsymbol{\xi}_+)$? The key observation is that the numbers δ and γ do not affect the formula for $\mathbf{F}(\varphi, \boldsymbol{\xi}_+)$, only its domain and range. Furthermore, contraction mappings have unique fixed points. If $\delta' < \delta$ and $\gamma' < \gamma$, then $X_{\delta'} \subset X_\delta$ and $Y_{\gamma'} \subset Y_\gamma$. When $\mathbf{h}(t, \boldsymbol{\xi}_+)$ is the fixed point in $X_{\delta'}$ for $\boldsymbol{\xi}_+ \in Y_{\gamma'}$, it is automatically the unique fixed point in X_δ for $\boldsymbol{\xi}_+ \in Y_\gamma$. Hence the function $\mathbf{h}$ on $Y_{\gamma'}$ is obtained by restricting the function $\mathbf{h}$ on Y_γ to the smaller set $Y_{\gamma'}$. In other words, $\mathbf{h}(t, \boldsymbol{\xi}_+)$ is uniquely determined for all δ and γ satisfying the conditions in *Proposition 8.18* so long as $|\boldsymbol{\xi}_+| < \gamma$.

For the rest of the chapter, $\mathbf{h}$ and $\boldsymbol{\theta}$ will be determined as above for some choice of δ and γ satisfying the conditions in *Proposition 8.18* or more restrictive ones. The immediate goal is to study the properties of the solutions $\boldsymbol{\theta}(t, \boldsymbol{\xi}_+)$.

Proposition 8.19 *Consider* $\dot{\mathbf{x}} = A\mathbf{x} + \mathbf{g}(\mathbf{x})$ *with* $A \in \mathcal{H}$ *and* $\mathbf{g}$ *in* $\mathcal{L}(\psi)$. *If*

$$\begin{aligned} \psi(2\delta) &< \frac{\alpha}{4KL} \\ \gamma &< \frac{\delta}{K}, \end{aligned}$$

then for every $\boldsymbol{\xi}_+$ *such that* $|\boldsymbol{\xi}_+| \le \gamma$ *there exists a solution* $\boldsymbol{\theta}(t, \boldsymbol{\xi}_+)$ *of* $\dot{\mathbf{x}} = A\mathbf{x} + \mathbf{g}(\mathbf{x})$ *defined for* $t \ge 0$ *with the following properties:*

(a) $P_+(\boldsymbol{\theta}(0, \boldsymbol{\xi}_+)) = \boldsymbol{\xi}_+$,

(b) $|\boldsymbol{\theta}(t, \boldsymbol{\xi}_+)| \le 2\delta$ *for* $t \ge 0$,

(c) $\boldsymbol{\theta}(t, \boldsymbol{\xi}_+) \equiv \mathbf{0}$ *if and only if* $\boldsymbol{\xi}_+ = \mathbf{0}$,

(d) $\boldsymbol{\xi}_+ \to \boldsymbol{\theta}(\cdot, \boldsymbol{\xi}_+)$ *is a continuous function from* Y_γ *into* $\mathcal{C}([0, \infty), \mathbb{R}^d)$.

Proof. For part (a), the improper integral in *(8.10)* converges to a point in E^-. Then setting $t = 0$ and applying P_+ to *(8.10)* proves that $P_+(\boldsymbol{\theta}(0, \boldsymbol{\xi}_+)) = \boldsymbol{\xi}_+$.

For the 2δ bound of $\boldsymbol{\theta}$ in part (b), it suffice to show that both $\mathbf{h}(t, \boldsymbol{\xi}_+)$ and $e^{At}\boldsymbol{\xi}_+$ are bounded by δ for $t \ge 0$. In the first two paragraphs of the proof of *Proposition 8.18*, it was shown that $|\mathbf{e}^{At}\boldsymbol{\xi}_+| \le \delta$ for $t \ge 0$ and $|\mathbf{F}(\varphi, \boldsymbol{\xi}_+)| \le \delta$. It follows from *(8.8)* that $|\mathbf{h}(t, \boldsymbol{\xi}_+)| < \delta$ for $t \ge 0$.

Turning to part (c), it is easy to verify that $\mathbf{h}(t, \mathbf{0}) \equiv \mathbf{0}$ using *(8.9)*. It follows from *(8.11)* that $\boldsymbol{\theta}(t, \mathbf{0}) \equiv \mathbf{0}$. The converse is a consequence of part (a).

To prove part (d), observe that for $\boldsymbol{\xi}_+$ and $\boldsymbol{\xi}'_+$ in Y_γ and $t \ge 0$ the definition of $\boldsymbol{\theta}(t, \boldsymbol{\xi})$ implies that

$$|\boldsymbol{\theta}(t, \boldsymbol{\xi}_+) - \boldsymbol{\theta}(t, \boldsymbol{\xi}'_+)| \le |\mathbf{h}(t, \boldsymbol{\xi}_+) - \mathbf{h}(t, \boldsymbol{\xi}'_+)| + |e^{At}\boldsymbol{\xi}_+ - e^{At}\boldsymbol{\xi}'_+|.$$

Obviously,

$$|e^{At}\boldsymbol{\xi}_+ - e^{At}\boldsymbol{\xi}'_+| \le Ke^{-\alpha t}|\boldsymbol{\xi}_+ - \boldsymbol{\xi}'_+| \le K|\boldsymbol{\xi}_+ - \boldsymbol{\xi}'_+|.$$

For the other term use, *(8.6)* with $\lambda = 1/2$ and *(8.7)* as follows:

$$\begin{aligned}|\mathbf{h}(t,\boldsymbol{\xi}_+) - \mathbf{h}(t,\boldsymbol{\xi}'_+)| &\leq \\ \frac{1}{1-(1/2)}\,|\mathbf{F}(\mathbf{h}(t,\boldsymbol{\xi}'_+),\boldsymbol{\xi}_+) - \mathbf{F}(\mathbf{h}(t,\boldsymbol{\xi}'_+),\boldsymbol{\xi}'_+)| &\leq \\ K|\boldsymbol{\xi}_+ - \boldsymbol{\xi}'_+|.\end{aligned}$$

Putting the two calculations together yields

$$|\boldsymbol{\theta}(t,\boldsymbol{\xi}_+) - \boldsymbol{\theta}(t,\boldsymbol{\xi}'_+)| \leq 2K|\boldsymbol{\xi}_+ - \boldsymbol{\xi}'_+| \tag{8.12}$$

to prove the continuity of $\boldsymbol{\xi}_+ \to \boldsymbol{\theta}(\cdot,\boldsymbol{\xi}_+)$. □

The construction of $\boldsymbol{\theta}(t,\boldsymbol{\xi}_+)$ using the fixed points of $\mathbf{F}(\varphi,\boldsymbol{\xi}_+)$ produces solutions that stay near $\mathbf{0}$ for $t \geq 0$ and some notation is needed for such solutions. The *local positively stable manifolds* at $\mathbf{0}$ or *local P-stable manifolds* for short are defined by

$$\mathcal{W}_\delta^+ = \{\boldsymbol{\xi} : |\mathbf{x}(t,\boldsymbol{\xi})| \leq \delta \text{ for } t \geq 0\}. \tag{8.13}$$

It is implicit in the definition of the local P-stable manifolds that when $\boldsymbol{\xi}$ is in $\mathcal{W}_\delta^+$, then $\mathbf{x}(t,\boldsymbol{\xi})$ is defined for all $t \geq 0$. Clearly, $\mathcal{W}_{\delta'}^+ \subset \mathcal{W}_\delta^+$ when $\delta' < \delta$.

Remark *If $\boldsymbol{\xi}$ is in $\mathcal{W}_\delta^+$, then $\mathbf{x}(\tau,\boldsymbol{\xi})$ is also in $\mathcal{W}_\delta^+$ when $\tau \geq 0$.*

Proof. Just note that $|\mathbf{x}(t,\mathbf{x}(\tau,\boldsymbol{\xi}))| = |\mathbf{x}(t+\tau,\boldsymbol{\xi})| \leq \delta$ when both $t \geq 0$ and $\tau \geq 0$. □

By interchanging the roles of E^+ and E^- in Section 2 and the preceding results in this section, there is a corresponding family of solutions staying near $\mathbf{0}$ for $t \leq 0$. Naturally, the *local N-stable manifolds* are defined by

$$\mathcal{W}_\delta^- = \{\boldsymbol{\xi} : |\mathbf{x}(t,\boldsymbol{\xi})| \leq \delta \text{ for } t \leq 0\} \tag{8.14}$$

and the constant solution $\mathbf{x}(t,\mathbf{0}) \equiv \mathbf{0}$ is in $\mathcal{W}_\delta^+ \cap \mathcal{W}_\delta^-$.

The wording here is a bit cumbersome and slightly nonstandard, but we want to emphasize that the properties of $\mathcal{W}_\delta^-$ for $t \leq 0$ mirror those of $\mathcal{W}_\delta^+$ for $t \geq 0$, and together they provide a local structure analogous to E^+ and E^-. The sets $\mathcal{W}_\delta^+$ and $\mathcal{W}_\delta^-$ are also often called the *local stable manifolds* and the *local unstable manifolds.*

The definition of the local P-stable manifolds is strictly in terms of the qualitative behavior of solutions and makes no reference to the construction of $\boldsymbol{\theta}(t,\boldsymbol{\xi}_+)$. The next task is to show that for small δ the local P-stable manifold $\mathcal{W}_\delta^+$ is synonymous with $\boldsymbol{\theta}(t,\boldsymbol{\xi}_+)$. Since $\boldsymbol{\theta}(t,\boldsymbol{\xi}_+)$ is parameterized by $\boldsymbol{\xi}_+$ in E^+ such that $|\boldsymbol{\xi}_+| < \gamma$, this provides a link between E^+ and $\mathcal{W}_\delta^+$.

Theorem 8.20 *Consider* $\dot{\mathbf{x}} = A\mathbf{x} + \mathbf{g}(\mathbf{x})$ *with* $A \in \mathcal{H}$ *and* $\mathbf{g}$ *in* $\mathcal{L}(\psi)$, *and suppose* δ *and* γ *satisfy the conditions*

$$\psi(4\delta) < \frac{\alpha}{4KL}$$
$$\gamma < \frac{\delta}{2K}.$$

(a) If $\boldsymbol{\xi}$ *is in* $\mathcal{W}_\delta^+$ *and* $|P_+(\boldsymbol{\xi})| < \gamma$, *then*

$$\mathbf{x}(t, \boldsymbol{\xi}) = \boldsymbol{\theta}(t, P_+(\boldsymbol{\xi})).$$

(b) The space

$$W = \mathcal{W}_\delta^P \cap \{\boldsymbol{\xi} : |P_+(\boldsymbol{\xi})| < \gamma\}$$

is homeomorphic to an open disk of dimension $\iota(A)$.

Proof. Suppose $\boldsymbol{\xi}$ is in the local stable manifold $\mathcal{W}_\delta^+$ so that $|\mathbf{x}(t, \boldsymbol{\xi})| \leq \delta$ for $t \geq 0$. As Usual,

$$|e^{At} P_+(\boldsymbol{\xi})| \leq K e^{-\alpha t} |P_+(\boldsymbol{\xi})| \leq K\gamma \leq \delta$$

for $t \geq 0$ because $|P_+(\boldsymbol{\xi})| < \gamma < \delta/2K$. It follows that

$$|\mathbf{x}(t, \boldsymbol{\xi}) - e^{At} P_+(\boldsymbol{\xi})| \leq 2\delta$$

for $t \geq 0$.

Because $\mathbf{x}(t, \boldsymbol{\xi})$ is a bounded solution of $\dot{\mathbf{x}} = A\mathbf{x} + \mathbf{g}(\mathbf{x})$, it satisfies the integral equation in *Theorem 8.10.* By writing $\mathbf{x}(t, \boldsymbol{\xi}) = \mathbf{x}(t, \boldsymbol{\xi}) - e^{At} P_+(\boldsymbol{\xi}) + e^{At} P_+(\boldsymbol{\xi})$, it is apparent that $\boldsymbol{\varphi}(t) = \mathbf{x}(t, \boldsymbol{\xi}) - e^{At} P_+(\boldsymbol{\xi})$ satisfies *(8.3)* with $\boldsymbol{\xi}_+ = P_+(\boldsymbol{\xi})$, and hence $\mathbf{x}(t, \boldsymbol{\xi}) - e^{At} P_+(\boldsymbol{\xi})$ is a fixed point of $\mathbf{F}(\boldsymbol{\varphi}, P_+(\boldsymbol{\xi}))$ in $X_{2\delta}$. It follows that

$$\mathbf{x}(t, \boldsymbol{\xi}) - e^{At} P_+(\boldsymbol{\xi}) = \mathbf{h}(t, P_+(\boldsymbol{\xi}))$$

for three reasons:

(a) The pair of numbers 2δ and γ satisfy the hypothesis of *Proposition 8.18.*

(b) $|P_+(\boldsymbol{\xi})| \leq \gamma$.

(c) Contractions have unique fixed points.

Therefore,

$$\mathbf{x}(t, \boldsymbol{\xi}) = \mathbf{h}(t, P_+(\boldsymbol{\xi})) + e^{At} P_+(\boldsymbol{\xi}) = \boldsymbol{\theta}(t, P_+(\boldsymbol{\xi}))$$

to complete the proof of the first part.

The projection P_+ is a continuous map of W into $V = \{\mathbf{v} \in E^+ : |\mathbf{v}| < \gamma\}$ and the latter is an open disk of the right dimension. The proof will be completed by showing that P_+ restricted to W is a homeomorphism onto V.

Because

$$\begin{aligned}\psi(\delta) &\leq \psi(4\delta) < \frac{\alpha}{4KL} \\ \gamma &< \frac{\delta}{2K} = \frac{\delta/2}{K}\end{aligned}$$

Proposition 8.19 applies to $\boldsymbol{\xi}_+$ in V using $\delta/2$ and γ. In particular, part (b) of it implies that $|\boldsymbol{\theta}(t, \boldsymbol{\xi}_+)| \leq \delta$ for $t \geq 0$ when $|\boldsymbol{\xi}_+| \leq \gamma$. Hence, $\boldsymbol{\theta}(0, \boldsymbol{\xi}_+)$ is in $\mathcal{W}_\delta^+$. Furthermore, $P_+(\boldsymbol{\theta}(0, \boldsymbol{\xi}_+)) = \boldsymbol{\xi}_+$ by part (a) of the same proposition making $Q : \boldsymbol{\xi}_+ \to \boldsymbol{\theta}(0, \boldsymbol{\xi}_+)$ the natural candidate for the continuous inverse of P_+ restricted to W.

The map $Q : \boldsymbol{\xi}_+ \to \boldsymbol{\theta}(0, \boldsymbol{\xi}_+)$ is continuous on V because

$$|\boldsymbol{\theta}(0, \boldsymbol{\xi}_+) - \boldsymbol{\theta}(0, \boldsymbol{\xi}'_+)| \leq 2K|\boldsymbol{\xi}_+ - \boldsymbol{\xi}'_+|$$

by *(8.12)*. It is obvious from the previous paragraph that $P_+(Q(\mathbf{v})) = \mathbf{v}$ for $\mathbf{v} \in V$. Part (a) of this proposition implies that $Q(P_+(\boldsymbol{\xi})) = \boldsymbol{\xi}$ for $\boldsymbol{\xi} \in W$, proving that Q is the required continuous inverse of P_+ restricted to W. □

It follows from this theorem that when $\iota(A) > 0$, the local P-stable manifolds $\mathcal{W}_\delta^+$ contain solutions other than the trivial solution $\mathbf{x}(t, \mathbf{0}) \equiv \mathbf{0}$. Moreover, for small δ and γ the solutions $\mathbf{x}(t, \boldsymbol{\xi})$ with $\boldsymbol{\xi} \in \mathcal{W}_\delta^+$ behave in a positively stable manner because they start within a δ distance from $\mathbf{0}$ and stay within a δ distance from $\mathbf{0}$ for $t \geq 0$. Surprisingly, they also exhibit the same exponential decay seen in E^+, but this is a much deeper result. The result depends only on the bound on the solutions and is independent of the existence of such solutions. (If they do not exist, they do not violate the conclusion.) Since the role of γ is primarily to prove that the local P-stable manifolds are nonempty, γ does not play a role in the next theorem.

Theorem 8.21 *Consider $\dot{\mathbf{x}} = A\mathbf{x} + \mathbf{g}(\mathbf{x})$ with $A \in \mathcal{H}$ and $\mathbf{g}$ in $\mathcal{L}(\psi)$, and let δ be positive a number satisfying:*

$$\psi(2\delta) < \frac{\alpha}{4KL}.$$

If $\boldsymbol{\xi}$ and $\boldsymbol{\xi}'$ are in $\mathcal{W}_\delta^+$, then

$$|\mathbf{x}(t, \boldsymbol{\xi}) - \mathbf{x}(t, \boldsymbol{\xi}')| \leq 2Ke^{-\alpha t/2}|P_+(\boldsymbol{\xi}) - P_+(\boldsymbol{\xi}')|$$

and

$$|\mathbf{x}(t, \boldsymbol{\xi})| \leq 2Ke^{-\alpha t/2}|P_+(\boldsymbol{\xi})|$$

for $t \geq 0$.

Proof. Since both $\boldsymbol{\xi}$ and $\boldsymbol{\xi}'$ are in $\mathcal{W}_\delta^+$, the function $|\mathbf{x}(t, \boldsymbol{\xi}) - \mathbf{x}(t, \boldsymbol{\xi}')|$ is bounded by 2δ for $t \geq 0$. The function $|\mathbf{x}(t, \boldsymbol{\xi}) - \mathbf{x}(t, \boldsymbol{\xi}')|$ also satisfies a functional inequality. Because both solutions are bounded for $t \geq 0$, *Theorem 8.10* can be applied to both of them and the differences estimated. Consequently,

$$\begin{aligned}
|\mathbf{x}(t,\boldsymbol{\xi}) - \mathbf{x}(t,\boldsymbol{\xi}')| &\leq e^{At}|P_+(\boldsymbol{\xi}) - P_+(\boldsymbol{\xi}')| \\
&+ \int_0^t \left| e^{A(t-s)} P_+ \left[\mathbf{g}\left(\mathbf{x}(s,\boldsymbol{\xi})\right) - \mathbf{g}\left(\mathbf{x}(s,\boldsymbol{\xi}')\right)\right]\right| ds \\
&+ \int_0^\infty \left| e^{-As} P_- \left[\mathbf{g}\left(\mathbf{x}(t+s,\boldsymbol{\xi})\right) - \mathbf{g}\left(\mathbf{x}(t+s,\boldsymbol{\xi}')\right)\right]\right| ds.
\end{aligned}$$

Using the properties of K, α, and $\mathcal{L}(\psi)$, produces the estimate

$$\begin{aligned}
|\mathbf{x}(t,\boldsymbol{\xi}) - \mathbf{x}(t,\boldsymbol{\xi}')| &\leq Ke^{-\alpha t}|P_+(\boldsymbol{\xi}) - P_+(\boldsymbol{\xi}')| \\
&+ KL\psi(2\delta)\int_0^t e^{-\alpha(t-s)}|\mathbf{x}(s,\boldsymbol{\xi}) - \mathbf{x}(s,\boldsymbol{\xi}')|\, ds \\
&+ KL\psi(2\delta)\int_0^\infty e^{-\alpha s}|\mathbf{x}(t+s,\boldsymbol{\xi}) - \mathbf{x}(t+s,\boldsymbol{\xi}')|\, ds
\end{aligned}$$

and then

$$\begin{aligned}
|\mathbf{x}(t,\boldsymbol{\xi}) - \mathbf{x}(t,\boldsymbol{\xi}')| &\leq Ke^{-\alpha t}|P_+(\boldsymbol{\xi}) - P_+(\boldsymbol{\xi}')| \\
&+ \frac{\alpha}{4}\int_0^t e^{-\alpha(t-s)}|\mathbf{x}(s,\boldsymbol{\xi}) - \mathbf{x}(s,\boldsymbol{\xi}')|\, ds \\
&+ \frac{\alpha}{4}\int_0^\infty e^{-\alpha s}|\mathbf{x}(t+s,\boldsymbol{\xi}) - \mathbf{x}(t+s,\boldsymbol{\xi}')|\, ds
\end{aligned}$$

because $KL\psi(2\delta) < \alpha/4$. By setting $u(t) = |\mathbf{x}(t,\boldsymbol{\xi}) - \mathbf{x}(t,\boldsymbol{\xi}')|$, the above estimate can be rewritten as

$$u(t) \leq KLe^{-\alpha t}|P_+(\boldsymbol{\xi}) - P_+(\boldsymbol{\xi}')| + \frac{\alpha}{4}\int_0^t e^{-\alpha(t-s)}u(s)\, ds + \frac{\alpha}{4}\int_0^\infty e^{-\alpha s}u(t+s)\, ds.$$

The above inequality suggests that Gronwall's inequality might be used, but doing so requires a rather technical argument that is best treated as a separate lemma.

Lemma *Suppose* a, b, M_1, M_2, and M_3 *are positive constants. Let* $u(t)$ *be a non negative continuous function defined on* $[0,\infty)$ *satisfying*

$$u(t) \leq M_1 e^{-at} + M_2\int_0^t e^{-a(t-s)}u(s)ds + M_3\int_0^\infty e^{-bs}u(t+s)ds.$$

If

$$\eta = \frac{M_2}{a} + \frac{M_3}{b} < 1,$$

then

$$u(t) \leq \frac{M_1}{1-\eta}\exp\left(-\left[a - \frac{M_2}{(1-\eta)}\right]t\right)$$

for $t \geq 0$.

Proof. The first step is to show that $\lim_{t\to\infty} u(t) = 0$ using the limit superior. The *limit superior* of a real-valued function $u(t)$ is defined by

$$\overline{\lim}_{t\to\infty} u(t) = \inf_{\mu>0}\left\{\sup_{t>\mu} u(t)\right\}.$$

Because $u(t)$ is a non negative function, a straightforward argument shows that $\lim_{t\to\infty} u(t) = 0$ if and only if $\overline{\lim}_{t\to\infty} u(t) = 0$. Similarly, $\overline{\lim}_{t\to\infty} u(t)$ is finite because $u(t)$ is a bounded function.

If $\lim_{t\to\infty} u(t) \neq 0$, then the limit superior $\overline{\lim}_{t\to\infty} u(t) = \rho > 0$. By the hypothesis of the lemma there exists σ such that $1 < \sigma < 1/\eta$. Since $\sigma\rho > \rho$, it follows from the definition of limit superior that there exists τ such that $u(t) \leq \sigma\rho$ for $t \geq \tau$. For $t \geq \tau$, the function $u(t)$ can be estimated as follows:

$$\begin{aligned}
u(t) &\leq M_1 e^{-at} + M_2 \int_0^\tau e^{-a(t-s)} u(s)ds \\
&+ M_2 \int_\tau^t e^{-a(t-s)} \sigma\rho\, ds + M_3 \int_0^\infty e^{-bs} \sigma\rho\, ds \\
&\leq M_1 e^{-at} + M_2 e^{-at} \int_0^\tau e^{as} u(s)ds \\
&+ \frac{M_2}{a}\sigma\rho + \frac{M_3}{b}\sigma\rho \\
&= M_1 e^{-at} + M_2 e^{-at} \int_0^\tau e^{as} u(s)ds + \eta\sigma\rho.
\end{aligned}$$

Notice, on the one hand, that the right side of the above estimate goes to $\eta\sigma\rho$ as t goes to infinity and that $\eta\sigma\rho < \rho$ by the choice of σ. Let $\varepsilon = \rho(1-\eta\sigma)/2$ Hence, for large t the right side is less than $\rho - \varepsilon$. On the other hand, by the definition of limit superior there exist arbitrarily large t such that $u(t) > \rho - \varepsilon$ for any $\varepsilon > 0$. Consequently, there must exist large t for which the left side of the estimate is greater than the right side. This contradiction completes the proof that $\lim_{t\to\infty} u(t) = 0$. (This is already enough to show that $\omega(\boldsymbol{\xi}) = \mathbf{0}$ when $\boldsymbol{\xi}$ is in $\mathcal{W}^S_\delta$ for small δ.)

Set $v(t) = \sup\{u(s) : s \geq t\}$. Clearly $u(s) \leq v(t)$ for all $s \geq t \geq 0$ and $v(t)$ is nonincreasing. It is easy to check that $v(t)$ is also continuous. The next step is to estimate $v(t)$ by breaking the first integral into two integrals similar to the way that $u(t)$ was estimated. Because $\lim_{t\to\infty} u(t) = 0$, given $t \geq 0$,

$$\tau = \sup\{s : u(t) = v(s) \text{ and } s \geq t\} < \infty.$$

(Actually τ is a function of t, but that dependency will be suppressed in the notation.) Obviously $u(s) \leq v(t)$ for $s \geq t$.

Note that $v(t) = u(\tau)$ and use it as the starting point for the following estimate:

$$v(t) \quad = \quad u(\tau)$$

$$
\begin{aligned}
&\leq M_1 e^{-a\tau} + M_2 \int_0^t e^{-a(\tau-s)} u(s) ds \\
&+ M_2 \int_t^\tau e^{-a(\tau-s)} u(s) + M_3 \int_0^\infty e^{-bs} u(\tau+s) ds \\
&\leq M_1 e^{-a\tau} + M_2 \int_0^t e^{-a(\tau-s)} v(s) ds \\
&+ v(t) \left[M_2 \int_t^\tau e^{-a(\tau-s)} ds + M_3 \int_0^\infty e^{-bs} ds \right] \\
&\leq M_1 e^{-a\tau} + M_2 \int_0^t e^{-a(\tau-s)} v(s) ds + \eta v(t).
\end{aligned}
$$

Collecting the $v(t)$ terms on the left produces

$$(1-\eta)v(t) \leq M_1 e^{-a\tau} + M_2 \int_0^t e^{-a(\tau-s)} v(s) ds.$$

The above must still be modified before Gronwall's inequality can be applied. First, multiply by e^{at} to obtain

$$(1-\eta)e^{at}v(t) = M_1 e^{a(t-\tau)} + M_2 e^{a(t-\tau)} \int_0^t e^{as} v(s) ds,$$

which simplifies to

$$(1-\eta)e^{at}v(t) \leq M_1 + M_2 \int_0^t e^{as} v(s) ds$$

because $a(t-\tau) < 0$. Notice that the function $v(t)$ now satisfies an inequality that is independent of τ and it is no longer relevant to the proof.

By setting $w(t) = e^{at}v(t)$, the above inequality becomes

$$w(t) \leq \frac{M_1}{1-\eta} + \int_0^t w(s) \frac{M_2}{1-\eta} ds$$

and Gronwall's inequality applies yielding

$$w(t) \leq \frac{M_1}{1-\eta} e^{M_2 t/(1-\eta)}.$$

Finally,

$$u(t) \leq v(t) = e^{-at}w(t) \leq \frac{M_1}{1-\eta} \exp\left(-\left[a - \frac{M_2}{(1-\eta)} \right] t \right)$$

to complete the proof. □

Returning to the proof of the theorem, we can now apply the lemma with $a = b = \alpha$, $M_1 = K|P_+(\boldsymbol{\xi}) - P_+(\boldsymbol{\xi}')|$ and $M_2 = M_3 = \alpha/4$. Hence,

$$\eta = \frac{M_2}{a} + \frac{M_3}{b} = \frac{1}{2}.$$

It follows that $1/(1-\eta) = 2$,

$$\frac{M_1}{1-\eta} = 2K|P_+(\boldsymbol{\xi}) - P_+(\boldsymbol{\xi}')|,$$

and

$$a - \frac{M_2}{1-\eta} = \alpha - 2\,\frac{\alpha}{4} = \frac{\alpha}{2}.$$

The result follows by substituting the appropriate values into the conclusion of the lemma and noting that the second inequality follows from the first by setting $\boldsymbol{\xi}' = \mathbf{0}$. □

Corollary *Consider $\dot{\mathbf{x}} = A\mathbf{x} + \mathbf{g}(\mathbf{x})$ with $A \in \mathcal{H}$ and $\mathbf{g}$ in $\mathcal{L}(\psi)$, and let δ be positive a number satisfying*

$$\psi(2\delta) < \frac{\alpha}{4KL}.$$

Then $\omega(\boldsymbol{\xi}) = \mathbf{0}$ if and only if there exists τ such that $\mathbf{x}(\tau, \boldsymbol{\xi})$ is in $\mathcal{W}_\delta^+$.

Proof. Exercise.

Proposition 8.22 *The local stable manifold $\mathcal{W}_\delta^+$ is tangent to E^+ at $\mathbf{0}$ in the sense that*

$$\frac{|P_-(\boldsymbol{\xi})|}{|P_+(\boldsymbol{\xi})|} \to 0$$

as $\boldsymbol{\xi}$ goes to $\mathbf{0}$ in $\mathcal{W}_\delta^+$.

Proof. Let $\boldsymbol{\xi}$ be a point in $\mathcal{W}_\delta^+$. To estimate $|P_-(\boldsymbol{\xi})|$, first use *Theorem 8.10* with $t = 0$ to show that

$$P_-(\boldsymbol{\xi}) = P_-(\mathbf{x}(0, \boldsymbol{\xi})) = -\int_0^\infty e^{-As} P_-[\mathbf{g}(\mathbf{x}(s, \boldsymbol{\xi}))]ds.$$

Use the second inequality in *Theorem 8.21* to see that $|\mathbf{x}(s, \boldsymbol{\xi})| \leq 2K\,|P_+(\boldsymbol{\xi})|$ for $s \geq 0$. Hence,

$$\begin{aligned} |P_-(\boldsymbol{\xi})| &\leq \int_0^\infty K e^{-\alpha s} L\psi\left(|\mathbf{x}(s, \boldsymbol{\xi})|\right) |\mathbf{x}(s, \boldsymbol{\xi})|\, ds \\ &\leq KL \int_0^\infty e^{-\alpha s} \psi(2K|P_+(\boldsymbol{\xi})|) 2K|P_+(\boldsymbol{\xi})|\, ds \\ &\leq \frac{2K^2 L\psi(2KL|\boldsymbol{\xi}|)\,|P_+(\boldsymbol{\xi})|}{\alpha}. \end{aligned}$$

Therefore,

$$\frac{|P_-(\boldsymbol{\xi})|}{|P_+(\boldsymbol{\xi})|} \leq \frac{2K^2 L\psi(2KL|\boldsymbol{\xi}|)}{\alpha}$$

and the right side goes to 0 as $\boldsymbol{\xi}$ goes to $\mathbf{0}$ in $\mathcal{W}_\delta^+$. □

It is time to broaden the discussion to include the local N-stable manifolds $\mathcal{W}_\delta^+$. All the choices of constants and so forth have been symmetric for positive and negative time. Consequently, all the results in this section have parallels for $t \leq 0$ without changing the choice of δ and γ. We will begin using those results without stating them formally.

Theorem 8.23 *Consider $\dot{\mathbf{x}} = A\mathbf{x} + \mathbf{g}(\mathbf{x})$ with $A \in \mathcal{H}$ and $\mathbf{g}$ in $\mathcal{L}(\psi)$, and let δ and γ be positive numbers satisfying:*

$$\begin{aligned} \psi(4\delta) &< \frac{\alpha}{4KL} \\ \gamma &< \frac{\delta}{2K}. \end{aligned}$$

There exists $\eta > 0$ such that $|\boldsymbol{\xi}| < \eta$ and $\boldsymbol{\xi}$ is in neither $\mathcal{W}_\delta^+$ nor $\mathcal{W}_\delta^-$ implies there exist $\tau_1 < 0 < \tau_2$ such that $|\mathbf{x}(\tau_j, \boldsymbol{\xi})| > \delta$ for $j = 1, 2$. Furthermore,

$$\mathcal{W}_\delta^+ \cap \mathcal{W}_\delta^- = \{\mathbf{0}\}.$$

Proof. Choose $\eta > 0$ such that $|\boldsymbol{\xi}| < \eta$ implies that both $|P_+(\boldsymbol{\xi})| < \gamma$ and $|P_+(\boldsymbol{\xi})| < \gamma$. If $|\mathbf{x}(t, \boldsymbol{\xi})| \leq \delta$ for all $t \geq 0$ or for all $t \leq 0$, then $\boldsymbol{\xi}$ would be in $\mathcal{W}_\delta^+$ or $\mathcal{W}_\delta^-$, respectively. Since $\boldsymbol{\xi}$ is in neither $\mathcal{W}_\delta^+$ nor $\mathcal{W}_\delta^-$ by hypothesis, the required τ_1 and τ_2 must exist.

Suppose $\boldsymbol{\xi}$ is in $\mathcal{W}_\delta^+ \cap \mathcal{W}_\delta^-$. Then $\mathbf{x}(t, \boldsymbol{\xi})$ is defined and satisfies $|\mathbf{x}(t, \boldsymbol{\xi})| \leq \delta$ for all $t \in \mathbb{R}$. Furthermore, $\mathbf{x}(\tau, \boldsymbol{\xi})$ is in $\mathcal{W}_\delta^+ \cap \mathcal{W}_\delta^-$ for every τ because $\mathbf{x}(t, \mathbf{x}(\tau, \boldsymbol{\xi})) = \mathbf{x}(t + \tau, \boldsymbol{\xi})$.

Let ε be an arbitrary positive number such that $0 < \varepsilon < \delta$. The version of *Theorem 8.21* for local N-stable manifolds implies that there exists $\tau < 0$ such that $|P_+(\mathbf{x}(\tau, \boldsymbol{\xi}))| < \varepsilon/(2K)$. Since $\mathbf{x}(\tau, \boldsymbol{\xi})$ is in $\mathcal{W}_\delta^+$, *Theorem 8.21* can be applied to $\mathbf{x}(t, \mathbf{x}(\tau, \boldsymbol{\xi}))$ for $t \geq 0$ to obtain the following estimate:

$$|\mathbf{x}(t, \mathbf{x}(\tau, \boldsymbol{\xi}))| \leq 2Ke^{-\alpha t/2}|P_+(\mathbf{x}(\tau, \boldsymbol{\xi}))| < \varepsilon.$$

In particular, when $t = -\tau > 0$, it follows that $|\boldsymbol{\xi}| = |\mathbf{x}(-\tau, \mathbf{x}(\tau, \boldsymbol{\xi}))| < \varepsilon$. Since ε can be arbitrarily small, $|\boldsymbol{\xi}| = 0$ and $\boldsymbol{\xi} = \mathbf{0}$. $\square$

The preceding results can now be pulled together into a single theorem that gives a remarkably complete picture for a smooth vector field of the phase portrait of $\dot{\mathbf{x}} = \mathbf{f}(\mathbf{x})$ near a fixed point $\mathbf{p}$ with $\mathbf{f}'(\mathbf{p}) \in \mathcal{H}$. Moreover, this phase portrait has the same basic structure as the phase portrait of a linear hyperbolic system near the origin.

Theorem 8.24 *Let $\mathbf{f}(\mathbf{x})$ be a smooth vector field on $\mathbb{R}^d$ with a fixed point at $\mathbf{p}$. If $\mathbf{f}'(\mathbf{p})$ is in $\mathcal{H}$, then there exist local stable manifolds $\mathcal{W}^+(\mathbf{p})$ and $\mathcal{W}^-(\mathbf{p})$ with the following properties:*

(a) *There exists an open sets V_1 and V_2 containing $\mathbf{p}$ such that $\mathcal{W}^+(\mathbf{p}) \cap V_1$ is homeomorphic to an open disk of dimension $\iota(\mathbf{f}'(\mathbf{p}))$ and $\mathcal{W}^-(\mathbf{p}) \cap V_2$ is homeomorphic to an open disk of dimension $d - \iota(\mathbf{f}'(\mathbf{p}))$.*

(b) *$\mathcal{W}^+(\mathbf{p})$ is positively invariant and $\mathcal{W}^-(\mathbf{p})$ is negatively invariant.*

(c) *If $\boldsymbol{\xi}$ is in $\mathcal{W}^+(\mathbf{p})$ or $\mathcal{W}^-(\mathbf{p})$, then $|\mathbf{x}(t,\boldsymbol{\xi}) - \mathbf{p}|$ decays exponentially as t goes to infinity or minus infinity, respectively.*

(d) *$\omega(\boldsymbol{\xi}) = \mathbf{p}$ or $\alpha(\boldsymbol{\xi}) = \mathbf{p}$ if and only if there exists τ such that $\mathbf{x}(\tau,\boldsymbol{\xi})$ is in $\mathcal{W}^+(\mathbf{p})$ or $\mathcal{W}^-(\mathbf{p})$, respectively.*

(e) *There exist $\delta > \eta > 0$ such that $|\boldsymbol{\xi}| < \eta$ and $\boldsymbol{\xi} \notin \mathcal{W}^+(\mathbf{p})$ or $\boldsymbol{\xi} \notin \mathcal{W}^-(\mathbf{p})$ implies that $\mathbf{x}(t,\boldsymbol{\xi}) \notin \{\mathbf{x} : |\mathbf{x}| < \delta\}$ for some $t > 0$ or $t < 0$, respectively.*

(f) *$\mathcal{W}^+(\mathbf{p}) \cap \mathcal{W}^-(\mathbf{p}) = \{\mathbf{p}\}$.*

Proof. Clearly, $\mathbf{f}(\mathbf{x}+\mathbf{p})$ is a smooth vector field with a fixed point at $\mathbf{0}$. Set $A = \mathbf{f}'(\mathbf{p})$ and $\mathbf{g}(\mathbf{x}) = \mathbf{f}(\mathbf{x}+\mathbf{p}) - A\mathbf{x}$. Then, $\mathbf{g}(\mathbf{0}) = \mathbf{0}$ and $\mathbf{g}'(\mathbf{0}) = \mathrm{O}$. Moreover, $\mathbf{g}(\mathbf{x})$ is the nonlinear part of $\mathbf{f}(\mathbf{x}+\mathbf{p})$ at $\mathbf{0}$ and $\mathbf{g}(\mathbf{x})$ is in $\mathcal{L}(\psi)$ for some ψ by *Proposition 8.11.* Thus all the prior results in this section can be applied to $\dot{\mathbf{x}} = A\mathbf{x} + \mathbf{g}(\mathbf{x})$.

Choose δ and γ satisfying the hypothesis of *Theorem 8.20.* Then δ and γ also satisfy the hypotheses of *Proposition 8.19*, *Theorem 8.21*, and *Theorem 8.23.* Set $\mathcal{W}^+(\mathbf{p}) = \mathcal{W}_\delta^+ + \mathbf{p}$ and $\mathcal{W}^-(\mathbf{p}) = \mathcal{W}_\delta^- + \mathbf{p}$. The proof is completed by observing that statements are just translations of the known properties of $\mathcal{W}_\delta^+$ and $\mathcal{W}_\delta^-$ to $\mathcal{W}^+(\mathbf{p})$ and $\mathcal{W}^-(\mathbf{p})$. □

It is now possible to define *global stable manifolds* at a fixed point $\mathbf{p}$ of a smooth vector field $\mathbf{f}$ such that $\mathbf{f}'(\mathbf{p})$ is a hyperbolic matrix. This is done as follows:

$$\begin{aligned} \mathcal{S}^+(\mathbf{p}) &= \left\{\boldsymbol{\xi} : \lim_{t\to\infty} \mathbf{x}(t,\boldsymbol{\xi}) = \mathbf{p}\right\} \\ &= \{\boldsymbol{\xi} : \omega(\boldsymbol{\xi}) = \{\mathbf{p}\}\} \\ &= \left\{\boldsymbol{\xi} : \mathbf{x}(t,\boldsymbol{\xi}) \in \mathcal{W}^+(\mathbf{p}) \text{ for some } t > 0\right\} \end{aligned}$$

and

$$\begin{aligned} \mathcal{S}^-(\mathbf{p}) &= \left\{\boldsymbol{\xi} : \lim_{t\to-\infty} \mathbf{x}(t,\boldsymbol{\xi}) = \mathbf{p}\right\} \\ &= \{\boldsymbol{\xi} : \alpha(\boldsymbol{\xi}) = \{\mathbf{p}\}\} \\ &= \left\{\boldsymbol{\xi} : \mathbf{x}(t,\boldsymbol{\xi}) \in \mathcal{W}^-(\mathbf{p}) \text{ for some } t < 0\right\}. \end{aligned}$$

Note that they are invariant sets by definition.

Stable manifolds are generally important pieces of the phase portrait of a smooth vector field, but they can fit together in complicated ways because of the ways they can intersect. For example, the intersection of $\mathcal{S}^+(\mathbf{p})$ and $\mathcal{S}^-(\mathbf{p})$ can contain solutions other than the fixed point and that does not happen with local stable manifolds. A point $\boldsymbol{\xi}$ is called a *homoclinic point* if $\boldsymbol{\xi} \in \mathcal{S}^+(\mathbf{p}) \cap \mathcal{S}^-(\mathbf{p})$ and $\boldsymbol{\xi} \neq \mathbf{p}$. A point $\boldsymbol{\xi}$ is called a *heteroclinic point* if $\boldsymbol{\xi} \in \mathcal{S}^+(\mathbf{p}) \cap \mathcal{S}^-(\mathbf{q})$ for distinct hyperbolic fixed points $\mathbf{p}$ and $\mathbf{q}$.

Stable manifold theorems occur in other settings such as periodic orbits of smooth flows and fixed points of diffeomorphisms. A number of these theorems can be found in [20].

EXERCISES

1. As used in the proof of the lemma for *Theorem 8.21*, let $u(t)$ be a continuous non-negative real-valued function such that $\lim_{t\to\infty} u(t) = 0$ and set $v(t) = \sup\{u(s) : s \geq t\}$. Show that $v(t)$ is a continuous non increasing function.

2. Define the *limit inferior* of a real-valued function $u(t)$ defined on $[0, \infty)$ by
$$\underline{\lim}_{t\to\infty} u(t) = \sup_{\mu>0} \left\{ \inf_{t>\mu} u(t) \right\},$$
and recall that the limit superior was defined in the proof of *Theorem 8.21*. Show that
$$\underline{\lim}_{t\to\infty} u(t) \leq \overline{\lim}_{t\to\infty} u(t)$$
and that $\lim_{t\to\infty} u(t)$ exists if and only if equality holds in the above expression.

3. Consider the planar system
$$\begin{aligned} \dot{x} &= x^3 - x \\ \dot{y} &= -y. \end{aligned}$$
Find the global stable manifolds for the hyperbolic fixed points and sketch the phase portrait.

4. Write $\ddot{x} - (2\mu - x^2)\dot{x} + x = 0$ as a planar system. Show that $\mathbf{0}$ is a hyperbolic fixed point for all $\mu \neq 0$ and use *Theorem 5.7* to determine $\mathcal{S}^+(\mathbf{0})$ and $\mathcal{S}^-(\mathbf{0})$ when $\mu < 0$.

5. Find the hyperbolic fixed points for the undamped pendulum $\ddot{x} + \sin x = 0$, and identify any homoclinic or heteroclinic orbits.

6. The usual second-order differential equation of a damped pendulum is $\ddot{x} + c\dot{x} + \sin x = 0$ with $c < 2$. Write it as a system, show that the solutions are defined for all time, and completely determine its phase portrait.

7. Consider the planar Hamiltonian system with $H(p, q) = q^2 - p^2 - 2p^3/3$. Sketch the phase portrait, determine the stable manifolds at the hyperbolic fixed points, and identify any homoclinic or heteroclinic orbits.

Chapter 9

Bifurcations

Studying changes in the qualitative behavior of solutions of a differential equation as a parameter varies is generally called bifurcation theory. The word bifurcate means to divide into two parts or branches. The most common usage of bifurcation in differential equations refers to the phenomena of a fixed point or periodic point splitting into several fixed points or periodic points at a critical value of the parameter.

When a real parameter μ in the autonomous differential equation $\dot{\mathbf{x}} = \mathbf{f}(\mathbf{x}, \mu)$ is varied, the phase portrait will often remain unchanged over long μ-intervals punctuated by occasional dramatic changes at specific values for μ. The values of μ at which significant changes occur in the phase portrait are called *bifurcation values*. Of course, the obvious problem is to find these bifurcation points and understand the transitions that occurs there. But an equally important question is to determine when fixed points or periodic points persist without bifurcating.

The question of persistence of fixed points leads naturally to the implicit function theorem. Because of the importance of the implicit function theorem and the inverse function theorem in analysis in general and differential equations and dynamics in particular, the first section is devoted to these two theorems and several applications. The inverse function theorem will be used to prove the implicit function theorem, and the implicit function theorem will then play a substantive role in the remaining two sections of this chapter.

The second section examines the persistence of periodic points in an autonomous vector field. In other words, when must a periodic point continue to be present under small variations of the parameter? Here, the algebraic multiplicity of the eigenvalue one for the derivative of the general solution with respect to the initial position comes into play and draws on material in sections 5.4, Linear Systems with Periodic Coefficients, and 7.2, Differentiation in Initial Conditions.

The third section is devoted to the formidable task of proving a Hopf bifurcation theorem. This deep result actually gives conditions under which a fixed point will spawn a curve of periodic points at a bifurcation value. Fortunately, there is also an array of simple examples of the bifurcation of a fixed point into a periodic orbit and a fixed point that can be used to illustrate Hopf bifurcations.

9.1 The Implicit Function Theorem

The implicit function theorem is essential to understanding where bifurcations can occur and where they cannot occur. Since the absence of bifurcations of fixed points in a parameterized system is almost synonymous with the implicit function theorem, the section begins with a few simple scalar examples of bifurcations of fixed points to set the stage for the implicit function theorem.

Consider an autonomous scalar differential equation $\dot{x} = f(x, \mu)$ with μ a real parameter and $f(x, \mu)$ a smooth function. For convenience, assume $f(x, \mu)$ is defined on $\mathbb{R}^2$. The phase portrait of $\dot{x} = f(x, \mu)$ for fixed μ is determined by the fixed points and the sign of $f(x, \mu)$ between fixed points. (See Chapter 2, page 64.) The bifurcation values are the values of μ at which there is a change in the number of fixed points. Thus the problem is to understand how the solutions of $f(x, \mu) = 0$ depend on μ.

For example, let $f(x, \mu) = x^2 - \mu$. Then $x^2 - \mu = 0$ if and only if $\mu = x^2$. So there are no fixed points when $\mu < 0$ and two fixed points when $\mu > 0$. Thus as μ crosses the value zero there is a dramatic change from no fixed points to two fixed points. At the bifurcation point, $\mu = 0$ there is one fixed point. Notice that for $\mu > 0$ there are two curves of fixed points, $x = \sqrt{\mu}$ and $x = -\sqrt{\mu}$. This type of bifurcation is usually called a *saddle-node bifurcation.*

Figure 9.1 is a graphical representation of this bifurcation. The slightly darker curve $\mu = x^2$ shows the location of the fixed points in a vertical phase space as μ changes. The vertical lines with arrows show the typical phase portraits for μ less than zero, equal to zero, and greater than zero. Note that the fixed points along the curve $x = -\sqrt{\mu}$ for $\mu > 0$ are positively asymptotically stable and those along the curve $x = \sqrt{\mu}$ are negatively asymptotically stable. At $\mu = 0$, there is a single fixed point that has no stable behavior.

A figure such as *Figure 9.1* is called a *bifurcation diagram.* Frequently, the vertical axis represents points in $\mathbb{R}^d$ not just $\mathbb{R}$ and the bifurcation diagram is more of a schematic devise.

For a second example, let $f(x, \mu) = x^3 - \mu x = x(x^2 - \mu)$. Now the situation is more complicated because $x = 0$ is a fixed point for all values of μ and the curve $\mu = x^2$ is overlaid on it. Once again, $\mu = 0$ is the only bifurcation point, but now one fixed point bifurcates into 3 instead of the appearance of fixed points where there had been none. For $\mu > 0$, the functions $x \equiv 0$, and $x = \pm\sqrt{\mu}$ give the locations of the fixed points. Moreover, the phase portraits for $\mu > 0$ are all the same; $x = 0$ is positively asymptotically stable and both $x = \pm\sqrt{\mu}$ are both negatively asymptotically stable. The bifurcation diagram for $\dot{x} = f(x, \mu) = x^3 - \mu x$ appears in *Figure 9.2* and is usually called a *pitch fork bifurcation.* Notice that the single fixed point is negatively asymptotically stable for $\mu < 0$ and splits into two negatively asymptotically stable fixed points at $\mu = 0$ creating a positively asymptotically stable fixed point for $\mu > 0$.

Both of these examples came down to solving $f(x, \mu) = 0$ for x as a function

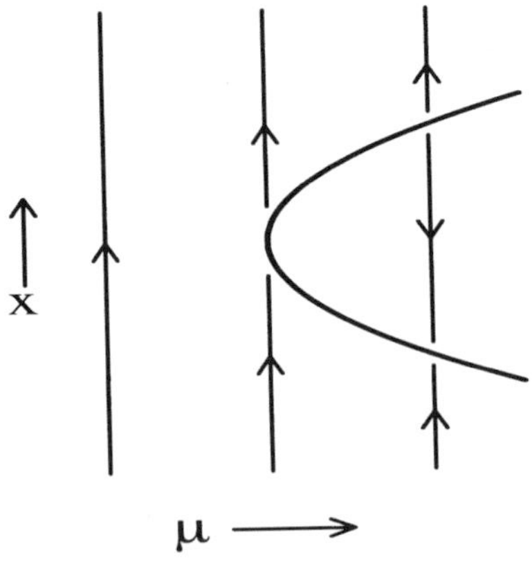

Figure 9.1: Bifurcation diagram for the saddle-node $\dot{x} = x^2 - \mu$.

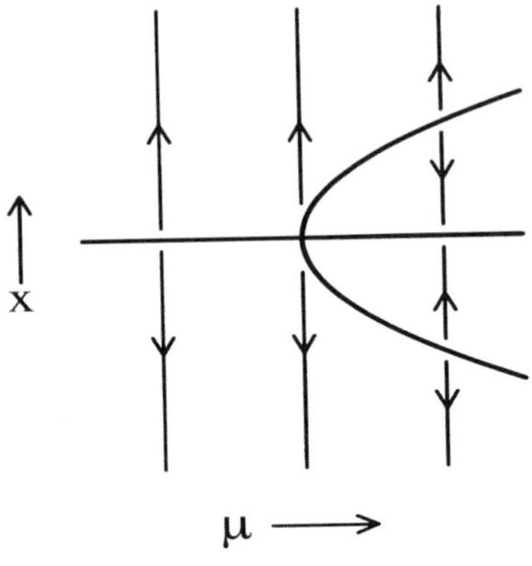

Figure 9.2: Bifurcation diagram for the pitch fork $\dot{x} = x^3 - \mu x$.

of μ. The fixed points persisted without dramatic changes when they were given by smooth functions of μ on an open interval and bifurcation points occurred otherwise. At non-bifurcation points, there were μ-intervals with zero, one, two, and three functions on them describing the locations of the fixed points of $\dot{x} = f(x, \mu)$. When it is not obvious, as it was in the above examples, how to solve $f(x, \mu) = 0$ for x as a function of μ, the key to proving that such functions exist is applying the implicit function theorem and that requires $\frac{\partial f}{\partial x} \neq 0$. Note that in both examples $\frac{\partial f}{\partial x} = 0$ precisely when $\mu = 0$ and the bifurcations occur.

Consequently, to establish the non-bifurcation or persistence of fixed points for the general smooth autonomous differential equation $\dot{\mathbf{x}} = \mathbf{f}(\mathbf{x}, \boldsymbol{\mu})$, the full power of the implicit function theorem must be applied to $\mathbf{f}(\mathbf{x}, \boldsymbol{\mu}) = \mathbf{0}$. Although the implicit function theorem for two or three variables is rather familiar fare, the general version is less familiar but more important. The best route to the implicit function theorem is through the inverse function theorem. The proof of the inverse function theorem will make good use of several ideas from the previous chapter.

Theorem 9.1 (Inverse Function Theorem) *Let* $\mathbf{f} : \Omega \to \mathbb{R}^d$ *be a smooth function on an open set* Ω *of* $\mathbb{R}^d$. *If* $\mathrm{D}\mathbf{f}(\mathbf{p})$ *is an invertible linear transformation, then there exist open sets* U *and* V *containing* $\mathbf{p}$ *and* $\mathbf{f}(\mathbf{p})$, *respectively, such that* $\mathbf{f}$ *restricted to* U *has a smooth inverse* $\mathbf{f}^{-1}$ *mapping* V *onto* U. *Furthermore,*

$$\mathrm{D}\mathbf{f}^{-1}(\mathbf{y}) = \left[\mathrm{D}\mathbf{f}(\mathbf{f}^{-1}(\mathbf{y}))\right]^{-1}$$

and

$$(\mathbf{f}^{-1})'(\mathbf{y}) = \left[\mathbf{f}'(\mathbf{f}^{-1}(\mathbf{y}))\right]^{-1}.$$

Proof. The proof begins with several reductions to simpler situations based on the chain rule (*Theorem 7.5* on page 217). Since the maps $\mathbf{x} \to \mathbf{x} + \mathbf{q}$ are smooth with smooth inverses on $\mathbb{R}^d$, it suffices to the prove the first part of the theorem with $\mathbf{p}$ replaced by $\mathbf{0}$ and the additional hypothesis that $\mathbf{f}(\mathbf{0}) = \mathbf{0}$. This modification of the function has no affect on the derivative of $\mathbf{f}$. Let $T = \mathrm{D}\mathbf{f}(\mathbf{0})$. By the chain rule the derivative, of $T^{-1} \circ \mathbf{f}$ at $\mathbf{0}$ is the linear map $\mathrm{I}(\mathbf{x}) = \mathbf{x}$. Since $T^{-1}(\mathbf{f}(\mathbf{0})) = \mathbf{0}$, there is no loss in generality by adding the assumption that $\mathrm{D}\mathbf{f}(\mathbf{0}) = \mathrm{I}$ to the assumption that $\mathbf{f}(\mathbf{0}) = \mathbf{0}$.

As a result of these reductions, $\mathbf{f}$ can be written as $\mathbf{f}(\mathbf{x}) = \mathbf{x} + \mathbf{g}(\mathbf{x})$ with $\mathbf{g} \in \mathcal{L}(\psi)$ by *Proposition 8.11*. For a given $\mathbf{y}$, the equation $\mathbf{f}(\mathbf{x}) = \mathbf{y}$ can be written as the fixed-point equation $\mathbf{y} - \mathbf{g}(\mathbf{x}) = \mathbf{x}$, and thus solving $\mathbf{f}(\mathbf{x}) = \mathbf{y}$ for $\mathbf{x}$ is the same as finding a fixed point of $\mathbf{F}(\mathbf{x}, \mathbf{y}) = \mathbf{y} - \mathbf{g}(\mathbf{x})$.

Let $X_\delta = \{\mathbf{x} \in \mathbb{R}^d : |\mathbf{x}| \le \delta\}$ and $Y_\gamma = \{\mathbf{y} \in \mathbb{R}^d : |\mathbf{y}| \le \gamma\}$, and set $\mathbf{F}(\mathbf{x}, \mathbf{y}) = \mathbf{y} - \mathbf{g}(\mathbf{x})$. Choose δ and γ such that $\psi(\delta) < 1/2$ and $\gamma < \delta/2$. Then for $(\mathbf{x}, \mathbf{y}) \in X_\delta \times Y_\gamma$,

$$|\mathbf{F}(\mathbf{x}, \mathbf{y})| = |\mathbf{y} - \mathbf{g}(\mathbf{x})| \le |\gamma| + \psi(\delta)\delta \le \delta.$$

Similarly,

$$|\mathbf{F}(\mathbf{x}_1, \mathbf{y}) - \mathbf{F}(\mathbf{x}_2, \mathbf{y})| = |\mathbf{g}(\mathbf{x}_2) - \mathbf{g}(\mathbf{x}_1)| \le \psi(\delta)|\mathbf{x}_2 - \mathbf{x}_1| \le \frac{1}{2}|\mathbf{x}_2 - \mathbf{x}_1|$$

for $\mathbf{x}_1, \mathbf{x}_2$ in X_δ and $\mathbf{y} \in Y_\gamma$. Obviously, $\mathbf{F}(\mathbf{x}, \mathbf{y})$ is a continuous function. Now, *Theorem 8.17* applies and there exists a continuous function $\mathbf{h} : Y_\gamma \to X_\delta$ such that

$$\mathbf{F}(\mathbf{h}(\mathbf{x}), \mathbf{y}) = \mathbf{h}(\mathbf{y}),$$

which can be rewritten and simplified as follows:

$$\begin{aligned} \mathbf{y} - \mathbf{g}(\mathbf{h}(\mathbf{y})) &= \mathbf{h}(\mathbf{y}) \\ \mathbf{h}(\mathbf{y}) + \mathbf{g}(\mathbf{h}(\mathbf{y})) &= \mathbf{y} \\ \mathbf{f}(\mathbf{h}(\mathbf{y})) &= \mathbf{y}. \end{aligned}$$

Set $V = \{\mathbf{y} \in \mathbb{R}^d : |\mathbf{y}| < \gamma\}$ and $U = \mathbf{f}^{-1}(V) \cap \{\mathbf{x} \in \mathbb{R}^d : |\mathbf{x}| < \delta\}$, which is open because $\mathbf{f}$ is continuous. For $\mathbf{x} \in U$, the equation $\mathbf{f}(\mathbf{x}) = \mathbf{x} + \mathbf{g}(\mathbf{x})$ is the same as $\mathbf{F}(\mathbf{x}, \mathbf{f}(\mathbf{x})) = \mathbf{x}$. Then by the uniqueness of the fixed point

$$\mathbf{h}(\mathbf{f}(\mathbf{x})) = \mathbf{x}.$$

Therefore, $\mathbf{h} : V \to U$ is at least a continuous inverse.

Because $\mathbf{f}$ is smooth, the function $\mathbf{f}' : \Omega \to \mathcal{M}_d(\mathbb{R})$ is continuous, and δ can be chosen so that $\mathbf{f}'(\mathbf{x})$ is invertible for all $\mathbf{x} \in X_\delta$. It follows that $\mathrm{D}\mathbf{f}(\mathbf{x})$ is an invertible linear transformation for all $\mathbf{x} \in U$.

To prove the differentiability of $\mathbf{h}$ at $\mathbf{y}_0 \in V$, we start with the differentiability of $\mathbf{f}$ at $\mathbf{x}_0 = \mathbf{h}(\mathbf{y}_0) \in U$. Let $T = \mathrm{Df}(\mathbf{x}_0)$ and write

$$\mathbf{f}(\mathbf{x}) - \mathbf{f}(\mathbf{x}_0) - T(\mathbf{x} - \mathbf{x}_0) = \mathbf{e}(\mathbf{x}),$$

where $\mathbf{e}(\mathbf{x})$ is an error term satisfying

$$\lim_{\mathbf{x} \to \mathbf{x}_0} \frac{|\mathbf{e}(\mathbf{x})|}{|\mathbf{x} - \mathbf{x}_0|} = 0$$

because $\mathbf{f}$ is differentiable at $\mathbf{x}_0$. Apply T^{-1} to the above to get

$$T^{-1}\big(\mathbf{f}(\mathbf{x}) - \mathbf{f}(\mathbf{x}_0)\big) - (\mathbf{x} - \mathbf{x}_0) = T^{-1}\big(\mathbf{e}(\mathbf{x})\big)$$

and then substitute $\mathbf{x} = \mathbf{h}(\mathbf{y})$ and $\mathbf{x}_0 = \mathbf{h}(\mathbf{y}_o)$ to get

$$T^{-1}\big(\mathbf{y} - \mathbf{y}_0\big) - (\mathbf{h}(\mathbf{y}) - \mathbf{h}(\mathbf{y}_0)) = T^{-1}\big(\mathbf{e}(\mathbf{h}(\mathbf{y}))\big).$$

It suffices to show that

$$\lim_{\mathbf{y} \to \mathbf{y}_0} \frac{\big|T^{-1}\big(\mathbf{e}(\mathbf{h}(\mathbf{y}))\big)\big|}{|\mathbf{y} - \mathbf{y}_0|} = 0,$$

which reduces to showing

$$\lim_{\mathbf{y} \to \mathbf{y}_0} \frac{|\mathbf{e}(\mathbf{h}(\mathbf{y}))|}{|\mathbf{y} - \mathbf{y}_0|} = 0$$

because $|T^{-1}(\mathbf{e}(\mathbf{h}(\mathbf{y})))| \le |T|\,|\mathbf{e}(\mathbf{h}(\mathbf{y}))|$.

To establish the above limit write

$$\frac{|\mathbf{e}(\mathbf{h}(\mathbf{y}))|}{|\mathbf{y} - \mathbf{y}_0|} = \frac{|\mathbf{e}(\mathbf{h}(\mathbf{y}))|}{|\mathbf{h}(\mathbf{y}) - \mathbf{h}(\mathbf{y}_0)|} \frac{|\mathbf{h}(\mathbf{y}) - \mathbf{h}(\mathbf{y}_0)|}{|\mathbf{y} - \mathbf{y}_0|}$$

noting that $\mathbf{h}(\mathbf{y}) \neq \mathbf{h}(\mathbf{y}_0)$ when $\mathbf{y} \neq \mathbf{y}_0$. Since $\mathbf{h}$ is continuous, $\mathbf{h}(\mathbf{y})$ goes to $\mathbf{h}(\mathbf{y}_0)$ as $\mathbf{y}$ goes to $\mathbf{y}_0$ and

$$\lim_{\mathbf{y} \to \mathbf{y}_0} \frac{|\mathbf{e}(\mathbf{h}(\mathbf{y}))|}{|\mathbf{h}(\mathbf{y}) - \mathbf{h}(\mathbf{y}_0)|} = 0.$$

To complete the proof that $\mathbf{h}$ is differentiable at $\mathbf{y}_0$, it suffices to show that

$$\frac{|\mathbf{h}(\mathbf{y}) - \mathbf{h}(\mathbf{y}_0)|}{|\mathbf{y} - \mathbf{y}_0|} \leq 2.$$

The above equation is an immediate consequence of *(8.6)* on page 281 because

$$|F(\mathbf{h}(\mathbf{y}_0), \mathbf{y}) - F(\mathbf{h}(\mathbf{y}_0), \mathbf{y}_0)| = |\mathbf{y} - \mathbf{g}(\mathbf{h}(\mathbf{y}_0)) - (\mathbf{y}_0 - \mathbf{g}(\mathbf{h}(\mathbf{y}_0))| = |\mathbf{y} - \mathbf{y}_0|.$$

Thus far it has been shown that if $\mathrm{D}\mathbf{f}(\mathbf{p})$ is an invertible linear transformation, then there exist open sets U and V containing $\mathbf{p}$ and $\mathbf{f}(\mathbf{p})$, respectively, such that $\mathbf{f}$ restricted to U has a differentiable inverse $\mathbf{f}^{-1} = \mathbf{h}$ mapping V onto U. It remains to prove that $\mathbf{f}^{-1}$ is smooth and that the formulas for the derivative of the inverse hold. This is best done back in the general setting of the theorem.

The two formulas are immediate consequences of the chain rule applied to $\mathbf{f} \circ \mathbf{f}^{-1} = \mathrm{I}$. The right-hand side of the second formula

$$(\mathbf{f}^{-1})'(\mathbf{y}) = \left[\mathbf{f}'(\mathbf{f}^{-1}(\mathbf{y}))\right]^{-1}$$

is continuous because $\mathbf{f}'$ is a continuous function by hypothesis and taking the inverse of a matrix is a continuous function. Therefore, $(\mathbf{f}^{-1})'(\mathbf{y})$ is continuous and $\mathbf{f}^{-1}$ is also a smooth function. □

When the inverse function theorem applies, it is worth noting that $\mathbf{f}$ restricted to the open set U is a homeomorphism of U onto the open set V because $\mathbf{f}$ is continuous on U and its inverse is continuous on V. It follows that $\mathbf{f}$ maps open sets contained in U to opens sets in V. (Recall from Section 8.3 that because U is open in $\mathbb{R}^d$, the open sets of U thought of as a metric space are just open sets of $\mathbb{R}^d$ contained in U. And homeomorphisms map open sets to open sets.)

Before moving on to the implicit function theorem, there is a simple application of the inverse function theorem that deserves mention. A fixed-point $\mathbf{p}$ of an autonomous differential equation $\dot{\mathbf{x}} = \mathbf{f}(\mathbf{x})$ is said to be *isolated fixed point* if there exists an open set U containing $\mathbf{p}$ such that $\mathbf{p}$ is the only fixed point in U.

Corollary *Let* $\mathbf{p}$ *be fixed point of a smooth vector field* $\mathbf{f}(\mathbf{x})$ *on an open set* Ω. *If* $\mathbf{f}'(\mathbf{p})$ *is invertible, then* $\mathbf{p}$ *is an isolated fixed point of* $\dot{\mathbf{x}} = \mathbf{f}(\mathbf{x})$.

Proof. The matrix $\mathbf{f}'(\mathbf{p})$ is invertible if and only if the linear transformation $\mathrm{D}\mathbf{f}(\mathbf{p})$ is an invertible. So the inverse function theorem applies. Since $\mathbf{f}$ must be one-to-one on U, there are no other points $\mathbf{x}$ in U such that $\mathbf{f}(\mathbf{x}) = \mathbf{0}$. □

If $\mathbf{G} : \mathbb{R}^m \to \mathbb{R}^m$ is a smooth function such that $\operatorname{Det} \mathbf{G}'(\mathbf{u})$ is always invertible, then the inverse function theorem implies that for a given $\mathbf{v} \in \mathbb{R}^m$ the set $\{\mathbf{u} : \mathbf{G}(\mathbf{u}) = \mathbf{v}\}$ is a set of isolated points in $\mathbb{R}^m$. When $\mathbf{G} : \mathbb{R}^m \to \mathbb{R}^n$ is a smooth function with $m \neq n$, things are not so simple. The implicit function theorem is an important tool for studying the structure of $\mathbf{G}^{-1}(\mathbf{v}) = \{\mathbf{u} : \mathbf{G}(\mathbf{u}) = \mathbf{v}\}$ for a smooth function $\mathbf{G} : \mathbb{R}^m \to \mathbb{R}^n$ when $m > n$. In this situation, it will be convenient to let $m = n + k$, to think of $\mathbb{R}^m$ as $\mathbb{R}^n \times \mathbb{R}^k$, and to write $(\mathbf{x}, \mathbf{y})$ for a generic point in $\Omega \subset \mathbb{R}^n \times \mathbb{R}^k$.

The notation for the matrix of first partial derivatives also requires some modification. The $n \times n$ matrix of first partial derivatives with respect to $x_1, \ldots, x_n$ will be denoted as before by

$$\frac{\partial \mathbf{G}}{\partial \mathbf{x}} = \left[\frac{\partial G_i}{\partial x_j}\right] = \begin{bmatrix} \frac{\partial G_1}{\partial x_1} & \cdots & \frac{\partial G_1}{\partial x_n} \\ \vdots & \ddots & \vdots \\ \frac{\partial G_n}{\partial x_1} & \cdots & \frac{\partial G_n}{\partial x_n}, \end{bmatrix}$$

but now it is only the first n columns of the entire matrix $n \times m$ matrix of first partial derivatives. Of course, the missing piece is $\frac{\partial \mathbf{G}}{\partial \mathbf{y}}$. Using this notation, the theorem can be stated as follows:

Theorem 9.2 (Implicit Function Theorem) *Let $\mathbf{G} : \Omega \to \mathbb{R}^n$ be a smooth function on Ω an open subset of $\mathbb{R}^{n+k}$, and let $(\mathbf{p}, \mathbf{q})$ be a point in Ω with $\mathbf{G}(\mathbf{p}, \mathbf{q}) = \mathbf{v}$. If the determinant of the $n \times n$ matrix*

$$\frac{\partial \mathbf{G}(\mathbf{p}, \mathbf{q})}{\partial \mathbf{x}} = \begin{bmatrix} \frac{\partial G_1(\mathbf{p},\mathbf{q})}{\partial x_1} & \cdots & \frac{\partial G_1(\mathbf{p},\mathbf{q})}{\partial x_n} \\ \vdots & \ddots & \vdots \\ \frac{\partial G_n(\mathbf{p},\mathbf{q})}{\partial x_1} & \cdots & \frac{\partial G_n(\mathbf{p},\mathbf{q})}{\partial x_n}, \end{bmatrix}$$

is nonzero, then there exist open sets $U_1 \subset \mathbb{R}^n$ and $U_2 \subset \mathbb{R}^k$ containing $\mathbf{p}$ and $\mathbf{q}$, respectively, and a smooth function $\mathbf{g} : U_2 \to U_1$ such that $\mathbf{G}(\mathbf{g}(\mathbf{y}), \mathbf{y}) = \mathbf{v}$. Furthermore, $\mathbf{G}(\mathbf{x}, \mathbf{y}) = \mathbf{v}$ for $(\mathbf{x}, \mathbf{y}) \in U_1 \times U_2$ if and only if $\mathbf{x} = \mathbf{g}(\mathbf{y})$.

Proof. Extend $\mathbf{G}$ to a smooth mapping from Ω into $\mathbb{R}^{n+k}$ by setting $\mathbf{F}(\mathbf{x}, \mathbf{y}) = (\mathbf{G}(\mathbf{x}, \mathbf{y}), \mathbf{y})$. Then the matrix

$$\mathbf{F}'(\mathbf{p}, \mathbf{q}) = \begin{bmatrix} \frac{\partial \mathbf{G}(\mathbf{p},\mathbf{q})}{\partial \mathbf{x}} & \frac{\partial \mathbf{G}(\mathbf{p},\mathbf{q})}{\partial \mathbf{y}} \\ \mathrm{O} & \mathrm{I} \end{bmatrix}$$

is invertible because its determinant is obviously nonzero, and the inverse function theorem applies to $\mathbf{F}$ at $(\mathbf{p}, \mathbf{q})$. So there exist open sets U and V and a smooth inverse of $\mathbf{F}$ restricted to U. By making U (and V) a little smaller, it can be assumed that $U = U_1 \times U_2$, where U_1 and U_2 are open subsets of $\mathbb{R}^n$ and $\mathbb{R}^k$, respectively. (See the comments immediately following the proof of the inverse function theorem.)

It is easy to see that $\mathbf{F}^{-1}$ on V has the form $\mathbf{F}^{-1}(\mathbf{x}, \mathbf{y}) = (\mathbf{h}(\mathbf{x}, \mathbf{y}), \mathbf{y})$ for a smooth function $\mathbf{h} : V \to U_1$. [Write $\mathbf{F}^{-1}$ in coordinate function form as $(h_1, \ldots, h_{n+k})$. Then, $\mathbf{F}(\mathbf{F}^{-1}(\mathbf{x}, \mathbf{y})) = (\mathbf{x}, \mathbf{y})$ implies that $h_{n+k}(\mathbf{x}, \mathbf{y}) = y_k$ for all $(\mathbf{x},\mathbf{y}) \in V$.] Let $\pi(\mathbf{x},\mathbf{y}) = \mathbf{x}$ be the projection of $\mathbb{R}^{n+k}$ onto $\mathbb{R}^n$ and observe that $\pi(\mathbf{F}(\mathbf{x},\mathbf{y})) = \mathbf{G}(\mathbf{x},\mathbf{y})$. It follows that

$$\begin{aligned} \mathbf{G}(\mathbf{h}(\mathbf{x},\mathbf{y}),\mathbf{y}) &= \pi(\mathbf{F}(\mathbf{h}(\mathbf{x},\mathbf{y}),\mathbf{y})) \\ &= \pi(\mathbf{F}(\mathbf{F}^{-1}(\mathbf{x},\mathbf{y}))) \\ &= \pi(\mathbf{x},\mathbf{y}) \\ &= \mathbf{x}. \end{aligned}$$

In particular, $\mathbf{G}(\mathbf{h}(\mathbf{v},\mathbf{y})) = \mathbf{v}$ and $\mathbf{g}(\mathbf{y}) = \mathbf{h}(\mathbf{v},\mathbf{y})$ is the required smooth function such that $\mathbf{G}(\mathbf{g}(\mathbf{y}),\mathbf{y})) = \mathbf{v}$. The form of $\mathbf{F}$ guarantees that $\{(\mathbf{v},\mathbf{y}) : \mathbf{y} \in U_2\} \subset V$ and $\mathbf{g}$ is defined on U_2.

The preceding shows that $(\mathbf{g}(\mathbf{y}),\mathbf{y})$ is in $U_1 \times U_2$ and $\mathbf{G}(\mathbf{g}(\mathbf{y}),\mathbf{y}) = \mathbf{v}$. For the converse, suppose that $\mathbf{G}(\mathbf{x},\mathbf{y}) = \mathbf{v}$ and $(\mathbf{x},\mathbf{y})$ is in $U_1 \times U_2$. Then, $\mathbf{F}(\mathbf{x},\mathbf{y}) = (\mathbf{v},\mathbf{y})$ and by the above $\mathbf{F}(\mathbf{g}(\mathbf{y}),\mathbf{y})) = (\mathbf{v},\mathbf{y})$. Since $\mathbf{F}$ is one-to-one on $U_1 \times U_2$, it follows that $\mathbf{x} = \mathbf{g}(y)$. $\square$

When Ω is an open set in $\mathbb{R}^d$, a *smooth parameterized vector field* on Ω will mean a smooth function $\mathbf{f} : \Omega \times W \to \mathbb{R}^d$, where W is an open set in $\mathbb{R}^k$ for some $k > 0$ and it will be written as $\mathbf{f}(\mathbf{x},\boldsymbol{\mu})$ with $\mathbf{x} \in \Omega$ and $\boldsymbol{\mu} \in W$. When $\mathbf{f}$ is a smooth parameterized vector field, the differential equation of interest will always be $\dot{\mathbf{x}} = \mathbf{f}(\mathbf{x},\boldsymbol{\mu})$. In this context, it will be convenient to think of $\boldsymbol{\mu}$ more as a constant than a variable and to use the notation

$$\mathbf{f}'_{\mathbf{x}}(\mathbf{x},\boldsymbol{\mu}) = \frac{\partial \mathbf{f}(\mathbf{x},\boldsymbol{\mu})}{\partial \mathbf{x}} = \begin{bmatrix} \frac{\partial f_1(\mathbf{x},\boldsymbol{\mu})}{\partial x_1} & \cdots & \frac{\partial f_1(\mathbf{x},\boldsymbol{\mu})}{\partial x_m} \\ \vdots & \ddots & \vdots \\ \frac{\partial f_n(\mathbf{x},\boldsymbol{\mu})}{\partial x_1} & \cdots & \frac{\partial f_n(\mathbf{x},\boldsymbol{\mu})}{\partial x_m}, \end{bmatrix}.$$

Corollary (Persistence of Fixed Points) *Let $\mathbf{f}$ be a smooth parameterized vector field on Ω. If $\mathbf{f}(\mathbf{p},\boldsymbol{\nu}) = \mathbf{0}$ and the determinant of $\mathbf{f}'_{\mathbf{x}}(\mathbf{p},\boldsymbol{\nu})$ is nonzero, then there exists a smooth function $\mathbf{g} : U_2 \to \mathbb{R}^d$ defined on an open set U_2 containing $\boldsymbol{\nu}$ such that $\mathbf{g}(\boldsymbol{\nu}) = \mathbf{p}$ and $\mathbf{g}(\boldsymbol{\mu})$ is an isolated fixed point of $\dot{\mathbf{x}} = \mathbf{f}(\mathbf{x},\boldsymbol{\mu})$ for each $\boldsymbol{\mu} \in U_2$.*

This corollary is little more than a restatement of the implicit function theorem, but it makes the point that a fixed point will not bifurcate in a smoothly parameterized vector field when the derivative of the vector field is invertible. The following variant of it is also of interest:

Corollary (Persistence of Hyperbolic Fixed Points) *Let $\mathbf{f}$ be a smooth parameterized vector field on Ω. If $\mathbf{f}(\mathbf{p},\boldsymbol{\nu}) = \mathbf{0}$ and $\mathbf{f}'_{\mathbf{x}}(\mathbf{p},\boldsymbol{\nu})$ is in $\mathcal{H}_j$, then there exists a smooth function $\mathbf{g} : U_2 \to \mathbb{R}^d$ defined on an open set U_2 containing $\boldsymbol{\nu}$ such that $\mathbf{g}(\boldsymbol{\nu}) = \mathbf{p}$ and $\mathbf{g}(\boldsymbol{\mu})$ is an isolated fixed point of $\dot{\mathbf{x}} = \mathbf{f}(\mathbf{x},\boldsymbol{\mu})$ and $\mathbf{f}'_{\mathbf{x}}(\mathbf{g}(\boldsymbol{\mu}),\boldsymbol{\mu})$ is in $\mathcal{H}_j$ for each $\boldsymbol{\mu} \in U_2$.*

Proof. Since $\mathcal{H}_j$ is an open set in $\mathcal{M}_d(\mathbb{R})$ and $\mathbf{f}'_{\mathbf{x}}$ is a continuous function with values in $\mathcal{M}_d(\mathbb{R})$, the open set U_2 can be chosen small enough to satisfy this condition because $\mathbf{f}'_{\mathbf{x}}(\mathbf{p}, \boldsymbol{\nu})$ is in $\mathcal{H}_j$. □

As an example, consider the system in $\mathbb{R}^2$ with μ a real parameter

$$\begin{aligned} \dot{x} &= x^2 - y^2 + \mu \\ \dot{y} &= xy. \end{aligned}$$

Then the critical matrix of first partial derivatives is

$$\begin{bmatrix} 2x & -2y \\ y & x \end{bmatrix},$$

which is invertible except when $x = y = 0$. In fact, there is a pair of saddle node bifurcations at the origin given by $y^2 = \mu$ and $x = 0$, and by $-x^2 = \mu$ and $y = 0$.

Thus far the examples have only involved a single real parameter, but the next example contains a pair of real parameters. Consider

$$\dot{x} = \alpha + \beta x - x^3$$

with α and β real parameters. In other words, $d = 1$, $k = 2$, and $\boldsymbol{\mu} = (\alpha, \beta)$. Here, $f'_x = \beta - 3x^2$ and to find the bifurcation points we must solve the system of equations

$$\begin{aligned} \alpha + \beta x - x^3 &= 0 \\ \beta - 3x^2 &= 0. \end{aligned}$$

Replacing β in the first equation with $3x^2$ leads to $\alpha = -2x^3$. It follows that at any bifurcation point, (α, β) must satisfy

$$\frac{\alpha^2}{4} = \frac{\beta^3}{27}$$

or

$$\beta = 3\left(\frac{\alpha}{2}\right)^{2/3}$$

as shown in *Figure 9.3* and then

$$\mathbf{x} = \sqrt[3]{-\alpha/2}$$

is the fixed point at which the bifurcation occurs.

It is an exercise to show that above the curve $\alpha^2/4 = \beta^3/27$ there are always three fixed points and below it there is only one. For values of α and β on the curve, there are two fixed points except at the cusp where there is only one fixed point. Moving from above the curve to below it two fix points coalesce into one on the curve and disappear below it except at the cusp that occurs at the origin. Here all three fixed points coalesce into one that continues below the curve.

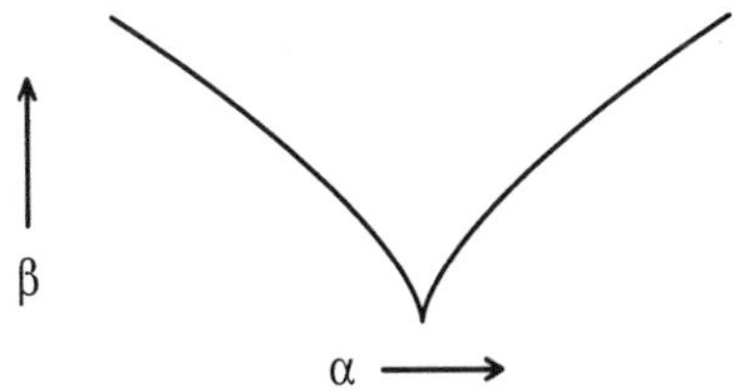

Figure 9.3: Graph of the bifurcation values for $\dot{x} = \alpha + \beta x + x^3$

To continue the discussion of the implicit function theorem and its applications to dynamics, let $\mathbf{G} : \Omega \to \mathbb{R}^n$ be a smooth function on an open set Ω contained in $\mathbb{R}^m$. Assume that $m > n$. If the derivative $\mathrm{D}\mathbf{G}(\mathbf{u}) : \mathbb{R}^{\mathrm{m}} \to \mathbb{R}^{\mathrm{n}}$ is onto, then there is a set of n columns of the matrix $\mathbf{G}'(\mathbf{u})$ that form a basis for $\mathbb{R}^n$. This is equivalent to requiring that the rank of $\mathbf{G}'(\mathbf{u})$ be n. The variables can be permuted so that the first n columns are linearly independent and labeled $\mathbf{x} = (x_1, \ldots, x_n)$, then the matrix

$$\begin{bmatrix} \frac{\partial g_1(\mathbf{u})}{\partial x_1} & \cdots & \frac{\partial g_1(\mathbf{u})}{\partial x_n} \\ \vdots & \ddots & \vdots \\ \frac{\partial g_n(\mathbf{u})}{\partial x_1} & \cdots & \frac{\partial g_n(\nu)}{\partial x_n}, \end{bmatrix}$$

is invertible and the implicit function theorem can be applied.

A value $\mathbf{v}$ of $\mathbf{G}$ is called a *regular value* if $\mathrm{D}\mathbf{G}(\mathbf{u})$ maps $\mathbb{R}^m$ onto $\mathbb{R}^n$ for all $\mathbf{u}$ such that $\mathbf{G}(\mathbf{u}) = \mathbf{v}$. Or equivalently the rank of $\mathbf{G}'(\mathbf{u})$ is n for all $\mathbf{u}$ such that $\mathbf{G}(\mathbf{u}) = \mathbf{v}$. When $\mathbf{v}$ is a regular value, the implicit function theorem applies at every $\mathbf{u}$ such that $\mathbf{G}(\mathbf{u}) = \mathbf{v}$.

There are usually lots of regular values when the range of $\mathbf{G}$ includes an open set. For example, when $\mathbf{G}$ has continuous partial derivatives of all orders, then Sard's theorem shows that almost all values are regular values. (For an elementary proof of Sard's theorem see [26].)

A metric space M is a *manifold of dimension* k if every point of M is contained in an open set that is homeomorphic to an open ball $\{\mathbf{x} \in \mathbb{R}^k : |\mathbf{x}| < r\}$ in $\mathbb{R}^k$. This definition is independent of the norm used. (For an introductory treatment of manifolds see [24].) It is, for example, easy to see using projections that $\{\mathbf{x} \in \mathbb{R}^m : \|\mathbf{x}\| = 1\}$ is a manifold of dimension $m-1$. The implicit function theorem is a wonderful source of manifolds.

Proposition 9.3 *Let* $\mathbf{G} : \Omega \to \mathbb{R}^n$ *be a smooth function on an open set* Ω *contained in* $\mathbb{R}^{n+k}$. *If* $\mathbf{v}$ *is a regular value of* $\mathbf{G}$, *then* $M = \{\mathbf{u} \in \Omega : \mathbf{G}(\mathbf{u}) = \mathbf{v}\}$ *is a manifold of dimension* k

Proof. Let $\mathbf{u}$ be a point in M. Since $\mathbf{v}$ is a regular value, the implicit function theorem can be applied to $\mathbf{u}$. Then, after rearranging the coordinates and writing $\mathbf{u} = (\mathbf{p}, \mathbf{q})$ as in *Theorem 9.2*, there exist open sets $U_1 \subset \mathbb{R}^n$ and

$U_2 \subset \mathbb{R}^k$ containing $\mathbf{p}$ and $\mathbf{q}$, respectively, and a smooth function $\mathbf{g} : U_2 \to U_1$ such that $\mathbf{G}(\mathbf{g}(\mathbf{y}), \mathbf{y}) = \mathbf{v}$ and such that $\mathbf{G}(\mathbf{x}, \mathbf{y}) = \mathbf{v}$ for $(\mathbf{x}, \mathbf{y}) \in U_1 \times U_2$ if and only if $\mathbf{x} = \mathbf{g}(\mathbf{y})$. The map $\mathbf{y} \to (\mathbf{g}(\mathbf{y}), \mathbf{y})$ is the required homeomorphism of U_2 onto the open set $(U_1 \times U_2) \cap M$ of M and its inverse is $(\mathbf{x}, \mathbf{y}) \to \mathbf{y}$. □

Even if $\mathbf{v}$ is not a regular value, the above argument still works at any points in M at which $\mathrm{D}\mathbf{g}(\mathbf{u})$ maps $\mathbb{R}^{n+k}$ onto $\mathbb{R}^n$. Consequently, M can appear to be a manifold at some points but not all points. For example, a figure eight fails to be a one-dimensional manifold only at the crossing point in the middle.

Actually, $M = \{\mathbf{u} \in \Omega : \mathbf{g}(\mathbf{u}) = \mathbf{v}\}$ is a differentiable manifold when $\mathbf{v}$ is a regular value. Roughly speaking, a real-valued function $h : M \to \mathbb{R}$ is said to be differentiable if $h(\mathbf{g}(\mathbf{y}), \mathbf{y})$ is differentiable for all the functions $\mathbf{g}$ given by the implicit function theorem. In this sense, the implicit function theorem is the origin of the study of differentiable manifolds and subsequently the dynamics of diffeomorphisms and differentiable flows on them. The books [6], [18], and [20] contain chapters devoted to differentiable dynamics on manifolds. Our interest in manifolds is, however, more modest.

Let $\mathbf{f}$ be a smooth vector field on Ω. A manifold M contained in Ω is an *invariant manifold* for $\dot{\mathbf{x}} = \mathbf{f}(\mathbf{x})$ if $\mathbf{x}(t, \boldsymbol{\xi})$ is in M for all t in its range, when $\boldsymbol{\xi}$ is in M. We have already encountered invariant manifolds, but their manifold structure was suppressed. The level sets of integrals are invariant and the previous proposition shows that they are also manifolds for regular values. Hence, we have the following:

Remark *Let $\mathbf{f}$ be a smooth vector field on Ω and let E be a smooth integral for $\dot{\mathbf{x}} = \mathbf{f}(\mathbf{x})$. If $\nabla \mathbf{f}(\mathbf{x}) \neq \mathbf{0}$ on $M = \{\mathbf{x} \in \Omega : \mathbf{g}(\mathbf{x}) = c\}$, then M is a $d-1$ dimensional invariant manifold.*

It is now possible to generalize the concept of an integral for a smooth vector field $\mathbf{f}$ on Ω an open subset in $\mathbb{R}^d$ to functions $\mathbf{E} : \Omega \to \mathbb{R}^k$ with $d > k$.

Proposition 9.4 *Let $\mathbf{f}$ be a smooth vector field on Ω and let $\mathbf{E} : \Omega \to \mathbb{R}^k$ with $d > k$ be a smooth function. If $\mathbf{f}(\mathbf{x})$ is in the null space of $\mathrm{D}\,\mathbf{E}(\mathbf{x})$ for $\mathbf{x} \in \Omega$ or equivalently $\mathbf{E}'(\mathbf{x})\mathbf{f}(\mathbf{x}) = \mathbf{0}$, then the level sets $M = \{\mathbf{x} \in \Omega : \mathbf{E}(\mathbf{x}) = \boldsymbol{\nu}\}$ are invariant. If, in addition, $\boldsymbol{\nu}$ is a regular value of $\mathbf{E}$, then M is a $d-k$ dimensional invariant manifold.*

Proof. Apply the chain rule to $\mathbf{E}(\mathbf{x}(t, \boldsymbol{\xi}))$. □

It is worth revisiting an earlier example in light of this proposition. Let A be the 4×4 matrix given by equation *(6.7)* on page 204. Recall that the functions $E_1(x, y, u, v) = x^2 + y^2$ and $E_2(x, y, u, v) = u^2 + v^2$ are both integrals of $\dot{\mathbf{x}} = A\mathbf{x}$ with $\mathbf{x} = (x, y, u, v)$. Set $\mathbf{E}(\mathbf{x}) = (E_1(\mathbf{x}), E_2(\mathbf{x}))$ and note that

$$\mathbf{E}'(\mathbf{x}) = \begin{bmatrix} 2x & 2y & 0 & 0 \\ 0 & 0 & 2u & 2v \end{bmatrix}.$$

An easy calculation shows that $\mathrm{D}\,\mathbf{E}(\mathbf{x})(\mathrm{A}\mathbf{x}) = \mathbf{E}'(\mathbf{x})\mathrm{A}\mathbf{x} = \mathbf{0}$ and that (r_1, r_2) is a regular value of $\mathbf{E}$ if and only if both r_1 and r_2 are positive. Thus by *Proposition 9.4* all the sets of the form $M = \{\mathbf{x} \in \Omega : \mathbf{E}(\mathbf{x}) = (r_1, r_2)\}$ with $r_1 > 0$ and $r_2 > 0$ are two-dimensional invariant manifolds for the flow determined by $\dot{\mathbf{x}} = A\mathbf{x}$. They are, of course, the invariant tori studied in Section 6.3.

EXERCISES

1. Find the bifurcation values and construct a bifurcation diagram for $\dot{x} = \mu x + x^2$. This example is a *transcritical bifurcation.*

2. Find the bifurcation values and construct a bifurcation diagram for $\dot{x} = \mu + x - x^3$.

3. Analyze the bifurcations of the general quadratic $\dot{x} = \alpha + \beta x + x^2$.

4. Find the bifurcation values and construct a bifurcation diagram for $\dot{x} = (x^2 - \mu)(x^2 - 2\mu)$.

5. Show that for $\dot{x} = \alpha + \beta x - x^3$ above the curve

$$\frac{\alpha^2}{4} = \frac{\beta^3}{27}$$

   shown in *Figure 9.3* there are always three fixed points and below it there is only one. Sketch the phase portraits in each case.

6. Find the bifurcation values for

$$\begin{aligned} \dot{x} &= x^2 - y^2 + \mu \\ \dot{y} &= xy \end{aligned}$$

   and determine the phase portraits near the persistent fixed points.

7. Consider the Lorenz system

$$\begin{aligned} \dot{x} &= -\sigma x + \sigma y \\ \dot{y} &= \rho x - y - xz \\ \dot{z} &= -\beta z + xy \end{aligned}$$

   for $\beta > 0$, $\rho > 0$, and $\sigma > 0$. The origin is a fixed point for all values of the parameters. For what values of the parameters is the origin a hyperbolic fixed point and what are the dimensions of the local stable manifolds? Are there any bifurcation values and additional fixed points that can be expressed as functions of the three parameters?

8. Show that the inverse function applies to the map $A \to e^A$ from $\mathcal{M}_d(\mathbb{R})$ to $GL_d(\mathbb{R})$ at $A = \mathrm{O}$, the matrix of all zeros.

9. Use the implicit function theorem to show that the special linear groups is a manifold of dimension $d^2 - 1$.

10. Let $\mathbf{f}(\mathbf{x})$ be a smooth vector field on Ω, an open subset of $\mathbb{R}^d$, and let $\mathbf{p}$ be a moving point (not a fixed point) for $\dot{\mathbf{x}} = \mathbf{f}(\mathbf{x})$. Using the notation of Section 6.1, let L be a local section at $\mathbf{p}$ of length 2η. Show that the first return time $\rho(\mathbf{v})$ (See *Theorem 6.4* on page 191 and the discussion preceding it.) is a smooth function on the open set

$$W = \{\mathbf{v} : \psi(\mathbf{v}) \in L^o,\ \rho(\mathbf{v}) \neq \infty, \text{ and } \mathbf{x}(\rho(\mathbf{v}), \psi(\mathbf{v})) \in L^o\}$$

by applying the implicit function theorem to $\mathbf{F}(t, \mathbf{v}) = (\mathbf{x}(t, \psi(\mathbf{v}) - \mathbf{p})) \cdot \mathbf{f}(\mathbf{p})$. Then show that the Poincaré return map $R(\mathbf{v})$ is also smooth on W.

9.2 Persistence of Periodic Points

Periodic points can also bifurcate as a parameter changes. In some cases, periodic points are part of a dramatic bifurcation of an entire system. A simple linear system provides a good example. Consider the parameterized 2×2 linear system $\dot{\mathbf{x}} = A(\mu)\mathbf{x}$ with

$$A(\mu) = \begin{bmatrix} \mu & 1 \\ -1 & \mu \end{bmatrix}.$$

The eigenvalues of $A(\mu)$ are $\mu \pm i$. So the origin is positively asymptotically stable for $\mu < 0$ and negatively asymptotically stable for $\mu > 0$. At $\mu = 0$, every point except the origin is periodic. Thus $\mu = 0$ is a rather dramatic bifurcation point at which for an instant every point is periodic and the system reverses its asymptotic behavior. What is happening is that $A(\mu)$ is crossing the boundary between $\mathcal{H}_2$ and $\mathcal{H}_0$ at $\mu = 0$. From this perspective, it is not surprising that the periodic orbits do not persist. Of course, similar bifurcations can occur in linear systems of all dimensions.

To prove a substantive result about the persistence of periodic points we need some preliminary facts about periodic points of smooth systems. Recall that for a smooth vector field the global solution $\mathbf{x}(t, \tau, \boldsymbol{\xi})$ of $\dot{\mathbf{x}} = \mathbf{f}(\mathbf{x})$ is a smooth function on its domain by *Theorem 7.9*. It follows that $\mathbf{x}(t, \boldsymbol{\xi}) = \mathbf{x}(t, 0, \boldsymbol{\xi})$ is also a smooth function. For periodic points, differentiability in initial conditions has a special implication.

Proposition 9.5 *Let* $\mathbf{f}$ *be a smooth vector field on* Ω*, an open subset of* $\mathbb{R}^d$*. If* $\mathbf{p}$ *is a periodic point for* $\dot{\mathbf{x}} = \mathbf{f}(\mathbf{x})$ *of period* T*, then one is an eigenvalue of the matrix*

$$\left[\frac{\partial \mathbf{x}(T, \mathbf{p})}{\partial \boldsymbol{\xi}}\right]$$

and $\mathbf{f}(\mathbf{p})$ *is an eigenvector for the eigenvalue one.*

Proof. Starting with the familiar equation

$$\mathbf{x}(s, \mathbf{x}(t, \mathbf{p})) = \mathbf{x}(s + t, \mathbf{p}),$$

set $s = T$ and differentiate

$$\mathbf{x}(T, \mathbf{x}(t, \mathbf{p})) = \mathbf{x}(T + t, \mathbf{p}),$$

with respect to t. By the chain rule, this produces

$$\left[\frac{\partial \mathbf{x}(T, \mathbf{x}(t, \mathbf{p}))}{\partial \boldsymbol{\xi}}\right] \dot{\mathbf{x}}(t, \mathbf{p}) = \dot{\mathbf{x}}(T + t, \mathbf{p})$$

or

$$\left[\frac{\partial \mathbf{x}(T, \mathbf{x}(t, \mathbf{p}))}{\partial \boldsymbol{\xi}}\right] \mathbf{f}(\mathbf{x}(t, \mathbf{p})) = \mathbf{f}(\mathbf{x}(T + t, \mathbf{p})) = \mathbf{f}(\mathbf{x}(t, \mathbf{p})).$$

Set $t = 0$ to obtain

$$\left[\frac{\partial \mathbf{x}(T, \mathbf{p})}{\partial \boldsymbol{\xi}}\right] \mathbf{f}(\mathbf{p}) = \mathbf{f}(\mathbf{p})$$

and complete the proof. □

In part (b) of *Theorem 7.10*, it was shown that the $d \times d$ matrix of first partial derivatives with respect to the initial position

$$\left[\frac{\partial \mathbf{x}(t, \boldsymbol{\xi})}{\partial \boldsymbol{\xi}}\right]$$

is the principal matrix solution at 0 of the linear matrix differential equation

$$\dot{W} = \left[\frac{\partial \mathbf{f}\big(\mathbf{x}(t, \boldsymbol{\xi})\big)}{\partial \mathbf{x}}\right] W,$$

which can be written more simply in the present context as

$$\dot{W} = \left[\mathbf{f}'\big(\mathbf{x}(t, \boldsymbol{\xi})\big)\right] W.$$

Again there are special consequences for periodic points.

Proposition 9.6 *Let* $\mathbf{f}$ *be a smooth vector field on* Ω*, an open subset of* $\mathbb{R}^d$*. If* $\mathbf{p}$ *is a periodic point for* $\dot{\mathbf{x}} = \mathbf{f}(\mathbf{x})$ *of period* T*, then there exists* $B \in \mathcal{M}_d(\mathbb{C})$ *and a continuous periodic function* $P : \mathbb{R} \to GL_d(\mathbb{R})$ *such that*

$$\left[\frac{\partial \mathbf{x}(t, \mathbf{p})}{\partial \boldsymbol{\xi}}\right] = P(t)e^{Bt}$$

and one is a characteristic multiplier of

$$\dot{W} = [\mathbf{f}'(\mathbf{x}(t, \mathbf{p}))] W.$$

Proof. Since $\mathbf{p}$ is a periodic point, the matrix $\mathbf{f}'(\mathbf{x}(t,\mathbf{p}))$ is periodic and Floquet's theorem (*Theorem 5.20* on page 177) applies. Thus $\partial\mathbf{x}(t,\mathbf{p})/\partial\boldsymbol{\xi}$ has the required form. Since it is a principal matrix at 0, it is also a monodromy matrix. (Characteristic values and eigenvalues are thus synonymous in this context.) Hence, one is a characteristic multiplier because one is an eigenvalue by the previous proposition. □

For a parameterized vector field, the global solution $\mathbf{x}(t,\boldsymbol{\xi},\boldsymbol{\mu})$ is also a smooth function on its domain by *Theorem 7.11.* The next result about the persistence of periodic solutions is an analogue of the corollary about the persistence of fixed points on page 305.

Theorem 9.7 (Poincaré) *Let $\mathbf{f}(\mathbf{x},\boldsymbol{\mu})$ be a smooth parameterized vector field on Ω, an open subset of $\mathbb{R}^d$, and suppose that $\mathbf{p}$ is a periodic point of period T for $\dot{\mathbf{x}} = \mathbf{f}(\mathbf{x},\boldsymbol{\nu})$. If the eigenvalue one of the matrix*

$$\left[\frac{\partial\,\mathbf{x}(T,\mathbf{p},\boldsymbol{\nu})}{\partial\,\boldsymbol{\xi}}\right]$$

has algebraic multiplicity one, then there exists $r > 0$ and differentiable functions

$$\mathbf{h} : \{\boldsymbol{\mu} \in \mathbb{R}^k : \|\boldsymbol{\mu} - \boldsymbol{\nu}\| < r\} \to \Omega$$

and

$$\theta : \{\boldsymbol{\mu} \in \mathbb{R}^k : \|\boldsymbol{\mu} - \boldsymbol{\nu}\| < r\} \to \mathbb{R}$$

such that $\mathbf{h}(\boldsymbol{\mu})$ is a periodic point for $\dot{\mathbf{x}} = \mathbf{f}(\mathbf{x},\boldsymbol{\mu})$ of period $\theta(\boldsymbol{\mu})$.

Proof. By translation of coordinates, it can be assumed that $\mathbf{p} = \mathbf{0}$. A linear change of variables of the form $\mathbf{x} = B\mathbf{y}$ with $B \in GL_d(\mathbb{R})$ can also be made without loss of generality. Under such a change in variables, $\dot{\mathbf{x}} = \mathbf{f}(\mathbf{x},\boldsymbol{\mu})$ becomes $\dot{\mathbf{y}} = B^{-1}\mathbf{f}(B\mathbf{y},\boldsymbol{\mu})$ by *Proposition 2.7* on page 70. Moreover, $B\varphi(t)$ is a solution of $\dot{\mathbf{x}} = \mathbf{f}(\mathbf{x},\boldsymbol{\mu})$ if and only if $\varphi(t)$ is a solution of $\dot{\mathbf{y}} = B^{-1}\mathbf{f}(B\mathbf{y},\boldsymbol{\mu})$. Since $B^{-1}\mathbf{f}(B\mathbf{0},\boldsymbol{\nu}) = B^{-1}\mathbf{f}(\mathbf{0},\boldsymbol{\nu})$, the matrix B can be chosen so that $B^{-1}\mathbf{f}(\mathbf{0},\boldsymbol{\nu}) = (0,\ldots,0,1) = \mathbf{e}_d$. This effectively makes the plane perpendicular to $\mathbf{e}_d$ a local cross section near $\mathbf{0}$.

As a result of the previous paragraph, it suffices to prove the theorem under the additional hypothesis that $\mathbf{p} = \mathbf{0}$ and $\mathbf{f}(\mathbf{0},\boldsymbol{\nu}) = \mathbf{e}_d$. Let $\mathbf{w}$ denote a variable in $\mathbb{R}^{d-1}$ so that $(\mathbf{w},0,\boldsymbol{\mu})$ is in $\mathbb{R}^{d+k}$, and set

$$\mathbf{F}(\mathbf{w},s,\boldsymbol{\mu}) = \mathbf{x}(T+s,(\mathbf{w},0),\boldsymbol{\mu}) - (\mathbf{w},0).$$

[The order of the variables in $\mathbf{F}(\mathbf{w},s,\boldsymbol{\mu})$ is designed to help make the matrix calculations easier to follow.] Then $\mathbf{F}$ is a well-defined smooth function on an open set V of the form

$$V = \{\mathbf{w} \in \mathbb{R}^{d-1} : \|\mathbf{w}\| < \beta\} \times (-\alpha,\alpha) \times \{\boldsymbol{\mu} \in \mathbb{R}^k : \|\boldsymbol{\mu}\| < \gamma\}$$

in $\mathbb{R}^{d+k}$ with values in $\mathbb{R}^d$.

The key observation about $\mathbf{F} : V \to \mathbb{R}^d$ is that $\mathbf{F}(\mathbf{w}, s, \boldsymbol{\mu}) = \mathbf{0}$ implies that $(\mathbf{w}, 0)$ is a periodic point of period $T + s$. Since $\mathbf{F}(\mathbf{0}, 0, \boldsymbol{\nu}) = \mathbf{0}$, the natural next step is to apply the implicit function theorem, but this requires further careful analysis.

To invoke the implicit function theorem it is necessary to show that the matrix

$$\begin{bmatrix} \frac{\partial F_1}{\partial w_1} & \cdots & \frac{\partial F_1}{\partial w_{d-1}} & \frac{\partial F_1}{\partial s} \\ \vdots & \vdots & & \vdots \\ \frac{\partial F_d}{\partial w_1} & \cdots & \frac{\partial F_d}{\partial w_{d-1}} & \frac{\partial F_d}{\partial s} \end{bmatrix}$$

has a nonzero determinant at $(\mathbf{0}, 0, \boldsymbol{\nu})$. The last column of this matrix is just

$$\dot{\mathbf{x}}(T + s, (\mathbf{w}, 0), \boldsymbol{\mu}) = \mathbf{f}(\mathbf{x}(T + s, (\mathbf{w}, 0), \boldsymbol{\mu}))$$

and at $(\mathbf{0}, 0, \boldsymbol{\nu})$ it simplifies to

$$\dot{\mathbf{x}}(T, \mathbf{0}, \boldsymbol{\nu}) = \mathbf{f}(\mathbf{x}(T, \mathbf{0}, \boldsymbol{\nu})) = \mathbf{f}(\mathbf{0}, \boldsymbol{\nu}) = \mathbf{e}_d.$$

Hence, it suffices to show that the matrix

$$\begin{bmatrix} \frac{\partial F_1}{\partial w_1} & \cdots & \frac{\partial F_1}{\partial w_{d-1}} \\ \vdots & & \vdots \\ \frac{\partial F_{d-1}}{\partial w_1} & \cdots & \frac{\partial F_{d-1}}{\partial w_{d-1}} \end{bmatrix}$$

has a nonzero determinant at $(\mathbf{0}, 0, \boldsymbol{\nu})$.

Next, observe that

$$\begin{aligned} \frac{\partial F_j}{\partial w_k} &= \frac{\partial x_j(T + s, (\mathbf{w}, 0), \boldsymbol{\mu})}{\partial \xi_j} - 1 & \text{when } j = k \\ \frac{\partial F_j}{\partial w_k} &= \frac{\partial x_j(T + s, (\mathbf{w}, 0), \boldsymbol{\mu})}{\partial \xi_k} & \text{when } j \neq k. \end{aligned}$$

Therefore, to apply the implicit function theorem it suffices to show that the matrix

$$\begin{bmatrix} \frac{\partial x_1}{\partial \xi_1} & \frac{\partial x_1}{\partial \xi_2} & \cdots & \frac{\partial x_1}{\partial \xi_{d-1}} \\ \vdots & \vdots & & \vdots \\ \frac{\partial x_{d-1}}{\partial \xi_1} & \frac{\partial x_{d-1}}{\partial \xi_2} & \cdots & \frac{\partial x_{d-1}}{\partial \xi_{d-1}} \end{bmatrix} - \begin{bmatrix} 1 & 0 & \cdots & 0 \\ \vdots & \vdots & & \vdots \\ 0 & 0 & \cdots & 1 \end{bmatrix}$$

has non zero determinant at $(\mathbf{0}, 0, \boldsymbol{\nu})$. In other words, it must be shown that one is not an eigenvalue of

$$\begin{bmatrix} \frac{\partial x_1}{\partial \xi_1} & \frac{\partial x_1}{\partial \xi_2} & \cdots & \frac{\partial x_1}{\partial \xi_{d-1}} \\ \vdots & \vdots & & \vdots \\ \frac{\partial x_{d-1}}{\partial \xi_1} & \frac{\partial x_{d-1}}{\partial \xi_2} & \cdots & \frac{\partial x_{d-1}}{\partial \xi_{d-1}} \end{bmatrix}$$

at $(\mathbf{0}, 0, \boldsymbol{\nu})$.

It follows from *Proposition 9.5* that

$$\left[\frac{\partial \mathbf{x}(T,\mathbf{0},\boldsymbol{\nu})}{\partial \xi}\right]\mathbf{f}(\mathbf{0},\boldsymbol{\nu}) = \mathbf{f}(\mathbf{0},\boldsymbol{\nu}).$$

Since $\mathbf{f}(\mathbf{0},\boldsymbol{\nu}) = \mathbf{e}_d$, the vector $\mathbf{e}_d$ is an eigenvector for the eigenvalue one and the last column of the matrix is

$$\begin{pmatrix} 0 \\ \vdots \\ 0 \\ 1 \end{pmatrix}.$$

Shifting to the complex numbers briefly, the hypothesis that the algebraic multiplicity of the eigenvalue one is one implies that the generalized eigenspace for the eigenvalue one is simply all complex scalar multiples of $\mathbf{e}_d$ or $\mathbb{C}\mathbf{e}_d$. If one is an eigenvalue for the first $d-1$ rows and columns of the above matrix and $\mathbf{w} \in \mathbb{C}^d$ is an eigenvector, then a straightforward calculation shows that $(\mathbf{w},0)$ is a generalized eigenvector of order at most two for the original $d \times d$ matrix and in the generalized eigenspace for the eigenvalue one. This contradicts the hypothesis of the theorem. Therefore, one cannot be an eigenvalue for the first $d-1$ rows and columns of the above matrix, and the implicit function theorem can be applied at $(0,\mathbf{0},\boldsymbol{\nu})$.

Specifically, there exists an open set U of $\mathbb{R}^k$ containing $\boldsymbol{\nu}$ and a smooth function $\mathbf{g}: U \to \mathbb{R}^d$ such that $\mathbf{g}(\boldsymbol{\nu}) = \mathbf{0}$ and $\mathbf{F}(g_1(\boldsymbol{\mu}),(g_2(\boldsymbol{\mu}),\ldots,g_d(\boldsymbol{\mu})),\boldsymbol{\mu}) = \mathbf{0}$. It follows that $\mathbf{h}(\boldsymbol{\mu}) = (g_2(\boldsymbol{\mu}),\ldots,g_{d-1}(\boldsymbol{\mu}),0)$ and $\theta(\boldsymbol{\mu}) = T + g_d(\boldsymbol{\mu})$ are the required smooth functions. □

Poincaré 's theorem can also be applied to perturbations. Suppose $\mathbf{p}$ is a periodic point of period T for the smooth vector field $\mathbf{f}$ on an open subset Ω of $\mathbb{R}^d$. By *Proposition 9.5*, one is an eigenvalue of

$$\left[\frac{\partial \mathbf{x}(T,\mathbf{p})}{\partial \xi}\right]$$

Suppose its algebraic multiplicity is one and let $\mathbf{g}(\mathbf{x},\boldsymbol{\mu})$ be a smooth parameterized vector field on Ω such that $\mathbf{g}(\mathbf{x},\mathbf{0}) \equiv \mathbf{0}$. Then there exists $r > 0$ and differentiable functions

$$\mathbf{h} : \{\boldsymbol{\mu} \in \mathbb{R}^k : \|\boldsymbol{\mu}\| < r\} \to \Omega$$

and

$$\theta : \{\boldsymbol{\mu} \in \mathbb{R}^k : \|\boldsymbol{\mu}\| < r\} \to \mathbb{R}$$

such that $\mathbf{h}(\boldsymbol{\mu})$ is a periodic point for $\dot{\mathbf{x}} = \mathbf{f}(\mathbf{x}) + \mathbf{g}(\mathbf{x},\boldsymbol{\mu})$ of period $\theta(\boldsymbol{\mu})$. In other words, when the algebraic multiplicity of the eigenvalue one is one, smooth perturbations cannot immediately destroy a periodic point.

Corollary *Let $\mathbf{f}(\mathbf{x}, \boldsymbol{\mu})$ be a smooth parameterized vector field on Ω, an open subset of $\mathbb{R}^2$, and suppose that $\mathbf{p}$ is a periodic point of period T for $\dot{\mathbf{x}} = \mathbf{f}(\mathbf{x}, \boldsymbol{\nu})$. If*

$$\int_0^T \operatorname{Tr} \mathbf{f}'_{\mathbf{x}}(\mathbf{x}(s, \mathbf{p}, \boldsymbol{\nu}))ds \neq 0,$$

then there exists $r > 0$ and differentiable functions

$$\mathbf{h} : \{\boldsymbol{\mu} \in \mathbb{R}^k : \|\boldsymbol{\mu} - \boldsymbol{\nu}\| < r\} \to \Omega$$

and

$$\theta : \{\boldsymbol{\mu} \in \mathbb{R}^k : \|\boldsymbol{\mu} - \boldsymbol{\nu}\| < r\} \to \mathbb{R}$$

such that $\mathbf{h}(\boldsymbol{\mu})$ is a periodic point for $\dot{\mathbf{x}} = \mathbf{f}(\mathbf{x}, \boldsymbol{\mu})$ of period $\theta(\boldsymbol{\mu})$.

Proof. Recall from *Proposition 5.23* on page 179 that

$$\exp\left(\int_0^T \operatorname{Tr} \mathbf{f}'_{\mathbf{x}}(\mathbf{x}(s, \mathbf{p}, \boldsymbol{\nu}))ds\right)$$

is the product of the characteristic multipliers. It follows that there must be a characteristic multiplier not equal to one. Since the dimension of the space is two, the algebraic multiplicity of the eigenvalue or characteristic multiplier one must be one. □

As a final example, consider the familiar system

$$\begin{aligned} \dot{x} &= -y + x(\mu - x^2 - y^2) \\ \dot{y} &= x + y(\mu - x^2 - y^2) \end{aligned}$$

and recall that $(\sqrt{\mu}\cos t, \sqrt{\mu}\sin t)$ is a periodic solution for $\mu > 0$. An easy calculation shows that the Jacobian matrix of the vector field with respect to x and y is

$$\begin{bmatrix} \mu - 3x^2 - y^2 & -1 - 2xy \\ 1 - 2xy & \mu - x^2 - 3y^2 \end{bmatrix}$$

and its trace is $2\mu - 4(x^2 + y^2)$. It follows for $\mu > 0$ that

$$\int_0^{2\pi} \mathbf{f}'_{\mathbf{x}}(\sqrt{\mu}\cos s, \sqrt{\mu}\sin s)\, ds = \int_0^{2\pi} -2\mu\, ds = -4\mu\pi \neq 0.$$

Hence, the algebraic multiplicity of the eigenvalue one is one and all these periodic points are persistent. For example, if $(g_1(x, y), g_2(x, y))$ is any smooth vector field on $\mathbb{R}^2$, then for $\mu > 0$ and $|\nu|$ small the system

$$\begin{aligned} \dot{x} &= -y + x(\mu - x^2 - y^2) + \nu g_1(x, y) \\ \dot{y} &= x + y(\mu - x^2 - y^2) + \nu g_2(x, y) \end{aligned}$$

will have a periodic point even when it is not readily calculated.

EXERCISES

1. Consider the planar system

$$\begin{aligned}\dot{x} &= -y + x\mu(1 - x^2 - y^2) \\ \dot{y} &= x + y\mu(1 - x^2 - y^2)\end{aligned}$$

and note that $\mathbf{p} = (1, 0)$ is a periodic point of period 2π. Show that one is an eigenvalue of algebraic multiplicity one for

$$\left[\frac{\partial \mathbf{x}(T, \mathbf{p}, \boldsymbol{\nu})}{\partial \boldsymbol{\xi}}\right]$$

if and only if $\mu \neq 0$. When $\mu \neq 0$, what is the other eigenvalue?

2. Rewrite the planar system

$$\begin{aligned}\dot{x} &= -y + x[\mu - (1 - x^2 - y^2)^2](1 - x^2 - y^2) \\ \dot{y} &= x + y[\mu - (1 - x^2 - y^2)^2](1 - x^2 - y^2)\end{aligned}$$

in polar coordinates. Show that there is a pitchfork bifurcation of a periodic orbit, and sketch the typical phase portraits according to the values of μ.

3. Describe the phase portraits and bifurcations of the parameterized planar system

$$\begin{aligned}\dot{x} &= -y + (\mu - \sin r^2)x \\ \dot{y} &= x + (\mu - \sin r^2)y,\end{aligned}$$

where as usual $r^2 = x^2 + y^2$.

9.3 Hopf Bifurcations

The example at the end of the previous section merits further discussion. Thus far the system

$$\begin{aligned}\dot{x} &= -y + x(\mu - x^2 - y^2) \\ \dot{y} &= x + y(\mu - x^2 - y^2)\end{aligned}$$

was examined for $\mu > 0$. It has a single periodic orbit for $\mu > 0$ that could not bifurcate if it wanted to by *Theorem 9.7.* Moreover, the origin is the only fixed point and it is negatively asymptotically stable.

It is clear from the polar form,

$$\begin{aligned}\dot{r} &= r(\mu - r^2) \\ \dot{\theta} &= 1,\end{aligned}$$

of this system that for $\mu < 0$ there are no periodic points and exactly one fixed point. In fact, not only is the origin a positively asymptotically stable hyperbolic fixed point, its positive stable manifold is the entire plane.

What is happening at $\mu = 0$? First the asymptotic stability is reversing and the eigenvalues are pure imaginary. Second, and more importantly, periodic points suddenly appear as μ crosses zero from negative-to-positive values. Thus the fixed point splits into a fixed point and a periodic point instead of splitting into several fixed points.

This phenomenon of a fixed point bifurcating into a curve of fixed points and a curve of periodic points is generally called a *Hopf bifurcation* and is often predictable. The goal of this section is to prove a Hopf bifurcation theorem, that is, a theorem that gives conditions under which periodic points will appear at the bifurcation of a fixed point. Once again the implicit function theorem will play a key role, but getting to the point where it can be applied requires considerable preliminary analysis. For simplicity, parameterized vector fields in this section will be limited to a single real parameter μ on an open interval $-c < \mu < c$.

A pair of pure imaginary eigenvalues of algebraic multiplicity one is one of the key hypotheses in predicting Hopf bifurcations. The algebraic multiplicity of one allows us to find a curve of complex eigenvalues depending on the parameter μ, but this requires differentiation with respect to a complex variable. So we begin with a short discussion of complex differentiation and complex polynomials.

Let $g : \mathbb{C} \to \mathbb{C}$ be a complex valued function of a complex variable. The function g is differentiable at ζ if

$$\lim_{z \to \zeta} \frac{g(z) - g(\zeta)}{z - \zeta} \tag{9.1}$$

exists and is denoted by $g'(\zeta)$ when it does. The proofs of most elementary results about differentiation of a real-valued function of a real variable carry over to a complex variable.

In particular, if $g(z)$ is differentiable at ζ then it is continuous at ζ, and the usual formulas for the derivative of a sum or product of two functions are valid for functions of a complex variable. It follows that a complex polynomial,

$$p(z) = \sum_{k=0}^{n} a_k z^k$$

is differentiable at every z and

$$p'(z) = \sum_{k=1}^{n} k a_k z^{k-1}.$$

The derivative of a polynomial can be used to test the multiplicity of a root.

Remark *Let ζ be a root of $p(z) = \sum_{k=0}^{n} a_k z^k$. Then ζ is a root of multiplicity one if and only if $p'(\zeta) \neq 0$.*

Proof. Exercise.

Using $z = x + iy$, the function $g(z)$ can be written in the form

$$g(z) = u(x, y) + iv(x, y)$$

where $u(x, y)$ and $v(x, y)$ are real-valued functions of two real variables. Letting $z = \zeta + h$ and $z = \zeta + hi$ with $h \in \mathbb{R}$ in *(9.1)*, produces two different formulas for $g'(\zeta)$. Specifically, letting $\zeta = \alpha + i\beta$

$$\begin{aligned} \lim_{h \to 0} \frac{g(\zeta + h) - g(\zeta)}{h} &= \frac{\partial u(\alpha, \beta)}{\partial x} + i \frac{\partial v(\alpha, \beta)}{\partial x} \\ \lim_{h \to 0} \frac{g(\zeta + hi) - g(\zeta)}{hi} &= -i \frac{\partial u(\alpha, \beta)}{\partial y} + \frac{\partial v(\alpha, \beta)}{\partial y}. \end{aligned}$$

Hence,

$$\begin{aligned} g'(\zeta) &= \frac{\partial u(\alpha, \beta)}{\partial x} + i \frac{\partial v(\alpha, \beta)}{\partial x} \\ &= \frac{\partial v(\alpha, \beta)}{\partial y} - i \frac{\partial u(\alpha, \beta)}{\partial y}. \end{aligned} \tag{9.2}$$

Since both of these expressions also equal $g'(\zeta)$, their real and complex parts must be equal. Consequently, if $g(z)$ is differentiable at every point in $\mathbb{C}$, then

$$\begin{aligned} \frac{\partial u}{\partial x} &\equiv \frac{\partial v}{\partial y} \\ -\frac{\partial u}{\partial y} &\equiv \frac{\partial v}{\partial x}, \end{aligned} \tag{9.3}$$

which are the *Cauchy-Riemann equations.* They will be used in the first proposition.

Proposition 9.8 *Let*

$$p(z, \mu) = \sum_{k=0}^{n} a_k(\mu) z^k$$

be a complex polynomial whose coefficients $a_0(\mu), a_1(\mu), \ldots, a_n(\mu)$ *are smooth functions of* μ *on* $-c < \mu < c$. *If* ζ *is a root of multiplicity one of* $p(z, 0)$, *then there exists a smooth complex valued function* $\rho(\mu)$ *defined on an open interval* $(-b, b)$ *such that* $\rho(\mu)$ *is a root of* $p(z, \mu)$ *with multiplicity one,* $\rho(0) = \zeta$, *and*

$$\rho'(\mu) = -\frac{\partial p(\rho(\mu), \mu)}{\partial \mu} \Big/ \frac{dp(\rho(\mu), \mu)}{dz}. \tag{9.4}$$

Proof. Write $p(z, \mu) = u(x, y, \mu) + iv(x, y, \mu)$ and consider the function $\mathbf{F} : \mathbb{R}^2 \times (-c, c) \to \mathbb{R}^2$ defined by $\mathbf{F}(x, y, \mu) = (u(x, y, \mu), v(x, y, \mu))$. Then, $\mathbf{F}(x, y, \mu) = \mathbf{0}$ if and only if $z = x + iy$ is a root of $p(z, \mu)$. To establish the existence of the function $\rho(\mu)$ the implicit function theorem will be applied at $(\zeta, 0) = (\alpha, \beta, 0)$. This requires first showing that the following determinant is not equal to zero at $(\alpha, \beta, 0) = (\zeta, 0)$:

$$\text{Det}\begin{bmatrix} \frac{\partial u}{\partial x} & \frac{\partial u}{\partial y} \\ \frac{\partial v}{\partial x} & \frac{\partial v}{\partial y} \end{bmatrix} = \frac{\partial u}{\partial x}\frac{\partial v}{\partial y} - \frac{\partial u}{\partial y}\frac{\partial v}{\partial x}.$$

Using equations *(9.3)* and *(9.2)*, the right side can be written as follows:

$$\frac{\partial u}{\partial x}\frac{\partial v}{\partial y} - \frac{\partial u}{\partial y}\frac{\partial v}{\partial x} = \left(\frac{\partial u}{\partial x}\right)^2 + \left(\frac{\partial v}{\partial x}^2\right) = |p'(z,\mu)|^2.$$

Since $(\zeta, 0)$ is a root of multiplicity one, the remark applies and $p'(\zeta, 0) \neq 0$. Hence, $|p'(\zeta, 0)|^2 \neq 0$ and the implicit function theorem implies the existence of real-valued functions $\rho_1(\mu)$ and $\rho_2(\mu)$ such that $\mathbf{F}(\rho_1(\mu), \rho_2(\mu), \mu) = \mathbf{0}$. Setting $\rho(\mu) = \rho_1(\mu) + i\rho_2(\mu)$ produces a continuous complex valued function on an interval $-b < \mu < b$ such that $\rho(0) = \zeta$ and $p'(\rho(\mu)) \neq 0$ because $p'(\zeta, 0) \neq 0$. Hence, $\rho(\mu)$ is a root of $p(z, \mu)$ of multiplicity one.

It remains to derive equation *(9.4)*. Using the chain rule to differentiate $\mathbf{F}(\rho_1(\mu), \rho_2(\mu), \mu) = \mathbf{0}$ with respect to μ produces

$$\begin{bmatrix} \frac{\partial u}{\partial x} & \frac{\partial u}{\partial y} & \frac{\partial u}{\partial \mu} \\ \frac{\partial v}{\partial x} & \frac{\partial v}{\partial y} & \frac{\partial v}{\partial \mu} \end{bmatrix} \begin{pmatrix} \rho_1' \\ \rho_2' \\ 1 \end{pmatrix} = \begin{pmatrix} 0 \\ 0 \end{pmatrix},$$

which can be written as

$$\begin{bmatrix} \frac{\partial u}{\partial x} & \frac{\partial u}{\partial y} \\ \frac{\partial v}{\partial x} & \frac{\partial v}{\partial y} \end{bmatrix} \begin{pmatrix} \rho_1' \\ \rho_2' \end{pmatrix} = -\begin{pmatrix} \frac{\partial u}{\partial \mu} \\ \frac{\partial v}{\partial \mu} \end{pmatrix}.$$

The above 2×2 system can be solved for ρ_1' and ρ_2' because it has already been shown that its determinant is $|p'(\rho(\mu), \mu)|^2$ which is nonzero. To complete the proof, the resulting formula for $\rho'(\mu) = \rho_1'(\mu) + i\rho_2'(\mu)$ must be simplified using the natural notation

$$\frac{\partial p(z,\mu)}{\partial \mu} = \frac{\partial u}{\partial \mu} + i\frac{\partial v}{\partial \mu}$$

and a little complex arithmetic. □

Proposition 9.9 *Let $A : (-c, c) \to \mathcal{M}_d(\mathbb{C})$ be a smooth function. If λ is an eigenvalue of $A(0)$ of algebraic multiplicity one, then there exist smooth functions $\rho : (-b, b) \to \mathbb{C}$ and $\mathbf{w} : (-b, b) \to \mathbb{C}^d$ satisfying the following conditions:*

(a) $\rho(\mu)$ is an eigenvalue of algebraic multiplicity one for $A(\mu)$ and $\rho(0) = \lambda$,

(b) $\mathbf{w}(\mu)$ is an eigenvector of $A(\mu)$ for $\rho(\mu)$.

Proof. The existence of a smooth function satisfying (a) is a consequence of the previous proposition applied to

$$p(z, \mu) = \text{Det } [A(\mu) - z\mathrm{I}]$$

with $\zeta = \lambda$.

Let $\mathbf{w}_0$ be an eigenvector for λ and note that $M(\lambda) = \{\gamma \mathbf{w}_0 : \gamma \in \mathbb{C}\} = \mathbb{C}\mathbf{w}_0$ because the algebraic multiplicity of λ is one. Set $V = \{\mathbf{z} \in \mathbb{C}^d : \mathbf{z} \cdot \mathbf{w}_0 = 0\}$. The dot product $\mathbf{z} \cdot \mathbf{w}$ for complex vectors is defined by *(4.7)* on page 139, and it follows from the lemma on page 139 that $\mathbb{C}^d = V \oplus \mathbb{C}\mathbf{w}_0$. Let $P : \mathbb{C}^d \to \mathbb{C}^d$ be the linear projection with null space $\mathbb{C}\mathbf{w}_0$ and range V, that is $P(\mathbf{v} + \gamma \mathbf{w}_0) = \mathbf{v}$ for $\mathbf{v} \in V$.

Define a parameterized family of linear transformations on $\mathbb{C}^d$ by setting

$$T_\mu(\mathbf{z}) = P\left([A(\mu) - \rho(\mu)\mathrm{I}]\, \mathbf{z}\right).$$

Clearly, the range of T_μ is contained in V for every μ. The strategy will be to show that for small μ, the restriction of T_μ to V is an isomorphism of V onto V and then to use T_μ^{-1} on V to calculate $\mathbf{w}(\mu)$.

Let K_μ be the null space or kernel of T_μ and show that $K_\mu \cap V = \mathbf{0}$ for small μ. If this is not the case, there exists a sequence μ_k converging to 0 and $\mathbf{v}_k \in K_\mu \cap V$ with $|\mathbf{v}_k| = 1$ converging to $\mathbf{v}_0$. Obviously, $\mathbf{v}_0$ is in V and $\mathbf{v}_0 \neq \mathbf{0}$. Since $A(\mu)$ and $\rho(\mu)$ are continuous functions, $P([A(0) - \rho(0)]\mathbf{v}_0) = \mathbf{0}$. Therefore, either $[A(0) - \rho(0)]\mathbf{v}_0 = \mathbf{0}$ or $[A(0) - \rho(0)]\mathbf{v}_0 = \gamma \mathbf{w}_0$, $\gamma \neq 0$. The first implies there is an eigenvector for $\rho(0) = \lambda$ in V, which is impossible because the algebraic multiplicity of λ is one. The second would imply that $\mathbf{v}_0$ is a generalized eigenvector of order two for λ, which is also impossible for the same reason. Since both cases led to contradictions, $K_\mu \cap V = \mathbf{0}$ for small μ.

Since $K_\mu \cap V = \mathbf{0}$ for small μ, the restriction of T_μ to V is an isomorphism of V onto V and T_μ is invertible on V but not on $\mathbb{C}^d$. Furthermore, none of the eigenvectors of $A(\mu)$ for $\rho(\mu)$ are in V for small μ, because they are in K_μ. Using the fact that the algebraic multiplicity of $\rho(\mu)$ is one, it follows that there exists a unique eigenvector of $A(\mu)$ for $\rho(\mu)$ of the form $\mathbf{w}(\mu) = \mathbf{v}(\mu) + \mathbf{w}_0$ for μ in a small interval $(-b, b)$ on which $\rho(\mu)$ satisfies part (a) and $K_\mu \cap V = \mathbf{0}$. The problem is now to show that $\mathbf{w}(\mu)$ is a smooth function.

Because $\mathbf{w}(\mu) = \mathbf{v}(\mu) + \mathbf{w}_0$ is an eigenvector,

$$[A(\mu) - \rho(\mu)\mathrm{I}]\,(\mathbf{v}(\mu) + \mathbf{w}_0) = \mathbf{0}$$

which implies that

$$T_\mu(\mathbf{v}(\mu) + \mathbf{w}_0) = \mathbf{0}$$

or

$$T_\mu(\mathbf{v}(\mu)) = -T_\mu(\mathbf{w}_0).$$

Because T_μ is invertible on V, the above equation can be solved for $\mathbf{v}(\mu)$. With a little care, it can also be shown that the solution is a smooth function.

Let $\mathbf{u}_1, \ldots, \mathbf{u}_{d-1}$ be a basis for V and let $B(\mu)$ be the matrix of the linear transformation $\mathbf{z} \to [A(\mu) - \rho(\mu)\mathrm{I}]\,\mathbf{z}$ with respect to the basis $\mathbf{u}_1, \ldots, \mathbf{u}_{d-1}, \mathbf{w}_0$.

Since the entries of $B(\mu)$ are linear combinations of the entries of $A(\mu)$ and $\rho(\mu)$, every entry of $B(\mu)$ is also a smooth function.

Let $C(\mu)$ be the $d-1 \times d-1$ matrix obtained by deleting the last row and column from $B(\mu)$. It is the matrix of T_μ restricted to V with respect to the basis $\mathbf{u}_1, \ldots, \mathbf{u}_{d-1}$. Therefore, $C(\mu)$ is invertible because T_μ restricted to V is invertible. Since the inverse of a matrix is a rational function of its entries, $C(\mu)^{-1}$ is a smooth function of μ.

The last column of $B(\mu)$ determines $T_\mu(\mathbf{w}_0)$. Specifically.

$$T_\mu(\mathbf{w}_0) = \sum_{k=1}^{d-1} b_{k\,d}(\mu)\mathbf{v}_k$$

and therefore, in the $\mathbf{u}_1, \ldots, \mathbf{u}_{d-1}$ coordinate system $\mathbf{v}(\mu)$ is given by

$$-\,C(\mu)^{-1} \begin{pmatrix} b_{1\,d}(\mu) \\ \vdots \\ b_{(d-1)\,d}(\mu) \end{pmatrix},$$

which is a smooth function. It follows that $\mathbf{w}(\mu) = \mathbf{v}(\mu) + \mathbf{w}_0$ is a smooth function in the standard coordinates. □

The following general, but probably not very well-known, result about difference quotients will also be needed for the proof of the Hopf bifurcation theorem.

Proposition 9.10 *If $F : \Omega \to \mathbb{R}$ is a smooth real-valued function on an open set Ω in $\mathbb{R}^m$ and*

$$G(\mathbf{x}, \sigma) = \begin{cases} \frac{F(\mathbf{x}+\sigma\mathbf{e}_k)-F(\mathbf{x})}{\sigma} & \text{if } \sigma \neq 0 \\ \frac{\partial F(\mathbf{x})}{\partial x_k} & \text{if } \sigma = 0 \end{cases}$$

on $\Omega' = \{(\mathbf{x}, \sigma) \in \mathbb{R}^{m+1} : \mathbf{x} \in \Omega$ and $(\mathbf{x} + \sigma\mathbf{e}_k) \in \Omega\}$, then $G(\mathbf{x}, \sigma)$ is a continuous function on Ω'. Furthermore, if $F(\mathbf{x})$ has continuous second partial derivatives on Ω, then $G(\mathbf{x}, \sigma)$ is a smooth function on Ω'.

Proof. Obviously, G is continuous at $(\mathbf{x}, \sigma)$ if $\sigma \neq 0$. Let $(\mathbf{x}_n, \sigma_n)$ be a sequence in Ω' converging to $(\mathbf{x}, 0)$. It suffices to show that $G(\mathbf{x}_n, \sigma_n)$ converges to $G(\mathbf{x}, 0)$. First, assume that $\sigma_n \neq 0$ for all n. For each n, the mean-value theorem can be applied to $F(\mathbf{x}_n + \sigma_n\mathbf{e}_k) - F(\mathbf{x})$ as a function of x_k and

$$\frac{F(\mathbf{x}_n + \sigma_n\mathbf{e}_k) - F(\mathbf{x}_n)}{\sigma_n} = \frac{\partial F(\mathbf{x}_n + \nu_n\mathbf{e}_k)}{\partial x_k}$$

for some ν_n such that $|\nu_n| < |\sigma_n|$. Because F is a smooth function

$$\lim_{n\to\infty} \frac{\partial F(\mathbf{x}_n + \nu_n\mathbf{e}_k)}{\partial x_k} = \frac{\partial F(\mathbf{x})}{\partial x_k} = G(\mathbf{x}, 0)$$

Secondly, if $\sigma_n = 0$ for all n, then the limit of $G(\mathbf{x}_n, 0)$ is also $G(\mathbf{x}, 0)$, because F is smooth. Therefore, G is continuous on Ω'. This completes the proof of the first statement in the lemma.

If F has continuous second partial derivatives, then

$$\frac{\partial G(\mathbf{x},\sigma)}{\partial x_j} = \begin{cases} \frac{1}{\sigma}\left(\frac{\partial F(\mathbf{x}+\sigma\mathbf{e}_k)}{\partial x_j} - \frac{\partial F(\mathbf{x})}{\partial x_j}\right) & \text{if } \sigma \neq 0 \\ \frac{\partial^2 f(\mathbf{x})}{\partial x_j \partial x_k} = \frac{\partial^2 f(\mathbf{x})}{\partial x_k \partial x_j} & \text{if } \sigma = 0. \end{cases}$$

Since $\frac{\partial F(\mathbf{x})}{\partial x_j}$ is now a smooth function, the first part of the proof applies to the above formula for $\frac{\partial G(\mathbf{x},\sigma)}{\partial x_j}$ and it is continuous.

It remains to calculate the partial of $G(\mathbf{x},\sigma)$ with respect to σ and show that it is also a continuous function. For $\sigma \neq 0$,

$$\frac{\partial G(\mathbf{x},\sigma)}{\partial \sigma} = \frac{\sigma \frac{\partial F(x+\sigma\mathbf{e}_k)}{\partial x_k} - (F(\mathbf{x}+\sigma\mathbf{e}_k) - F(\mathbf{x}))}{\sigma^2}$$

and

$$\begin{aligned} \frac{\partial G(\mathbf{x},0)}{\partial \sigma} &= \lim_{\sigma\to 0} \frac{\frac{f(\mathbf{x}+\sigma\mathbf{e}_k)-f(\mathbf{x})}{\sigma} - \frac{\partial F(\mathbf{x})}{\partial x_k}}{\sigma} \\ &= \lim_{\sigma\to 0} \frac{F(\mathbf{x}+\sigma\mathbf{e}_k) - F(\mathbf{x}) - \sigma\frac{\partial F(\mathbf{x})}{\partial x_k}}{\sigma^2}. \end{aligned}$$

Using Taylor's theorem in the variable x_k, we can write

$$F(\mathbf{x}+\sigma\mathbf{e}_k) = F(\mathbf{x}) + \sigma\frac{\partial F(\mathbf{x})}{\partial x_k} + \frac{\sigma^2}{2}\frac{\partial^2 F(\mathbf{x}+\nu\mathbf{e}_k)}{\partial^2 x_k}$$

with $|\nu| < |\sigma|$. It follows that

$$\frac{\partial G(\mathbf{x},0)}{\partial \sigma} = \frac{1}{2}\lim_{\sigma\to 0}\frac{\partial^2 F(\mathbf{x}+\nu\mathbf{e}_k)}{\partial^2 x_k} = \frac{1}{2}\frac{\partial^2 F(\mathbf{x})}{\partial^2 x_k},$$

and the first partial of G with respect to σ exists throughout Ω.

To show that $\frac{\partial G(\mathbf{x},\sigma)}{\partial \sigma}$ is continuous, it suffices again to consider a sequence $(\mathbf{x}_n, \sigma_n)$ converging to $(\mathbf{x}, 0)$ with $\sigma_n \neq 0$ for all n because F has continuous second partial derivatives. Using the Taylor's formula with remainder again

$$\begin{aligned} \lim_{n\to\infty}\frac{\partial G(\mathbf{x}_n,\sigma_n)}{\partial \sigma} &= \lim_{n\to\infty}\frac{\sigma_n\frac{\partial F(x_n+\sigma_n\mathbf{e}_k)}{\partial x_k} - (F(\mathbf{x}_n+\sigma_n\mathbf{e}_k) - F(\mathbf{x}_n))}{\sigma_n^2} \\ &= \lim_{n\to\infty}\frac{\frac{\partial F(x_n+\sigma_n\mathbf{e}_k)}{\partial x_k} - \frac{\partial F(x_n)}{\partial x_k}}{\sigma_n} - \frac{1}{2}\frac{\partial^2 F(\mathbf{x}_n+\nu_n\mathbf{e}_k)}{\partial^2 x_k} \\ &= \frac{1}{2}\frac{\partial^2 F(\mathbf{x})}{\partial^2 x_k}, \end{aligned}$$

and completes the proof of the proposition. □

The stage is now set for the main theorem of the section.

Theorem 9.11 (Hopf Bifurcation Theorem) *Let* $\mathbf{f}(\mathbf{x}, \mu)$ *be a parameterized vector field on* Ω *with* $-c < \mu < c$. *Suppose* $\mathbf{f}$ *satisfies the following conditions:*

(a) $\mathbf{f}(\mathbf{x}, \mu)$ *has continuous second partial derivatives on its domain* $\Omega \times (-c, c)$,

(b) $\mathbf{f}(\mathbf{0}, \mu) \equiv \mathbf{0}$,

(c) $\mathbf{f}'_{\mathbf{x}}(\mathbf{0}, 0)$ *has a pair of pure imaginary eigenvalues of algebraic multiplicity one denoted by* $\pm\theta i$,

(d) $k\theta i$, $k \in \mathbb{Z}$ *and* $k \neq \pm 1$, *is not an eigenvalue of* $\mathbf{f}'_{\mathbf{x}}(\mathbf{0}, \mu)$,

(e) letting

$$p(z, \mu) = \text{Det}\,[\mathbf{f}'_{\mathbf{x}}(\mathbf{0}, \mu) - z\mathrm{I}]$$

with z *complex, the real part of the complex number*

$$\frac{\partial p}{\partial \mu} \Big/ \frac{dp}{dz}$$

at $(\theta i, 0)$ *is nonzero.*

Then there exist smooth functions $\mu(\sigma)$, $\tau(\sigma)$, *and* $\boldsymbol{\xi}(\sigma)$ *defined on an open interval* $(-b, b)$ *and having the following properties:*

i. $\mathbf{x}(2\pi\tau(\sigma), \boldsymbol{\xi}(\sigma), \mu(\sigma)) = \boldsymbol{\xi}(\sigma)$,

ii. $\mathbf{f}(\boldsymbol{\xi}(\sigma), \mu(\sigma)) \neq 0$ *for* $\sigma \neq 0$,

iii. $\tau(0) = 1/\theta$, $\boldsymbol{\xi}(0) = \mathbf{0}$, *and* $\mu(0) = 0$.

Proof. The proof proceeds in three stages. First, a linear change in variables is made to more conveniently position the variables for the implicit function theorem. The second step is to construct a functional equation whose solutions will be the desired periodic points. The final stage is to show that the implicit function theorem applies to the functional equation and produces the required smooth functions.

For convenience, let $A(\mu) = \mathbf{f}'_{\mathbf{x}}(\mathbf{0}, \mu)$ and let $\theta i, -\theta i, \lambda_3, \ldots, \lambda_q$ denote the distinct eigenvalues of $A(0)$. The dimension of $M(\pm\theta i)$ is one by the hypothesis. If $\mathbf{w}$ is an eigenvector for θi, then $\overline{\mathbf{w}}$ is an eigenvector for $-\theta i$, and the vectors $\mathbf{v}_1 = \mathbf{w} + \overline{\mathbf{w}}$ and $\mathbf{v}_2 = -i(\mathbf{w} - \overline{\mathbf{w}})$ are linearly independent vectors in $\mathbb{R}^d$. Moreover, $A(0)\mathbf{v}_1 = -\theta\mathbf{v}_2$ and $A(0)\mathbf{v}_2 = \theta\mathbf{v}_1$. Set

$$\begin{aligned} V_1 &= \{\alpha\mathbf{v}_1 + \beta\mathbf{v}_2 : \alpha \in \mathbb{R} \text{ and } \beta \in \mathbb{R}\} \\ &= \left\{\mathbf{u} \in \mathbb{R}^d : \mathbf{u} \in M(\theta i) \oplus M(-\theta i)\right\} \end{aligned}$$

and

$$V_2 = \left\{\mathbf{u} \in \mathbb{R}^d : \mathbf{u} \in M(\lambda_3) \oplus \ldots \oplus M(\lambda_q)\right\}.$$

Because $A(0)$ is a real matrix, it is easy to see as in the proof of *Theorem 8.1* that $V_1 \oplus V_2 = \mathbb{R}^d$ and both V_1 and V_2 are $A(0)$ invariant.

By a linear change in variables of the form $\mathbf{x} = B\mathbf{y}$, it can be assumed that

$$\begin{aligned} \mathbf{v}_1 &= (1, 0, 0, \ldots, 0) \\ \mathbf{v}_2 &= (0, 1, 0 \ldots, 0) \\ V_2 &= \{(0, 0, x_3, \ldots, x_d) : x_k \in \mathbb{R}\}. \end{aligned}$$

More importantly such a change in variables will not affect the original hypothesis because the Jacobian matrix for $\dot{\mathbf{y}} = B^{-1}\mathbf{f}(B\mathbf{y}, \mu)$ with respect to $\mathbf{y}$ at $\mathbf{0}$ is $B^{-1}A(\mu)B$ by the chain rule. For the rest of the proof, it will be assumed without loss of generality that the original problem now has the above structure.

Making such a change of variables, puts $A(0)$ in the block form

$$A(0) = \begin{bmatrix} A_1 & \mathrm{O} \\ \mathrm{O} & A_2 \end{bmatrix}$$

with

$$A_1 = \begin{bmatrix} 0 & \theta \\ -\theta & 0 \end{bmatrix}.$$

This completes the first stage of the proof.

The smooth function $\mathbf{g}(\mathbf{x}, \mu) = \mathbf{f}(\mathbf{x}, \mu) - A(\mu)\mathbf{x}$ has the following properties:

$$\begin{aligned} \mathbf{g}(\mathbf{0}, \mu) &= \mathbf{0} \\ \mathbf{g}'_{\mathbf{x}}(\mathbf{0}, \mu) &= \mathrm{O} \end{aligned}$$

The generalized variation of constants formula (*Proposition 5.18* on page 174) can be applied to $\mathbf{f}(\mathbf{x}, \mu) = A(\mu) + \mathbf{g}(\mathbf{x}, \mu)$, so that solutions $\mathbf{x}(t, \boldsymbol{\xi}, \mu)$ of $\dot{\mathbf{x}} = \mathbf{f}(\mathbf{x}, \mu)$ can be expressed as

$$\mathbf{x}(t, \boldsymbol{\xi}, \mu) = e^{tA(\mu)}\boldsymbol{\xi} + \int_0^t e^{(t-s)A(\mu)}\mathbf{g}(\mathbf{x}(s, \boldsymbol{\xi}, \mu), \mu)\, ds.$$

Given $\sigma > 0$, it follows that $2\pi\sigma$ is a period for $\boldsymbol{\xi}$ if and only if

$$\left[e^{2\pi\sigma A(\mu)} - \mathrm{I}\right]\boldsymbol{\xi} + \int_0^{2\pi\sigma} e^{(2\pi\sigma - s)A(\mu)}\mathbf{g}(\mathbf{x}(s, \boldsymbol{\xi}, \mu), \mu)\, ds = \mathbf{0}.$$

Before the implicit function theorem can be applied, the number of variables must be reduced by one. To take advantage of the fact that one periodic point determines the entire periodic orbit, we will try to determine where periodic orbits cross either the hyperplane $x_1 = 0$ or the hyperplane $x_2 = 0$.

Since it is expected that periodic points will be born or die near at $\mu = 0$, we cannot expect to specify the location of the periodic points as a function of μ. To get around this difficulty a new parameter, σ, will be introduced along with some special notation to facilitate working with it.

Let $\hat{\boldsymbol{\xi}} = (\xi_3, \ldots, \xi_d)$ denote a generic variable in $\mathbb{R}^{d-2}$ and for convenience set $\boldsymbol{\xi} = (1, 0, \xi_3, \ldots, \xi_d)$, which is a slightly more restrictive use of the variable $\boldsymbol{\xi}$ than usual. Define $\mathbf{h} : W \to \mathbb{R}^d$ on an open subset W of $\mathbb{R}^{d+1}$ by setting

$$\mathbf{h}(\mu, \tau, \hat{\boldsymbol{\xi}}, y) = \left[e^{2\pi\tau A(\mu)} - \mathrm{I}\right] y\,\boldsymbol{\xi} + \int_0^{2\pi\tau} e^{(2\pi\tau - s)A(\mu)} \mathbf{g}(\mathbf{x}(s, y\,\boldsymbol{\xi}, \mu), \mu)\, ds.$$

The function $\mathbf{h}(\mu, \tau, \hat{\boldsymbol{\xi}}, y)$ also has continuous second partial derivatives on its domain as a consequence of the Leibnitz rules for differentiating under an integral sign. The condition $\mathbf{f}(\mathbf{0}, \mu) \equiv \mathbf{0}$ implies that $\mathbf{x}(t, \mathbf{0}, \mu) \equiv \mathbf{0}$ and $\mathbf{g}(\mathbf{0}, \mu) \equiv \mathbf{0}$. Hence, $\mathbf{h}(\mu, \tau, \hat{\boldsymbol{\xi}}, 0) \equiv \mathbf{0}$. Since $\mathbf{g}'_{\mathbf{x}}(\mathbf{0}, \mu)$ is the zero matrix, it follows by the chain rule that

$$\frac{\partial\, \mathbf{h}(\mu, \tau, \hat{\boldsymbol{\xi}}, 0)}{\partial\, y} = \left[e^{2\pi\tau A(\mu)} - \mathrm{I}\right] \boldsymbol{\xi}.$$

Furthermore, $2\pi\tau$ is a period for $y\,\boldsymbol{\xi}$ if and only if $\mathbf{h}(\mu, \tau, \hat{\boldsymbol{\xi}}, y) = \mathbf{0}$, but again there are difficulties. The implicit function theorem cannot be applied directly to $\mathbf{h}(\mu, \tau, \hat{\boldsymbol{\xi}}, y)$ because too many of its partial derivatives are zero.

Proposition 9.10 can be applied, however, to the y variable in each coordinate function of $\mathbf{h}$ to construct a new smooth function $\mathbf{G}(\mu, \tau, \hat{\boldsymbol{\xi}}, y, \sigma)$. Therefore,

$$\mathbf{H}(\mu, \tau, \hat{\boldsymbol{\xi}}, \sigma) = \mathbf{G}(\mu, \tau, \hat{\boldsymbol{\xi}}, 0, \sigma)$$

is also smooth and using $\mathbf{h}(\mu, \tau, \hat{\boldsymbol{\xi}}, 0) \equiv \mathbf{0}$ can expressed in terms of $\mathbf{h}$ as follows:

$$\begin{aligned} \mathbf{H}(\mu, \tau, \hat{\boldsymbol{\xi}}, \sigma) &= \frac{\mathbf{h}(\mu, \tau, \hat{\boldsymbol{\xi}}, \sigma)}{\sigma} && \text{if } \sigma \neq 0 \\ \mathbf{H}(\mu, \tau, \hat{\boldsymbol{\xi}}, \sigma) &= \frac{\partial\, \mathbf{h}(\mu, \tau, \hat{\boldsymbol{\xi}}, 0)}{\partial\, y} && \text{if } \sigma = 0, \end{aligned}$$

and hence $\mathbf{H}(\mu, \tau, \hat{\boldsymbol{\xi}}, \sigma)$ equals

$$\left[e^{2\pi\tau A(\mu)} - \mathrm{I}\right] \boldsymbol{\xi} + \frac{1}{\sigma} \int_0^{2\pi\tau} e^{(2\pi\tau - s)A(\mu)} \mathbf{g}(\mathbf{x}(s, \sigma\boldsymbol{\xi}, \mu), \mu)\, ds$$

or

$$\left[e^{2\pi\tau A(\mu)} - \mathrm{I}\right] \boldsymbol{\xi}$$

according as $\sigma \neq 0$ or $\sigma = 0$. Thanks to *Proposition 9.10* we already know that the above formulas define a smooth function.

When $\sigma = 0$, the point $\sigma\boldsymbol{\xi} = \mathbf{0}$ is a fixed point for all μ. So a true periodic point will not occur for $\sigma = 0$. When $\sigma \neq 0$, it is follows that $\mathbf{H}(\mu, \tau, \hat{\boldsymbol{\xi}}, \sigma) = \mathbf{0}$ if and only if $\mathbf{h}(\mu, \tau, \hat{\boldsymbol{\xi}}, \sigma) = \mathbf{0}$ if and only if $2\pi\tau$ is a period for $\sigma\boldsymbol{\xi}$. Therefore, the function $\mathbf{H}$ can be used to detect periodic points.

The final stage of the proof is to apply the implicit function theorem to $\mathbf{H}(\mu, \tau, \hat{\boldsymbol{\xi}}, \sigma)$ at the point $(0, 1/\theta, \mathbf{0}, 0)$ to show that μ, τ, and $\boldsymbol{\xi}$ are functions of σ on an interval $-b < \sigma < b$. Since σ will be zero in the calculation of the first partial derivatives of $\mathbf{H}(\mu, \tau, \hat{\boldsymbol{\xi}}, \sigma)$ with respect to $\mu, \tau, \xi_3, \ldots,$ and ξ_d, applying the implicit function theorem is a matter of calculating the first partial derivatives of $[e^{2\pi\tau A(\mu)} - \mathrm{I}]\boldsymbol{\xi}$ with respect to $\mu, \tau, \xi_3, \ldots, \xi_d$. Because these variables enter in quite different ways, this will involve several distinct calculations.

For convenience, set $\eta = 1/\theta$, so that $\eta\theta = 1$. The $d \times d$ matrix of partial derivatives of $[e^{2\pi\tau A(\mu)} - \mathrm{I}]\boldsymbol{\xi}$ with respect to $\mu, \tau, \xi_3, \ldots, \xi_d$ at $(0, \eta, \mathbf{0})$ will be written and analyzed in the following form:

$$\begin{bmatrix} C & E \\ F & D \end{bmatrix}$$

where C is a 2×2 matrix, D is a $(d-2) \times (d-2)$ matrix, and so forth. Rather than stringing the pieces together in one long argument showing that the determinant of the above matrix is nonzero, the major pieces of the argument will be separated out as lemmas.

Lemma A *The following formulas hold:*

$$\begin{aligned} D &= e^{2\pi\eta A_2} - \mathrm{I} \\ E &= \mathrm{O}. \end{aligned}$$

Proof of Lemma. The map $\hat{\boldsymbol{\xi}} \to [e^{2\pi\eta A(0)} - \mathrm{I}]\boldsymbol{\xi}$ is the composition of two linear maps and a translation. Specifically, $\hat{\boldsymbol{\xi}} \to (\mathbf{0}, \hat{\boldsymbol{\xi}})$ is linear, $\mathbf{v} \to \mathbf{v} + \mathbf{e}_1$ is a translation, and $\mathbf{u} \to [e^{2\pi\eta A(0)} - \mathrm{I}]\mathbf{u}$ is linear. Since the derivative of a translation is the identity map, the derivative of $\hat{\boldsymbol{\xi}} \to [e^{2\pi\eta A(0)} - \mathrm{I}]\boldsymbol{\xi}$ is the linear map

$$\hat{\boldsymbol{\xi}} \to \begin{bmatrix} e^{2\pi\eta A_1} & \mathrm{O} \\ \mathrm{O} & e^{2\pi\eta A_2} \end{bmatrix} \begin{pmatrix} \mathbf{0} \\ \hat{\boldsymbol{\xi}} \end{pmatrix} = \begin{bmatrix} \mathrm{O} \\ e^{2\pi\eta A_2} \end{bmatrix} \hat{\boldsymbol{\xi}}$$

and obviously $D = e^{2\pi\eta A_2} - \mathrm{I}$ and $E = 0$. □

Therefore, the matrix F can be ignored and it suffices to show that both C and D have none zero determinants.

Lemma B Det $[e^{2\pi\eta A_2} - \mathrm{I}] \neq 0$.

Proof of Lemma. If Det $[e^{2\pi\eta A_2} - \mathrm{I}] = 0$, then 1 is an eigenvalue of $e^{2\pi\eta A_2}$ and $2\pi ki$ is an eigenvalue of $2\pi\eta A_2$ for some $k \in \mathbb{Z}$. Therefore, for some eigenvalue λ of A_2 it must be true that $2\pi\eta\lambda = 2\pi ki$ or $\lambda = k\theta i$, which is excluded by part (d) of the hypothesis of the theorem. □

To apply the implicit function theorem, it now suffices to show that the determinant of C is not zero. The matrix C consists of the first partial derivatives of the first two coordinate functions of $\mathbf{H}$ with respect to μ and τ evaluated at $(0, \eta, \mathbf{0})$.

It is now necessary to work for awhile with $A(\mu)$ as a complex matrix because θi is complex and its eigenvectors lie in $\mathbb{C}^d$. To start, note that

$$\mathbf{w}_L = \left(\frac{1}{\sqrt{2}}, \frac{-i}{\sqrt{2}}, 0, \ldots, 0\right)$$

and

$$\mathbf{w}_R = \left(\frac{1}{\sqrt{2}}, \frac{i}{\sqrt{2}}, 0, \ldots, 0\right)$$

are left and right eigenvectors of $A(0)$ for θi, respectively. By *Proposition 9.9*, there exist smooth complex valued function $\rho(\mu)$ and smooth vector valued functions $\mathbf{w}_L(\mu)$ and $\mathbf{w}_R(\mu)$ such that $\rho(\mu)$ is an eigenvalue of algebraic multiplicity one for $A(\mu)$ with $\rho(0) = \theta i$, and $\mathbf{w}_L(\mu)$ and $\mathbf{w}_R(\mu)$ are left and right eigenvectors of $A(\mu)$ for $\rho(\mu)$. In the calculations that follow, $\mathbf{w}_L(\mu)$ and $\mathbf{w}_R(\mu)$ will be thought of as $1 \times d$ and $d \times 1$ matrices, respectively.

The matrix product $\mathbf{w}_L(\mu)\mathbf{w}_R(\mu)$ is defined and produces 1×1 matrix or just a number. Since $\mathbf{w}_L(0)\mathbf{w}_R(0) = 1$, they can be normalized so that $\mathbf{w}_L(\mu)\mathbf{w}_R(\mu) = 1$ for all μ. Clearly, $\mathbf{w}_L(\mu)$ and $\mathbf{w}_R(\mu)$ are also left and right eigenvectors of $B(\mu) = e^{2\pi\eta A(\mu)}$ for the eigenvalue $e^{2\pi\eta\rho(\mu)}$. In particular, 1 is an eigenvalue for $B(0)$ because $2\pi\eta\rho(0) = 2\pi i$.

Lemma C *The following formulas hold:*

$$\begin{aligned}
\mathbf{w}_L(\mu)B(\mu)\mathbf{w}_R(\mu) &= e^{2\pi\eta\rho(\mu)} && (9.5)\\
\mathbf{w}_L(\mu)B'(\mu)\mathbf{w}_R(\mu) &= 2\pi\eta\rho'(\mu)e^{2\pi\eta\rho(\mu)} && (9.6)\\
\mathbf{w}_L(0)B'(0)\mathbf{w}_R(0) &= 2\pi\eta\rho'(0). && (9.7)
\end{aligned}$$

Proof of Lemma. The following calculation proves the first equation:

$$\begin{aligned}
\mathbf{w}_L(\mu)B(\mu)\mathbf{w}_R(\mu) &= \big(\mathbf{w}_L(\mu)B(\mu)\big)\mathbf{w}_R(\mu)\\
&= e^{2\pi\eta\rho(\mu)}\mathbf{w}_L(\mu)\mathbf{w}_R(\mu)\\
&= e^{2\pi\eta\rho(\mu)}.
\end{aligned}$$

Differentiating the first equation with respect to μ yields

$$\begin{aligned}
\mathbf{w}'_L(\mu)B(\mu)\mathbf{w}_R(\mu) &+ \mathbf{w}_L(\mu)B'(\mu)\mathbf{w}_R(\mu) +\\
\mathbf{w}_L(\mu)B(\mu)\mathbf{w}'_R(\mu) &= 2\pi\eta\rho'(\mu)e^{2\pi\eta\rho(\mu)}.
\end{aligned}$$

So it suffices to show that

$$\mathbf{w}'_L(\mu)\big(B(\mu)\mathbf{w}_R(\mu)\big) + \big(\mathbf{w}_L(\mu)B(\mu)\big)\mathbf{w}'_R(\mu) = 0.$$

As in the proof of the first equation, the left side simplifies to

$$e^{2\pi\eta\rho(\mu)}\big(\mathbf{w}'_L(\mu)\mathbf{w}_R(\mu) + \mathbf{w}_L(\mu)\mathbf{w}'_R(\mu)\big)$$

and

$$\mathbf{w}'_L(\mu)\mathbf{w}_R(\mu) + \mathbf{w}_L(\mu)\mathbf{w}'_R(\mu) = 0$$

because $\mathbf{w}_L(\mu)\mathbf{w}_R(\mu) \equiv 1$.

The third equation comes from the second by substituting $\mu = 0$ and using $2\pi\eta\rho(0) = 2\pi\eta\theta i = 2\pi i$ to simplify it. $\square$

If the upper left 2×2 sub matrix of $B'(0)$ is

$$\begin{bmatrix} a & b \\ c & d \end{bmatrix},$$

then the third equation in the previous lemma simplifies to

$$(a+d) + (b-c)i = 2\pi\eta\rho'(0).$$

Since the real part of $\rho'(0)$ is not zero by condition (d) in the hypothesis of the theorem, it follows that $a + d \neq 0$, so a and d are not both zero. Our choice of $\boldsymbol{\xi}$ corresponds to $a \neq 0$. If $a = 0$, then one sets $\boldsymbol{\xi} = (0, 1, \xi_3, \ldots, \xi_d)$. Assume $a \neq 0$ for the rest of the proof.

Calculating the partial derivatives of $[e^{2\pi\tau A(\mu)} - \mathrm{I}]\boldsymbol{\xi}$ with respect to μ at $(0, \eta, \mathbf{0})$ is the same as calculating the first column of $B'(0)$, because $\boldsymbol{\xi} = \mathbf{e}_1$ when $\hat{\boldsymbol{\xi}} = \mathbf{0}$. Hence, the first column of the 2×2 matrix C is

$$\begin{pmatrix} a \\ c \end{pmatrix}.$$

Information about the second column of C will come from the first partial derivatives of the first two coordinate functions of $\mathbf{H}$ with respect to τ.

Lemma D *If $a \neq 0$, then the second column of C equals*

$$\begin{pmatrix} 0 \\ -2\pi\theta \end{pmatrix}.$$

Proof of Lemma. Note that

$$e^{2\pi\tau A_1} = \begin{bmatrix} \cos(2\pi\tau\theta) & \sin(2\pi\tau\theta) \\ -\sin(2\pi\tau\theta) & \cos(2\pi\tau\theta) \end{bmatrix}$$

and its derivative with respect τ is

$$\begin{bmatrix} -2\pi\theta\sin(2\pi\tau\theta) & 2\pi\theta\cos(2\pi\tau\theta) \\ -2\pi\theta\cos(2\pi\tau\theta) & -2\pi\theta\sin(2\pi\tau\theta) \end{bmatrix}.$$

Because $\boldsymbol{\xi} = \mathbf{e}_1$, when $\hat{\boldsymbol{\xi}} = \mathbf{0}$, evaluating the first column of this matrix at $\tau = \eta$ shows that the second column of C is

$$\begin{pmatrix} 0 \\ -2\pi\theta \end{pmatrix}$$

as required. $\square$

Therefore, the determinant of C is $2a\pi\theta$, which is not equal to zero. Along with lemmas A and B, this proves that the implicit function theorem applies at $(0, \eta, \mathbf{0}, 0)$ and there exist smooth functions $\mu(\sigma)$, $\tau(\sigma)$, and $\hat{\boldsymbol{\xi}}(\sigma)$ defined on an open interval containing 0 satisfying

$$\mathbf{H}(\mu(\sigma), \tau(\sigma), \hat{\boldsymbol{\xi}}(\sigma), \sigma) = \mathbf{0}.$$

For $\sigma \neq 0$, it follows that

$$\mathbf{h}(\mu(\sigma), \tau(\sigma), \hat{\boldsymbol{\xi}}(\sigma), \sigma) \quad = \quad \mathbf{0}.$$

Before writing this formula in terms of the definition of $\mathbf{h}(\mu, \tau, \hat{\boldsymbol{\xi}}, \sigma)$ set

$$\boldsymbol{\xi}(\sigma) = \sigma\widehat{\hat{\boldsymbol{\xi}}(\sigma)}$$

and avoid writing the right-hand side more than once. Going back to the definition of $\mathbf{h}(\mu, \tau, \hat{\boldsymbol{\xi}}, y)$ the formula $\mathbf{h}(\mu(\sigma), \tau(\sigma), \hat{\boldsymbol{\xi}}(\sigma), \sigma) = \mathbf{0}$ can be written as

$$\left[e^{2\pi\tau(\sigma)A(\mu(\sigma))} - \mathrm{I}\right] \boldsymbol{\xi}(\sigma) +$$

$$\int_0^{2\pi\tau(\sigma)} e^{(2\pi\tau(\sigma)-s)A(\mu(\sigma))}\mathbf{g}(\mathbf{x}(s, \boldsymbol{\xi}(\sigma), \mu(\sigma)), \mu(\sigma))\, ds = \mathbf{0}.$$

This can be rewritten as

$$\boldsymbol{\xi}(\sigma) = e^{2\pi\tau(\sigma)A(\mu)}\boldsymbol{\xi}(\sigma) +$$

$$\int_0^{2\pi\tau(\sigma)} e^{(2\pi\tau(\sigma)-s)A(\mu(\sigma))}\mathbf{g}(\mathbf{x}(s, \boldsymbol{\xi}((\sigma)), \mu(\sigma)), \mu(\sigma))\, ds =$$

$$\mathbf{x}(2\pi\tau(\sigma), \boldsymbol{\xi}(\sigma), \mu(\sigma))$$

to establish the first conclusion of the theorem. The third conclusion follows from the fact that the implicit function theorem was applied at the point $(0, \eta, \mathbf{0}, 0)$. The second requires a little more work.

It remains to be shown that $\mathbf{f}(\boldsymbol{\xi}(\sigma), \mu(\sigma)) \neq \mathbf{0}$ when $\sigma \neq 0$ and lies in a small open interval $-b < \sigma < b$. Part (d) of the hypothesis implies that 0 is not an eigenvalue of $\mathbf{f}'_{\mathbf{x}}(\mathbf{0}, 0)$, and thus $\mathbf{f}'_{\mathbf{x}}(\mathbf{0}, 0)$ is invertible. It follows that for small μ the point $(\mathbf{0}, \mu)$ is an isolated fixed point of the vector field $\mathbf{f}(\mathbf{x}, \mu(\sigma))$ by the corollary to the inverse function theorem on page 303. If σ is small and $\mathbf{f}(\boldsymbol{\xi}(\sigma), \mu(\sigma)) = \mathbf{0}$, then $(\boldsymbol{\xi}(\sigma), \mu(\sigma)) = (\mathbf{0}, \mu(\sigma))$ because $\mu(\sigma)$ and $\boldsymbol{\xi}(\sigma)$ are continuous functions satisfying $\mu(0) = 0$ and $\boldsymbol{\xi}(0) = \mathbf{0}$. It is impossible that $\boldsymbol{\xi}(\sigma) = \mathbf{0}$ when $\sigma \neq 0$ because the first coordinate of $\boldsymbol{\xi}(\sigma) = \sigma$. $\square$

In the last paragraph, it was shown that the function $\boldsymbol{\xi}(\sigma)$ moves away from the origin for $\sigma \neq 0$. Of course, $\tau(\sigma) \neq 0$ for small σ because $\tau(0) = 1/\theta \neq 0$. The first and second conclusions of the Hopf bifurcation theorem along with

$\tau(\sigma) \neq 0$ guarantee that the points $\boldsymbol{\xi}(\sigma)$ are genuine periodic points. The behavior of the function $\mu(\sigma)$ can, however, be more pathological.

To illustrate the conclusions of the Hopf bifurcation theorem consider the planar system

$$\begin{aligned} \dot{x} &= -y + h(r^2, \mu)x \\ \dot{y} &= x + h(r^2, \mu)y \end{aligned} \tag{9.8}$$

with $r^2 = x^2 + y^2$ and $h(s, \mu)$ a real-valued function.

The dynamics of this system will be examined for several specific functions $h(s, \mu)$. In these examples, the periodic points can be found without using the Hopf bifurcation theorem. What is interesting about them is that the functions $\mu(\sigma)$, $\tau(\sigma)$ and $\boldsymbol{\xi}(\sigma)$ can be rather explicitly determined and thus illustrate different Hopf bifurcation patterns that can occur.

Some assumptions need to be made about $h(s, \mu)$ to ensure that the Hopf bifurcation theorem applies. Assume that $h(s, \mu)$ has continuous second partial derivatives and that $h(0, 0) = 0$. Let $\mathbf{f}(x, y, \mu) = (-y + h(r^2, \mu)x, x + h(r^2, \mu)y)$ and note that $\mathbf{f}(0, 0, \mu) \equiv 0$. The matrix of first partial derivatives of $\mathbf{f}$ with respect to x and y is given by

$$\begin{bmatrix} h(r^2, \mu) + 2\frac{\partial h(r^2, \mu)}{\partial s}x^2 & -1 + 2\frac{\partial h(r^2, \mu)}{\partial s}xy \\ 1 + 2\frac{\partial h(r^2, \mu)}{\partial s}xy & h(r^2, \mu) + 2\frac{\partial h(r^2, \mu)}{\partial s}y^2 \end{bmatrix}.$$

Using the same notation as in the proof of the Hopf bifurcation theorem,

$$A(\mu) = \begin{bmatrix} h(0, \mu) & 1 \\ -1 & h(0, \mu) \end{bmatrix}$$

and the eigenvalues of $A(\mu)$ are $h(0, \mu) \pm i$. Thus $A(0)$ has a pair of pure imaginary eigenvalues because $h(0, 0) = 0$. We can use

$$\begin{aligned} \rho(\mu) &= h(0, \mu) + i \\ \rho'(\mu) &= \frac{\partial h(0, \mu)}{\partial \mu} \end{aligned}$$

and assume that $\frac{\partial h(0,0)}{\partial \mu} > 0$ to ensure that part (e) of the hypothesis is satisfied.

To summarize, if the real-valued function $h(s, \mu)$ defined on $\mathbb{R}^2$ satisfies the following conditions:

(a) $h(s, \mu)$has continuous second partial derivatives,

(b) $h(0, 0) = 0$,

(c) $\frac{\partial h(0,0)}{\partial \mu} > 0$,

then the Hopf bifurcation theorem applies to the system *(9.8)* and it has periodic points. Those periodic points can be better understood from the perspective of polar coordinates.

Using *Proposition 2.7* to change to polar coordinates, the above system becomes

$$\begin{pmatrix} \dot{r} \\ \dot{\theta} \end{pmatrix} = \begin{bmatrix} \cos\theta & \sin\theta \\ -r^{-1}\sin\theta & r^{-1}\cos\theta \end{bmatrix} \begin{pmatrix} -r\sin\theta + h(r^2,\mu)r\cos\theta \\ r\cos\theta + h(r^2,\mu)r\sin\theta \end{pmatrix}$$

and this simplifies to

$$\begin{pmatrix} \dot{r} \\ \dot{\theta} \end{pmatrix} = \begin{pmatrix} rh(r^2,\mu) \\ 1 \end{pmatrix}.$$

Therefore, the origin is the only fixed point, every other solution spirals around it, and the periodic orbits are circles of radius r such that $h(r^2,\mu) = 0$. The periodic points all have period 2π and the function $\tau(\sigma) \equiv 1$ and can be ignored. Since $\frac{\partial h(0,0)}{\partial \mu} > 0$, the $h(r^2,\mu) = 0$ equation can be solved for μ as a function of r^2, and r can play the role of σ in the theorem. The graph of $h(r^2,\mu) = 0$ with $r \geq 0$ provides a graphic representation of the Hopf bifurcation at the origin. Now for some specific examples.

First, let $h(s,\mu) = \mu(1-s)$ so that

$$\begin{aligned} \dot{r} &= \mu r(1-r^2) \\ \dot{\theta} &= 1. \end{aligned}$$

This example was carefully analyzed at the end of Section 2.2. It has exactly one periodic orbit, the circle $r = 1$, for $\mu \neq 0$ and every orbit is periodic for $\mu = 0$. Solving $h(r^2,\mu) = 0$ for μ as a function of r when $-1 < r < 1$ and $\frac{\partial h(r^2,\mu)}{\partial \mu} > 0$, yields $\mu(r) \equiv 0$. The graph of $h(r^2,\mu) = 0$ for $0 < r < 1$ lies on the r-axis and the birth of the periodic points is for one value of μ only. In this case, $\boldsymbol{\xi}(r) = (r,0)$ works, and it gives the same periodic orbit for $\pm r$. The curve of periodic points $\boldsymbol{\xi}(r) = (r,0)$ intersects the curve of periodic points $(1,\mu)$, which is not part of the Hopf bifurcation.

For a second example, let $h(s,\mu) = \mu - s$. Then the system is

$$\begin{aligned} \dot{r} &= r(\mu - r^2) \\ \dot{\theta} &= 1, \end{aligned}$$

which was discussed at the beginning of this section. Now $\mu = r^2$ and $\boldsymbol{\xi}(r) = (r,0)$ works, and for every positive μ there is one periodic orbit. Thus the periodic orbits are born when $\mu = 0$ and persist for all $\mu > 0$, but there are no periodic points for $\mu < 0$. This is an example of a *supercritical bifurcation.*

Lastly, consider $h(s,\mu) = \mu + 1 - (s-1)^2$ and the system

$$\begin{aligned} \dot{r} &= r(\mu + 1 - (r^2-1)^2) \\ \dot{\theta} &= 1. \end{aligned}$$

Here, $\mu = (r^2-1)^2 - 1$, which is a quartic. Its minimum for positive r occurs when $r = 1$ and $\mu = -1$, and it has a local maximum at $r = 0$ and $\mu = 0$. Consequently, as r increases from 0 to 1, the values of μ at which a periodic point occurs are decreasing until they reach -1. Then they increase and grow infinitely large. This type of bifurcation is usually called a *subcritical bifurcation.* Thus for values for $-1 < \mu < 0$ there are two values of r such that $r = (r, 0)$ is a periodic point. This is illustrated in *Figure 9.4*. The horizontal line is the line $r = 0$ and represents the fixed point $(0, 0)$ for all μ. The darker curve provides the value(s) of r for which the circle of radius r is a periodic orbit and touches the horizontal line at the point where the Hopf bifurcation occurs, namely, $r = 0$ and $\mu = 0$.

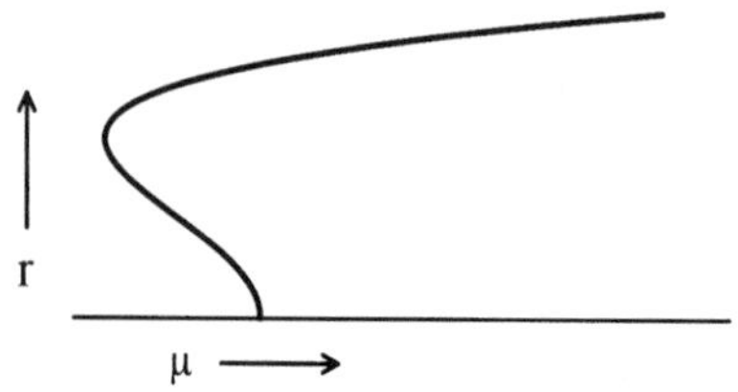

Figure 9.4: Subcritical Hopf bifurcation

Note that $\mu = -1$ is also a bifurcation value because there are no periodic points for $\mu < -1$, but one appears at $\mu = -1$. In fact, this is a saddle-node bifurcation for periodic orbits. The theorem of Poincaré does not apply because one is an eigenvalue of algebraic multiplicity two for the matrix $\left[\frac{\partial \mathbf{x}(2\pi,(1,0),-1)}{\partial \boldsymbol{\xi}}\right]$ because $\mathrm{Tr}\left[\mathbf{f}'(\cos t, \sin t, -1\right] \equiv 0$.

Bifurcations have been extensively studied for the last 50 years and there is a wealth of results and examples. For further reading, see [16] and [29].

EXERCISES

1. Prove the following generalization of the remark on page 317: Let ζ be a root of $p(z) = \sum_{k=0}^{n} a_k z^k$. Then ζ is a root of multiplicity k if and only if $p^{(j)}(\zeta) = 0$ for $j = 1, \ldots, k-1$ and $p^{(k)}(\zeta) \neq 0$.

2. Let $g(z) = u(x, y) + iv(x, y)$ be a complex valued function of the complex variable $z = x + iy$. Show that if $g(z)$ is differentiable at $\zeta = \alpha + i\beta$ as defined by *(9.1)*, then $F(x, y) = (u(x, y), v(x, y))$ is differentiable at (α, β) as defined on page 214 but not conversely.

3. Describe the bifurcation behavior of

$$\begin{aligned} \dot{x} &= -y + x(\mu + x^2 + y^2) \\ \dot{y} &= x + y(\mu + x^2 + y^2). \end{aligned}$$

4. Complete the calculation of the formula for $\rho'(\mu)$ given by *(9.4)* in the proof of *Proposition 9.9*.

5. Construct a bifurcation diagram for the planar system

$$\begin{aligned}\dot{x} &= -y + x(\mu - r^2)(\mu - 2r^2)\\ \dot{y} &= x + y(\mu - r^2)(\mu - 2r^2)\end{aligned}$$

and show that the Hopf bifurcation theorem (*Theorem 9.11*) does not apply to this system.

6. Show that the Hopf bifurcation theorem can be applied at $\mu = \pm 1$ to the system

$$\begin{aligned}\dot{x} &= -y + x(\mu + 1 - r^2)(\mu - 1 - r^2)\\ \dot{y} &= x + y(\mu + 1 - r^2)(\mu - 1 + r^2)\end{aligned}$$

and construct a bifurcation diagram in the $\mu\ r$ plane.

7. Discuss the bifurcation behavior of

$$\begin{aligned}\dot{x} &= -y + x(\sin(\mu) - x^2 - y^2)\\ \dot{y} &= x + y(\sin(\mu) - x^2 - y^2).\end{aligned}$$

8. Show that $\ddot{x} + (x^2 - 2\mu)\dot{x} + x = 0$ has periodic solutions for small positive μ. (*Exercise 4* on page 297 provides information about the phase portraits for $\mu < 0$.)

Bibliography

[1] Ralph Abraham and Jerrold E. Marsden. *Foundations of Mechanics.* Benjamin/Cummings Publ. Co., Reading, 1977.

[2] Kathleen T. Alligood, Tim D. Sauer, and James A. Yorke. *Chaos: An Introduction to Dynamical Systems.* Textbooks in Mathematical Sciences. Springer, New York, 1996.

[3] Joesph Auslander. *Minimal Flows and their Extensions*, volume 153 of *North-Holland Mathematics Studies.* Elsevier Science Publishers B.V., Amsterdam, 1988.

[4] Paul W Berg and James L McGregor. *Elementary Partial Differential Equations.* Holden-Day Series in Mathematics. Holden-Day, Oakland, 1966.

[5] Harald Bohr. *Almost Periodic Functions.* Chelsea Publishing Company, New York, 1947.

[6] Michael Brin and Garrett Stuck. *Introduction to Dynamical Systems.* Cambridge University Press, Canbridge, 2002.

[7] Earl A. Coddington and Norman Levinson. *Theory of Ordinary Differential Equations.* International Series in Pure and Applied Mathematics. McGraw-Hill Book Company, New York, 1955.

[8] Kevin R. Coombs, Brian R. Hunt, Ronald L. Lipsman, John E. Osborn, and Garrett J. Stuck. *Differential Equations with Maple.* John Wiley & Sons, Inc., New York, second edition, 1996.

[9] Kevin R. Coombs, Brian R. Hunt, Ronald L. Lipsman, John E. Osborn, and Garrett J. Stuck. *Differential Equations with Mathematica.* John Wiley & Sons, Inc., New York, second edition, 1998.

[10] Kevin R. Coombs, Brian R. Hunt, Ronald L. Lipsman, John E. Osborn, and Garrett J. Stuck. *Differential Equations with Matlab.* John Wiley & Sons, Inc., New York, 2000.

[11] Jane Cronin. *Differential Equation Introduction and Qualitative Theory,* volume 54 of *Pure and Applied Mathematics.* Marcel Dekker, New York, 1980.

[12] Robert L. Devaney. *An Introduction to Chaotic Dynamical Systems.* Addison-Wesley Publishing Company, Redwood City, second edition, 1989.

[13] C. William Gear. *Numerical Initial Value Problems in Ordinary Differential Equations.* Prentice-Hall Series in Automatic Computation. Prentice-Hall, Englewood Cliffs, 1971.

[14] Walter Helbig Gottschalk and Gustav Arnold Hedlund. *Topological Dynamics*, volume 36 of *Colloquium Publications.* American Mathematical Society, Providence, 1955.

[15] Jack K. Hale. *Ordinary Differential Equations*, volume XXI of *Pure and Applied Mathematics.* John Wiley & Sons, Inc., New York, 1969.

[16] Jack K. Hale and Hüseyin Kǫcak. *Dynamics and Bifurcations*, volume 3 of *Texts in Applied Mathematics.* Springer-Verlag, New York, 1991.

[17] Philip Hartman. *Ordinary Differential Equations.* BirkHauser, Boston, second edition, 1982.

[18] Boris Hasselblatt and Anatole Katok. *A First Course in Dynamics.* Cambridge University Press, Cambridge, 2003.

[19] John G. Hocking and Gail S. Young. *Toplogy.* Addison-Wesley Series in Mathematics. Addison-Wesley Publishing Company, Reading, 1961.

[20] M. C. Irwin. *Smooth DynamicalSystems*, volume 94 of *Pure and Applied Mathematics.* Academic Press, London, 1980.

[21] Frank Jones. *Lebesque Integration on Euclidean Space.* Jones and Bartlett, Boston, 1993.

[22] Mohamed A. Khamsi and Wiliam A Kirk. *An Introduction to Metric Spaces and Fixed Point Theory.* Pure and Applied Mathematics. John Wiley & Sons, Inc., New York, 2001.

[23] Rainer Kress. *Numerical Analysis*, volume 181 of *Graduate Texts in Mathematics.* Springer, New York, 1998.

[24] John M. Lee. *Introduction to Topological Manifolds*, volume 202 of *Graduate Texts in Mathematics.* Springer, New York, 2000.

[25] K. R. Meyer and G. R. Hall. *Introduction to Hamiltonian Dynamical Systems and the N-body Problem*, volume 90 of *Applied Mathematical Sciences.* Springer-Verlag, New York, 1992.

[26] John W. Milnor. *Topology from the Differentiable Viewpoint.* The University Press of Virginia, Charlottesville, 1965.

[27] V. V. Nemytskii and V. V. Stepanov. *Qualitative Theory of Differential Equations.* Princeton University Press, Princeton, 1960.

[28] Helena E. Nusse and James A. Yorke. *Dynamics: Numerical Explorations*, volume 101 of *Applied Mathematical Sciences*. Springer, New York, second edition edition, 1998.

[29] Lawarence Perko. *Differential Equations and Dynamical Systems*, volume 7 of *Texts in Applied Mathematics*. Springer-Verlag, New York, 1991.

[30] Harry Pollard. *Mathematical Introduction to Celestial Mechanics*. Prentice-Hall Mathematics Series. Prentice-Hall, Englewood Cliffs, 1966.

[31] H. L. Royden. *Real Analysis*. Macmillan Publishing Company, New York, third edition, 1988.

[32] Walter Rudin. *Principles of Mathematical Analysis*. International Series in Pure and Applied Mathematics. McGraw-Hill Book Company, New York, second edition, 1964.

[33] Arthur A. Sagle and Ralph E. Walde. *Introduction to Lie Groups and Lie Algebras*, volume 51 of *Pure and Applied Mathematics*. Academic Press, New York, 1973.

[34] Edwin H. Spanier. *Algebraic Toplogy*. McGraw-Hill Series in Higher Mathematics. McGraw-Hill Book Company, New York, 1966.

[35] Michael Spivak. *Calculus on Manifolds*. Mathematical Monograph Series. W. A, Benjamin, New York, 1965.

[36] Peter Walters. *An Introduction to Ergodic Theory*, volume 79 of *Graduate Texts in Mathematics*. Springer-Verlag, New York, 1982.

Index

PURE AND APPLIED MATHEMATICS

A Wiley-Interscience Series of Texts, Monographs, and Tracts

Founded by RICHARD COURANT

Editors: MYRON B. ALLEN III, DAVID A. COX, PETER LAX

Editors Emeriti: PETER HILTON, HARRY HOCHSTADT, JOHN TOLAND

ADÁMEK, HERRLICH, and STRECKER—Abstract and Concrete Catetories
ADAMOWICZ and ZBIERSKI—Logic of Mathematics
AINSWORTH and ODEN—A Posteriori Error Estimation in Finite Element Analysis
AKIVIS and GOLDBERG—Conformal Differential Geometry and Its Generalizations
ALLEN and ISAACSON—Numerical Analysis for Applied Science
*ARTIN—Geometric Algebra
AUBIN—Applied Functional Analysis, Second Edition
AZIZOV and IOKHVIDOV—Linear Operators in Spaces with an Indefinite Metric
BERG—The Fourier-Analytic Proof of Quadratic Reciprocity
BERMAN, NEUMANN, and STERN—Nonnegative Matrices in Dynamic Systems
BERKOVITZ—Convexity and Optimization in $\mathbb{R}^n$
BOYARINTSEV—Methods of Solving Singular Systems of Ordinary Differential Equations
BURK—Lebesgue Measure and Integration: An Introduction
*CARTER—Finite Groups of Lie Type
CASTILLO, COBO, JUBETE, and PRUNEDA—Orthogonal Sets and Polar Methods in Linear Algebra: Applications to Matrix Calculations, Systems of Equations, Inequalities, and Linear Programming
CASTILLO, CONEJO, PEDREGAL, GARCIÁ, and ALGUACIL—Building and Solving Mathematical Programming Models in Engineering and Science
CHATELIN—Eigenvalues of Matrices
CLARK—Mathematical Bioeconomics: The Optimal Management of Renewable Resources, Second Edition
COX—Galois Theory
†COX—Primes of the Form $x^2 + ny^2$: Fermat, Class Field Theory, and Complex Multiplication
*CURTIS and REINER—Representation Theory of Finite Groups and Associative Algebras
*CURTIS and REINER—Methods of Representation Theory: With Applications to Finite Groups and Orders, Volume I
CURTIS and REINER—Methods of Representation Theory: With Applications to Finite Groups and Orders, Volume II
DINCULEANU—Vector Integration and Stochastic Integration in Banach Spaces
*DUNFORD and SCHWARTZ—Linear Operators
Part 1—General Theory
Part 2—Spectral Theory, Self Adjoint Operators in Hilbert Space
Part 3—Spectral Operators
FARINA and RINALDI—Positive Linear Systems: Theory and Applications
FOLLAND—Real Analysis: Modern Techniques and Their Applications
FRÖLICHER and KRIEGL—Linear Spaces and Differentiation Theory
GARDINER—Teichmüller Theory and Quadratic Differentials

*Now available in a lower priced paperback edition in the Wiley Classics Library.
†Now available in paperback.

GILBERT and NICHOLSON—Modern Algebra with Applications, Second Edition
*GRIFFITHS and HARRIS—Principles of Algebraic Geometry
GRILLET—Algebra
GROVE—Groups and Characters
GUSTAFSSON, KREISS and OLIGER—Time Dependent Problems and Difference Methods
HANNA and ROWLAND—Fourier Series, Transforms, and Boundary Value Problems, Second Edition
*HENRICI—Applied and Computational Complex Analysis
Volume 1, Power Series—Integration—Conformal Mapping—Location of Zeros
Volume 2, Special Functions—Integral Transforms—Asymptotics—Continued Fractions
Volume 3, Discrete Fourier Analysis, Cauchy Integrals, Construction of Conformal Maps, Univalent Functions
*HILTON and WU—A Course in Modern Algebra
*HOCHSTADT—Integral Equations
JOST—Two-Dimensional Geometric Variational Procedures
KHAMSI and KIRK—An Introduction to Metric Spaces and Fixed Point Theory
*KOBAYASHI and NOMIZU—Foundations of Differential Geometry, Volume I
*KOBAYASHI and NOMIZU—Foundations of Differential Geometry, Volume II
KOSHY—Fibonacci and Lucas Numbers with Applications
LAX—Functional Analysis
LAX—Linear Algebra
LOGAN—An Introduction to Nonlinear Partial Differential Equations
MARKLEY—Principles of Differential Equations
MORRISON—Functional Analysis: An Introduction to Banach Space Theory
NAYFEH—Perturbation Methods
NAYFEH and MOOK—Nonlinear Oscillations
PANDEY—The Hilbert Transform of Schwartz Distributions and Applications
PETKOV—Geometry of Reflecting Rays and Inverse Spectral Problems
*PRENTER—Splines and Variational Methods
RAO—Measure Theory and Integration
RASSIAS and SIMSA—Finite Sums Decompositions in Mathematical Analysis
RENELT—Elliptic Systems and Quasiconformal Mappings
RIVLIN—Chebyshev Polynomials: From Approximation Theory to Algebra and Number Theory, Second Edition
ROCKAFELLAR—Network Flows and Monotropic Optimization
ROITMAN—Introduction to Modern Set Theory
*RUDIN—Fourier Analysis on Groups
SENDOV—The Averaged Moduli of Smoothness: Applications in Numerical Methods and Approximations
SENDOV and POPOV—The Averaged Moduli of Smoothness
*SIEGEL—Topics in Complex Function Theory
Volume l—Elliptic Functions and Uniformization Theory
Volume 2—Automorphic Functions and Abelian Integrals
Volume 3—Abelian Functions and Modular Functions of Several Variables
SMITH and ROMANOWSKA—Post-Modern Algebra
STAKGOLD—Green's Functions and Boundary Value Problems, Second Editon
*STOKER—Differential Geometry
*STOKER—Nonlinear Vibrations in Mechanical and Electrical Systems
*STOKER—Water Waves: The Mathematical Theory with Applications

*Now available in a lower priced paperback edition in the Wiley Classics Library.
†Now available in paperback.

WATKINS—Fundamentals of Matrix Computations, Second Edition
WESSELING—An Introduction to Multigrid Methods
†WHITHAM—Linear and Nonlinear Waves
†ZAUDERER—Partial Differential Equations of Applied Mathematics, Second Edition

*Now available in a lower priced paperback edition in the Wiley Classics Library.
†Now available in paperback.